DYNAMICAL SYSTEMS
METHOD AND APPLICATIONS

DYNAMICAL SYSTEMS METHOD AND APPLICATIONS

Theoretical Developments and Numerical Examples

Alexander G. Ramm

Department of Mathematics
Kansas State University

Nguyen S. Hoang

Department of Mathematics
University of Oklahoma

A JOHN WILEY & SONS, INC., PUBLICATION

Published by John Wiley & Sons, Inc., Hoboken, New Jersey.
Published simultaneously in Canada.

For general information on our other products and services please contact our Customer Care Department within the United States at (800) 762-2974, outside the United States at (317) 572-3993 or fax (317) 572-4002.

Wiley also publishes its books in a variety of electronic formats. Some content that appears in print, however, may not be available in electronic formats. For more information about Wiley products, visit our web site at www.wiley.com.

Library of Congress Cataloging-in-Publication Data:

Ramm, A. G. (Alexander G.)
  Dynamical systems : methods and applications : theoretical developments and numerical examples / Alexander G. Ramm, Nguyen S. Hoang.
    p. cm.
  Includes bibliographical references and index.
  ISBN 978-1-118-02428-7 (cloth)
1. Differentiable dynamical systems. I. Hoang, Nguyen S., 1980- II. Title.
  QA614.8.R35 2011
  515'.35—dc23   2011036456

Printed in the United States of America.

10 9 8 7 6 5 4 3 2 1

To our families

CONTENTS IN BRIEF

CONTENTS

LIST OF FIGURES

LIST OF TABLES

PREFACE

In this monograph a general method for solving operator equations, especially nonlinear and ill-posed, is developed. The method is called the Dynamical Systems Method (DSM). Suppose one wants to solve an operator equation:

$$F(u) = f, \qquad (0.1)$$

where F is a nonlinear or linear map in a Hilbert or Banach space. We assume that equation (0.1) is solvable, possibly nonuniquely. The DSM for solving equation (0.1) consists of finding a map Φ such that the Cauchy problem

$$\dot{u} = \Phi(t, u), \quad u(0) = u_0; \qquad \dot{u} = \frac{du}{dt} \qquad (0.2)$$

has a unique global solution; that is, solution $u(t)$ defined for all $t \geq 0$, there exists $u(\infty) = \lim_{t \to \infty} u(t)$, and $F(u(\infty)) = 0$:

$$\exists! u \quad \forall t \geq 0; \quad \exists u(\infty); \quad F(u(\infty)) = f. \qquad (0.3)$$

If (0.3) holds, we say that DSM is justified for equation (0.1). Thus the dynamical system in this book is a synonym to an evolution problem (0.2). This explains the name DSM. The choice of the initial data $u(0)$ will be discussed for various classes of equations (0.1). It turns out that for many classes of equations (0.1) the initial approximation u_0 can be chosen arbitrarily, and, nevertheless, (0.3) holds, while for some problems the choice of u_0, for which (0.3) can be established, is restricted to some neighborhood of a solution to equation (0.1).

We describe various choices of Φ in (0.2) for which it is possible to justify (0.3). It turns out that the scope of DSM is very wide. To describe it, let us introduce some notions. Let us call problem (0.1) *well-posed* if

$$\sup_{u\in B(u_0,R)} \|[F'(u)]^{-1}\| \le m(R), \tag{0.4}$$

where $B(u_0, R) = \{u : \|u - u_0\| \le R\}$, $F'(u)$ is the Fréchet derivative (F-derivative) of the operator-function F at the point u, and the constant $m(R) > 0$ may grow arbitrarily as R grows. If (0.4) fails, we call problem (0.1) *ill-posed*. If problem (0.1) is ill-posed, we often assume that noisy data f_δ are given in place of f, $\|f_\delta - f\| \le \delta$. Although the equation $F(u) = f$ is solvable, the equation $F(u) = f_\delta$ may have no solutions.

The problem is:

Given $\{\delta, f_\delta, F\}$, find a stable approximation u_δ to a solution u of the equation $F(u) = f$; that is, find u_δ such that

$$\lim_{\delta\to 0} \|u_\delta - u\| = 0. \tag{0.5}$$

In Part I of this book, unless otherwise stated, we assume that

$$\sup_{u\in B(u_0,R)} \|F^{(j)}(u)\| \le M_j(R), \quad 0 \le j \le 2, \tag{0.6}$$

where $M_j(R)$ are some constants. In other words, we assume that the nonlinearity is C_{loc}^2, but the rate of its growth, as R grows, is not restricted. This assumption is dropped later on (see Chapter 20). We will obtain many results assuming only that $F'(u)$ is continuous with respect to u.

Let us now describe briefly the scope of the DSM.

Any well-posed problem (0.1) can be solved by a DSM which converges at an exponential rate, that is,

$$\|u(\infty) - u(t)\| \le re^{-c_1 t}, \quad \|F(u(t))\| \le \|F_0\|e^{-c_1 t}, \tag{0.7}$$

where $r > 0$ and $c_1 > 0$ are some constants, and $F_0 := F(u_0)$.

For ill-posed problems, in general, it is not possible to estimate the rate of convergence even for linear problems; depending on the data f, this rate can be arbitrarily slow. To estimate the rate of convergence for an ill-posed problem, one has to make some additional assumptions about the data f. Remember that by "any" we mean throughout any solvable problem (0.1).

Any solvable linear equation

$$F(u) = Au = f, \tag{0.8}$$

where A is a closed, linear, densely defined operator in a Hilbert space H, can be solved stably by a DSM. If noisy data f_δ are given, $\|f_\delta - f\| \le \delta$, then DSM yields a stable solution u_δ for which (0.5) holds, provided that one stops at a suitable time, the "stopping time" t_δ.

We derive stopping rules, that is, rules for choosing $t(\delta) := t_\delta$, the time at which $u_\delta(t_\delta) = u_\delta$ should be calculated, using f_δ in place of f, in order for (0.5) to hold.

For linear problems (0.8) the convergence of a suitable DSM is global with respect to u_0; that is, DSM converges to the unique minimal-norm solution y of (0.8) for any choice of u_0.

We prove similar results for equations (0.1) with monotone operators F : $H \to H$. Recall that F is called monotone if

$$\langle F(u) - F(v), u - v \rangle \ge 0 \quad \forall u, v \in H, \tag{0.9}$$

where H is a Hilbert space and $\langle \cdot, \cdot \rangle$ denotes the inner product in H. For hemicontinuous monotone operators the set $\mathcal{N} = \{u : F(u) = 0\}$ is closed and convex, and such sets in a Hilbert space have a unique minimal-norm element. A map F is called hemicontinuous if the function $\langle F(u + \lambda v), w \rangle$ is continuous with respect to $\lambda \in [0, \lambda_0)$ for any $u, v, w \in H$, where $\lambda_0 > 0$ is a number.

DSM is justified for any solvable equation (0.1) with monotone operators with continuous $F'(u)$, and a version of DSM is justified for continuous F without any smoothness of F assumed. Note that no restrictions on the growth of $M_j(R)$ as R grows are imposed, so the nonlinearity is C^2_{loc} but may grow arbitrarily fast. For monotone operators we will drop assumption (0.6) and construct a convergent DSM.

We justify DSM for an arbitrary solvable equation (0.1) in a Hilbert space with C^2_{loc} nonlinearity under a very weak assumption:

$$F'(y) \ne 0, \tag{0.10}$$

where y is a solution to equation (0.1).

We justify DSM for operators satisfying the following spectral assumption:

$$\|(F'(u) + \varepsilon)^{-1}\| \le \frac{c}{\varepsilon}, \quad 0 < \varepsilon \le \varepsilon_0, \quad \forall u \in H, \tag{0.11}$$

where $\varepsilon_0 > 0$ is an arbitrary small fixed number. Assumption (0.11) is satisfied, for example, for operators $F'(u)$ whose regular points, that is, points $z \in \mathbb{C}$ such that $(F'(u) - z)^{-1}$ is a bounded linear operator for

$$|z| < \varepsilon_0, \quad |\arg z - \pi| \le \varphi_0, \tag{0.12}$$

where $\varphi_0 > 0$ is an arbitrary small fixed number. Monotone operators with a continuous Fréchet derivative satisfy assumption (0.11) with $c = 1$ and any $\epsilon_0 > 0$. We also prove the existence of a solution to the equation

$$F(u) + \varepsilon u = 0, \tag{0.13}$$

provided that (0.6) and (0.11) hold.

We discuss DSM for equations (0.1) in Banach spaces. In particular, we discuss some singular perturbation problems for equations of the type (0.13): under what conditions a solution u_ε to equation (0.13) converges to a solution of equation (0.1) as $\varepsilon \to 0$.

In Newton-type methods, for example,

$$\dot{u} = -[F'(u)]^{-1}[F(u) - f], \quad u(0) = u_0, \tag{0.14}$$

the most difficult and time-consuming part is the inversion of the derivative $F'(u)$.

We justify a DSM method which avoids the inversion of the derivative.

For example, for well-posed problem (0.1) such a method is

$$\begin{aligned} \dot{u} &= -Q[F(u) - f], \quad u(0) = u_0, \\ \dot{Q} &= -TQ + A^*, \quad Q(0) = Q_0, \end{aligned} \tag{0.15}$$

where

$$A := F'(u), \quad T = A^* A, \tag{0.16}$$

A^* is the adjoint to A operator, and u_0 and Q_0 are suitable initial approximations.

We also give a similar DSM scheme for solving ill-posed problem (0.1).

We justify DSM for some classes of operator equations (0.1) with unbounded operators, for example, for operators $F(u) = Au + g(u)$, where A is a linear, densely defined, closed operator in a Hilbert space H and g is a nonlinear C^2_{loc} map.

We justify DSM for equation (0.1) with some nonsmooth operators, for example, with monotone and hemicontinuous operators, defined on all of H.

We show that the DSM can be used as a theoretical tool for proving conditions sufficient for the surjectivity of a nonlinear map or for this map to be a global homeomorphism.

One of our motivations is to develop a general method for solving operator equations, especially nonlinear and ill-posed. The other motivation is to develop a general approach to constructing convergent iterative processes for solving these equations.

The idea of this approach is straightforward: If the DSM is justified for solving equation (0.1), that is, (0.3) holds, then one considers a discretization of (0.2), for example, the explicit Euler method:

$$u_{n+1} = u_n + h_n \Phi(t_n, u_n), \quad u_0 = u_0, \quad t_{n+1} = t_n + h_n, \tag{0.17}$$

and if one can prove convergence of (0.17) to the solution of (0.2), then (0.17) is a convergent iterative process for solving equation (0.1).

We prove that any solvable linear equation (0.8) (with bounded or unbounded operator A) can be solved by a convergent iterative process which converges to the unique minimal-norm solution of (0.8) for any initial approximation u_0.

We prove a similar result for solvable equation (0.1) with monotone operators.

For general nonlinear equation (0.1), under suitable assumptions, a convergent iterative process is constructed. The initial approximation in this process does not have to be in a suitable neighborhood of a solution to (0.1).

We give several numerical examples of applications of the DSM. A detailed discussion of the problem of stable differentiation of noisy functions is given.

Among new technical tools, which we often use in this book, are some novel differential inequalities.

The first of these deals with the functions satisfying the following inequality:

$$\dot{g} \leq -\gamma(t)g(t) + \alpha(t)g^p(t) + \beta(t), \quad p > 1, \quad t \geq t_0 \geq 0, \qquad (0.18)$$

where g, γ, α, and β are nonnegative functions, and γ, α, and β are continuous on $[t_0, \infty)$. We assume that there exists a positive function $\mu \in C^1[t_0, \infty)$, such that

$$\frac{\alpha(t)}{\mu^p(t)} + \beta(t) \leq \frac{1}{\mu(t)}\left[\gamma(t) - \frac{\dot{\mu}(t)}{\mu(t)}\right], \qquad (0.19)$$

$$\mu(t_0)g(t_0) \leq 1, \qquad (0.20)$$

and prove that under the above assumptions, any nonnegative solution $g(t)$ to (0.18) is defined on $[t_0, \infty)$ and satisfies the following inequality:

$$0 \leq g(t) \leq \frac{1}{\mu(t)}. \qquad (0.21)$$

A more general inequality than (0.18) is also introduced, investigated, and applied. Namely,

$$\dot{g} \leq \gamma(t)g(t) + \alpha(t, g) + \beta(t), \qquad t \geq t_0,$$

where $\alpha(t, g) \geq 0$ is a locally Lipschitz-continuous function of g, $g \geq 0$, which is continuous with respect to t for any fixed g. The assumptions on $\gamma(t)$ and $\beta(t)$ are as above. We prove that if

$$\alpha(t, \mu^{-1}(t)) + \beta(t) \leq \mu^{-1}(t)\left(\gamma(t) - \frac{\dot{\mu}(t)}{\mu(t)}\right), \qquad \mu(t_0)g(t_0) \leq 1,$$

then $g(t)$ exists for all $t \geq t_0$, and

$$0 \leq g(t) \leq \mu^{-1}(t), \qquad t \geq t_0.$$

In particular, if $\lim_{t \to \infty} \mu(t) = \infty$, then

$$\lim_{t \to \infty} g(t) = 0.$$

Another inequality that we use is an operator version of the Gronwall inequality. Namely, assume that

$$\dot{Q} = -T(t)Q(t) + G(t), \quad Q(0) = Q_0, \tag{0.22}$$

where $T(t)$ and $G(t)$ are linear bounded operators on a Hilbert space depending continuously on a parameter $t \in [0, \infty)$. If there exists a continuous positive function $\varepsilon(t)$ on $[0, \infty)$ such that

$$\langle T(t)h, h \rangle \geq \varepsilon(t)\|h\|^2, \quad \forall h \in H, \tag{0.23}$$

then the solution to (0.22) satisfies the inequality

$$\|Q(t)\| \leq e^{-\int_0^t \varepsilon(x)dx} \left[\|Q_0\| + \int_0^t \|G(s)\| e^{\int_0^s \varepsilon(x)dx} ds \right]. \tag{0.24}$$

This inequality shows that $Q(t)$ is a bounded linear operator whose norm is bounded uniformly with respect to t if

$$\sup_{t \geq 0} \int_0^t \|G(s)\| e^{-\int_s^t \varepsilon(x)dx} ds < \infty. \tag{0.25}$$

We also study the following inequality:

$$y(t) - y(s) \leq \int_s^t h(\xi) \, d\xi, \quad 0 \leq s \leq t, \tag{0.26}$$

where $y(t) \geq 0$ is a continuous function on $[0, \infty)$, and $h(t) \geq 0, \forall t \in [0, \infty)$. Let $\omega(t) \geq 0$ be a nondecreasing continuous function, $w(t) > 0, \forall t > 0$. We prove that if

$$\int_0^\infty \omega\big(y(t)\big) \frac{1}{(1+t)^\alpha} dt < \infty, \quad 0 < \alpha \leq 1, \tag{0.27}$$

and

$$A := \limsup_{t \to \infty} h(t)t^\alpha < \infty, \tag{0.28}$$

then

$$\lim_{t \to \infty} y(t) = 0. \tag{0.29}$$

The following inequality is studied in Chapter 5:

$$y(t) - y(s) \leq \int_s^t [g(\xi)\varphi(y(\xi)) + h(\xi)] \, d\xi, \quad 0 \leq s \leq t, \tag{0.30}$$

where g and h are nonnegative locally integrable functions on $[0, \infty)$, $\varphi \geq 0$ is a continuous function on $[0, \infty)$, and the functions $\int_0^t g(x) \, dx$ and $\int_0^t h(x) \, dx$

are uniformly continuous with respect to t on $[0, \infty)$. Let $\omega(t) \geq 0$ be a nondecreasing continuous function $w(t) > 0, \forall t > 0$. We prove that if

$$\int_0^\infty \omega(y(t)) \, dt < \infty, \tag{0.31}$$

then

$$\lim_{t \to \infty} y(t) = 0. \tag{0.32}$$

Other useful inequalities are established in Chapter 5.

The DSM is shown to be useful as a tool for proving theoretical results (see Chapter 13).

The DSM is used in Chapter 14 for construction of convergent iterative processes for solving operator equation.

In Chapter 15 some numerical problems are discussed—in particular, the problem of stable differentiation of noisy data.

In Part II of the book, which consists of eight chapters, the Discrepancy Principle (DP) is introduced and justified for various versions of the DSM. These principles serve as stopping rules for integrating evolution problem (0.2) when the data are noisy. We prove that every solvable linear operator equation $Au = f$ with a closed densely defined operator A in H can be stably solved by a DSM if noisy data f_δ are known. Similar results are established for operator equations with monotone operators, and for more general classes of equations. Solving evolution problem (0.2) is a important component of the DSM. Global existence of a solution to problem (0.2) is an important and non-trivial question. Local existence of a solution $u(t)$ to problem (0.2) is usually established under the assumption that the non-linear operator $\Phi(t, u)$ satisfies a local Lipschitz condition with respect to u. If the local existence of $u(t)$ is established, then its global existence can be often derived from an a priori estimate

$$\sup_{t \geq t_0} \|u(t)\| \leq c < \infty.$$

We have developed a novel approach to these problems in Chapter 20.

In Part III the emphasis is on examples of numerical applications of the DSM to a number of problems of general interest. For example, the following numerical problems of practical interest are discussed: stable solution of linear ill-conditioned algebraic systems, stable solution of Fredholm and Volterra integral equations of the first kind, stable solution of Hammerstein nonlinear integral equations, image restoration by various versions of the DSM, and inversion of the Laplace transform of a compactly supported signal from a compact subset of the real axis.

The results in this book will be useful for scientists and engineers dealing with solving operator equations, linear and nonlinear, especially ill-posed, and for students in mathematics, computational mathematics, engineering, and physics.

Part I of this book is based on the monograph [151] and the book is based mostly on the papers, published by the authors and referenced in Appendix B: Bibliographical Notes. This book is essentially self-contained, and it requires a relatively modest background of the reader.

In Appendix A various auxiliary material is presented. Together with some known results, available in the literature, some less known results are included: for example, conditions for compactness of embedding operators and conditions for the continuity of the solutions to operator equations with respect to a parameter.

The table of contents gives a detailed list of topics discussed in this book.

ACKNOWLEDGMENTS

The authors thank their families for support.

This book is based mostly on the authors's published papers and the earlier published monograph [151]. The authors thank Elsevier for permission to use this monograph.

A. G. Ramm thanks the Max Planck Institute for Mathematics in the Sciences, Leipzig, for hospitality during the Summer of 2011.

PART I

CHAPTER 1

INTRODUCTION

1.1 WHAT THIS BOOK IS ABOUT

This book is about a general method for solving operator equations

$$F(u) = 0. \tag{1.1}$$

Here F is a nonlinear map in a Hilbert space H. Later on we consider maps F in Banach spaces as well. The general method, which we develop in this book and call the Dynamical Systems Method (DSM), consists of finding a nonlinear map $\Phi(t, u)$ such that the Cauchy problem

$$\dot{u} = \Phi(t, u), \quad u(0) = u_0 \tag{1.2}$$

has a unique global solution $u(t)$, that is, the solution defined for all $t \geq 0$, this solution has a limit $u(\infty)$:

$$\lim_{t \to \infty} \|u(\infty) - u(t)\| = 0, \tag{1.3}$$

and this limit solves equation (1.1):

$$F(u(\infty)) = 0. \tag{1.4}$$

Dynamical Systems Method and Applications: Theoretical Developments and Numerical Examples, First Edition. A. G. Ramm and N. S. Hoang.
Copyright © 2012 John Wiley & Sons, Inc.

Let us write these three conditions as

$$\exists! u(t) \quad \forall t \geq 0; \quad \exists u(\infty); \quad F(u(\infty)) = 0. \tag{1.5}$$

If (1.5) holds for the solution to (1.2), then we say that a DSM is justified for solving equation (1.1). There may be many choices of $\Phi(t, u)$ for which DSM can be justified. A number of such choices will be given in Chapter 3 and in other chapters. It should be emphasized that we do not assume that equation (1.1) has a unique solution. Therefore the solution $u(\infty)$ depends on the initial approximation u_0 in (1.2). The choice of u_0 in some cases is not arbitrary and in many cases this choice is arbitrary—for example, for problems with linear operators or with nonlinear monotone operators, as well as for a wide class of general nonlinear problems (see Chapters 4, 6, 7–9, 11–12, 14).

The existence and uniqueness of the local solution to problem (1.2) is guaranteed, for example, by a Lipschitz condition imposed on Φ:

$$\|\Phi(t, u) - \Phi(t, v)\| \leq L\|u - v\|, \qquad \forall u, v \in B(u_0, R), \tag{1.6}$$

where the constant L does not depend on $t \in [0, \infty)$ and

$$B(u_0, R) = \{u : \|u - u_0\| \leq R\}$$

is a ball, centered at the element $u_0 \in H$ and of radius $R > 0$.

1.2   WHAT THE DSM (DYNAMICAL SYSTEMS METHOD) IS

The DSM for solving equation (1.1) consists of finding a map $\Phi(t, u)$ and an initial element u_0 such that conditions (1.5) hold for the solution to the evolution problem (1.2).

If conditions (1.5) hold, then one solves Cauchy problem (1.2) and calculates the element $u(\infty)$. This element is a solution to equation (1.1). The important question one faces after finding a nonlinearity Φ, for which (1.5) holds, is the following one: How does one solve Cauchy problem (1.2) numerically? This question has been studied much in the literature. If one uses a projection method, that is, looks for the solution of the form

$$u(t) = \sum_{j=1}^{J} u_j(t) f_j, \tag{1.7}$$

where $\{f_j\}$ is an orthonormal basis of H, and $J > 1$ is an integer, then problem (1.2) reduces to a Cauchy problem for a system of J nonlinear ordinary differential equations for the scalar functions $u_j(t)$, $1 \leq j \leq J$, if the right-hand side of (1.2) is projected onto the J-dimensional subspace spanned by $\{f_j\}_{1 \leq j \leq J}$. This system is

$$\begin{aligned}
\dot{u}_j &= \left\langle \Phi\left(\sum_{m=1}^{J} u_m(t) f_m, t\right), f_j \right\rangle, & 1 \leq j \leq J, \\
u_j(0) &= \langle u_0, f_j \rangle, & 1 \leq j \leq J.
\end{aligned} \tag{1.8}$$

Numerical solution of the Cauchy problem for systems of ordinary differential equations has been much studied in the literature.

In this book the main emphasis is on the possible choices of Φ which imply properties (1.5).

1.3 THE SCOPE OF THE DSM

One of our aims is to show that DSM is applicable to a very wide variety of problems.

Specifically, we prove in this book that the DSM is applicable to the following classes of problems:

1. *Any well-posed solvable problem (1.1) can be solved by DSM.*

   By a *well-posed problem* (1.1) we mean the problem with the operator F satisfying the following assumptions:

   $$\sup_{u \in B(u_0, R)} \|[F'(u)]^{-1}\| \le m(R) \tag{1.9}$$

   and

   $$\sup_{u \in B(u_0, R)} \|F^{(j)}(u)\| \le M_j(R), \quad 0 \le j \le 2, \tag{1.10}$$

   where $F^{(j)}(u)$ is the jth Fréchet derivative of F.

   If assumption (1.9) does not hold but (1.10) holds, we call problem (1.1) *ill-posed*. This terminology is not quite standard. The standard notion of an ill-posed problem is given in Section 2.1.

   We prove that for any solvable well-posed problem, not only the DSM can be justified (i.e., Φ can be found such that for problem (1.2) conclusions (1.5) hold), but, in addition, the convergence of $u(t)$ to $u(\infty)$ is exponentially fast:

   $$\|u(t) - u(\infty)\| \le r e^{-c_1 t}, \tag{1.11}$$

   where $r > 0$ and $c_1 > 0$ are constants, and

   $$\|F(u(t))\| \le \|F_0\| e^{-c_1 t}, \quad F_0 := F(u_0). \tag{1.12}$$

2. *Any solvable linear ill-posed problem can be solved by DSM.*

   A linear problem (1.1) is a problem

   $$Au = f, \tag{1.13}$$

   where A is a linear operator. We always assume this operator to be closed and densely defined. Its null space is denoted

   $$\mathcal{N}(A) := \{u : Au = 0\},$$

its domain is denoted $D(A)$, and its range is denoted $R(A)$.

For a linear ill-posed problems a DSM can be justified and Φ can be found such that convergence (1.3) holds for *any* initial approximation u_0 in (1.2) and $u(\infty)$ is the unique minimal-norm solutions y to (1.5). However, in general, one cannot estimate the rate of convergence: It can be as slow as one wishes if f is chosen suitably. To obtain a rate of convergence for an ill-posed problem, one has to make additional assumptions on f. One can give a stable approximation to the minimal-norm solution y to problem (1.5) using DSM. This stable approximation u_δ should be found from the noisy data $\{f_\delta, \delta\}$, where f_δ, the noisy data, is an arbitrary element satisfying the inequality

$$\|f_\delta - f\| \le \delta \tag{1.14}$$

and $\delta > 0$ is a small number. The stable approximation is the approximation for which one has

$$\lim_{\delta \to 0} \|u_\delta - y\| = 0. \tag{1.15}$$

When one uses a DSM for stable solution of an ill-posed problem (1.5), or, more generally, of a nonlinear problem

$$F(u) = f, \tag{1.16}$$

then one solves the Cauchy problem (1.2), where Φ depends on the noisy data f_δ and one stops the calculation of the corresponding solution $u_\delta(t)$ at a time t_δ, which is called the stopping time. The stopping time should be chosen so that

$$\lim_{\delta \to 0} \|u_\delta(t_\delta) - u(\infty)\| = 0, \tag{1.17}$$

where $u(\infty)$ is the limiting value of the solution $u(t)$ to problem (1.2) corresponding to the exact data f. In Chapters 4, 6, and 7 we give some methods for choosing the stopping times for solving ill-posed problems.

3. *Any solvable ill-posed problem (1.16) with a monotone operator F, satisfying (1.10), can be solved stably by a DSM.*

   If the operator F in problem (1.16) is monotone, that is,

   $$\langle F(u) - F(v), u - v \rangle \ge 0, \tag{1.18}$$

   and assumption (1.10) holds, then one can find such Φ that (1.5) holds. Moreover, *convergence (1.3) holds for any initial approximation u_0 in (1.2), and $u(\infty)$ is the unique minimal-norm solution y to (1.16).*

   If noisy data $f_\delta, \|f_\delta - f\| \le \delta$, are given in place of the exact data f, then one integrates the Cauchy problem (1.2) with Φ corresponding to

f_δ and calculates the corresponding solution $u_\delta(t)$ at a suitably chosen stopping time t_δ.

If $u_\delta := u_\delta(t_\delta)$, then

$$\lim_{\delta \to 0} \|u_\delta - y\| = 0. \tag{1.19}$$

Some methods for finding the stopping time are discussed in Chapter 6.

4. *Any solvable ill-posed problem (1.16), such that*

$$F(y) = f, \quad F'(y) \neq 0,$$

and (1.10) holds, can be solved stably by a DSM.

5. *Any solvable ill-posed problem (1.16) with a monotone, hemicontinuous, defined on all of H operator F can be solved stably by a DSM.*

   For such operators, assumption (1.10) is dropped. One can choose such Φ that convergence (1.3) holds for any initial approximation u_0 in (1.2) and $u(\infty) = y$, where y is the unique minimal-norm solution to (1.16).

6. *If $F = L + g$, where L is a linear, closed, densely defined operator, g is a nonlinear operator satisfying (1.10), and equation (1.16) is solvable, then it can be solved by a DSM, provided that L^{-1} exists and is bounded and*

$$\sup_{u \in B(u_0, R)} \left\| \left(I + L^{-1} g'(u) \right)^{-1} \right\| \leq m(R). \tag{1.20}$$

   Thus DSM can be used for some equations (1.1) with unbounded operators F.

7. *DSM can be used for proving theoretical results.*

   For example:

   A map $F : H \to H$ is surjective if (1.9)–(1.10) hold and

$$\sup_{R > 0} \frac{R}{m(R)} = \infty. \tag{1.21}$$

   A map $F : H \to H$ is a global homeomorphism of H onto H if (1.10) holds and

$$\|[F'(u)]^{-1}\| \leq h(\|u\|), \tag{1.22}$$

   where $h(s) > 0$ is a continuous function on $[0, \infty)$ such that

$$\int_0^\infty h^{-1}(s) \, ds = \infty. \tag{1.23}$$

8. *DSM can be used for solving nonlinear well-posed and ill-posed problems (1.1) without inverting the derivative $F'(u)$.*

For example, if assumptions (1.9)–(1.10) hold and problem (1.1) is solvable, then the DSM

$$
\begin{aligned}
\dot{u} &= -QF(u), \\
\dot{Q} &= -TQ + A^*, \\
u(0) &= u_0, \qquad Q(0) = Q_0
\end{aligned}
\tag{1.24}
$$

converges to a solution of problem (1.1) as $t \to \infty$, and (1.5) holds. Here Q is an operator,

$$
A := F'(u), \quad T := A^*A,
\tag{1.25}
$$

and A^* is the adjoint to A operator.

Note that a Newton-type method for solving equation (1.1) by a DSM is of the form

$$
\dot{u} = -[F'(u)]^{-1}F(u), \quad u(0) = u_0.
\tag{1.26}
$$

This method is applicable to the well-posed problems only, because it requires $F'(u)$ to be boundedly invertible. Its regularized versions are applicable to many ill-posed problems also, as we demonstrate in this book. In practice the numerical inversion of $F'(u)$ is the most difficult and time-consuming part of the solution of equation (1.1) by the Newton-type methods. The DSM (1.24) avoids completely the inversion of the derivative $F'(u)$. Convergence of this method is proved in Chapter 10, where a DSM scheme, similar to (1.24), is constructed for solving ill-posed problems (1.1).

9. *DSM can be used for solving equations (1.1) in Banach spaces.*

   In particular, if $F : X \to X$ is an operator in a Banach space X and the following spectral assumption holds:

   $$
   \|A_\varepsilon^{-1}\| \le \frac{c}{\varepsilon}, \quad 0 < \varepsilon < \varepsilon_0,
   \tag{1.27}
   $$

   where $c > 0$ is a constant,

   $$
   A_\varepsilon := A + \varepsilon I, \quad \varepsilon = const > 0,
   \tag{1.28}
   $$

   and $\varepsilon_0 > 0$ is an arbitrary small fixed number, then the DSM can be used for solving the equation

   $$
   F(u) + \varepsilon u = 0.
   \tag{1.29}
   $$

10. *DSM can be used for construction of convergent iterative schemes for solving equation (1.1).*

    The general idea is simple. Suppose that a DSM is justified for equation (1.1). Consider a discretization of (1.2)

    $$
    u_{n+1} = u_n + h_n \Phi(t_n, u_n), \quad u_0 = u_0, \quad t_{n+1} = t_n + h_n,
    \tag{1.30}
    $$

or some other discretization scheme. Assume that the scheme (1.30) converges:

$$\lim_{n\to\infty} u_n = u(\infty). \tag{1.31}$$

Then (1.30) is a convergent iterative scheme for solving equation (1.1) because $F(u(\infty)) = 0$.

It is clear now that the DSM has a very wide range of applicability. The author hopes that some numerical schemes for solving operator equations (1.1), which are based on the DSM, will be more efficient than some of the currently used numerical methods.

1.4   A DISCUSSION OF DSM

The reader may ask the following question:

Why would one like to solve problem (1.2) in order to solve a simpler looking problem (1.1)?

The answer is:

First, one may think that problem (1.1) is simpler than problem (1.2), but, in fact, this thinking may not be justified. Indeed, if problem (1.1) is ill-posed and nonlinear, then there is no general method for solving this problem, while one may try to solve problem (1.2) by using a projection method and solving the Cauchy problem (1.8).

Secondly, there is no clearly defined measure of the notion of the simplicity of problem (1.1) as compared with problem (1.2). As we have mentioned in Section 1.2, the numerical methods for solving (1.8) have been studied in the literature extensively (see, e.g., [39]).

The attractive features of the DSM are: its wide applicability, its flexibility [there are many choices of Φ for which one can justify DSM, i.e., prove (1.5), and many methods for solving the Cauchy problem (1.2)], and its numerical efficiency (we show some evidences of this efficiency in Chapter 15). In particular, one can solve such classical problems as stable numerical differentiation of noisy data, solving ill-conditioned linear algebraic systems, and other problems more accurately and efficiently by a DSM than by traditional methods.

1.5   MOTIVATIONS

The motivations for the development of the DSM in this book are the following ones.

First, we want to develop a general method for solving linear and, especially, nonlinear operator equations. This method is developed especially, but not exclusively, for solving nonlinear ill-posed problems.

Secondly, we want to develop a general method for constructing convergent iterative methods for solving nonlinear ill-posed problems.

CHAPTER 2

ILL-POSED PROBLEMS

In this chapter we discuss various methods for solving ill-posed problems.

2.1 BASIC DEFINITIONS. EXAMPLES

Consider an operator equation

$$F(u) = f, \qquad (2.1)$$

where $F : X \to Y$ is an operator from a Banach space X into a Banach space Y.

Definition 2.1.1 *Problem (2.1) is called well-posed (by J. Hadamard) if F is injective and surjective and has continuous inverse. If the problem is not well-posed, then it is called ill-posed.*

Ill-posed problems are of great interest in applications. Let us give some examples of ill-posed problems which are of interest in applications.
   Example 2.1

Dynamical Systems Method and Applications: Theoretical Developments and Numerical Examples, First Edition. A. G. Ramm and N. S. Hoang.
Copyright © 2012 John Wiley & Sons, Inc.

Solving linear algebraic systems with ill-conditioned matrices.
Let

$$Au = f \qquad (2.2)$$

be a linear algebraic system in $\mathbb{R}^n$, $u, f \in \mathbb{R}^n$, $A = (a_{ij})_{1 \le i,j \le n}$ is an ill-conditioned matrix; that is, the condition number $\kappa(A) = \|A\| \, \|A^{-1}\|$ is large. This definition of the condition number preassumes that A is nonsingular, that is, $\mathcal{N}(A) = \{0\}$. If A is singular, i.e., $\mathcal{N}(A) \neq \{0\}$, then formally $\kappa(A) = \infty$ because $\|A^{-1}\| = \infty$. Indeed,

$$\|A^{-1}\| = \sup_{f \neq 0} \frac{\|A^{-1}f\|}{\|f\|} = \sup_{u = A^{-1}f \neq 0} \frac{1}{\frac{\|Au\|}{\|u\|}} = \frac{1}{\inf_{u \neq 0} \frac{\|Au\|}{\|u\|}} = \infty, \qquad (2.3)$$

because $Au = 0$ for some $u \neq 0$ if $\mathcal{N}(A) \neq \{0\}$.

Problem (2.2) is practically ill-posed if $\mathcal{N}(A) \neq \{0\}$ but $\kappa(A) \gg 1$; that is, $\kappa(A)$ is very large. Indeed, in this case small variations Δf of f may cause large variations Δu of the solution u. One has

$$\frac{\|\Delta u\|}{\|u\|} = \frac{\|A^{-1}\Delta f\|}{\|A^{-1}f\|} \le \frac{\|\Delta f\| \|A^{-1}\|}{\|f\| \|A\|^{-1}} = \kappa(A) \frac{\|\Delta f\|}{\|f\|}, \qquad (2.4)$$

where we have used the inequality $\|A^{-1}f\| \ge \|f\| \|A\|^{-1}$. If the equality sign is achieved in (2.4), then a relative error $\frac{\|\Delta f\|}{\|f\|}$ causes $\kappa(A) \frac{\|\Delta f\|}{\|f\|}$ relative error $\frac{\|\Delta u\|}{\|u\|}$ of the solution. If $\kappa(A) = 10^6$, then the relative error of the solution is quite large.

An example of an ill-conditioned matrix is Hilbert's matrix.

$$h_{ij} = \frac{1}{1 + i + j}, \quad 0 \le i, j \le n. \qquad (2.5)$$

Its condition number is of order 10^{13} for $n = 9$. A 2×2 matrix

$$A = \begin{pmatrix} 4.1 & 2.8 \\ 9.7 & 6.6 \end{pmatrix} \qquad (2.6)$$

has condition number $2,249.5$. Equation (2.2) with A defined by (2.6) and $u = \begin{pmatrix} 1 \\ 0 \end{pmatrix}$ is satisfied if $f = \begin{pmatrix} 4.1 \\ 9.7 \end{pmatrix}$. If $f_\delta = \begin{pmatrix} 4.11 \\ 9.70 \end{pmatrix}$, then the corresponding solution $u_\delta = \begin{pmatrix} 0.34 \\ 0.97 \end{pmatrix}$. One can see that a small perturbation of f produces a large perturbation of the solution.

The Hilbert matrix for all $n \ge 0$ is positive-definite, because it is a Gramian of a system of linearly independent functions:

$$h_{ij} = \int_0^1 x^{i+j} dx.$$

Example 2.2
Stable summation of the Fourier series and integrals with randomly perturbed coefficients.

Suppose that

$$f = \sum_{j=1}^{\infty} c_j h_j(x), \tag{2.7}$$

where $\langle h_i, h_j \rangle = \delta_{ij}$, where $\delta_{ij} = \begin{cases} 0, & i \neq j, \\ 1, & i = j, \end{cases}$ and $c_j = \langle f, h_j \rangle$, where $\langle f, h \rangle$ is the inner product in a Hilbert space $H = L^2(D)$.

Let us assume that $\{c_{j\delta}\}_{1 \leq j < \infty}$ are given, and $\sup_j |c_{j\delta} - c_j| \leq \delta$. The problem is to estimate f, given the set $\{\delta, c_{j\delta}\}_{1 \leq j < \infty}$.

If f_δ is an estimate of f and $\lim_{\delta \to 0} \|f_\delta - f\| = 0$, then the estimate is called stable. Here $\|f\| = \langle f, f \rangle^{1/2}$.

Methods for calculating a stable estimate of f given noisy data will be discussed later.

Example 2.3
Stable numerical differentiation of noisy data.

Suppose that $f \in C^2([0,1])$ is not known, but the noisy data $f_\delta \in L^\infty([0,1])$ are given and it is assumed that $\|f - f_\delta\| \leq \delta$, where the norm is L^∞-norm.

The problem is:

Given the noisy data $\{\delta, f_\delta\}$, estimate stably f'.

We prove that this problem, as stated, does not have a solution. In order to solve this problem, one has to have additional information, namely one has to assume an a priori bound

$$\|f^{(a)}\| \leq M_a, \quad a > 1, \tag{2.8}$$

where $f^{(a)}$ is the derivative of order a. If a is not an integer, one defines M_a as follows. Let $a = m + b$, where m is an integer and $0 < b < 1$.

Then

$$\|f^{(a)}\| = \|f^{(m)}\| + \sup_{x,s \in [0,1]} \frac{|f^{(m)}(x) - f^{(m)}(s)|}{|x - s|^b}. \tag{2.9}$$

One can prove that the data $\{\delta, f_\delta, M_a\}$ with any fixed $a > 1$ allow one to construct a stable approximation of f and to estimate the error of this approximation. For example, this error is $O(\delta^{\frac{1}{2}})$ if $a = 2$ and is $O(\delta^{\frac{b}{1+b}})$ if $a = 1 + b$, $0 < b < 1$.

Usually in the literature the stable approximation of f is understood as an estimate $R_\delta f_\delta$ such that

$$\lim_{\delta \to 0} \|R_\delta f_\delta - f'\| = 0. \tag{2.10}$$

A. G. Ramm had introduced ([128]) a new definition of the stable approximation of f namely, the following one:

Let us call the estimate $R_\delta f_\delta$ of f' stable if

$$\lim_{\delta \to 0} \sup_{f \in K(\delta,a)} \|R_\delta f_\delta - f'\| = 0, \qquad (2.11)$$

where

$$K(\delta, a) := \{f : \|f^{(a)}\| \le M_a, \quad \|f - f_\delta\| \le \delta\}. \qquad (2.12)$$

The new definition (2.11)–(2.12) has an advantage over the standard definition because in the new definition there is no dependence on f (and remember that f is unknown), in contrast with the standard definition, which uses the unknown f.

The estimate $R_\delta f_\delta$ has to be constructed on the basis of the known data $\{\delta, f_\delta, M_a\}$ only. These data may correspond to any f in the set $K(\delta, a)$. Since f is not known and can be any element from the set (2.12), it is more natural to define the stable approximation of f by formula (2.11) rather than by formula (2.10).

A detailed study of the practically important problem of stable numerical differentiation will be given in Section 15.2.

Example 2.4
Stable solution of Fredholm integral equations of the first kind.
Consider the equation

$$Au = f, \quad Au = \int_D A(x,y)u(y)\,dy, \qquad (2.13)$$

where $D \subset \mathbb{R}^n$ is a bounded domain and the function $A(x,y) \in L^2(D \times D)$ or

$$\sup_{x \in D} \int_D |A(x,y)|dy \le M. \qquad (2.14)$$

If $A \in L^2(D \times D)$, then the operator A in (2.13) is compact in $H = L(D)$. If (2.14) holds and

$$\lim_{h \to 0} \sup_{|x-s| \le h} \int_D |A(x,y) - A(s,y)|\,dy = 0, \qquad (2.15)$$

then the operator A in (2.13) is compact in $X = C(D)$.
Indeed, if $A \in L^2(D \times D)$, then

$$\int_D \int_D |A(x,y)|^2 dx dy < \infty. \qquad (2.16)$$

In this case the operator A in (2.13) is a Hilbert–Schmidt (HS) operator, which is known to be compact in $H = L^2(D)$.

For convenience of the reader, let us prove the following known [69] result:

Theorem 2.1.1 *Integral operator (2.13) is compact as an operator from $L^p(D)$ into $L^q(D)$ if*

$$\int_D \int_D |A(x,y)|^{r'} dx dy \le M^{r'}, \qquad r = \min(p,q), \tag{2.17}$$

where $M = const > 0$, $q' = \frac{q}{q-1}$, $r' = \frac{r}{r-1}$, $p \ge 1, q \ge 1$, and

$$\|A\|_{L^p \to L^q} := \|A\| \le M|D|^{\frac{q-r(q-1)}{qr}}, \qquad |D| := \text{meas } D. \tag{2.18}$$

Remark 2.1.2 If $p = q = r = r' = 2$, then (2.18) yields $\|A\| \le M$, so meas D can be infinite in this case, but $\|A\| \le M$.

Proof: If $p \ge r$, then one has

$$|(Au)(x)| \le \left(\int_D |A(x,y)|^{r'} dy \right)^{\frac{1}{r'}} \|u\|_{L^r} \le \left(\int_D |A(x,y)|^{r'} dy \right)^{\frac{1}{r'}} \|u\|_{L^p},$$

where $L^s = L^s(D)$.

Using Hölder's inequality, one gets

$$\left(\int_D |(Au)(x)|^q dx \right)^{\frac{1}{q}} \le \|u\|_{L^p} \left(\int_D dx \right)^{\frac{q-r(q-1)}{qr}} \left(\int_D dx \int_D |A(x,y)|^{r'} dy \right)^{\frac{1}{r'}}.$$

Thus, estimate (2.18) is proved.

To prove the compactness of A, note that estimate (2.17) implies that $A(x,y) \in L^2(D \times D)$. Therefore there is a finite-rank kernel $A_m(x,y) = \sum_{j,k=1}^{m} a_j(x) b_k(y)$, which approximates $A(x,y)$ in the $L^{r'}(D \times D)$ with arbitrary accuracy, provided that m is sufficiently large. Therefore

$$\lim_{m \to \infty} \|A - A_m\|_{L^p \to L^q} = 0, \tag{2.19}$$

by estimate (2.18). Since $A_m(x,y)$ is a finite-rank kernel, the corresponding operator A with this kernel is compact. Thus the operator A is compact being the limit of a sequence of compact operators A in the operator norm.

Theorem 2.1.1 is proved. ∎

It is well known that a linear compact operator in an infinite-dimensional Banach space X cannot have a bounded inverse. Indeed if A is a linear compact operator in X and B is its bounded inverse, then $BA = I$, where I is the identity operator, and $I = BA$ is compact as a product of a compact and bounded operator. But the identity operator is compact only in a finite-dimensional Banach space. Therefore B cannot be bounded if it exists. Consequently, problem (2.13) is ill-posed.

Some methods for stable solution of equation (2.13), given the noisy data, are developed in this chapter.

Example 2.5

Analytic continuation.

Let f be an analytic function in a domain D on a complex plane. Assume that f is known on a set $E \subset C$, which has a limit point inside D. Then, by the well-known uniqueness theorem, the function f is uniquely determined everywhere in D. The problem of analytical continuation of f from the set E to D is ill-posed. Indeed, if the noisy data f are given on the set E, such that $\sup_{z \in E} |f_\delta(z) - f(z)| \leq \delta$, then $f_\delta(z)$ may not be an analytic function in D, and in this case it may be not defined in D, or, if $f_\delta(z)$ is analytic in D, its values in D can differ very much from the values of f. Consider a simple example:

$$f(z) = e^z, \quad D = \{z : |z| < 1\}, \quad E = \{z : |z| \leq a\}.$$

Let $a = 10^{-5}$, $f_\delta(z) = \frac{1}{1-z}$. Then one has

$$\max_{|z| \leq a} \left| e^z - \frac{1}{1-z} \right| = \max_{|z| \leq a} \left| \sum_{n=2}^{\infty} z^n \left(\frac{1}{n!} - 1 \right) \right| \leq \frac{a^2}{2} + \frac{a^3}{1-a} := \delta.$$

If $a \leq 10^{-5}$, then $\delta < 10^{-10}$. However, at $z_0 = 1 - 10^{-5}$ one has

$$\left| e^{z_0} - \frac{1}{1 - z_0} \right| < 10^5.$$

Example 2.6

The Cauchy problem for elliptic equations.

Suppose that

$$\Delta u = 0 \text{ in } D \subset \mathbb{R}^n, \quad u|_S = f, \quad u_N|_S = h, \tag{2.20}$$

where D is a domain, S is its boundary, and N is the unit outer normal to S.

Finding u from the data $\{f, h\}$ is an ill-posed problem: Small perturbation of the data $\{f, h\}$ may lead to the pair $\{f_\delta, h_\delta\}$, which does not correspond to any harmonic function in D. The function h in the data (2.20) cannot be chosen arbitrarily. One knows that f alone determines u in (2.20) uniquely. Therefore f determines h uniquely as well. The map

$$\Lambda : f \to h$$

is called the Dirichlet-to-Neumann map. This map is injective and its properties are known.

Example 2.7

Minimization problems.

Let

$$f(u) \geq m > -\infty$$

be a continuous functional in a Banach space X. Consider the problem of finding its global minimum

$$m = \inf_u f(u)$$

and its global minimizer y,

$$f(y) = m.$$

We assume that the global minimizer exists and is unique and that $m > -\infty$. While the problem of finding global minimum is well-posed, the problem of finding global minimizer is ill-posed. Let us explain these claims. Consider $f_\delta(u) = f(u) + g_\delta(u)$, where $\sup_{u \in X} |g_\delta(u)| \leq \delta$. One has

$$\inf_u [f(u) + g_\delta(u)] \leq \inf_u f(u) + \sup g_\delta(u) \leq m + \delta$$

and

$$m - \delta \leq \inf_u f(u) - \sup_u |g_\delta| \leq \inf_u [f(u) + g_\delta(u)].$$

Thus

$$m - \delta \leq \inf_u [f(u) + g_\delta(u)] \leq m + \delta, \tag{2.21}$$

provided that

$$\sup_u |g_\delta(u)| \leq \delta.$$

This proves that small perturbations of f lead to small perturbations of the global minimum.

The situation with global minimizer is much worse: Small perturbation of f can lead to large perturbations of the global minimizer. For instance, consider the function

$$f(x) = -\cos x + \varepsilon x^2 e^{-x^2}, \quad x \in \mathbb{R}.$$

This function has a unique global minimizer $x = 0$, and the global minimum m is -1 for any fixed value of $\varepsilon > 0$. If $g_\delta(x)$ is a continuous function, such that

$$\sup_{x \in \mathbb{R}} |g_\delta(x)| \leq \delta, \quad g_\delta(0) > 0,$$

then one can choose $g_\delta(x)$ so that the global minimizer will be as far from $x = 0$ as one wishes.

Example 2.8

Inverse scattering problem in quantum mechanics [137]

Let

$$[\nabla^2 + k^2 - q(x)]u = 0 \quad \text{in } \mathbb{R}^3, \quad k = const > 0, \tag{2.22}$$

$$u = e^{ik\alpha \cdot x} + v, \quad \alpha \in S^2, \tag{2.23}$$

$$\lim_{r \to \infty} \int_{|x|=r} |\frac{\partial v}{\partial |x|} - ikv|^2 ds = 0, \tag{2.24}$$

where S^2 is the unit sphere in $\mathbb{R}^3$, α is a given unit vector, the direction of the incident plane wave, and q is a real-valued function, which is called potential and which we assume compactly supported:

$$q \in Q_a := \{q : q(x) = 0 \text{ if } |x| > a, \quad q(x) \in L^2(B_a), \quad q = \bar{q}\}, \tag{2.25}$$

where $B_a = \{x : |x| \leq a\}$. One can prove that the scattered field v is of the form

$$v = A(\alpha', \alpha, k)\frac{e^{ikr}}{r} + o\left(\frac{1}{r}\right), \quad r := |x| \to \infty, \quad \frac{x}{r} = \alpha'. \tag{2.26}$$

The coefficient $A = A(\alpha', \alpha, k)$ is called the scattering amplitude. The scattering problem (2.22)–(2.24) is uniquely solvable under the above assumptions (and even under less restrictive assumptions on the rate of decay of q at infinity; see, e.g., [137]). Therefore, the scattering amplitude $A = A_q$ is uniquely determined by the potential q. The inverse scattering problem of quantum mechanics consists of finding the potential from the knowledge of the scattering amplitude A on some subset of $S^2 \times S^2 \times \mathbb{R}_+$.

A detailed discussion of this problem in the case when the above subset is $S_1^2 \times S_2^2 \times k_0$, $k_0 = const > 0$, and S_1^2 and S_2^2 are arbitrary small open subsets of S^2, that is, in the case of fixed-energy data, is given in [137]. The inverse scattering problem, formulated above, is ill-posed: A small perturbation of the scattering amplitude may be a function $A(\alpha', \alpha, k_0)$, which is not a scattering amplitude corresponding to a potential from the class Q_a or even from a larger class of potentials.

A. G. Ramm [111, 112, 169] has established the uniqueness of the solution to inverse scattering problem with fixed energy data, gave a characterization of the class of functions which are the scattering amplitudes at a fixed energy of a potential $q \in Q_a$, and gave an algorithm for recovery of a q from $A(\alpha', \alpha) := A(\alpha', \alpha, k_0)$ known for all $\alpha \in S^2$ and all $\alpha' \in S^2$ at a fixed $k = k_0 > 0$ (see also [137]).

The error of this algorithm is also given in [137], see also [124]. Also a stable estimate of a $q \in Q_a$ is obtained in [137] when the noisy data $A_\delta(\alpha', \alpha)$ are given:

$$\sup_{\alpha, \alpha' \in S^2} |A_\delta(\alpha', \alpha) - A(\alpha', \alpha)| \leq \delta. \tag{2.27}$$

Recently [167, 168] A. G. Ramm has formulated and solved the following inverse scattering-type problem with fixed $k = k_0 > 0$ and fixed $\alpha = \alpha_0$ data $A(\beta) := A(\beta, \alpha_0, k_0)$, known for all $\beta \in S^2$. The problem consists in finding a potential $q \in L^2(D)$, such that the corresponding scattering amplitude $A_q(\beta, \alpha_0, k_0) := A(\beta)$ would approximate an arbitrary given function $f(\beta) \in L^2(S^2)$ with arbitrary accuracy:

$$\|f(\beta) - A(\beta)\|_{L^2(S^2)} < \epsilon,$$

where $\epsilon > 0$ is an a priori given, arbitrarily small, fixed number. In [167] it is proved that this problem has a (nonunique) solution, and an analytic formula is found for one of the potentials, which solve this problem. The domain $D \in \mathbb{R}^3$ in the above problem is an arbitrary bounded domain.

More recently A. G. Ramm [170]–[172] has proved uniqueness of the solution to two non-overdetermined inverse scattering problems in $\mathbb{R}^3$. One of them is the inverse problem with back-scattering data $A(-\beta, \beta, k)$, known for all β in an open subset of S^2 and all $k \in (k_1, k_2)$, where $0 \le k_1 < k_2$. The other one is the inverse problem with the scattering data $A(\beta, \alpha_0, k)$, known for a fixed $\alpha_0 \in S^2$, all $\beta \in S_1^2$, and all $k \in (k_1, k_2)$, $0 \le k_1 < k_2$. These problems were open for many decades.

Example 2.9

Inverse obstacle scattering.

Consider the scattering problem:

$$(\nabla^2 + k^2)u = 0 \text{ in } D' := \mathbb{R}^3 \setminus D, \tag{2.28}$$

$$u|_S = 0, \tag{2.29}$$

$$u = e^{ik\alpha \cdot x} + A(\alpha', \alpha)\frac{e^{ikr}}{r} + o\left(\frac{1}{r}\right), \quad r := |x| \to \infty, \quad \alpha' = \frac{x}{r}, \tag{2.30}$$

where D is a bounded domain with boundary S, $k = const > 0$ is fixed, $\alpha \in S^2$ is given, and the coefficient $A(\alpha', \alpha)$ is called the scattering amplitude.

Existence and uniqueness of the solution to problem (2.28)–(2.30), where D is an arbitrary bounded domain is proved in [137], where some references concerning the history of this problem can be found. In [137], one also finds proofs of the existence and uniqueness of the solution to similar problems with boundary conditions of Neumann type

$$u_N|_S = 0, \tag{2.31}$$

where N is the unit exterior normal to the surface S and of Robin type

$$(u_N + hu)|_S = 0, \tag{2.32}$$

under minimal assumptions on the smoothness of the boundary S. In (2.32), $h \ge 0$ is an $L^\infty(S)$ function. If the Neumann conditions holds, then S is assumed to be such that the imbedding operator

$$i_1 : H^1(D'_\mathbb{R}) \to L^2(D'_\mathbb{R})$$

is compact. Here D'_1 is an open subset of D', $D'_1 = D' \cap B_R$, where B_R is some ball containing D.

If the Robin condition holds, then we assume that i_1 and i_2 are compact, where i_1 has been defined above and

$$i_2 : H^1(D'_\mathbb{R}) \to L^2(S).$$

Here $L^2(S)$ is the L^2 space with the Hausdorff $(n-1)$-measure on it. The Hausdorff d-measure (d-dimensional measure) is defined as follows. If S is a set in $\mathbb{R}^n$, consider various coverings of this set by countably many balls of radii $r_j \le r$. Let

$$h(r) := B(d) \inf \sum_j r_j^d,$$

where $B(d)$ is the volume of a unit ball in $\mathbb{R}^d$ and the infimum is taken over all the coverings of S. Clearly $h(r)$ is a nonincreasing function, so that it is nondecreasing as $r \to 0$. Therefore there exists the limit (finite or infinite)

$$\lim_{r \to 0} h(r) := \Lambda(S).$$

This limit $\Lambda(S)$ is called d-dimensional Hausdorff measure of S. The restriction on the smoothness of S, which is implied by the compactness of the imbedding operators i_1 and i_2, are rather weak: Any Lipschitz boundary S satisfies these restrictions, but Lipschitz boundaries form a small subset of the boundaries for which i_1 and i_2 are compact (see [36] and [37]).

The existence and uniqueness of the solution to the obstacle scattering problem imply that the scattering amplitude $A(\alpha', \alpha)$ is uniquely defined by the boundary S and by the boundary condition on S (the Dirichlet condition (2.29), the Neumann condition (2.31), or the Robin one (2.32)).

The inverse obstacle scattering problem consists of finding S and the boundary condition (the Dirichlet, Neumann, or Robin) on S, given the scattering amplitude on a subset of $S^2 \times S^2 \times \mathbb{R}_+$. The first basic uniqueness theorem for this inverse problem has been obtained by M. Schiffer in 1964 (see [107] and [137]; M.Schiffer did not publish his beautiful proof). He assumed that the Dirichlet condition (2.29) holds and that $A(\alpha', \alpha, k)$ is known for a fixed $\alpha = \alpha_0$, all $\alpha' \in S$ and all $k > 0$.

The second basic uniqueness theorem was obtained in 1985 [107] by A. G. Ramm, who did not preassume the boundary condition on S and proved the following uniqueness theorem:

The scattering data $A(\alpha', \alpha)$, given at an arbitrary fixed $k = k_0 > 0$ for all $\alpha' \in S_1^2$ and $\alpha \in S_2^2$, determine uniquely the surface S and the boundary condition on S of Dirichlet, Neumann, or Robin type.

Here S_1^2 and S_2^2 are arbitrarily small fixed open subsets of S^2 (solid angles), and the boundary condition is either Dirichlet, or Neumann, or Robin type. It is still an open problem to prove the uniqueness theorem for the inverse obstacle scattering problem if $A(\alpha') := A(\alpha', \alpha_0, k_0)$ is known for all $\alpha' \in S^2$, a fixed $\alpha = \alpha_0 \in S^2$, and a fixed $k = k_0 > 0$.

A recent result ([144] in this direction is a uniqueness theorem under additional assumptions on the geometry of S (convexity of S and nonanalyticity of S).

The inverse obstacle scattering problem is ill-posed by the same reason as the inverse potential scattering problem in Example 2.8: Small perturbation of the scattering amplitude may throw it out of the set of scattering amplitudes. A characterization of the class of scattering amplitudes is given in [110]; see also [109] and [114].

The absolute majority of the practically interesting inverse problems are ill-posed.

Let us mention some of these problems in addition to the two inverse scattering problems mentioned above.

Example 2.10

Inverse problem of geophysics.

Let

$$[\nabla^2 + k^2 + k^2 v(x)]u = -\delta(x - y) \text{ in } \mathbb{R}^3, \qquad (2.33)$$

where $k = const > 0$, $v(x)$ is a compactly supported function, $v \in L^2(D)$, and $\bar{D} = \operatorname{supp} v \subset \mathbb{R}^3_-$, where supp v is the support of v, and $\mathbb{R}^3_- := \{x : x_3 < 0\}$. We assume that u satisfies the radiation condition

$$\frac{\partial u}{\partial |x|} - iku = o\left(\frac{1}{|x|}\right), \quad |x| \to \infty, \qquad (2.34)$$

uniformly in directions $\frac{x}{|x|}$. One may think that $P = \{x : x_3 = 0\}$ is the surface of the earth, $v(x)$ is the inhomogeneity in the velocity profile, u is the acoustic pressure, and y is the position of the point source of this pressure.

The simplest model inverse problem of geophysics consists of finding $v(x)$ from the knowledge of $u(x, y, k)$ for all $x \in P_1$, all $y \in P_2$, and a fixed $k > 0$ (or for all $k \in (0, k_0)$, where $k_0 > 0$ is an arbitrarily small fixed number; in this case the data $u(x, y, k)$, $k \in (0, k_0)$, are called low-frequency data).

Here P_1 and P_2 are open sets in P. A more realistic model allows one to replace equation (2.30) with

$$[\nabla^2 + k^2 n(x) + k^2 v(x)]u = \delta(x - y) \quad \text{in} \quad \mathbb{R}^3, \qquad (2.35)$$

where the nonconstant background refraction coefficient $n(x)$ is known. It can be a fairly arbitrary function [114].

In geophysical modeling, one often assumes that $n_0(x) = 1$ for $x_3 > 0$ (in hot air) and $n(x) = n_0 = const$ for $x_3 < 0$ (homogeneous earth).

The inverse geophysical problem is ill-posed: A small perturbation of the data $u(x, y, k)$, $x, y \in P_1 \times P_2$, may lead to a function which is not the value of the solution to problem (2.33)–(2.34) for $v \in L^2(D)$.

Example 2.11

Finding small subsurface inhomogeneities from surface scattering data.

The inverse problem can be formulated as the inverse problem of geophysics in Example 2.10, with the additional assumptions

$$D = \cup_{j=1}^J D_j, \quad \operatorname{diam} D_j := a_j, \quad \max_j a_j := a_j, \quad ka \ll 1, \quad kd \gg 1, \quad (2.36)$$

where

$$d = \min_{i \neq j} \operatorname{dist}(D_i, D_j).$$

The inverse problem is:

Given $u(x, y, k)$ for $x \in P_1, y \in P_2$, where P_1 and P_2 are the same as in Example 2.10 and $k = k_0 > 0$ fixed, find the positions of D_j, their number J, and their intensities $V_j := \int_{D_j} v(x) \, dx$.

This problem is ill-posed by the same reasons as the inverse geophysical problem. A method for solving this problem is given in [140].

Example 2.12

Antenna synthesis problem.

Let an electric current be flowing in a region D. This current creates an electromagnetic field according to the Maxwell's equations

$$\nabla \times E = iw\mu H, \quad \nabla \times H = -iw\varepsilon E + j, \tag{2.37}$$

where w is the frequency, ε and μ are dielectric and magnetic parameters, and j is the current. If ε and μ are constants, then one can derive the following formula ([137, p. 11]):

$$E = -iw\mu \frac{e^{ikr}}{r}[\alpha', [\alpha', J]] + o\left(\frac{1}{r}\right), \quad r = |x|, \quad \alpha' = \frac{x}{r}, \tag{2.38}$$

where $k = w\sqrt{\varepsilon\mu}$, $[a, b]$ is the cross product and

$$J = \frac{1}{4\pi} \int_D e^{-ik\alpha' \cdot y} j(y)\, dy, \tag{2.39}$$

where the integral is taken over the support D of the current $j(y)$. We assume that D is bounded.

The inverse problem, which is called the antenna synthesis problem, consists of finding $j(x)$ from the knowledge of the radiation pattern

$$[\alpha', [\alpha', J]]$$

for all $\alpha' \in S^2$.

This problem is ill-posed and, in general, it may have many solutions. One has to restrict the admissible currents to obtain an inverse problem which has at most one solution. For example, let us assume that

$$j(x) = j(x_3)\delta(x_1)\delta(x_2)e_3, \quad x_3 := z, \quad -a \le z \le a,$$

where $\delta(x_j)$ is the delta function and e_3 is the unit vector along the z-axis. Thus, we deal with the linear antenna. The radiation pattern in this case is

$$\frac{1}{4\pi}[\alpha', [\alpha', e_3]] \int_{-a}^{a} e^{-ikzu} j(z)\, dz, \quad u = \cos\theta = e_3 \cdot \alpha',$$

and the problem of linear antenna synthesis consists of finding $j(z)$ from the knowledge of the desired diagram $f(u)$:

$$\int_{-a}^{a} e^{-ikzu} j(z)\, dz = f(u), \quad -1 \le u \le 1. \tag{2.40}$$

Equation (2.40) has at most one solution. Indeed, if $f(u) = 0$, then the left-hand side of (2.40) vanishes on the set $-1 \le u \le 1$ and is an entire function of u.

By the uniqueness theorem for analytic functions, one concludes that

$$\int_{-a}^{a} e^{-ikzu} j(z) \, dz = 0 \quad \text{for all } u \in \mathbb{C},$$

and, in particular, for all $u \in \mathbb{R}$. This and the injectivity of the Fourier transform imply $j(z) = 0$. So, the uniqueness of the solution to (2.40) is proved. Solving equation (2.40) is an ill-posed problem because small perturbations of f may be nonanalytic, and equation (2.40) with any analytic right-hand side f has no solution. There is an extensive literature on antenna synthesis problems [9, 81, 96, 97, 99, 100].

Example 2.13

Inverse problem of potential theory.

Consider the gravitational potential generated by a mass with density ρ distributed in a domain D:

$$\int_{D} \frac{\rho(y)}{4\pi|x-y|} \, dy = u(x). \tag{2.41}$$

The inverse problem of potential theory consists of finding $\rho(y)$ from the measurements of the potential $u(x)$ outside D. This problem is not uniquely solvable, in general. For example, a point mass distribution, $\rho(y) = M\delta(x)$, generates the potential $\frac{M}{4\pi|x|}$ in the region $|x| > 0$, which is the same as the mass the uniformly distributed in a ball $B_a = \{x : |x| \leq a\}$ with the density

$$\rho(y) = \frac{3M}{4\pi a^3},$$

so that the total mass of the ball is equal to M.

However, if one assumes a priori that the mass density ρ is identically equal to 1, then the domain D, which is star-shaped with respect to a point $x \in D$, is uniquely determined by the knowledge of the potential $u(x)$ in the region $|x| > R$, where the ball B_R contains D.

Example 2.14

Tomography and integral geometry problems.

Suppose there is a family of curves, and the integrals of a function over every curve from this family are known. The problem consists of recovery of f from the knowledge of these integrals. An important example is tomography. If the family of straight lines $l_{\alpha p}$, where

$$l_{\alpha p} = \{x : \alpha \cdot x = p\}, \qquad \alpha \in S^1,$$

where S^1 is the unit sphere in $\mathbb{R}^2$, $x \in \mathbb{R}^2$, $p \geq 0$, is a number, then the knowledge of the family of integrals

$$\hat{f}(\alpha, p) = \int_{l_{\alpha p}} f(x) \, ds, \tag{2.42}$$

allows one to recover f uniquely. Analytical inversion formulas are known [118]. The function $\hat{f}(\alpha, p)$ is called the Radon transform of f. In applications $\hat{f}(\alpha, p)$ is called a tomogram. Finding f from $\hat{f}$ is an ill-posed problem: If $\hat{f}$ is the Radon transform of a compactly supported continuous function f (or L^2 function f), and if $\hat{f}$ is perturbed a little in the sup-norm (or L^2-norm), then the resulting function $\hat{f} + h$, $\|h\| \leq \delta$, may be not the Radon transform of any compactly supported L^2-function.

In practice, a dominant role is played by the local tomography.

In local tomography, one has only the local tomographic data, that is, the values $\hat{f}(\alpha, p)$ for α, p which satisfy the inequality

$$|\alpha \cdot x - p| \leq \varepsilon, \tag{2.43}$$

where $x_0 \in \mathbb{R}^2$ is a given fixed point in a region of interest, for example, in a region of human body around which one thinks a tumor is possible, and $\varepsilon > 0$ is a small number. One of the basic motivations for the theory of local tomography, originated in [116] and [117], and developed in [118], was the desire to minimize the X-ray radiation of patients. From the knowledge of local tomographic data, one cannot find the function f, because the inversion formula in $\mathbb{R}^2$ is nonlocal (e.g., see [118, p. 31]):

$$f(x) = \frac{1}{4\pi} \int_{S^1} d\alpha \int_{-\infty}^{\infty} \frac{\hat{f}_p(\alpha, p)}{\alpha \cdot x - p} \, dp. \tag{2.44}$$

Here

$$\hat{f}(\alpha, p) = \hat{f}(-\alpha, -p), \qquad \forall p \in \mathbb{R},$$

S^1 is the unit sphere in $\mathbb{R}^2$, that is, the set $|\alpha| = 1$, $\alpha \in \mathbb{R}^2$, and $\hat{f}_p = \frac{\partial \hat{f}}{\partial p}$. Formula (2.44) is proved in [118] for smooth rapidly decaying functions f, but it remains valid for $f \in H_0^1(\mathbb{R}^2)$, where $H_0^1(\mathbb{R}^2)$ is the set of functions in the Sobolev space $H^1(\mathbb{R})$ with compact support.

Nonlocal nature of the inversion formula (2.44) means that one has to know $\hat{f}(\alpha, p)$ for all $\alpha \in S^1$ and all $p \in \mathbb{R}$ in order to recover $f(x)$ at the point x. A. G. Ramm has posed the following question:

If one cannot recover $f(x)$ from local tomographic data, what practically useful information can one recover from these data?

The answer, given in [116]–[118], is:

One can recover the discontinuity curves of f and the sizes of the jumps of f across the discontinuity curves.

Example 2.15

Inverse spectral problem.

Consider the problem

$$lu := -u'' + q(x)u + \lambda u, \quad 0 \leq x \leq 1, \tag{2.45}$$

$$u(0) = u(1) = 0. \tag{2.46}$$

Assume that

$$q = \bar{q}, \quad q \in L^1(0,1). \tag{2.47}$$

Then problem (2.45)–(2.46) has a discrete spectrum

$$\lambda_1 < \lambda_2 \le ..., \quad \lim_{n \to} \lambda_n = \infty. \tag{2.48}$$

A natural question is:

Does the set of eigenvalues $\{\lambda_j\}_{j=1,2,...}$ determine $q(x)$ uniquely?

The answer is:

It does not, in general.

The set of $\{\lambda_j\}_{\forall j}$ determines roughly speaking half of the potential $q(x)$ in the following sense (see, e.g., [137]): If $q(x)$ is unknown on half of the interval $0 \le x \le \frac{1}{2}$, then the knowledge of all $\{\lambda_j\}_{\forall j}$ determines $q(x)$ on the remaining half of the interval $\frac{1}{2} < x < 1$ uniquely.

There is an exceptional result, however, due to Ambarzumian (1929) [8], which says that if $u'(0) = u'(1) = 0$, then the set of the corresponding eigenvalues $\{\mu_j\}_{\forall j}$ determines q uniquely [137]. A multidimensional generalization of this old result is given in [137].

Let us define the spectral function of the self-adjoint Dirichlet operator $l = -\frac{d^2}{dx^2} + q(x)$ in $L(\mathbb{R}_+)$, $\mathbb{R}_+ = [0, \infty)$. This operator can be defined as the closure of the symmetric operator l_0 defined on twice continuously differentiable functions u vanishing at $x = 0$ and near infinity and such that $lu \in L(\mathbb{R}_+)$.

It is not trivial to prove that l_0 is densely defined in $H = L^2(\mathbb{R}_+)$ if one assumes that $q \in L^1(\mathbb{R}_+)$ or if

$$q \in L_{1,1} := \{q : q = \bar{q}, \int_0^\infty x|q(x)| \, dx < \infty\}.$$

The idea of the proof is as follows (cf. [86]): The operator

$$l_0 + a^2 = \frac{d^2}{dx^2} + q(x) + a^2$$

is symmetric on its domain of definition $D(l_0) \subset H$, as one can check easily by integration by parts; the equation

$$lu + a^2u = -u'' + q(x)u + a^2u = f, \quad u(0) = 0, \tag{2.49}$$

is uniquely solvable for any sufficiently large $a > 0$ and for any $f \in H$, and its solution $u \in H^1(\mathbb{R}_+)$; if $h \in H$ and $h \perp D(l_0)$, then $h = lv$ and $\langle v, l_0 w + a^2 w \rangle = 0$, $\forall w \in D(l_0)$, where $\langle u, w \rangle$ is the inner product in $L^2(0, \infty)$; consequently, $v \in H$ solves homogeneous ($f = 0$) problem (2.49); if $a > 0$ is sufficiently large, this problem has only the initial the trivial solution $v = 0$; thus $h = 0$, and the claim is proved: $D(l_0)$ is dense in H.

With the self-adjoint Dirichlet operator l one associates the spectral function $d\rho(\lambda)$. This is the function for which

$$\int_0^\infty |f(x)|^2 \, dx + \sum_{j=1}^J \langle f, \varphi_j \rangle \varphi_j = \int_{-\infty}^\infty |\tilde{f}(\lambda)|^2 \, d\rho(\lambda), \qquad (2.50)$$

for any $f \in L^2(\mathbb{R}_+)$. Here

$$\tilde{f}(\lambda) = \int_0^\infty f(x)\varphi(x,\lambda) \, dx, \quad l\varphi = \lambda\varphi, \quad \varphi(0,\lambda) = 0, \quad \varphi'(0,\lambda) = 1,$$

and

$$l\varphi_j = \lambda_j \varphi_j, \quad \varphi_j(0) = 0, \quad \|\varphi_j\|_H = 1, \qquad (2.51)$$

i.e. φ_j are the normalized eigenfunctions of l, corresponding to possibly existing negative spectrum of l. If $q \in L_{1,1}$ this spectrum is known to be finite (e.g., see [35]).

If $q \in L_{1,1}$ then there is a unique spectral function corresponding to the self-adjoint operator l.

The inverse spectral problem consists of finding $q(x)$ from the knowledge of $\rho(\lambda)$. This problem is ill-posed: Small perturbations of $\rho(\lambda)$ may lead to a function which is not a spectral function.

Example 2.16

Inverse problems for wave equation.

Consider the wave equation

$$\frac{1}{c^2(x)} u_{tt} = \Delta u + \delta(x - y) \quad \text{in} \quad \mathbb{R}^3, \qquad (2.52)$$

$$u|_{t=0} = 0, \quad u_t|_{t=0} = 0. \qquad (2.53)$$

The function $c(x)$ is the wave velocity. Assume that

$$c^{-2}(x) = c_0^{-2}(x)[1 + v(x)],$$

where $c_0(x)$ is the wave velocity in the background medium and $v(x)$ is the inhomogeneity in the wave velocity. Assume that $v(x)$ is unknown, compactly supported in $\mathbb{R}^3_- := \{x : x_3 < 0\}$.

The inverse problem is:

Given the measurements of $u(x,y,t)$, the solution to (2.52)–(2.53) for all values of $x \in P_1$, $y \in P_2$ and $t \in [0,T]$, where $T > 0$ is some number, find $v(x)$, provided that $c_0(x)$ is known. Here P_1 and P_2 are open sets on the plane $x_3 = 0$, which is the surface of the Earth in the geophysical model.

This inverse problem is ill-posed: Small perturbations of $u(x,y,t)$ may lead to a function which is not a restriction of the solution to problem (2.52)–(2.53) to the plane $x_3 = 0$.

Example 2.17

Inverse problems for the heat equation.
Consider the problem

$$u_t = \Delta u - q(x)u, \quad t \geq 0, \quad x \in D, \tag{2.54}$$

$$u|_{t=0} = 0, \tag{2.55}$$

$$u|_S = f, \quad S := \partial D. \tag{2.56}$$

This problem has a unique solution $u(x,t)$.

Let the flux

$$u_N|_S = h(s,t), \tag{2.57}$$

be measured for any $f \in H^{\frac{3}{2}}(S)$, where $H^m(S)$ is the Sobolev space and N is the exterior unit normal to S.

The inverse problem is:

Given the set of pairs $\{f, h\}$ for all $t \in [0,T]$, where $T > 0$ is a number, find $q(x)$.

This problem is ill-posed: Small perturbations of h may lead to a function which is not a normal derivative of a solution to problem of the type (2.54)–(2.56).

Example 2.18
Inverse conductivity problem.

This problem is also called impedance tomography problem. Consider the stationary problem

$$\nabla \cdot (a(x)\nabla u) = 0 \text{ in } D, \quad u|_S = f, \tag{2.58}$$

where

$$0 < a_0 \leq a(x) \leq a_1 < \infty, \quad a \in H^2(D), \tag{2.59}$$

and assume that $S = \partial D$ is sufficiently smooth. Problem (2.58) has a unique solution, so the function

$$u_N|_S = h, \tag{2.60}$$

where N is the unit exterior normal to S, is uniquely defined if $a(x)$ and f are known.

The inverse problem is:

Given the set $\{f, h\}$, find $a(x)$.

This problem has at most one solution but it is ill-posed (see, e.g., [114] and [137]).

Example 2.19
Deconvolution and other imaging problems.

Consider, for instance, the problem

$$\int_0^t k(t,s)u(s)\, ds = f(t). \tag{2.61}$$

The deconvolution problem consists of finding $u(s)$ from the knowledge of f and $k(t, s)$. This problem is ill-posed as was explained in Example 2.4, where a more general Fredholm-type equation of the first kind was discussed.

Example 2.20

Heat equation with reversed time.

Consider the problem:

$$u_t = u_{xx}, \quad t \geq 0, \quad 0 \leq x \leq x_b; \quad u(0, t) = u(x_b, t) = 0, \quad u(x, 0) = f(x).$$

This problem has a unique solution:

$$u(x, t) = \sum_{j=1}^{\infty} e^{-\lambda_j t} f_j \varphi_j(x); \quad f = \langle f, \varphi_j \rangle_{L^2(0, \pi)} := \langle f, \varphi_j \rangle,$$

$$\varphi_j(x) = \sqrt{\frac{2}{\pi}} \sin(jx); \quad \lambda_j = j^2; \quad \langle \varphi_j, \varphi_m \rangle = \delta_{j,m}.$$

Consider the following problem:

Given the function $g(x)$ and a number $T > 0$, can one find $f(x) = u(x, 0)$ such that $u(x, T) = g(x)$?

In general, the answer is no: Not every $g(x)$ can be the value of the solution $u(x, t)$ of the heat equation at $t = T > 0$. For example, $g(x)$ has to be infinitely differentiable. However, given an arbitrary $g \in L^2(0, \pi)$ and an arbitrary small $\varepsilon > 0$, one can find f, such that $\|u(x, T) - g(x)\| \leq \varepsilon$, where $u(x, t)$ is the solution to the heat equation with $u(x, 0) = f(x)$. This is easy to see from the formula

$$u(x, T) = \sum_{j=1}^{\infty} e^{-\lambda_j T} f_j \varphi_j(x).$$

The inequality

$$\|u(x, T) - g\| \leq \varepsilon$$

holds if

$$\sum_{j=1}^{\infty} |e^{-\lambda_j T} f_j - g_j|^2 \leq \varepsilon^2.$$

If $\sum_{j \geq j(\varepsilon)} |g_j|^2 < \frac{\varepsilon^2}{2}$, then one may take

$$f_j = 0 \quad \text{for} \quad j \geq j(\varepsilon)$$

and

$$f_j = g_j e^{\lambda_j T} \quad \text{for} \quad j < j(\varepsilon)$$

and get

$$\|u(x, T) - g\| \leq \varepsilon.$$

Note that if $T > 0$ and $j(\varepsilon)$ are large, then the coefficients f_j are extremely large.

Thus, the above problem is highly ill-posed: small perturbations of g may lead to arbitrary large perturbations of f, or throw g out of the set of functions, which are the values of $u(x, T)$. Similar results hold in multidimensional problems for the heat equation.

The number of examples of ill-posed problems, which are of interest in applications, can be easily increased.

Let us define the notion of regularizer.

Definition 2.1.2 *An operator R_δ is called a regularizer for the problem*

$$Au = f \tag{2.62}$$

if

$$\lim_{\delta \to 0} \|R_\delta f_\delta - u\| = 0, \tag{2.63}$$

where R_δ is a bounded operator defined on the whole space; the relation (2.63) should hold for any $f \in R(A)$ and some solution u of equation (2.62).

In the literature it is often assumed that A is injective, and then the solution to equation (2.63) is unique. However, one may also consider the case when A is not injective. Usually the operator R_δ is constructed as a two-parameter family $R_{\delta,a}$, where parameter a is called a regularization parameter, and for a suitable choice of $a = a(\delta)$ the operator $R_{\delta,a(\delta)} := R_\delta$ satisfies Definition 2.1.2; methods for constructing of $R_{\delta,a}$ and for choosing $a = a(\delta)$ are discussed in many papers and books (e.g., see [64], [137], and [193] and references therein).

Sometimes the requirement (2.63) is replaced by

$$\lim_{\delta \to 0} \sup_{f_\delta \in B(f,\delta)} \|R_\delta f_\delta - u\| = 0, \tag{2.64}$$

where $B(f, \delta) = \{g : \|g - f\| \leq \delta\}$.

A. G. Ramm has proposed a different definition of the regularizer R_δ (see [137]):

Definition 2.1.3 *An operator R_δ is a regularizer for the problem $Au = f$ if*

$$\lim_{\delta \to 0} \sup_{f \in B(f_\delta, \delta),\; f = Au} \|R_\delta f_\delta - u\| = 0, \tag{2.65}$$

where $B(f_\delta, \delta) = \{g : \|g - f_\delta\| \leq \delta\}$.

The motivation for this new definition is natural: The given data are $\{f_\delta, \delta\}$ and f is unknown. This unknown f may be any element of the set $B(f_\delta, \delta) \cap R(A)$. Therefore the regularizer must recover u corresponding to any such f. There is a considerable practical difference between the two definitions (2.64) and (2.65). For instance, one may be able to find a regularizer in the sense (2.64), but this regularizer will not be a regularizer in the sense (2.65). Consider a particular example. Let

$$R_\delta f_\delta = \frac{f_\delta(x + h(\delta)) - f_\delta(x - h(\delta))}{2h(\delta)}. \tag{2.66}$$

This is a regularizer for the problem of stable numerical differentiation. It was first proposed in [98] and then used extensively (see [137]). The choice of $h(\delta)$ depends on the a priori information about the unknown function f whose derivative we want to estimate stably given the data $\{f_\delta, \delta, M_a\}$, where $\|f_\delta - f\|_\infty \leq \delta$, the norm is $L^\infty(a,b)$-norm, the interval (a,b) is arbitrary. Without loss of generality one may take $a = 0$, $b = 1$, which we will do below. The number M_a is the a priori known upper bound for the derivative of f of order $a > 0$. If $a = j + b$, where j is an integer, $j > 0$, and $b \in (0,1)$, then

$$M_a = \sup_{x \in [0,1]} \{|f^{(j)}(x)| + |f(x)|\} + \sup_{x,y \in [0,1]} \frac{|f^{(j)}(x) - f^{(j)}(y)|}{|x - y|^b}. \tag{2.67}$$

It is proved in [137] that no regularizer (linear or nonlinear) can be found for the problem of stable numerical differentiation of noisy data if the regularizer is understood in the sense (2.65) and $a \leq 1$, and that such a regularizer can be found in the form (2.66) with a suitable $h(\delta)$ provided that $a > 1$. A regularizer in the sense (2.64) can be found for the problem of stable numerical differentiation with $a = 1$; however, practically this regularizer is of no use. A detailed discussion of this problem is given in Section 15.2.

2.2 VARIATIONAL REGULARIZATION

Consider a linear ill-posed problem

$$Au = f, \tag{2.68}$$

where A is a closed, densely defined linear operator in a Hilbert space H.

We assume that problem (2.68) is ill-posed, that $Ay = f$, where

$$y \perp \mathcal{N}, \quad \mathcal{N} := \mathcal{N}(A) := \{u : Au = 0\},$$

and that f_δ is given in place of f,

$$\|f_\delta - f\| \leq \delta, \quad \|f_\delta\| > c\delta, \quad c = const, \quad c \in (1,2).$$

$$\mathcal{F}(u) := \|Au - f_\delta\|^2 + a\|u\|^2 = \min, \tag{2.69}$$

where $a > 0$ is regularization parameter.

The method of variational regularization (VR) for stable solution of equation (2.68), that is, for finding the regularizer R_δ such that (2.63) holds, consists of finding the functional (2.69) and then choosing $a = a(\delta)$ such that

$$\lim_{\delta \to 0} a(\delta) = 0, \tag{2.70}$$

and (2.63) holds with $R_\delta f_\delta = u_\delta := u_{a(\delta),\delta}$.

Let us show that the global minimizer of (2.69) exists and is unique. Indeed, functional (2.69) is quadratic, so a necessary condition for its minimizer is a linear equation, the Euler equation. Assume first that A is bounded. Then the Euler equation for the functional (2.69) is

$$A^*Au + au = A^*f_\delta. \tag{2.71}$$

Let us denote

$$A^*A := T, \quad T_a : T + aI. \tag{2.72}$$

Since $T = T^* \geq 0$, the operator T has a bounded inverse, $\|T_a^{-1}\| \leq \frac{1}{a}$, so equation (2.71) has a solution $u_{a,\delta} = T_a^{-1}A^*f_\delta$ and this solution is unique. One has

$$\mathcal{F}(u_{a,\delta}) \leq \mathcal{F}(u). \tag{2.73}$$

Indeed, one can check that

$$\mathcal{F}(u_{a,\delta}) \leq \mathcal{F}(u_{a,\delta} + v), \quad \forall v \in H,$$

and the equation sign is attained if and only if $v = 0$.

Let us choose $a = a(\delta)$ so that $u_\delta := u_{a(\delta),\delta}$ would satisfy the relation

$$\lim_{\delta \to 0} \|u_\delta - y\| = 0. \tag{2.74}$$

To do so, we estimate:

$$\|T_a^{-1}A^*f_\delta - y\| \leq \|T_a^{-1}A^*(f_\delta - f)\| + \|T_a^{-1}A^*f - y\| := J_1 + J_2. \tag{2.75}$$

Note that

$$\|T_a^{-1}A^*\| \leq \frac{1}{2\sqrt{a}}. \tag{2.76}$$

To prove (2.76) one uses the commutation formula

$$T_a^{-1}A^* = A^*Q_a^{-1}, \quad Q := AA^*. \tag{2.77}$$

This formula is easy to check: Multiply (2.77) by Q_a from the right and by T_a from the left and get an obvious relation:

$$A^*(AA^* + aI) = (A^*A + aI)A^*.$$

Reversing the above derivation, one gets (2.77). Using the polar decomposition

$$A^* = U(AA^*)^{\frac{1}{2}}, \tag{2.78}$$

where U is a partial isometry, one gets, using the spectral theorem,

$$\|T_a^{-1}A^*\| = \|A^*Q_a^{-1}\| \leq \|Q^{\frac{1}{2}}Q_a^{-1}\| = \sup_{s \geq 0} \frac{s^{\frac{1}{2}}}{s + a} = \frac{1}{2\sqrt{a}}, \tag{2.79}$$

so formula (2.76) is verified. Thus

$$J_1 \le \frac{\delta}{2\sqrt{a}}. \tag{2.80}$$

To estimate J_2 in (2.75), one uses the spectral theorem again and gets

$$J_2^2 = \|T_a^{-1}Ty - y\|^2 = a^2\|T_a^{-1}y\| = \int_0^{\|T\|} \frac{a^2 d\langle E_s y, y\rangle}{(a+s)^2} := \beta^2(a). \tag{2.81}$$

One has

$$\lim_{a\to 0} \beta^2(a) = \|P_{\mathcal{N}} y\|^2 = 0, \tag{2.82}$$

because $y \perp \mathcal{N}$ by the assumption, and

$$\mathcal{N} = (E_0 - E_{-0})H.$$

From (2.75) and (2.80)–(2.82), one gets

$$\|u_{a,\delta} - y\| \le \frac{\delta}{2\sqrt{a}} + \beta(a). \tag{2.83}$$

Taking any $a(\delta)$ such that

$$\lim_{\delta\to 0} a(\delta) = 0, \quad \lim_{\delta\to 0} \frac{\delta}{\sqrt{a(\delta)}} = 0 \tag{2.84}$$

and setting $u_{a(\delta),\delta} := u_\delta$, one obtains (2.74).

Let us summarize our result.

Theorem 2.2.1 *Assume that $a = a(\delta)$ and (2.84) holds. Then the element $u_\delta = T_{a(\delta)}^{-1} A^* f_\delta$ satisfies (2.74).*

Remark 2.2.2 Without additional assumptions on y, it is impossible to estimate the rate of decay of $\beta(a)$ as $a \to 0$. Therefore, it is impossible to get a rate of convergence in (2.74): The convergence can be as slow as one wishes for some y. The usual assumption which would guarantee some rate of decay of β and, therefore, of $\|u_\delta - y\|$ is the following one:

$$y = T^b z, \quad 0 < b \le 1. \tag{2.85}$$

If (2.85) holds, then

$$\beta^2(a) = \int_0^{\|T\|} \frac{a^2 s^{2b} d\langle E_s z, z\rangle}{(a+s)^2} \le a^{2b}(1-b)^{2-2b} b^{2b} \|z\|^2. \tag{2.86}$$

If $\|T\| < \infty$, then one can give a rate for $b > 1$ as well, but the rate will be $O(a)$ as $a \to 0$, so that for $b \ge 1$ there is a saturation. The case $\|T\| = \infty$ is discussed below separately.

Thus,

$$\beta(a) \leq c\|z\|a^b, \quad 0 < b \leq 1, \tag{2.87}$$

where $c = (1-b)^{1-b}b^b$.

If one minimizes for a small fixed $\delta > 0$ the function

$$\frac{\delta}{2\sqrt{a}} + c\|z\|a^b = \min \tag{2.88}$$

in the region $a > 0$, then one gets

$$a = a(\delta) = c_1\delta^{\frac{2}{2b+1}}, \quad c_1 = \left(\frac{1}{4bc\|z\|}\right)^{\frac{2}{2b+1}}. \tag{2.89}$$

Thus, under the condition (2.87), one gets

$$\|u_\delta - y\| = O(\delta^{\frac{2}{2b+1}}), \quad \delta \to 0. \tag{2.90}$$

Let us now generalize the above theory to include unbounded, closed, densely defined operators A. The main difficulty is the following one: The element f_δ in (2.71) may not belong to the domain $D(A^*)$ of A^*. Our result is stated in the following theorem.

Theorem 2.2.3 *The operator $T_a^{-1}A^*$, defined originally on $D(A^*)$, is closable. Its closure denoted again $T_a^{-1}A^*$ is defined on all of H and is a bounded linear operator with the norm bounded as in (2.76) for any $a > 0$. The relation (2.77) holds.*

Proof: Let us first check that the operator $T_a^{-1}A^*$ with the domain $D(A^*)$ is closable. Recall that a densely defined linear operator B in a Hilbert space is closable if it has a closed extension. This happens if and only if the closure of the graph of B is again a graph. In other words, if $u_n \to 0$ and $Bu_n \to f$, then $f = 0$.

The operator B is called closed if $u_n \to u$ and $Bu_n \to f$ implies $u_n \in D(B)$ and $Bu = f$. The operator B is closed if and only if its graph is a closed subset in $H \times H$, that is, the set $\{u, Bu\}$ is a closed linear subspace in $H \times H$.

Let us check that the operator $T_a^{-1}A^*$ with domain $D(A^*)$ is closable. Let $u_n \in D(A^*)$ and $u_n \to 0$, and assume that $T_a^{-1}A^*u_n \to g$, as $n \to \infty$. We wish to prove that $g = 0$. For any $u \in H$ one gets

$$\langle g, u \rangle = \lim_{n \to \infty} \langle T_a^{-1}A^*u_n, u \rangle = \lim_{n \to \infty} \langle u_n, AT_a^{-1}u \rangle = 0, \tag{2.91}$$

where we have used the closedness of A.

Since u is arbitrary, equation (2.91) implies $g = 0$. So the operator $T_a^{-1}A^*$ with the domain $D(A^*)$ is closable.

Estimate (2.76) and formula (2.77) remain valid and their proofs are essentially the same.

Theorem 2.2.3 is proved.   ∎

Theorem 2.2.4 *The conclusion of Theorem 2.2.1 remains valid for unbounded, closed, densely defined operator A under the same assumptions as in Theorem 2.2.1.*

Proof of Theorem 2.2.4 is the same as that of Theorem 2.2.1 after Theorem 2.2.3 is established.

Let us now discuss an a posteriori choice of $a(\delta)$, the discrepancy principle. Let $u_{a,\delta} = T_a^{-1} A^* f_\delta$. For a fixed $\delta > 0$, consider the equation

$$\|Au_{a,\delta} - f_\delta\| = \|AT_a^{-1} A^* f_\delta - f_\delta\| = c\delta, \quad c \in (1, 2), \tag{2.92}$$

as an equation for $a = a(\delta)$. Here c is a constant and we assume that $\|f_\delta\| > c\delta$. We prove the following result.

Theorem 2.2.5 *Equation (2.92) has a unique solution $a = a(\delta)$ for every sufficiently small $\delta > 0$, provided that $\|f_\delta - f\| \le \delta$, $\|f_\delta\| > c\delta$, and $f = Ay$. One has $\lim_{\delta \to 0} a(\delta) = 0$ and (2.74) holds with $u_\delta := T_{a(\delta)}^{-1} A^* f_\delta$.*

Proof: Using formula (2.77) and the spectral theorem for the self-adjoint operator Q, one gets

$$\|AT_a^{-1} A^* f_\delta - f_\delta\|^2 = \|[QQ_a^{-1} - I]f_\delta\|^2 = \int_0^\infty \frac{a^2 d\langle E_s f_\delta, f_\delta\rangle}{(a+s)^2} := h(\delta, a). \tag{2.93}$$

The function $h(\delta, a)$ is continuous with respect to a on the interval $(0, \infty)$ for any fixed $\delta > 0$. One has

$$h(\delta, \infty) = \int_0^\infty d\langle E_s f_\delta, f_\delta\rangle = \|f_\delta\|^2 > c^2\delta^2, \tag{2.94}$$

and

$$h(\delta, +0) = \|P_{\mathcal{N}^*} f_\delta\|^2 \le \delta^2, \tag{2.95}$$

where

$$\mathcal{N}^* = \mathcal{N}(A^*) = \mathcal{N}(Q),$$

P is the orthogonal projector onto $\mathcal{N} = \mathcal{N}(A)$, and we have used the following formulas:

$$\lim_{a \to 0} \int_0^\infty \frac{a^2 d\langle \mathcal{E}_s f_\delta, f_\delta\rangle}{(a+s)^2} = \|(\mathcal{E}_0 - \mathcal{E}_{-0})f_\delta\|^2 = \|P_{\mathcal{N}^*} f_\delta\|^2, \tag{2.96}$$

where $\mathcal{E}_s$ is the resolution of the identity corresponding to Q, and

$$\|P_{\mathcal{N}^*} f_\delta\| \le \|P_{\mathcal{N}^*} f\| + \|P_{\mathcal{N}^*} (f_\delta - f)\| \le \delta, \tag{2.97}$$

because $P_{\mathcal{N}^*}(A) = 0$ and $f \in \mathcal{R}(A)$.

From (2.94), (2.95), and the continuity of $h(\delta, a)$ it follows that equation (2.92) has a solution. This solution is unique because $h(\delta, a)$ is a monotonically growing function of a for a fixed $\delta > 0$.

Also,

$$\lim_{\delta \to 0} a(\delta) = 0,$$

because $\lim_{\delta \to 0} h(\delta, a(\delta)\mathcal{R}) = 0$, and $h(\delta, a(\delta)) \geq c_1 > 0$ if $\lim_{\delta \to 0} a(\delta) \geq c_2 > 0$, where c_1 and c_2 are some constants.

Let us now check that (2.74) holds with $u_\delta = T_a^{-1} A^* f_\delta$. We have

$$\mathcal{F}(u_\delta) = c^2 \delta^2 + a(\delta) \|u_\delta\|^2 \leq \mathcal{F}(y) = \delta^2 + a(\delta) \|y\|^2. \tag{2.98}$$

Since $c > 1$, one gets

$$\|u_\delta\| \leq \|y\|. \tag{2.99}$$

Thus

$$\limsup_{\delta \to 0} \|u_\delta\| \leq \|y\|. \tag{2.100}$$

It follows from (2.99) that there exists a weakly convergent subsequence $u_\delta \rightharpoonup u$ as $\delta \to 0$, where we have denoted this subsequence also u_δ. Thus

$$\|u\| \leq \liminf_{\delta \to 0} \|u_\delta\|. \tag{2.101}$$

From (2.100) and (2.101) it follows that $\|u\| \leq \|y\|$. Let us prove that $Au = f$. Since $\|u\| \leq \|y\|$ and the minimal-norm solution to equation (2.68) is unique, we conclude that $u = y$. To verify the equation $Au = f$, we argue as follows: from (2.92) it follows that $\lim_{\delta \to 0} \|Au_\delta - f\| = 0$, so, for any $g \in D(A^*)$, one gets

$$\langle f, g \rangle = \lim_{\delta \to 0} \langle Au_\delta, g \rangle = \lim_{\delta \to 0} \langle u_\delta, A^* g \rangle = \langle u, A^* g \rangle. \tag{2.102}$$

Since $A^{**} = \overline{A} = A$, where the overbar denotes the closure of A, this implies

$$\langle f - Au, g \rangle = 0, \qquad \forall g \in D(A^*).$$

Since $D(A^*)$ is dense, it follows that $Au = f$, as claimed. The density of $D(A^*)$ follows from the assumption that A is closed and densely defined.

Let us now prove (2.74). We have already proved that $u_\delta \rightharpoonup y$ and $\|u_\delta\| \leq \|y\|$. This implies (2.74). Indeed

$$\|u_\delta - y\|^2 = \|u_\delta\|^2 + \|y\|^2 - 2\mathrm{Re}\langle u_\delta, y \rangle \to 0 \quad \text{as} \quad \delta \to 0.$$

Theorem 2.2.5 is proved. ∎

We have assumed in Theorem 2.2.5 that $u_{a,\delta} := T_a^{-1} A^* f_\delta$ is the exact minimizer of the functional (2.69). Suppose that $w_{a,\delta}$ is an approximate minimizer of (2.69) in the following sense:

$$\mathcal{F}(w_{a,\delta}) \leq m + (c^2 - 1 - b)\delta^2, \quad c^2 > 1 + b, \tag{2.103}$$

where $c \in (1,2)$ is a constant, $b > 0$ is a constant, and $m := \inf_u \mathcal{F}(u)$.

The problem is

Will the equation

$$\|Aw_{a,\delta} - f_\delta\| = c\delta \tag{2.104}$$

be solvable for a for any $\delta > 0$ sufficiently small?

Will the element $w_\delta := w_{a(\delta),\delta}$ converge to y?

Our result is stated in the following new discrepancy principle.

Theorem 2.2.6 *Assume that assumption (2.103) holds and that the following assumptions hold:*

$$\|f_\delta\| > c\delta, \quad c = const \in (1,2), \quad \|f_\delta - f\| \le \delta, \quad f = Ay, \quad y \perp \mathcal{N}.$$

Then, for any $w_{a,\delta}$, satisfying (2.103) and depending continuously on a, equation (2.104) has a solution $a = a(\delta)$, such that $w_\delta = w_{a(\delta),\delta}$ converges to y:

$$\lim_{\delta \to 0} \|w_\delta - y\| = 0. \tag{2.105}$$

Proof: Let $H(\delta,a) := \|Aw_{a,\delta} - f_\delta\|$. Then $H(\delta,a)$ is continuous with respect to a because $w_{a,\delta}$ is continuous with respect to a. Let us verify that

$$H(\delta, +0) < c\delta, \quad H(\delta, \infty) > c\delta. \tag{2.106}$$

Then there exists $a = a(\delta)$ which solves (2.104).

As $a \to \infty$, we have

$$a\|w_{a,\delta}\|^2 \le \mathcal{F}(w_{a,\delta}) \le m + (c - 1 - b)\delta^2 \le \mathcal{F}(0) + (c^2 - 1 - b)\delta^2. \tag{2.107}$$

Since $\mathcal{F}(0) = \|f_\delta\|^2$, one gets from (2.107)

$$\|w_{a,\delta}\| \le \frac{c}{\sqrt{a}}, \quad a \to \infty. \tag{2.108}$$

Therefore

$$\lim_{a \to \infty} \|w_{a,\delta}\| = 0,$$

so

$$H(\delta, \infty) = \|A0 - f_\delta\| = \|f_\delta\| > c\delta. \tag{2.109}$$

Let $a \to 0$. Then

$$H^2(\delta, a) \le \mathcal{F}(w_{a,\delta}) \le m + (c^2 - 1 - b)\delta^2 \le \mathcal{F}(y) + (c^2 - 1 - b)\delta^2.$$

One has

$$\mathcal{F}(y) = \delta^2 + a\|y\|^2.$$

So

$$H^2(\delta, a) \le (c^2 - b)\delta^2 + a\|y\|^2.$$

Thus

$$H(\delta, +0) \le (c^2 - b)^{\frac{1}{2}}\delta < c\delta. \tag{2.110}$$

From (2.110) and (2.109), one gets (2.106). So, the existence of the solution $a = a(\delta)$ of equation (2.104) is proved.

Let us prove (2.105). One has

$$\mathcal{F}(w_\delta) = \|Aw_\delta - f_\delta\|^2 + a(\delta)\|w_\delta\|^2 = c^2\delta^2 + a(\delta)\|w_{a,\delta}\|^2 \le m + (c^2 - 1 - b)\delta^2.$$

Thus, using the inequality $m \le \delta^2 + a(\delta)\|y\|^2$, one gets

$$c^2\delta^2 + a(\delta)\|w_\delta\|^2 \le a(\delta)\|y\|^2 + (c^2 - b)\delta^2.$$

Therefore

$$\|w_\delta\| \le \|y\|. \tag{2.111}$$

As in the proof of Theorem 2.2.5, inequality (2.111) and equation (2.104) imply $w_\delta \rightharpoonup w$, $Aw_\delta \longrightarrow f$ as $\delta \to 0$, and $Aw = f$. Since $\|w\| \le \|y\|$, as follows from (2.111), it follows that $w = y$, because y is the unique minimal-norm solution to equation (2.68). Thus, $w_\delta \rightharpoonup y$ and $\|w_\delta\| \le \|y\|$. This implies (2.105), as was shown in the proof of Theorem 2.2.5. Theorem 2.2.6 is proved. ∎

Let us discuss variational regularization for nonlinear equations. Let A be a possibly nonlinear, injective, closed map from a Banach space X into a Banach space Y. Let

$$F(u) := \|A(u) - f_\delta\| + \delta g(u),$$

where $g(u) \ge 0$ is a functional. Assume that the set $\{u : g(u) \le c\}$ is precompact in X. Assume that

$$A(y) = f, \quad \|f_\delta - f\| \le \delta, \quad \text{and} \quad D(A) \subset D(g),$$

so that $y \in D(g)$.

Theorem 2.2.7 *Under the above assumptions let u_δ be any sequence such that $F(u_\delta) \le c\delta$, $c := 2 + g(y)$. Then $\lim_{\delta \to 0} \|u_\delta - y\| = 0$.*

Proof: Let

$$m := \inf_{u \in D(A)} F(u), \quad F(u_n) \le m + \frac{1}{n},$$

where $n = n(\delta)$ is the smallest positive integer satisfying the inequality $\frac{1}{n} \le \delta$. One has

$$m \le F(y) = \delta[1 + g(y)]$$

and

$$F(u_n) \le c\delta, \quad c = 2 + g(y).$$

Thus, $g(u_n) \leq c$. Consequently, one can select a convergent subsequence u_δ, $\|u_\delta - u\| \to 0$ as $\delta \to 0$. One has

$$0 = \lim_{\delta \to 0} F(u_\delta) = \lim_{\delta \to 0} \{\|A(u_\delta) - f_\delta\| + \delta g(u_\delta)\} = \lim_{\delta \to 0} \|A(u_\delta) - f\|.$$

Thus $u_\delta \to u$ and $A(u_\delta) \to f$. Since A is closed, this implies $A(u) = f$. Since A is injective and $A(y) = f$, one gets $u = y$. Thus

$$\lim_{\delta \to 0} \|u_\delta - y\| = 0.$$

Theorem 2.2.7 is proved.   ∎

Let us prove the following theorem.

Theorem 2.2.8 *Functional (2.69) has a unique global minimizer*

$$u_{a,\delta} = A^*(Q + aI)^{-1} f_\delta$$

for any $f_\delta \in H$, where $Q := AA^*$, $a = const > 0$.

Proof: Consider the equation

$$(Q + aI)w_{a,\delta} = (AA^* + aI)w_{a,\delta} = f_\delta.$$

It is uniquely solvable:
$$w_{a,\delta} = (Q + aI)^{-1} f_\delta.$$

Define $u_{a,\delta} := A^* w_{a,\delta}$. Then

$$Au_{a,\delta} - f_\delta = -aw_{a,\delta}.$$

One has

$$\begin{aligned} F(u + v) =& \|Au - f_\delta\|^2 + a\|u\|^2 + \|Av\|^2 + a\|v\|^2 \\ &+ 2\mathrm{Re}[\langle Au - f_\delta, Av \rangle + a\langle u, v \rangle], \qquad \forall v \in D(A). \end{aligned}$$

Let $u = u_{a,\delta}$. Then

$$\langle Au_{a,\delta} - f_\delta, Av \rangle + a\langle u_{a,\delta}, v \rangle = -a\langle w_{a,\delta}, Av \rangle + a\langle u_{a,\delta}, v \rangle = 0,$$

because
$$\langle w_{a,\delta}, Av \rangle = \langle A^* w_{a,\delta}, v \rangle = \langle u_{a,\delta}, v \rangle.$$

Consequently,

$$F(u_{a,\delta} + v) = F(u_{a,\delta}) + a\|v\|^2 + \|Av\|^2 \geq F(u_{a,\delta}),$$

and $F(u_{a,\delta} + v) = F(u_{a,\delta})$ if and only if $v = 0$. Therefore

$$u_{a,\delta} = A^*(Q + aI)^{-1} f_\delta$$

is the unique global minimizer of $F(u)$.

Theorem 2.2.8 is proved.   ∎

Let us show how the results can be extended to the case when not only f is known with an error δ, but the bounded operator A is also known with an error δ, that is, A_δ is known, $\|A_\delta - A\| \leq \delta$, and A is unknown.

Consider, for instance, an analog of Theorem 2.2.1. Let us define

$$u_\delta := T_{a(\delta),\delta}^{-1} A_\delta^* f_\delta, \qquad (2.112)$$

where

$$T_{a,\delta} := A_\delta^* A_\delta + aI. \qquad (2.113)$$

The element

$$u_{a,\delta} := T_{a,\delta}^{-1} A_\delta^* f_\delta \qquad (2.114)$$

solves the equation

$$(T_\delta + aI)u_{a,\delta} = A_\delta^* f_\delta. \qquad (2.115)$$

We have

$$\|T_\delta - T\| \leq \|(A_\delta^* - A^*)A_\delta\| + \|A^*(A_\delta - A)\| \leq 2\delta(\|A\| + \delta). \qquad (2.116)$$

Equation (2.115) can be written as

$$\begin{aligned} u_{a,\delta} =& T_a^{-1} A^* f + T_a^{-1}(A_\delta^* - A^*)f_\delta \\ &+ T_a^{-1} A^*(f_\delta - f) + T_a^{-1}(A^* A - A_\delta^* A_\delta)u_{a,\delta}. \end{aligned} \qquad (2.117)$$

We have

$$\|T_a^{-1}(A^* A - A_\delta^* A_\delta)\| \leq \frac{2\delta(\|A\| + \delta)}{a}. \qquad (2.118)$$

Assume that $a(\delta)$ satisfies the conditions

$$\lim_{\delta\to 0} \frac{\delta}{a(\delta)} = 0, \quad \lim_{\delta\to 0} a(\delta) = 0. \qquad (2.119)$$

Then (2.118) implies that equation (2.117) is uniquely solvable for $u_{a(\delta),\delta} := u_\delta$, and

$$\|u_\delta - T_{a(\delta)}^{-1} A^* f\| \leq c\left(\frac{\delta}{a(\delta)}\|f\| + \frac{\delta}{2\sqrt{a(\delta)}}\right) \xrightarrow[\delta\to 0]{} 0, \qquad (2.120)$$

where c is an upper bound of the norm of the operator $[I - (A^* A - A_\delta^* A_\delta)]^{-1}$. Thus, $c = c(\delta)$ and $\lim_{\delta\to 0} c(\delta) = 1$.

We have proved the following result.

Theorem 2.2.9 *Assume that f_δ and A_δ are given,*

$$\|f_\delta - f\| \leq \delta, \quad \|A_\delta - A\| \leq \delta, \quad Ay = f, \quad y \perp \mathcal{N}(A),$$

u_δ is defined in (2.112), and (2.119) holds. Then $\lim_{\delta \to 0} \|u_\delta - y\| = 0$.

The discrepancy principle can also be generalized to the case when A_δ is given in place of A. This principle for choosing $a(\delta)$ can be formulated as the equation

$$\|A_\delta T_{a,\delta}^{-1} A_\delta^* f_\delta - f_\delta\| = c\delta, \qquad c = const \in (R, R + 1), \tag{2.121}$$

where $R \geq 1 + \|y\|$ is a constant, and we assume that

$$\|f_\delta\| > c\delta. \tag{2.122}$$

Equation (2.121) is uniquely solvable for $a = a(\delta)$, and $\lim_{\delta \to 0} = 0$. This is proved as in the proof of Theorem 2.2.5. Condition (2.121) allows one to get the estimate for $u_\delta := u_{a(\delta),\delta}$, where $a(\delta)$ solves (2.121). This estimate

$$\|u_\delta\| \leq \|y\| \tag{2.123}$$

is similar to the estimate (2.99). It allows one to derive the relation (2.74). The constant $1 + \|y\|$ is a lower bound for R in (2.121) because the inequality

$$\|A_\delta u_{a,\delta} - f_\delta\|^2 + a\|u_{a,\delta}\| \leq \|A_\delta y - f_\delta\|^2 + a\|y\|^2 \tag{2.124}$$

and the relation (2.121) imply inequality (2.123), provided that

$$c \geq \|y\| + 1. \tag{2.125}$$

This follows from the estimate

$$\begin{aligned} \|A_\delta y - f_\delta\| &\leq \|(A_\delta - A)y\| + \|Ay - f_\delta\| \\ &\leq \delta\|y\| + \delta = \delta(\|y\| + 1). \end{aligned} \tag{2.126}$$

Thus, we have proved the following result similar to Theorem 2.2.5.

Theorem 2.2.10 *Equation (2.121) is uniquely solvable for $a = a(\delta)$, provided that (2.92) holds. The relation $\lim_{\delta \to 0} a(\delta) = 0$ holds. The element $u_\delta := u_{a(\delta),\delta}$ satisfies the relation (2.74), provided that (2.125) holds.*

Although the solution y is unknown, an upper bound on y is often known a priori as a part of a priori information about the unkown solution.

2.3   QUASI-SOLUTIONS

Let us assume that $A : X \to Y$ is a continuous operator from a Banach space X into a Banach space Y, $K \subset X$ is a compact set.

Definition 2.3.1 *An element $z \in K$ is called a quasi-solution of the equation $Au = f$ if*

$$\|Az - f\| = \inf_{u \in K} \|Au - f\|.$$

If A is continuous, then the functional $\|Au - f\|$ is continuous. Every continuous functional achieves on a compact set its infimum. Therefore quasi-solutions are well-defined for any $f \in Y$.

Under suitable assumptions, one can prove that the quasi-solution not only exists, but is unique and depends continuously on f. To formulate these assumptions, let us recall some geometrical notions.

Definition 2.3.2 *A Banach space is called strictly convex if $\|u + v\| = \|u\| + \|v\|$ implies $u = \lambda v$, $\lambda = const \in \mathbb{R}$.*

Definition 2.3.3 *Let $M \subset X$ be a convex set, that is, $u, v \in M$ implies $\lambda u + (1 - \lambda)v \in M$ for all $\lambda \in (0,1)$. Then metric projection of an element $w \in X$ onto M is an element $Pw = v \in M$, such that*

$$\|w - v\| = \inf_{z \in M} \|w - z\|.$$

Lemma 2.3.1 *In a strictly convex Banach space X the metric projection onto a convex set M is unique.*

Proof: Suppose that v_1 and v_2 are metric projections of $u \notin M$ onto M, so that

$$\|u - v_1\| = \|u - v_2\| = \inf_{z \in M} \|u - z\| := m > 0.$$

Then, since M is convex, $\frac{v_1 + v_2}{2} \in M$, and we get

$$m \leq \|u - \frac{v_1 + v_2}{2}\| \leq \frac{1}{2}(\|u - v_1\| + \|u - v_2\|) = m.$$

Thus

$$\|u - v_1\| = \|u - v_2\| = \|\frac{u - v_1 + u - v_2}{2}\| > 0.$$

Denote $u - v_1 = p$, $u - v_2 = q$. Then $\|p\| = \|q\| = \frac{1}{2}\|p + q\| > 0$, so $\|p+q\| = \|p\| + \|q\|$. Since X is strictly convex, it follows that $p = \lambda q$, $|\lambda| = 1$, so $\lambda = 1$ or $\lambda = -1$, because λ is real-valued. If $\lambda = -1$, then $\|p + q\| = 0$, a contradiction. So $\lambda = 1$. Thus $p = q$, so $v_1 = v_2$. Lemma 2.3.1 is proved. ∎

Remark 2.3.2 Hilbert space is strictly convex, Lebesgue's spaces $L^p(D)$, $1 < p < \infty$, are strictly convex, but $C(D)$ and $L^1(D)$ are not strictly convex.

Lemma 2.3.3 *The operator P of metric projection onto a convex compact set M of a strictly convex Banach space is continuous.*

Proof: We claim that the distance

$$d(f, M) = \inf_{z \in M} \|f - z\|$$

is a continuous function of f for any set M, not necessarily convex or compact.
Indeed,

$$d(f_1, M) \leq d(f_1, z) \leq d(f_1, f_2) + d(f_2, z),$$

so

$$d(f_1, M) - d(f_2, M) \leq d(f_1, f_2).$$

By symmetry,

$$d(f_2, M) - d(f_1, M) \leq d(f_1, f_2).$$

Thus

$$|d(f_1, M) - d(f_2, M)| \leq d(f_1, f_2),$$

as claimed.

Let us prove the continuity of $P = P_M$. Let $\lim_{n \to \infty} \|f_n - f\| = 0$, and
assume that $\|q_n - q\| > \varepsilon > 0$, where $q_n = P_M f_n$ and $q = P_M f$. Since M is
compact and q is bounded (by the above claim), one may select a convergent
subsequence, which we denote again by q_n,

$$q_n \to v \in M, \quad \|v - q\| \geq \varepsilon > 0.$$

One has $\|f - q\| \leq \|f - v\|$ and

$$\|f - v\| \leq \|f - f_n\| + \|f_n - q_n\| + \|q_n - v\|.$$

Note that

$$\lim_{n \to \infty} \|f - f_n\| = 0, \quad \lim_{n \to \infty} \|q_n - v\| = 0,$$

and, by the claim,

$$\lim_{n \to \infty} \|f_n - q_n\| = \lim_{n \to \infty} d(f_n, M) = d(f, M) = \|f - q\|.$$

Therefore

$$\|f - q\| = \|f - v\|.$$

Lemma 2.3.3 is proved.   ∎

From Lemmas 2.3.1 and 2.3.3 the following result follows.

Theorem 2.3.4 *Assume that A is a linear continuous injection, $M \subset X$ is
convex and compact, and X is strictly convex. Then the quasi-solution to
equation $A(u) = f$ exists, is unique, and depends continuously on f.*

Proof: Existence of the quasi-solution follows from compactness of M and
continuity of A, as was explained below Definition 2.3.1. Uniqueness of the
quasi-solution follows from the injectivity of A. The quasi-solution depends
continuously on f by Lemma 2.3.3 because the set AM is convex and compact,
so the map $f \to P_{AM} f$ is continuous, and the quasi-solution $u = A^{-1} P_{AM} f$
is continuous on the set AM if M is compact.   ∎

The continuity of A^{-1} on AM is a consequence of the following lemma.

Lemma 2.3.5 *Assume that $A : M \to X$ is a possibly nonlinear, injective, closed map from a compact set M of a complete metric space X into X. Then the inverse map A^{-1} is continuous on AM.*

Proof: Let $A(u_n) = f_n$, $f_n \to f$ as $n \to \infty$, $u_n \in M$. Since M is compact, one can select a convergent subsequence, which we denote u_n again, $u_n \to u \in M$. Since A is closed, the convergence $u_n \to u$, $A(u_n) \to f$ implies $A(u) = f$. Since A is injective, the equation $A(u) = f$ defines uniquely $u = u_f = A^{-1}(f)$. Thus $A^{-1}(f_n) \to A^{-1}(f)$. Since the limit of every subsequence u_n is the same, $A^{-1}(f)$, the sequence $u_n = A^{-1}(f_n)$ converges to $A^{-1}(f)$. Lemma 2.3.5 is proved. ∎

Remark 2.3.6 Lemma 2.3.5 differs from the well-known result [29, Lemma I.5.8], because the continuity of A is replaced by the closedness of A.

Let us now consider quasi-solutions for nonlinear equations. Let $A : X \to Y$ be an injective, possibly nonlinear, closed map from a Banach space X into a Banach space Y. Let $K \subset D(A) \subseteq X$ be a compact set. Assume that

$$A(y) = f, \qquad \|f_\delta - f\| \leq \delta.$$

Let

$$m := \inf_{u \in K} \|A(u) - f_\delta\|.$$

Choose a minimizing sequence u_n such that

$$\|A(u_n) - f - \delta\| \leq m + \frac{1}{n}, \quad n = 1, 2....$$

Let $n = n(\delta)$ be the smallest positive integer such that $\frac{1}{n} \leq \delta$, so that

$$\|A(u_n) - f_\delta\| \leq m + \delta.$$

Since

$$m \leq \|A(y) - f_\delta\| = \delta,$$

one has

$$\|A(u_n) - f_\delta\| \leq 2\delta.$$

Thus

$$\|A(u_n) - f\| \to 0 \quad \text{as} \quad \delta \to 0.$$

Since $u \in K$ and K is compact, one can select a convergent subsequence $u_{\delta_j} := v_j$, $v_j \to v$ as $j \to \infty$, $v \in K$. Thus $A(v_j) \to f$, $v_j \to v$, so $A(v) = f$, because A is closed. Since A is injective, the element v is uniquely determined by f. Therefore $v = y$, and

$$\lim_{\delta \to 0} \|u_\delta - y\| = 0.$$

We have proved the following theorem.

Theorem 2.3.7 *Let $A : X \to Y$ be an injective, closed, possibly nonlinear, map from a Banach space X into a Banach space Y. Let $K \subset D(A)$ be a compact set. Assume that $A(y) = f$, and $\|f_\delta - f\| \leq \delta$. Let $u_\delta \in K$ be any sequence such that $\|A(u_\delta) - f_\delta\| \leq 2\delta$. Then $\lim_{\delta \to 0} \|u_\delta - y\| = 0$.*

2.4   ITERATIVE REGULARIZATION

The main result in this section is the following theorem.

Theorem 2.4.1 *Every solvable linear equation $Au = f$ with a closed densely defined operator A in a Hilbert space H can be solved by a convergent iterative process.*

Proof: Let $T = A^*A$ and $Q = AA^*$ be nonnegative self-adjoint operators, $a = const > 0$, $T_a = T + aI$, I is the identity operator, $B = aT_a^{-1}$ we have proved (see Section 2.2, proof of Theorem 2.2.3) that $T_a^{-1}A^*$ is a bounded operator defined on all of H, $\|T_a^{-1}A^*\| \leq \frac{1}{2\sqrt{a}}$. Consider an iterative process

$$u_{n+1} = Bu_n + T_a^{-1}A^*f, \quad u_1 \perp \mathcal{N}, \quad B = aT_a^{-1}, \qquad (2.127)$$

where $\mathcal{N} = \mathcal{N}(A) = \mathcal{N}(T)$. Denote

$$w_n := u_n - y, \qquad n \geq 1,$$

where $Ay = f$, $y \perp \mathcal{N}$. One has

$$y = By + T_a^{-1}A^*f. \qquad (2.128)$$

Thus

$$w_{n+1} = Bw_n, \quad w_1 = u_1 - y \perp \mathcal{N}. \qquad (2.129)$$

Therefore,

$$w_{n+1} = B^n w_1, \quad w_1 \perp \mathcal{N}. \qquad (2.130)$$

Let E_s be the resolution of the identity corresponding to the self-adjoint operator T. Then

$$\lim_{n \to \infty} \|w_{n+1}\|^2 = \lim_{n \to \infty} \int_0^\infty \frac{a^{2n}}{(a+s)^{2n}} d\langle E_s w_1, w_1 \rangle$$
$$= \|(E_0 - E_{-0})w_1\|^2 = \|P_N w_1\|^2 = 0, \qquad (2.131)$$

where P_N is the orthoprojector onto $\mathcal{N}$.   ∎

Remark 2.4.2 If A is a bounded operator, then the operator T_a^{-1} can be easily computed if $a > \|T\| = \|A\|^2$. Indeed

$$T_a^{-1} = (T + aI)^{-1} = a(I + a^{-1}T)^{-1}$$

and $\|T\|a^{-1} < 1$ if $a > \|T\|$, so that

$$(I + a^{-1}T)^{-1} = \sum_{j=0}^{\infty}(-1)^j a^{-j}T^j,$$

and the series converges at the rate of geometrical series.

Remark 2.4.3 One cannot give a rate of convergence in (2.131) without extra assumptions on w_1. If, for example, $w_1 = Tz$, then the integral in (2.131) can be written as

$$\int_0^{\infty} \frac{a^{2n}s^2}{(a+s)^{2n}} d\langle E_s z, z\rangle.$$

One can check that

$$\max_{s \geq 0} \frac{a^{2n}s^2}{(a+s)^{2n}} \leq c\frac{a^2}{n^2},$$

where $c = const > 0$ does not depend on a and n,

$$c = \max \frac{n^2}{(n-1)^2} \frac{1}{(1 + \frac{1}{n-1})^{2n}} \leq 4.$$

Thus, if the extra assumption on w_1 is $w_1 = Tz$, then the rate of decay in (2.131) is

$$\|w_{n+1}\|^2 \leq c\frac{a^2}{n^2}\|z\|^2.$$

A similar calculation can be made if $w_1 = T^b z$, where $b > 0$ is a constant.

Remark 2.4.4 An idea, similar to the one used in the proof of Theorem 2.4.1, can be applied to the equation

$$Au = f, \tag{2.132}$$

where $A = A^*$ is self-adjoint not necessarily bounded operator, and we assume that

$$Ay = f, \quad y \perp \mathcal{N}.$$

Let $a > 0$ be a constant. Equation (2.73) is equivalent to

$$u = iaA_{ia}^{-1}u + A_{ia}^{-1}f, \quad A_{ia} := A + iaI. \tag{2.133}$$

Consider the iterative process

$$u_{n+1} = iaA_{ia}^{-1}u_n + A_{ia}^{-1}f, \quad u_1 \perp \mathcal{N}. \tag{2.134}$$

Theorem 2.4.5 *If $A = A^*$ and $a > 0$, then $\lim_{n\to\infty}\|u_n - y\| = 0$, provided that $u_1 \perp \mathcal{N}$.*

Proof: Let $v_n := u_n - y$. Since

$$y = ia A_{ia}^{-1} y + A_{ia}^{-1} f,$$

one gets

$$v_{n+1} = ia A_{ia}^{-1} v_n, \quad v_1 = u_1 - y \perp \mathcal{N}. \tag{2.135}$$

Thus $v_{n+1} = (ia)^n A_{ia}^{-1} v_1$, so

$$\lim_{n \to \infty} \|v_{n+1}\|^2 = \lim_{n \to \infty} \int_{-\infty}^{\infty} \frac{a^{2n} d\langle E_s v_1, v_1 \rangle}{(a^2 + s^2)^n} = \|P_{\mathcal{N}} v_1\|^2 = 0, \tag{2.136}$$

where E is the resolution of the identity, corresponding to A, and $\mathcal{N} = \mathcal{N}(A)$. Theorem 2.4.5 is proved. ∎

Let us discuss another iterative process for solving equation $Au = f$. We assume that this equation is solvable, and $y \perp \mathcal{N}$ is its minimal-norm solution. Let us also assume that A is bounded. Let $T = A^* A$. If $Au = f$, then $Tu = A^* f$. Conversely, if equation $Au = f$ is solvable, then every solution to the equation $Tu = A^* f$ solves equation $Au = f$. Indeed, write $f = Ay$. Then $Tu = Ty$. Multiply this equation by $u - y$ and get $A(u - y) = 0$. Thus $Au = Ay = f$.

The iterative process for solving linear solvable equation $Au = f$ can be written as follows:

$$u_{n+1} = u_n - (T u_n - A^* f), \quad u_1 = 0, \tag{2.137}$$

provided that

$$\|T\| \leq 1. \tag{2.138}$$

Note that (2.138) is not really a restriction because one can always divide the equation $Tu = A^* f$ by $\|T\| > 0$ and denote $\frac{T}{\|T\|}$ by T_1. Then $\|T_1\| = 1$, so T_1 satisfies the restriction (2.138).

Theorem 2.4.6 *One has $\lim_{n \to \infty} \|u_n - y\| = 0$, where $Ay = f$, $y \perp \mathcal{N}$, (2.138) is assumed, and u_n is defined by (2.137).*

Proof: Let $u_n - y := z_n$. One has

$$y = y - (Ty - A^* f).$$

Thus

$$z_{n+1} = z_n - T z_n = (I - T)^n z_1. \tag{2.139}$$

Let E_s be the resolution of the identity corresponding to T. Then

$$\lim_{n \to \infty} \|z_{n+1}\|^2 = \lim_{n \to \infty} \int_0^1 (1 - s)^{2n} d\langle E_s z_1, z_1 \rangle = \|P_N z_1\|^2 = 0, \tag{2.140}$$

because $z_1 = u_1 - y = -y \perp \mathcal{N}$.

Theorem 2.4.6 is proved. ∎

2.5 QUASI-INVERSION

The quasi-inversion method (see [78]) was applied to the ill-posed problem for the heat equation considered in Example 2.20. Following [78], we consider a more general setting. Let

$$\dot{u} + Au = 0, \quad u(0) = f, \tag{2.141}$$

where $A = A^* \geq cI > 0$ is a selfadjoint operator in a Hilbert space H. The problem is:

   Given $g \in H$, a number $T > 0$, and an arbitrary small $\varepsilon > 0$, find f such that $\|u(T) - g\| \leq \varepsilon$.

   The solution $u(t) = u(t; f)$ to (2.141) exists and is unique. It can be written as

$$u = e^{-tA}f = \int_c^\infty e^{-ts}dE_s f,$$

where E_s is the resolution of the identity for A.

   Note that if there is an f such that $u(T; f) = g$, then such an f is unique. Indeed, if there are two such f, f_1, and f_2, then $0 = \int_c^\infty e^{-Ts}dE_s f$, where $f := f_1 - f_2$, so

$$0 = \int_c^\infty e^{-Ts} d\langle E_s f, f \rangle \qquad c > 0.$$

Since $(E_s f, f)$ is a monotonically nondecreasing function, it follows that $f = 0$.

   It is also easy to check that, for any fixed $g \in H$, one has

$$\inf_{f \in H} \|u(T; f) - g\| = 0. \tag{2.142}$$

Indeed, otherwise one would have a nonzero element g_1 orthogonal to the span of $u(T; f)$ for all $f \in H$, i.e., $\langle u(T; f), g_1 \rangle = 0, \forall f \in H$. Taking $f = g_1$, one gets

$$\int_c^\infty e^{-Ts}d\langle E_s g_1, g_1 \rangle = 0.$$

This implies $g_1 = 0$. Thus, (2.142) is proved.

   The problem

$$\dot{u} + Au = 0, \quad u(T) = g \tag{2.143}$$

can be reduced to the problem

$$-\frac{du}{d\tau} + Au = 0, \quad u(0) = g, \quad \tau = T - t. \tag{2.144}$$

   Consider the set

$$M := \{u : \|u(T)\| \leq c\},$$

where u solves (2.144).

Lemma 2.5.1 *If $u \in M$ and $\|g\| \leq \varepsilon$, then*

$$\|u(\tau)\| \leq c^{\frac{\tau}{T}} \varepsilon^{\frac{T-\tau}{T}}, \quad 0 \leq \tau \leq T. \tag{2.145}$$

Proof: Let $h := \|u(\tau)\|^2$. Then

$$\dot{h} := \frac{dh}{d\tau} = 2\mathrm{Re}\langle \dot{u}, u \rangle, \quad \ddot{h} = 2\|\dot{u}\|^2 + 2\mathrm{Re}\langle \ddot{u}, u \rangle.$$

One has

$$\langle \ddot{u}, u \rangle = \langle Au, Au \rangle = \|\dot{u}\|^2,$$

so

$$\ddot{h} = 4\|\dot{u}\|^2.$$

Define $p(\tau) := \ln h$. Then

$$\ddot{p} = \frac{\ddot{h}h - \dot{h}^2}{h^2} = \frac{4\|\dot{u}\|^2 h - 4|Re\langle \dot{u}, u \rangle|^2}{h^2} \geq 0.$$

Therefore $p(\tau)$ is a convex function. Thus

$$p(\tau) \leq \frac{T-\tau}{T}p(0) + \frac{\tau}{T}p(T).$$

This implies

$$h(\tau) \leq [h(0)]^{\frac{T-\tau}{T}}[h(T)]^{\frac{\tau}{T}},$$

or

$$\|u(\tau)\| \leq c^{\frac{\tau}{T}}\varepsilon^{\frac{T-\tau}{T}}.$$

Lemma 2.5.1 is proved.   ∎

Estimate (2.145) gives a continuous dependence of the solution to (2.144) or (2.143) on g under a priori assumption $u \in M$.

Given g, the quasi-inversion method for finding f in the sense

$$\|g - e^{-TA}f\| \leq \eta, \tag{2.146}$$

where $\eta > 0$ is an arbitrary small given number, consists of solving the problem

$$\dot{u}_\varepsilon + Au_\varepsilon - \varepsilon A^2 u_\varepsilon = 0, \quad u(T) = g, \quad t \geq T, \tag{2.147}$$

and then finding $f = f_\eta = u_\varepsilon(0)$. Here $\varepsilon = const > 0$. If $\varepsilon > 0$ is sufficiently small, then

$$\|g - e^{-TA}u_\varepsilon(0)\| \leq \eta.$$

Let us justify the above claim. The unique solution to (2.147) is

$$u_\varepsilon(t) = e^{-(t-T)(A-\varepsilon A^2)}g, \tag{2.148}$$

where the element $u_\varepsilon(t)$,

$$u_\varepsilon(t) = e^{-(t-T)(A-\varepsilon A^2)}g = \int_c^\infty e^{-(t-T)(s-\varepsilon s^2)}dE_s g,$$

is well-defined for any $g \in H$ and for $t \in [0, T)$, provided that $\varepsilon > 0$. Let

$$f = f_\varepsilon := u_\varepsilon(0) = e^{T(A-\varepsilon A^2)}g. \tag{2.149}$$

Then

$$\|g - e^{-TA}f_\varepsilon\|^2 = \|g - e^{-\varepsilon TA^2}g\|^2 = \int_c^\infty |1 - e^{-\varepsilon Ts^2}|^2 d\langle E_s g, g \rangle. \tag{2.150}$$

Thus, for any fixed $g \in H$, one can pass to the limit $\varepsilon \to 0$ in (2.150) and get

$$\lim_{\varepsilon \to 0} \|g - e^{-TA}f_\varepsilon\| = 0. \tag{2.151}$$

Therefore, for any $g \in H$ and any $\eta > 0$, however small, one can find f_ε such that

$$\|g - e^{-TA}f_\varepsilon\| \le \eta.$$

If $g = e^{-TA}h$ for some $h \in H$, that is, g is the value of a solution to equation (2.141) at $t = T$ with $u(0) = h$, then

$$\|g - e^{-TA}f_\varepsilon\| \le \|e^{-TA}\|\|h - f_\varepsilon\| \le \|h - f_\varepsilon\|.$$

Therefore, if $g = e^{-TA}h$, $h \in H$, then taking any f satisfying the inequality $\|f_\varepsilon - h\| \le \eta$ implies $\|g - e^{-TA}f_\varepsilon\| \le \eta$.

Remark 2.5.2 In [78] the case when $A = A(t)$ is treated under suitable assumptions.

Remark 2.5.3 There are infinitely many elements f satisfying (2.146). Formula (2.149) gives one concrete such an element.

Remark 2.5.4 The quasi-inversion method yields stable solution of the above problem: If g_δ is given, $\|g_\delta - g\| \le \delta$, then

$$\|e^{T(A-\varepsilon A^2)}(g_\delta - g)\| \le \|e^{T(A-\varepsilon A^2)}\|\delta = N(\varepsilon, T)\delta, \tag{2.152}$$

where

$$N(\varepsilon, T) = \sup_{s \ge c} e^{T(s-\varepsilon s^2)} = e^{\frac{T}{4\varepsilon}}. \tag{2.153}$$

Remark 2.5.5 In [78] the quasi-inversion method is applied to solving Cauchy problems for elliptic equations and to other ill-posed problems.

2.6   DYNAMICAL SYSTEMS METHOD (DSM)

The idea of a DSM has been described in Section 1.1.

We want to solve an operator equation

$$F(u) - f = 0 \tag{2.154}$$

in a Hilbert space H. We assume that this equation has a solution y, that

$$\sup_{u \in B(u_0, R)} \|F^{(j)}(u)\| \leq M_j(R), \quad j = 0, 1, 2, \tag{2.155}$$

where $F^{(j)}(u)$ are Fréchet derivatives of F, that $u_0 \in H$ is some element,

$$B(u_0, R) = \{u : \|u - u_0\| \leq R, \ u \in H\},$$

and that $M_j(R)$ are some positive constants. We do not restrict the growth of these constants as $R \to \infty$. This means that the nonlinearity F can grow arbitrarily fast as $\|u\|$ grows, but is locally twice Fréchet differentiable.

Definition 2.6.1 *We will call problem (2.154) well-posed (WP) if*

$$\sup_{u \in B(u_0, R)} \|[F'(u)]^{-1}\| \leq m(R) \tag{2.156}$$

and will call it ill-posed (IP) otherwise.

Condition (2.156) implies that F is a local homeomorphism in a neighborhood of the point u.

If $F(u) = Au$, where A is a bounded linear operator, then $F'(u) = A$ for any $u \in H$, and condition (2.156) implies that A is an isomorphism.

If $F(u) = Au$ so that $F'(u) = A$, and A is not boundedly invertible, that is, (2.156) fails, then either A is not injective, or A is not surjective, or A^{-1} is not bounded. If A is not injective, that is, $\mathcal{N} = \mathcal{N}(A) \neq \{0\}$, then one can consider A as a mapping from the factor space $H/\mathcal{N}$ into H, and then this mapping is injective, or one can consider A as a mapping from $H_1 = H \ominus \mathcal{N}$, and this mapping is injective.

If A is not surjective but its range $\mathcal{R}(A)$ is closed, then A as a mapping from H_1 into $\mathcal{R}(A)$ is injective and surjective and has a bounded inverse, so in this case the problem of solving equation $Au - f = 0$ may be reduced to a well-posed problem:

If $\|f_\delta - f\| \leq \delta$, but $f \notin \mathcal{R}(A)$, then one projects f_δ onto $\mathcal{R}(A)$, and the solution $A^{-1}P_{\mathcal{R}(A)}f_\delta$ depends continuously on f_δ.

However, if the range $\mathcal{R}(A)$ is not closed, $\mathcal{R}(A) \neq \overline{\mathcal{R}(A)}$, then small perturbations f_δ, $\|f_\delta - f\| \leq \delta$, of f may lead to large perturbations of the solution, or f_δ may be out of $\mathcal{R}(A)$, in which case the equation $Au = f_\delta$ is not solvable. Thus, if $\mathcal{R}(A) \neq \overline{\mathcal{R}(A)}$, then the problem of solving equation $Au = f$ is ill-posed.

Calculating the null space $\mathcal{N}$ of a linear operator A is an ill-posed problem: A small perturbation of A may transform A into an injective operator.

If F is a nonlinear mapping, then condition (2.156), in general, does not imply injectivity or surjectivity globally. For example, if $F(u) = e^u$ is a mapping of $\mathbb{R}$ into $\mathbb{R}$, then

$$F'(u) = e^u, \quad |[F'(u)]^{-1}| = |e^{-u}| \leq m(R), \quad |u| \leq R;$$

however, equation $e^u = 0$ has no solution. An example of noninjectivity: Let

$$F : \mathbb{R}^2 \to \mathbb{R}^2, \quad F(u) = \begin{pmatrix} e^{u_1} \cos u_2 \\ e^{u_1} \sin u_2 \end{pmatrix}.$$

Then

$$F'(u) = \begin{pmatrix} \dfrac{\partial F_1}{\partial u_1} & \dfrac{\partial F_1}{\partial u_2} \\[2mm] \dfrac{\partial F_2}{\partial u_1} & \dfrac{\partial F_2}{\partial u_2} \end{pmatrix} = \begin{pmatrix} e^{u_1} \cos u_2 & -e^{u_1} \sin u_2 \\ e^{u_1} \sin u_2 & e^{u_1} \cos u_2 \end{pmatrix},$$

$$[F'(u)]^{-1} = e^{-u_1} \begin{pmatrix} \cos u_2 & \sin u_2 \\ -\sin u_2 & \cos u_2 \end{pmatrix},$$

so

$$\|[F'(u)]^{-1}\| \leq e^{-u_1} \leq e^R, \quad |u_1|^2 + |u_2|^2 \leq R^2.$$

Condition (2.156) holds with any $R > 0$, but F is not injective: If equation $F(u) = f$ has a solution $\begin{pmatrix} u_1 \\ u_2 \end{pmatrix}$, then it has solutions

$$\begin{pmatrix} cu_1 \\ u_2 + 2n\pi \end{pmatrix}, \qquad n = \pm 1, \pm 2, \dots.$$

The DSM method for solving equation (2.154) consists of finding a nonlinear mapping $\Phi(t, u)$ such that the Cauchy problem

$$\dot{u} = \Phi(t, u), \quad u(0) = u_0 \tag{2.157}$$

has a unique global solution, that there exists $u(\infty)$, and that $u(\infty)$ solves equation (2.154):

$$\exists! u(t) \quad \forall t \geq 0; \quad \exists u(\infty); \quad F(u(\infty)) - f = 0. \tag{2.158}$$

If F is nonlinear, then the global existence of a solution to (2.157) is a difficult problem by itself. To guarantee the local existence and uniqueness of the solution to (2.157), we assume that Φ satisfies a Lipschitz condition

$$\|\Phi(t, u) - \Phi(t, v)\| \leq K\|u - v\|, \qquad \forall u, v \in B(u_0, R), \tag{2.159}$$

where $K = $ constant > 0 does not depend on t, u, v but may depend on R. We also assume that

$$\sup_{t \in (t_0, t_0 + b)} \sup_{u \in B(u_0, R)} \|\Phi(t, u)\| \leq K_0(R) := K_0, \tag{2.160}$$

where $K_0 > 0$ and $b > 0$ are constants. Such an assumption on Φ will be satisfied in most of our applications. Assumption (2.159) does not restrict the growth of nonlinearity of Φ because K may grow rapidly as R grows.

Problem (2.157), in general, may have no global solution: The solution may blow up in a finite time. The maximal interval $[0, T)$ of the existence of the solution may be finite, that is, $T < \infty$. In all the problems we deal with in this monograph, the global existence of the solution to (2.157) will be proved by establishing an a priori bound for the norm of the solution

$$\sup_{t>0} \|u(t)\| \leq c, \tag{2.161}$$

where $c = const > 0$ does not depend on t and the supremum is taken over all $t \in [0, T)$.

Lemma 2.6.1 *If (2.159) and (2.161) hold, then $T = \infty$.*

Proof: From the classical local existence and uniqueness theorem for the solution to (2.157) under the assumptions (2.159)–(2.160) one obtains that the solution to (2.157) with the initial condition $u(t_0) = u_0$ exists and is unique on the interval $|t - t_0| < \tau$, where

$$\tau = \min\left(\frac{1}{K}, \frac{R}{K_0}, b\right). \tag{2.162}$$

This result is well known, but we prove it for the convenience of the reader (see also Section 16.2). Using this result, we complete the proof of Lemma 2.6.1.

Assume that (2.161) holds but $T < \infty$. Let $u \in B(u_0, R)$ for $t \in [0, T - \frac{\tau}{2}]$. Take $t_0 = T - \frac{\tau}{2}$. Then the solution $u(t)$ to (2.157) with the initial condition $u(t_0) = u(T - \frac{\tau}{2})$ exists on the interval $(T - \frac{\tau}{2}, T + \frac{\tau}{2})$, so T is not the maximal interval of existence. Condition (2.161) guarantees that the Lipschitz constant K in (2.159) remains bounded, $K = K(c)$, the constant K_0 in (2.160) remains bounded, $K_0 = K_0(c)$, and the radius of the ball $B(u_0, R)$, within which the solution stays, remains bounded, $R = R(c)$. Thus, $\tau = \tau(c)$ does not decrease as the initial point gets close to T. This and the local existence and uniqueness theorem for the solution to (2.157) imply $T = \infty$. Lemma 2.6.1 is proved. ∎

In Theorem 2.6.2 below we assume that $\Phi(t, u)$ is continuous operator function of t.

Theorem 2.6.2 *(local existence and uniqueness theorem).*
Assume (2.159)–(2.160). Then problem (2.157) with $u(t_0) = u_0$, $t_0 > 0$, has a unique solution $u(t) \in B(u_0, R)$ on the interval $t \in (t, t + \tau)$, where τ is defined in (2.162).

Proof: Problem (2.157) with $u(t_0) = u_0$ is equivalent to the equation

$$u = u_0 + \int_{t_0}^{t} \Phi(s, u(s)) \, ds := B(u), \tag{2.163}$$

because we have assumed that Φ is continuous with respect to t. Let us check that operator B maps the ball $B(u_0, R)$ into itself and is a contraction mapping on this ball if (2.162) holds. This and the contraction mapping principle yield the conclusion of Theorem 2.6.2.

We have

$$\sup_{t \in (t_0, t_0+\tau)} \|u(t) - u_0\| \leq \sup_{s \in (t_0, t_0+\tau),\ u \in B(u_0, R)} \|\Phi(s, u(s))\|$$

$$\leq \tau K_0 \leq R. \tag{2.164}$$

Furthermore,

$$\sup_{t \in (t_0, t_0+\tau)} \|B(u) - B(v)\| \leq \tau \sup_{s \in (t_0, t_0+\tau),\ u \in B(u_0, R)} \|\Phi(s, u(s)) - \Phi(s, v(s))\|$$

$$\leq \tau K \sup_{s \in (t_0, t_0+\tau)} \|u(s) - v(s)\|. \tag{2.165}$$

Since $\tau K < 1$, the operator B is a contraction on the set $B(u_0, R)$. Theorem 2.6.2 is proved. ∎

2.7 VARIATIONAL REGULARIZATION FOR NONLINEAR EQUATIONS

Consider the equation

$$F(u) = f, \tag{2.166}$$

where F is a nonlinear map in a Hilbert space H.

Assume that there is a solution y to equation (2.166), $F(y) = f$.

We also assume that F is a weakly continuous (wc) map, that is,

$$u_n \rightharpoonup u \quad \Rightarrow \quad F(u_n) \rightharpoonup F(u). \tag{2.167}$$

Let $\|f_\delta - f\| \leq \delta$ and

$$g(u) = \|F(u) - f_\delta\| + b\delta\varphi(u), \quad b = const > 0, \tag{2.168}$$

where $\varphi : H_1 \to \mathbb{R}_+ := [0, \infty)$ is a functional, the set $\{u : \varphi(u) \leq c\}$ is bounded in H_1, $H_1 \subset H$ is a Hilbert space, $\|u\|_1 \geq \|u\|$, the embedding $i : H_1 \to H$ is compact, that is, the set $\{u : \|u\|_1 \leq c\}$ contains a convergent in H subsequence, φ is weakly lower semicontinuous (wlsc), that is,

$$u_n \rightharpoonup u \quad \Rightarrow \quad \lim_{n \to \infty} \varphi(u_n) \geq \varphi(u), \tag{2.169}$$

and $y \in H_1$. From (2.167) and (2.169) it follows that g is wlsc.

Lemma 2.7.1 *If (2.167) and (2.169) hold, then the problem*

$$g(u) = min \tag{2.170}$$

has a solution.

Proof: Since $g(u) \geq 0$, there exists $m := \inf_u g(u)$. Let $g(u_n) \to m$. Then $0 \leq g(u_n) \leq m + \delta$ for all $n \geq n(\delta)$, so

$$b\delta\varphi(u_n) \leq m + \delta. \tag{2.171}$$

We have

$$m \leq g(y) \leq \delta(1 + b\varphi(y)). \tag{2.172}$$

From (2.171) and (2.172) we get

$$\varphi(u_n) \leq \frac{2 + b\varphi(y)}{b} := c. \tag{2.173}$$

Thus, there exists a convergent subsequence of u_n, which is denoted u_n again:

$$\lim_{n \to \infty} u_n = u. \tag{2.174}$$

Since g is weakly lower semicontinuous, we obtain

$$m = \lim_{n \to \infty} g(u_n) \leq g(u) \leq m. \tag{2.175}$$

Thus, the lemma is proved.   ∎

Remark 2.7.2 The proof remains valid if we assume that the set $\{u : \varphi(u) \leq c\}$ is weakly precompact—that is, contains a weakly convergent subsequence $u_n \rightharpoonup u$, rather than the strongly convergent one. However, the assumption about compactness of the embedding $i : H_1 \to H$ will be used later.

Theorem 2.7.3 *Assume that F is injective, the embedding $i : H_1 \to H$ is compact, $f = F(y)$, (2.167) holds, $\|f_\delta - f\| \leq \delta$, and $g(u_\delta) = \inf_{u \in H} g(u)$. Then*

$$\|u_\delta - y\| \leq \eta(\delta) \to 0 \quad as \quad \delta \to 0. \tag{2.176}$$

Proof: Using the equation $F(y) = f$, (2.172), and (2.173), we obtain

$$\|F(u_\delta) - F(y)\| \leq \|F(u_\delta) - f_\delta\| + \|f_\delta - f\| \leq \delta(2 + b\varphi(y)) \tag{2.177}$$

and

$$\varphi(u_\delta) \leq c. \tag{2.178}$$

Therefore, (2.176) follows. Indeed, assuming that $\|u_\delta - y\| \geq \varepsilon > 0$, choose a subsequence

$$u_n := u_{\delta_n} \to u \quad as \quad \delta_n \to 0.$$

This is possible because i is compact and (2.178) holds. Then we get

$$\varepsilon \leq \lim_{n \to \infty} \|u_n - y\| = \|u - y\|. \tag{2.179}$$

Let us prove that
$$F(u) = y. \tag{2.180}$$

It follows from (2.177) that
$$\lim_{n \to \infty} \|F(u_n) - f\| = 0.$$

Assumption (2.167) and weak lower semicontinuity of the norm in H imply
$$0 = \lim_{n \to \infty} \|F(u_n) - f\| \geq \|F(u) - f\|. \tag{2.181}$$

Thus (2.180) is verified.

Since F is injective, it follows from (2.180) that $u = y$. This contradicts to the inequality (2.179).

Theorem 2.7.3 is proved. ∎

Let us discuss now a *new discrepancy principle*.

Earlier the discrepancy principle for finding the regularization parameter $a = a(\delta)$ was proposed and justified for linear equations in the following form. One finds a minimizer $u_{a,\delta}$ to the functional
$$G(u) := \|F(u) - f_\delta\|^2 + a\|u\|^2.$$

Then $a = a(\delta)$ is found as the solution (unique if F is a linear operator) of the equation
$$\|F(u_{a,\delta}) - f_\delta\| = c\delta, \quad \|f_\delta\| > c\delta, \quad c = const \in (1,2). \tag{2.182}$$

The justification of the discrepancy principle consisted of the proof of the unique solvability of equation (2.182) for $a = a(\delta)$ and of the proof of the relation
$$\lim_{\delta \to 0} \|u_\delta - y\| = 0, \quad u_\delta := u_{a(\delta),\delta}. \tag{2.183}$$

Equation (2.182) is a nonlinear equation for a even in the case when $F(u) = Au$ is a linear operator.

To avoid solving this equation, we propose to take $a(\delta) = b\delta$, $b = const > 0$.

We have proved that for any fixed $b = const > 0$, problem (2.170) has a solution u_δ and (2.176) holds, provided that F is injective, (2.167) holds, and i is compact.

Therefore one may use the following relaxed version of the discrepancy principle. The usual discrepancy principle requires us to solve the variational problem
$$\|F(u) - f_\delta\|^2 + a\varphi(u) = \min, \tag{2.184}$$

to find a minimizer $u_{a,\delta}$, and then to find the regularization parameter $a = a(\delta)$ by solving the nonlinear equation for a:
$$\|F(u_{a,\delta}) - f_\delta\| = c\delta, \quad c = const \in (1,2). \tag{2.185}$$

Finally, one has to prove that

$$u_\delta := u_{a(\delta),\delta}$$

satisfies (2.183).

The relaxed version of the discrepancy principle does not require solving nonlinear equation (2.185) for a, but allows one to take

$$a(\delta) = b\delta$$

with an arbitrary fixed $b = const > 0$.

One can also choose some b such that

$$\delta \leq \|F(u_{b\delta,\delta}) - f_\delta\| \leq c_1\delta, \qquad c_1 = const > 0, \qquad (2.186)$$

where $c_1 > 1$ is an arbitrary fixed constant. It is easier numerically to find b such that (2.186) holds than to solve equation (2.185) for $a = a(\delta)$.

CHAPTER 3

DSM FOR WELL-POSED PROBLEMS

In this chapter it is shown that every well-posed problem can be solved by a DSM which converges exponentially fast.

3.1 EVERY SOLVABLE WELL-POSED PROBLEM CAN BE SOLVED BY DSM

In this chapter we prove that every solvable equation

$$F(u) = 0, \tag{3.1}$$

where F satisfies assumptions (2.155) and (2.156), can be solved by a DSM (1.2) so that the three conditions (1.5) hold, and, in addition, the convergence of this DSM is exponentially fast:

$$\|u(t) - u(\infty)\| \le re^{-c_1 t}, \qquad \|F(u(t))\| \le \|F_0\|e^{-c_1 t}, \tag{3.2}$$

where $c_1 = const > 0$, $r = const > 0$ and $F_0 = F(u(0))$. We will specify c_1 and r later.

Dynamical Systems Method and Applications: Theoretical Developments and Numerical Examples, First Edition. A. G. Ramm and N. S. Hoang.
Copyright © 2012 John Wiley & Sons, Inc.

First, let us establish a general framework for a study of well-posed problems.

Assume that

$$\langle F'(u)\Phi(t, u), F(u)\rangle \leq -g_1(t)\|F(u)\|^2, \qquad \forall u \in H, \tag{3.3}$$

and

$$\|\Phi(t, u)\| \leq g_2(t)\|F(u)\|, \qquad \forall u \in H, \tag{3.4}$$

where $g_1(t)$ and $g_2(t)$ are positive functions, defined on $\mathbb{R}_+ = [0, \infty)$, g_2 is continuous, and $g_1 \in L^1(\mathbb{R}_+)$.

The assumption (3.3) can be generalized:

$$\|\Phi(t, u)\| \leq g_2(t)\|F(u)\|^b, \qquad b = const > 0.$$

Define the following function:

$$G(t) := g_2(t)e^{-\int_0^t g_1(s)ds}, \qquad t \geq 0.$$

Assume:

$$\int_0^\infty g_1(s)\, ds = \infty, \qquad G \in L^1(\mathbb{R}_+), \tag{3.5}$$

and

$$\|F(u_0)\| \int_0^\infty G(t)\, dt \leq R. \tag{3.6}$$

Let us formulate our first result.

Theorem 3.1.1 *Assume that (3.3)–(3.6) hold. Also assume that (2.155) holds for $j \leq 1$ and (2.159) holds. Then problem (2.157) has a unique global solution $u(t)$, there exists $u(\infty)$, $F(u(\infty)) = 0$, and the following estimates hold:*

$$\|u(t) - u(\infty)\| \leq \|F(u_0)\| \int_t^\infty G(s)\, ds \tag{3.7}$$

and

$$\|F(u(t))\| \leq \|F(u_0)\|e^{-\int_0^t g_1(s)\, ds}. \tag{3.8}$$

Proof: From the assumption (2.159) it follows that there exists a unique local solution to problem (2.157). To prove that this solution is global, we use Lemma 2.6.1 and establish estimate (2.161) for this solution. Let $g(t) := \|F(u(t))\|$ and $\dot{g} = \frac{dg}{dt}$. Then, using (2.155) with $j = 1$, we get

$$g\dot{g} = \langle F'(u)\dot{u}, F\rangle = \langle F'(u)\Phi(t, u), F\rangle \leq -g_1(t)g^2. \tag{3.9}$$

Since $g \geq 0$, we obtain

$$g(t) \leq g_0 e^{-\int_0^t g_1(s)\, ds}, \qquad g_0 = \|F(u_0)\|, \tag{3.10}$$

which is the inequality (3.8). From equation (2.157), inequality (3.4), and estimate (3.8) we derive the estimate

$$\|u(s) - u(t)\| \le \|F(u_0)\| \int_t^s G(p)\, dp. \tag{3.11}$$

Since $G \in L^1(\mathbb{R}_+)$, it follows from (3.11) that estimate (2.161) holds, the limit

$$u(\infty) = \lim_{t \to \infty} u(t)$$

exists, and estimate (3.7) holds. Taking $t \to \infty$ in (3.8), using the first assumption (3.5) and the continuity of F, one obtains the relation $F(u(\infty)) = 0$. Theorem 3.1.1 is proved. ∎

Let us replace assumption (3.3) by a more general one:

$$\langle F'\Phi, F \rangle \le -g_1(t)\|F(u)\|^a, \qquad a \in (0, 2). \tag{3.12}$$

Arguing as in the proof of Theorem 3.1.1, one gets the inequality

$$g^{1-a}\dot{g} \le -g_1(t),$$

so

$$0 \le g(t) \le \left[g^{2-a}(0) - (2-a) \int_0^t g_1(s)\, ds \right]^{\frac{1}{2-a}}. \tag{3.13}$$

If the first assumption (3.5) holds, then (3.13) implies that $g(t) = 0$ for all $t \ge T$, where T is defined by the equation

$$\int_0^T g_1(s)\, ds = \frac{g^{2-a}(0)}{2-a}, \qquad 0 < a < 2. \tag{3.14}$$

Thus

$$\|F(u(t))\| = 0, \qquad t \ge T. \tag{3.15}$$

Therefore $u(t)$ solves the equation $F(u) = 0$, and

$$\|u(T) - u(0)\| \le \|F(u_0)\| \int_0^T G(p)\, dp \le \|G(u_0)\| \int_0^T g_2(s)\, ds. \tag{3.16}$$

From (3.16) and (3.6) it follows that $u(t) \in B(u_0, R)$ for all $t \ge 0$.

Let us formulate the results we have proved:

Theorem 3.1.2 *Assume (3.12), (3.6), (3.5), (2.159), (2.155) with $j \le 1$, and let T be defined by equation (3.14). Then equation $F(u) = 0$ has a solution $u \in B(u_0, R)$, problem (2.157) has a unique global solution $u(t) \in B(u_0, R)$, and $F(u(t)) = 0$ for $t \ge T$.*

Assume now that the inequality (3.12) holds with $a > 2$ and the first assumption (3.5) holds. Then, arguing as in the proof of Theorem 3.1.1, one gets

$$0 \leq g(t) \leq \left[\frac{1}{g^{a-2}(0)} + (a-2) \int_0^t g_1(s)\, ds \right]^{-\frac{1}{a-2}} := h(t), \qquad (3.17)$$

where $\lim_{t \to \infty} h(t) = 0$ because of (3.5).

Assume that

$$\int_0^\infty g_2(s) h(s)\, ds \leq R. \qquad (3.18)$$

Then (2.157) and (3.4) imply

$$\|u(t) - u(0)\| \leq R, \quad \|u(t) - u(\infty)\| \leq \int_t^\infty g_2(s) h(s)\, ds, \qquad (3.19)$$

so

$$\lim_{t \to \infty} \|u(t) - u(\infty)\| = 0. \qquad (3.20)$$

Thus, the following result is obtained.

Theorem 3.1.3 *Assume that (3.12) holds with $a > 2$ and that (2.159), (3.4), (3.5), and (3.18) hold. Then there exists a unique global solution $u(t)$ to (2.157), $u(t) \in B(u_0, R)$, there exists $u(\infty)$ and $F(u(\infty)) = 0$.*

In the remaining sections of this chapter we will use the above results in a number of problems of interest in applications.

In particular, we will use the following consequence of Theorem 3.1.1.

Assume that $g_1(t) = c_1 = const > 0$ and $g_2(t) = c_2 = const > 0$, so that (3.3) and (3.4) take the form

$$\langle F'(u)\Phi(t, u), F(u) \rangle \leq -c_1 \|F(u)\|^2, \qquad \forall u \in H, \qquad (3.21)$$

and

$$\|\Phi(t, u)\| \leq c_2 \|F(u)\|, \qquad \forall u \in H. \qquad (3.22)$$

Then conditions (3.5) are trivially satisfied, (3.6) takes the form

$$\|F(u_0)\| \frac{c_2}{c_1} := r \leq R, \qquad (3.23)$$

(3.11) implies

$$\|u(\infty) - u(t)\| \leq r e^{-c_1 t}, \qquad (3.24)$$

and

$$\|u(t) - u(0)\| \leq r, \qquad (3.25)$$

and (3.10) yields

$$\|F(t)\| \leq \|F(u_0)\| e^{-c_1 t}. \qquad (3.26)$$

Note that (3.2) follows from (3.24) and (3.26).

Let us formulate what we have just demonstrated.

Theorem 3.1.4 *Assume (3.21)–(3.23), (2.155), and (2.156). Then problem (2.157) has a unique global solution $u(t)$, and this solution satisfies inequalities (3.24)–(3.26), so that there exists $\lim_{t \to \infty} u(t) := u(\infty)$, and $F(u(\infty)) = 0$. Thus, under the above assumptions, equation $F(u) = 0$ has a solution $u(\infty) = u(\infty; u_0)$, possibly nonunique.*

By a *global* solution to (2.157) we mean the solution which exists for all $t \geq 0$.

Note that if $u(\infty; u_0)$ is taken as the initial data, then $u(\infty; u(\infty; u_0))$ does not have to be equal to $u(\infty; u_0)$.

Example 3.1

Let

$$\dot{u} = c(1 + u^2)e^{-t}, \qquad u(0) = u_0 = 1.$$

Then

$$\arctan u(t) - \arctan u(0) = c(1 - e^{-t}),$$
$$u(t) = \tan(\arctan u_0 + c - ce^{-t}).$$

Thus

$$u(\infty; u_0) = \tan(\arctan 1 + c) = \tan\left(\frac{\pi}{4} + c\right)$$

and

$$u(\infty; u(\infty; u_0)) = \tan\left(\arctan \tan\left(\frac{\pi}{4} + c\right) + c\right) = \tan\left(\frac{\pi}{4} + 2c\right) \neq u(\infty; u_0)$$

provided that $\frac{\pi}{4} + c \neq \frac{\pi}{4} + 2c + n\pi$.

In order that $u(\infty; u_0) = u(\infty; u(\infty; u_0))$ for the problem

$$\dot{u} = \Phi(t, u), \qquad u(0) = u_0,$$

it is sufficient that $u(\infty; u_0)$ solves the equation $\Phi(\infty; u(\infty; u_0)) = 0$.

3.2   DSM AND NEWTON-TYPE METHODS

Consider the equation

$$F(u) = 0 \tag{3.27}$$

and the DSM for solving (3.27) of the form

$$\dot{u} = -[F'(u)]^{-1}F(u), \qquad u(0) = u_0. \tag{3.28}$$

This is a particular case of (2.157) which makes sense if (2.156) holds, that is, if the problem is well-posed in our terminology. Assumption (2.159) follows from (2.155) and (2.156) due to the differentiation formula:

$$([F'(u)]^{-1})' = -[F'(u)]^{-1} F''(u)[F'(u)]^{-1}, \tag{3.29}$$

which can be derived easily by differentiating the identity

$$[F'(u)]^{-1} F'(u) = I \tag{3.30}$$

with respect to u. We assume throughout the rest of this chapter that assumptions (2.155)–(2.156) hold. Let us apply Theorem 3.1.4 to problem (3.28). We have

$$\langle F'\Phi, F \rangle = -\|F\|^2, \tag{3.31}$$

so that $c_1 = 1$ in condition (3.21). Furthermore, (3.22) holds with $c_2 = m(R)$, where $m(R)$ is the constant from (2.156). Finally, to ensure the existence of a solution to equation (3.27), let us assume that (3.23) holds:

$$\|F(u_0)\|\frac{c_2}{c_1} := r \leq R, \quad \text{i.e.,} \quad \|F(u_0)\|m(R) \leq R. \tag{3.32}$$

Then Theorem 3.1.4 implies the following result.

Theorem 3.2.1 *Assume (2.155), (2.156), and (3.32). Then equation (3.27) has a solution in $B(u_0, R)$, problem (3.28) has a unique global solution $u(t) \in B(u_0, R)$, and there exists the limit*

$$u(\infty) = u(\infty; u_0) \in B(u_0, R);$$

this limit solves the equation

$$F(u(\infty)) = 0,$$

and the DSM method (3.28) converges exponentially fast in the sense that estimates (3.24) and (3.26) hold.

Remark 3.2.2 Condition (3.32) is always satisfied for a suitable u_0 if one knows a priori that equation (3.27) has a solution y, $F(y) = 0$. Indeed, in this case one can choose u_0 sufficiently close to y and, by the continuity of F, condition (3.32) will be satisfied because $\lim_{u_0 \to y} \|F(u_0)\| = \|F(y)\| = 0$. Most of the classical theorems about convergence of the discrete Newton-type methods contain the assumption that u_0 is sufficiently close to a solution to equation (3.27).

Remark 3.2.3 If one formally discretizes equation (3.28), one gets an iterative scheme

$$u_{n+1} = u_n - h_n [F'(u_n)]^{-1} F(u_n), \quad u_0 = u_0, \tag{3.33}$$

$$t_{n+1} = t_n + h_n, \quad u_n = u(t_n). \tag{3.34}$$

If $h_n = 1$, the process, (3.33) reduces to the classical Newton's method.

$$u_{n+1} = u_n - [F'(u_n)]^{-1}F(u_n), \quad u_0 = u_0. \tag{3.35}$$

Various conditions sufficient for the convergence of this process are known (see [25] and [69]).

For a comparison with Theorem 3.2.1 let us formulate a theorem (see [25, p. 157]) about convergence of the classical Newton's method.

Theorem 3.2.4 *Assume*

$$\|[F'(u_0)]^{-1}F(u_0)\| \le a, \qquad \|[F'(u_0)]^{-1}\| \le b,$$

$$\|F'(u) - F'(v)\| \le k\|u - v\|, \qquad u, v \in B(u_0, R),$$

$$q := 2kab < 1, \qquad 2a < R.$$

Then F has a unique zero $y \in B(u_0, R)$, and

$$\|u_n - y\| \le \frac{a}{2^{n-1}} q^{2^n - 1}, \qquad 0 < q < 1. \tag{3.36}$$

One can find a proof of Theorem 3.2.4 in [25, p. 158]. The proof is considerably more complicated than the proof of Theorem 3.2.1. The basic feature of the classical process (3.35) is its quadratic rate of convergence. This means that

$$\|u_{n+1} - y\| \le c\|u_n - y\|^2$$

for all sufficiently large n. This rate is achieved for the process (3.33) with a constant step size $h_n = h$ only if $h = 1$. For the continuous analog (3.28) of the method (3.35) the rate of convergence is exponential, and it is slower than quadratic.

3.3   DSM AND THE MODIFIED NEWTON'S METHOD

Consider equation (3.27) and the following DSM method for solving this equation:

$$\dot{u} = -[F'(u_0)]^{-1}F(u), \quad u(0) = u_0. \tag{3.37}$$

Let us check conditions of Theorem 3.1.4. Condition (3.22) is satisfied with $c_2 = m(R)$, where $m(R)$ is the constant from (2.156). Let us check condition (3.21). We have

$$-\langle F'(u)[F'(u_0)]^{-1}F(u), F(u)\rangle$$

$$= -\left\langle [F'(u) - F'(u_0)][F'(u_0)]^{-1}F(u), F(u)\right\rangle - \|F(u)\|^2, \tag{3.38}$$

and

$$\left| \left\langle [F'(u) - F'(u_0)][F'(u_0)]^{-1} F(u), F(u) \right\rangle \right| \leq M_2 \|u - u_0\| m \|F(u)\|^2, \quad (3.39)$$

where M_2 is the constant from (2.155).

Let us assume that $\|u - u_0\| \leq R$, that is, $u \in B(u_0, R)$, and choose R such that

$$m M_2 R = \frac{1}{2}. \quad (3.40)$$

Then (3.38)–(3.40) imply that $c_1 = \frac{1}{2}$. Condition (3.23) takes the form

$$2\|F(u_0)\| m \leq \frac{1}{2 m M_2},$$

that is,

$$4 m^2 M_2 \|F(u_0)\| \leq 1. \quad (3.41)$$

Theorem 3.1.4 yields the following result.

Theorem 3.3.1 *Assume (2.155)–(2.156) and (3.41). Then equation (3.27) has a solution in $B(u_0, R)$, problem (3.37) has a unique global solution $u(t) \in B(u_0, R)$, there exists $u(\infty) \in B(u_0, R)$, $F(u(\infty)) = 0$, and the DSM method (3.37) converges to the solution $u(\infty)$ exponentially fast in the sense that estimates (3.24) and (3.26) hold.*

3.4   DSM AND GAUSS–NEWTON-TYPE METHODS

Consider equation (3.27) and the following DSM method for solving this equation

$$\dot{u} = -T^{-1} A^* F(u), \quad u(0) = u_0, \quad (3.42)$$

where

$$A := F'(u), \quad T := A^* A, \quad (3.43)$$

A^* is the operator adjoint to A.

Let us apply Theorem 3.1.4. As always in this chapter, we assume (2.155) and (2.156). Condition (3.21) takes the form

$$-\left\langle F' T^{-1} A^* F, F \right\rangle = -\|F\|^2, \quad (3.44)$$

because $A(A^* A)^{-1} A^* = I$, since $F'(u) := A$ is a boundedly invertible operator (see (2.156)). Thus, $c_1 = 1$.

Let us find c_2. We have

$$\|T^{-1} A^* F\| \leq \|T^{-1}\| M_1 \|F\|, \quad (3.45)$$

where $\|A^*\| = \|A\| \leq M_1$ by (2.155). Finally

$$\|T^{-1}\| \leq \|A^{-1}\| \|(A^*)^{-1}\| \leq m^2, \quad (3.46)$$

where m is the constant from (2.156). Thus $c_2 = M_1 m^2$. Condition (3.23) takes the form

$$\|F(u_0)\| M_1 m^2 \leq R. \tag{3.47}$$

Theorem 3.1.4 yields the following result.

Theorem 3.4.1 *Assume (2.155), (2.156), and (3.47). Then equation (3.27) has a solution in $B(u_0, R)$, problem (3.42) has a unique global solution $u(t)$, there exists $u(\infty)$, $F(u(\infty)) = 0$, and the DSM method (3.42) converges to $u(\infty)$ exponentially fast in the sense that estimates (3.24) and (3.26) hold.*

3.5 DSM AND THE GRADIENT METHOD

Again we want to solve equation (3.27). The DSM method we use is the following one:

$$\dot{u} = -A^* F(u), \quad u(0) = u_0. \tag{3.48}$$

As before, $A = F'(u)$ and A^* is the adjoint to A. Condition (3.21) takes the form

$$-\langle AA^*F, F \rangle = -\|A^*F\|^2 \leq -c_1 \|F\|^2, \tag{3.49}$$

where $c_1 = m^{-2}$. Here we have used the following estimates:

$$\|A^*u\| \geq \|(A^*)^{-1}\|^{-1} \|u\|, \quad \|(A^*)^{-1}\| = \|A^{-1}\| = m, \tag{3.50}$$

where m is the constant from (2.156). Condition (3.22) holds with $c_2 = M_1$, where M_1 is the constant from (2.155) and we have used the relation $\|A^*\| = \|A\|$. Condition (3.23) takes the form

$$\|F(u_0)\| m^2 M_1 \leq R. \tag{3.51}$$

Theorem 3.1.4 yields the following result.

Theorem 3.5.1 *Assume (2.155), (2.156), and (3.51). Then equation (3.27) has a solution in $B(u_0, R)$, problem (3.48) has a unique global solution $u(t) \in B(u_0, R)$, there exists $u(\infty)$, $F(u(\infty)) = 0$, and the DSM method (3.48) converges exponentially fast in the sense that estimates (3.24) and (3.26) hold.*

3.6 DSM AND THE SIMPLE ITERATIONS METHOD

We want to solve equation (3.27) by the following DSM method:

$$\dot{u} = -F(u), \quad u(0) = u_0. \tag{3.52}$$

Let us assume that

$$F'(u) \geq c_1(R) > 0, \quad u \in B(u_0, R). \tag{3.53}$$

Condition (3.21) takes the form

$$-\langle F'F, F \rangle \leq -c_1 \|F\|^2, \tag{3.54}$$

so $c_1 := c_1(R)$ is the constant from (3.53). Condition (3.22) holds with $c_2 = 1$. Condition (3.23) takes the form

$$\|F(u_0)\| \frac{1}{c_1} \leq R. \tag{3.55}$$

Theorem 3.1.4 yields the following result.

Theorem 3.6.1 *Assume (2.155), (2.156), and (3.55). Then equation (3.27) has a solution in $B(u_0, R)$, problem (3.52) has a unique global solution $u(t) \in B(u_0, R)$, there exists $u(\infty)$, $F(u(\infty)) = 0$, and the DSM method (3.52) converges to $u(\infty)$ exponentially fast; that is, inequalities (3.24) and (3.26) hold.*

3.7   DSM AND MINIMIZATION METHODS

Let $f : H \to \mathbb{R}_+$ be twice Fréchet differentiable function, $f \in C^2_{\text{loc}}$. We want to use DSM for global minimization of f. In many cases this gives a method for solving operator equation $F(u) = 0$. Indeed, if this equation has a solution y, $F(y) = 0$, then y is the global minimizer of the functional $f(u) = \|F(u)\|^2$.

Let us consider the following DSM scheme:

$$\dot{u} = -\frac{h}{\langle f'(u), h \rangle} f(u), \quad u(0) = u_0, \tag{3.56}$$

where $h \in H$ is some element for which $|\langle f'(u(t)), h \rangle| > 0$ for $t \geq 0$, where $u(t)$ solves (3.56).

Condition (3.21) takes the form

$$-\frac{\langle f', h \rangle}{\langle f', h \rangle} f^2 = -f^2,$$

so $c_1 = 1$. Condition (3.22) is

$$\frac{\|h\|}{|\langle f'(u), h \rangle|} f(u) \leq c_2 f(u),$$

where c_2 depends on the choice of h and on $f(u)$. Choose

$$h = f'(u(t)).$$

Then (3.56) becomes

$$\dot{u} = -\frac{f'(u(t))}{\|f'(u(t))\|^2} f(u(t)), \quad u(0) = u_0. \tag{3.57}$$

Condition (3.22) is not satisfied, in general, and we assume that

$$\frac{f(u)}{\|f'(u)\|} \leq af^b, \tag{3.58}$$

where a and b are positive constants.

Then (3.57) implies

$$\dot{f} = f'(u)\dot{u} = -f,$$

so

$$f(u(t)) = f_0 e^{-t}, \quad f_0 := f(u_0) \tag{3.59}$$

and

$$\|\dot{u}\| \leq \frac{f}{\|f'(u(t))\|} \leq af_0^b e^{-bt} := c_3 e^{-bt}. \tag{3.60}$$

One has

$$\|u\|\dot{} \leq |\dot{u}\| \leq c_3 e^{-bt}, \tag{3.61}$$

so

$$\|u(t) - u_0\| \leq \frac{c_3}{b}, \quad \|u(t) - u(\infty)\| \leq \frac{c_3}{b} e^{-bt}. \tag{3.62}$$

Condition (3.23) takes the form

$$\frac{c_3}{b} \leq R, \quad \text{i.e.} \quad \frac{af_0^b}{b} \leq R. \tag{3.63}$$

We have proved the following result.

Theorem 3.7.1 *Assume that $f \in C_{loc}^2$ and that (3.58) and (3.63) hold. Then equation $f(u) = 0$ has a solution in $B(u_0, R)$, problem (3.57) has a unique global solution $u(t)$, $f(u(\infty)) = 0$, and estimates (3.62) hold.*

Remark 3.7.2 If $f(u) = \|F(u)\|^2$ and H is the real Hilbert space, then $f'(u) = 2[F'(u)]^*F(u)$. Assume (2.156). Denote $A := F'(u)$. Then

$$\|[A^*]^{-1}\| \leq m(R).$$

One has

$$\|A^*F\| \geq \|[A^*]^{-1}\|^{-1}\|F(u)\| = m^{-1}\|F(u)\|.$$

Thus, if

$$\Phi = -\frac{f'}{\|f'\|^2}f, \quad f = \|F(u)\|^2,$$

then

$$\|\Phi\| \leq \frac{1}{\|f'(u)\|}f \leq \frac{m}{2\|F\|}f = \frac{m}{2}f^{\frac{1}{2}},$$

so that (3.57) holds with $b = \frac{1}{2}$. Therefore assumption (2.156) ensures convergence of the DSM (3.57) for global minimization of the functional $f(u) = \|F(u)\|^2$ with an operator satisfying assumption (2.156), and the DSM converges exponentially fast.

3.8 ULM'S METHOD

In [33] and [189] the following method for solving equation $F(u) = 0$ is discussed under some assumptions on F, of which one is the existence of a bounded linear operator B in a Hilbert (or Banach) space such that

$$\|I - BF'(u_0)\| < 1, \tag{3.64}$$

where $u_0 \in H$ is some element. The method consists of using the following iterative process:

$$u_{n+1} = u_n - B_{n+1}F(u_n), \quad B_{n+1} = 2B_n - B_n F'(u_n)B_n. \tag{3.65}$$

As the initial approximation, one takes u_0 and $B_0 = B$ from the assumption (3.64).

The aim of this short Section is to prove the convergence of a simplified continuous analog of the above method using Theorem 3.1.1. The continuous analog is of the form

$$\dot{u} = -BF(u), \quad u(0) = u_0. \tag{3.66}$$

We have simplified the method by not considering the updates for B and assuming that

$$\|I - F'(u)B\| \le q < 1, \quad u \in B(u_0, R), \tag{3.67}$$

where $q \in (0, 1)$ is a number independent of u, and $R > 0$ is some number. Thus, B is an approximation to the inverse operator $[F'(u)]^{-1}$ in some sense. We wish to prove the convergence of the DSM (3.66). Our Φ from DSM (1.2) is

$$\Phi = -BF,$$

and we wish to verify conditions of Theorem 3.1.1.

The DSM (3.66) is similar to the Newton method. The main difference between the two methods is in taking an approximate inverse B in the sense (3.67) in place of the exact inverse $[F'(u)]^{-1}$ in the Newton method.

Let us verify the conditions of Theorem 3.1.1. Condition (3.3) is easy to verify:

$$-\langle F'BF, F \rangle \le -(1 - q)\|F\|^2,$$

where we have used the assumption (3.67). Thus,

$$g_1(t) = c_1 = 1 - q > 0,$$

and condition (3.3) holds.

Condition (3.4) obviously holds with

$$g_2(t) = \|B\| := c_2.$$

Condition (3.6) takes the form (3.23), that is,

$$\|F(u_0)\|\frac{c_2}{c_1} \leq R. \tag{3.68}$$

If condition (3.68) holds, then, by Theorem 3.1.1, the DSM (3.66) is justified, that is, the conclusions (1.5) are valid.

If equation (1.1) has a solution, then condition (3.68) is always satisfied if the initial approximation u_0 is taken sufficiently close to the solution.

CHAPTER 4

DSM AND LINEAR ILL-POSED PROBLEMS

In this chapter we prove that any solvable linear operator equation can be solved by a DSM and by a convergent iterative process.

4.1 EQUATIONS WITH BOUNDED OPERATORS

Consider the equation

$$Au = f, \tag{4.1}$$

where A is a bounded linear operator in a Hilbert space H. Assume that there exists a solution to (4.1), and that $\mathcal{N} = \mathcal{N}(A)$ is the null space of A, and denote by y the unique minimal-norm solution, that is, the solution $y \perp \mathcal{N}$. Let $T := A^*A$ and $T_a := T + aI$, where I is the identity operator. We assume that problem (4.1) is ill-posed in the sense that the range $\mathcal{R}(A)$ of the operator A is not closed.

If $\mathcal{R}(A)$ is closed, that is, $\mathcal{R}(A) = \overline{\mathcal{R}(A)}$, but A is not injective, then one may define an operator A_1, from $H_1 := H \ominus \mathcal{N}$ onto $\mathcal{R}(A)$. This operator will be continuous and injective and, by the closed graph theorem, the inverse operator A_1^{-1} is continuous from $\mathcal{R}(A)$ into H_1. Thus, the ill-posedness which

Dynamical Systems Method and Applications: Theoretical Developments and Numerical Examples, First Edition. A. G. Ramm and N. S. Hoang.
Copyright © 2012 John Wiley & Sons, Inc.

is due to the lack of injectivity is not a problem if $\mathcal{R}(A)$ is closed. Similarly, if A is not surjective but $\mathcal{R}(A)$ is closed, then small perturbations f_δ of f which do not throw f_δ out of $\mathcal{R}(A)$ (i.e., $f_\delta \in \mathcal{R}(A) = \overline{\mathcal{R}(A)}$, $\|f_\delta - f\| \le \delta$) lead to a small perturbation of the minimal-norm solution because A_1^{-1} is continuous. However, if $\mathcal{R}(A) \ne \overline{\mathcal{R}(A)}$, then the inverse operator A_1^{-1} from $\mathcal{R}(A)$ into H_1 is unbounded. Indeed, if A_1^{-1} is bounded then

$$\|u\| = \|A^{-1}Au\| \le c\|Au\|, \quad c = \|A^{-1}\| = const < \infty.$$

If

$$\|Au_n - Au_m\| \to 0, \quad m, n \to \infty,$$

then the above inequality implies $\|u_n - u_m\| \to 0$, $m, n \to \infty$, so $u_n \to u$, and, by continuity of A, $Au_n \to Au$. Therefore $\mathcal{R}(A)$ is closed, contrary to our assumption.

The case when $\mathcal{R}(A)$ is not closed leads to difficulties in solving equation (4.1) because small perturbations of f may lead to large perturbations of u or may lead to an equation which has no solutions.

Let us assume that $\mathcal{R}(A) \ne \overline{\mathcal{R}(A)}$, f_δ is given, $\|f_\delta - f\| \le \delta$, and

$$Ay = f, \quad y \perp \mathcal{N}.$$

How does one calculate by a DSM a stable approximation u_δ to y? That is, how does one calculate u_δ such that:

$$\lim_{\delta \to 0} \|u_\delta - y\| = 0. \tag{4.2}$$

There are several versions of DSM for solving ill-posed equation (4.1). One of these versions is

$$\dot{u} = -u + T_{a(t)}^{-1} A^* f, \quad u(0) = u_0, \tag{4.3}$$

where

$$a(t) > 0; \quad a(t) \searrow 0 \text{ as } t \to \infty, \tag{4.4}$$

and $\searrow$ denotes monotone decay to zero.

Let us motivate the choice of the DSM (4.3). Equation $Au = f$ with a bounded linear operator A is equivalent to the equation $Tu = A^* f$, where $T \ge 0$ is a monotone operator. In Chapter 6 the Newton-type DSM (6.34) is justified for monotone operators (see Theorem 6.2.1). The DSM (4.3) is identical to method (6.34), because the Fréchet derivative of a linear operator T is equal to T, so

$$-T_a^{-1}(Tu + au - A^* f) = -u + T_a^{-1} A^* f,$$

which is the right-hand side of equation (4.3). Therefore the DSM (4.3) is the Newton-type DSM for a linear equation $Tu = A^* f$.

Theorem 4.1.1 *Problem (4.3) has a unique global solution $u(t)$, there exists $u(\infty)$, $A(u(\infty)) = f$, $u(\infty) = y \perp \mathcal{N}$.*

The case of noisy data f_δ will be discussed after the proof of Theorem 4.1.1.

Proof of Theorem 4.1.1. The unique solution of (4.3) is

$$u(t) = u_0 e^{-t} + \int_0^t e^{-(t-s)} T_{a(s)}^{-1} A^* Ay \ ds.$$

Clearly, $\lim_{t \to \infty} \|u_0 e^{-t}\| = 0$. We claim that

$$\lim_{t \to \infty} \int_0^t e^{-(t-s)} T_{a(s)}^{-1} Ty \ ds = y. \tag{4.5}$$

This claim follows from two lemmas.

Lemma 4.1.2 *If g is a continuous function on $[0, \infty)$ with values in H and $g(\infty)$ exists, then*

$$\lim_{t \to \infty} \int_0^t e^{-(t-s)} g(s) \ ds = g(\infty). \tag{4.6}$$

Lemma 4.1.3 *One has*

$$\lim_{a \to 0} T_a^{-1} Ty = y, \quad a > 0, \quad y \perp \mathcal{N}. \tag{4.7}$$

Proof of Lemma 4.1.2. For any fixed τ, however large, one has

$$\lim_{t \to \infty} \int_0^\tau e^{-(t-s)} g(s) \ ds = 0.$$

If τ is sufficiently large, then $|g(s) - g(\infty)| < \eta$, $\forall s \geq \tau$, where $\eta > 0$ is an arbitrary small number. The conclusion (4.6) follows now from the relation $\lim_{t \to \infty} \int_\tau^t e^{-(t-s)} g(\infty) \ ds = g(\infty)$.

Lemma 4.1.2 is proved. ∎

Proof of Lemma 4.1.3. Using the spectral theorem for the self-adjoint operator T and denoting by E_s the resolution of the identity, corresponding to T, one gets

$$\lim_{a \to 0} \|T_a^{-1} Ty - y\|^2 = \lim_{a \to 0} \int_0^{\|T\|} \frac{a^2 d\langle E_s y, y \rangle}{(a + s)^2} = \|P_{\mathcal{N}} y\|^2 = 0, \tag{4.8}$$

where

$$P_{\mathcal{N}} = E_0 - E_{-0}$$

is the orthoprojector onto the null space $\mathcal{N}$ of operator T. Because $\mathcal{N}(T) = \mathcal{N}(A)$ we use the same letter $\mathcal{N}$ as for the null space of A.

Lemma 4.1.3 is proved. ∎

Since $\lim_{t \to \infty} a(t) = 0$, Theorem 4.1.1 is proved. ∎

Consider now the case of noisy data f_δ. Then we solve the problem

$$\dot{v} = -v + T_{a(t)}^{-1} A^* f_\delta, \quad v(0) = u_0, \tag{4.9}$$

stop integration at a stopping time t_δ, denote $v(t_\delta) := u_\delta$, and prove (4.2) for a suitable choice of t_δ. Let $u(t) - v(t) := w(t)$. Then

$$\dot{w} = -w + T_{a(t)}^{-1} A^* (f - f_\delta), \quad w(0) = 0. \tag{4.10}$$

By (2.79) we have

$$\|T_a^{-1} A^*\| \leq \frac{1}{2\sqrt{a}}. \tag{4.11}$$

Since

$$w(t) = \int_0^t e^{-(t-s)} T_{a(s)}^{-1} A^* (f - f_\delta) \, ds,$$

one has

$$\|w(t)\| \leq \int_0^t e^{-(t-s)} \frac{\delta}{2\sqrt{a(s)}} \, ds \leq \frac{\delta}{2\sqrt{a(t)}}. \tag{4.12}$$

Therefore, if t_δ is chosen so that

$$\lim_{\delta \to 0} \frac{\delta}{2\sqrt{a(t_\delta)}} = 0, \quad \lim_{\delta \to 0} t_\delta = \infty, \tag{4.13}$$

then (4.12) and Theorem 4.1.1 imply the following.

Theorem 4.1.4 *Assume that (4.13) holds and equation (4.1) is solvable. Then $u_\delta := v(t_\delta)$, where $v(t)$ is the unique solution to (4.9), satisfies (4.2).*

Remark 4.1.5 There are many a priori choices of the stopping times t_δ satisfying (4.13). No optimization of the choice of t_δ has been made. It is an open problem to propose such an optimization.

Let us propose an a posteriori choice of t_δ based on a discrepancy-type principle. We choose t_δ from the equation

$$\|A T_{a(t)}^{-1} A^* f_\delta - f_\delta\| = c\delta, \quad c = const \in (1, 2). \tag{4.14}$$

Denote $AA^* := Q$. We have

$$a^2(t) \int_0^{\|Q\|} \frac{1}{[s + a(t)]^2} d\langle F_s f_\delta, f_\delta \rangle = c^2 \delta^2, \tag{4.15}$$

where F_s is the resolution of the identity corresponding to the selfadjoint operator Q. Equation (4.15) for a has a unique solution a_δ and t_δ is found uniquely from the equation

$$a_\delta = a(t). \tag{4.16}$$

This equation has a unique solution $t = t_\delta$ because $a(t)$ is monotone. Equation (4.15) considered as an equation for a has a unique solution $a = a_\delta$. This was proved in Theorem 2.2.6. Since $\lim_{\delta \to 0} a_\delta = 0$, it follows that $\lim_{\delta \to 0} t_\delta = \infty$.

This allows us to prove the following result.

Theorem 4.1.6 *Equation (4.14) has a unique solution $t = t_\delta$, and $\lim_{\delta \to 0} t_\delta = \infty$. The element $u_\delta := v(t_\delta)$, where v solves (4.9) and t_δ is the solution of (4.14), satisfies (4.2), provided that*

$$\lim_{t \to \infty} \frac{\dot{a}(t)}{a(t)} = 0. \tag{4.17}$$

Proof: By Theorem 2.2.6 we have

$$\lim_{\delta \to 0} \|T_{a_\delta}^{-1} A^* f_\delta - y\| = 0. \tag{4.18}$$

Denote

$$w_\delta(t) := T_{a(t)}^{-1} A^* f_\delta.$$

Thus

$$T_{a(t)} w_\delta(t) = A^* f_\delta.$$

Differentiating this equation with respect to t one obtains

$$T_{a(t)} \dot{w}_\delta(t) + \dot{a}(t) w_\delta(t) = 0. \tag{4.19}$$

This implies

$$\|\dot{w}_\delta(t)\| \le |\dot{a}(t)| \|T_{a(t)}^{-1}\| \|w_\delta(t)\| \le \frac{|\dot{a}(t)|}{a(t)} \|w_\delta(t)\|. \tag{4.20}$$

We have

$$\|v(t_\delta) - y\| \le \|v(t_\delta) - w_\delta(t_\delta)\| + \|w_\delta(t_\delta) - y\|. \tag{4.21}$$

By (4.18),

$$\lim_{\delta \to 0} \|w_\delta(t_\delta) - y\| = 0. \tag{4.22}$$

Let us check that

$$\lim_{\delta \to 0} \|v(t_\delta) - w_\delta(t_\delta)\| = 0. \tag{4.23}$$

We have

$$v(t_\delta) = u_0 e^{-t_\delta} + \int_0^{t_\delta} e^{-(t_\delta - s)} w_\delta(s)\, ds, \quad w_\delta(s) := T_{a(s)}^{-1} A^* f_\delta. \tag{4.24}$$

Since

$$\lim_{\delta \to 0} t_\delta = \infty, \tag{4.25}$$

we have

$$\lim_{\delta \to 0} \|u_0\| e^{-t_\delta} = 0. \tag{4.26}$$

Furthermore,

$$\int_0^t e^{-(t-s)} w_\delta(s)\, ds = e^{-(t-s)} w_\delta(s)\Big|_0^t - \int_0^t e^{-(t-s)} \dot{w}_\delta(s)\, ds$$

$$= w_\delta(t) - e^{-t} w_\delta(0) - \int_0^t e^{-(t-s)} \dot{w}_\delta(s)\, ds. \qquad (4.27)$$

Therefore,

$$\lim_{\delta \to 0} \|v(t_\delta) - w_\delta(t_\delta)\| = \lim_{\delta \to 0} \Big\| \int_0^{t_\delta} e^{-(t_\delta - s)} \dot{w}_\delta(s)\, ds \Big\|$$

$$\leq \lim_{\delta \to 0} \int_0^{t_\delta} e^{-(t_\delta - s)} \|\dot{w}_\delta(s)\|\, ds$$

$$\leq \lim_{\delta \to 0} \int_0^{t_\delta} e^{-(t_\delta - s)} \frac{|\dot{a}(s)|}{a(s)} \|w_\delta(s)\|\, ds \qquad (4.28)$$

From (4.14) and Theorem 2.2.5 one concludes that the following inequality holds for all sufficiently small $\delta > 0$:

$$\|w_\delta(t_\delta)\| \leq \|y\| + \|w_\delta(t_\delta) - y\| \leq \|y\| + 1. \qquad (4.29)$$

It follows from (4.19) that

$$\frac{d}{dt} \|w_\delta(t)\|^2 = 2\operatorname{Re}\langle w_\delta(t), \dot{w}_\delta(t)\rangle = -\dot{a}(t) \operatorname{Re}\langle w_\delta(t), T_{a(t)}^{-1} w_\delta(t)\rangle \geq 0, \qquad (4.30)$$

for all $t \geq 0$. We have used the fact that $T_{a(t)}^{-1} > 0$, which follows from $T_{a(t)} \geq a(t)I > 0$, and the inequality $\dot{a}(t) \leq 0$. Thus

$$\|w_\delta(t)\| \leq \|w_\delta(t_\delta)\|, \qquad 0 \leq t \leq t_\delta. \qquad (4.31)$$

From (4.28), (4.29), and (4.31) one obtains

$$\lim_{\delta \to 0} \|v(t_\delta) - w_\delta(t_\delta)\| \leq \lim_{\delta \to 0} e^{-t_\delta} \int_0^{t_\delta} e^s \frac{|\dot{a}(s)|}{a(s)}\, ds (1 + \|y\|). \qquad (4.32)$$

Using L'Hospital's rule and (4.17), one can easily prove that

$$\lim_{t \to \infty} e^{-t} \int_0^t e^s \frac{|\dot{a}(s)|}{a(s)}\, ds = 0. \qquad (4.33)$$

This, (4.32) and (4.25) imply (4.23).

Theorem 4.1.6 is proved.   ■

Remark 4.1.7 Our proof of Theorem 4.1.6 is valid for a closed, densely defined in H, linear operator A, which may be unbounded.

Assume that

$$0 < a(t) \searrow 0, \qquad \lim_{t \to \infty} \frac{\dot{a}(t)}{a(t)} = 0. \tag{4.34}$$

Remark 4.1.8 Condition (4.17) holds, for instance, if

$$a(t) = \frac{c_0}{(c_1 + t)^b}, \quad c_1, c_0 > 0, \quad b > 0. \tag{4.35}$$

Remark 4.1.9 Let $Q := AA^*$, $Q_a := Q + aI$, and assume that A is bounded. Then

$$\|f_\delta\| \le \|Q_{a(t)}\| \|Q_{a(t)}^{-1} f_\delta\| \le (\|Q\| + a(t)) \|Q_{a(t)}^{-1} f_\delta\|.$$

Therefore,

$$\|Q_{a(t)}^{-1} f_\delta\| \ge \frac{\|f_\delta\|}{\|Q\| + a(0)}, \qquad \forall t \ge 0. \tag{4.36}$$

From (4.34) one derives

$$\lim_{t \to \infty} e^t a(t) = \infty. \tag{4.37}$$

This and (4.36) imply

$$\lim_{t \to \infty} e^t a(t) \left\| Q_{a(t)}^{-1} f_\delta \right\| = \infty, \tag{4.38}$$

Remark 4.1.10 One has

$$-a(t) Q_{a(t)}^{-1} f_\delta = (Q Q_{a(t)}^{-1} - I) f_\delta = (A T_{a(t)}^{-1} A^* f_\delta - f_\delta). \tag{4.39}$$

Here, we have used the commutation formula $Q Q_{a(t)}^{-1} = A T_{a(t)}^{-1} A^*$. It follows from (4.39), from the relation $\lim_{t \to \infty} a(t) = 0$, from equation (2.93), and from inequality (2.95) that

$$\lim_{t \to \infty} a(t) \|Q_{a(t)}^{-1} f_\delta\| \le \delta. \tag{4.40}$$

Let us formulate another version of the discrepancy principle, assuming that A is bounded.

Theorem 4.1.11 *Let $a(t)$ satisfy (4.34), $v(t)$ be the solution to (4.9), and*

$$v(0) = u_0 := T_{a(0)}^{-1} A^* f_\delta, \qquad \|A u_0 - f_\delta\| > c\delta. \tag{4.41}$$

Then there exists a unique $t_\delta > 0$ such that

$$\|A v(t_\delta) - f_\delta\| = c\delta, \quad \|A v(t) - f_\delta\| > c\delta, \qquad \forall t \in [0, t_\delta). \tag{4.42}$$

This t_δ satisfies the relation $\lim_{\delta \to 0} t_\delta = \infty$, and

$$\lim_{\delta \to 0} \|v_\delta - y\| = 0, \qquad v_\delta := v(t_\delta). \tag{4.43}$$

Proof: The uniqueness of t_δ follows from its definition (4.42).

   Let us prove the existence of t_δ satisfying (4.42). From (4.9) one gets

$$v(t) = e^{-t}u_0 + e^{-t}\int_0^t e^s T_{a(s)}^{-1} A^* f_\delta \, ds. \tag{4.44}$$

Thus,

$$\begin{aligned}
Av(t) - f_\delta &= e^{-t}Au_0 + e^{-t}\int_0^t e^s A T_{a(s)}^{-1} A^* f_\delta \, ds - f_\delta \\
&= e^{-t}(Au_0 - f_\delta) + e^{-t}\int_0^t e^s (QQ_{a(s)}^{-1} - I) f_\delta \, ds \\
&= e^{-t}(Au_0 - f_\delta) - \int_0^t e^{-(t-s)} a(s) Q_{a(s)}^{-1} f_\delta \, ds. \tag{4.45}
\end{aligned}$$

   Let us prove that

$$\lim_{t\to\infty} \frac{\left\| e^{-t}\int_0^t e^s a(s) Q_{a(s)}^{-1} f_\delta \, ds \right\|}{a(t)\|Q_{a(t)}^{-1} f_\delta\|} = 1. \tag{4.46}$$

By the spectral theorem one gets

$$\left\| \int_0^t e^{s-t} a(s) Q_{a(s)}^{-1} f_\delta \, ds \right\|^2 = \int_0^{\|Q\|} \left(\int_0^t \frac{e^{s-t}a(s)}{a(s)+\lambda} \, ds \right)^2 d\langle F_\lambda f_\delta, f_\delta \rangle, \tag{4.47}$$

$$\|a(t)Q_{a(t)}^{-1} f_\delta\|^2 = \int_0^{\|Q\|} \frac{a^2(t)}{(a(t)+\lambda)^2} d\langle F_\lambda f_\delta, f_\delta \rangle. \tag{4.48}$$

Since $0 < a(t) \searrow 0$ one has

$$\frac{a(t)}{a(s)} \leq \frac{a(t)+\lambda}{a(s)+\lambda} \leq \frac{a(t)+\|Q\|}{a(s)+\|Q\|}, \qquad 0 \leq s \leq t,\, 0 \leq \lambda \leq \|Q\|. \tag{4.49}$$

Let

$$g_\lambda(t) = \frac{\int_0^t \frac{e^s a(s)}{a(s)+\lambda} \, ds}{\frac{e^t a(t)}{a(t)+\lambda}}, \qquad t \geq 0,\, \lambda \geq 0. \tag{4.50}$$

From (4.49) one gets

$$g_0(t) = \frac{e^t - 1}{e^t} \leq g_\lambda(t) \leq g_{\|Q\|}(t) = \frac{\int_0^t \frac{e^s a(s)}{a(s)+\|Q\|} \, ds}{\frac{e^t a(t)}{a(t)+\|Q\|}}, \qquad 0 \leq \lambda \leq \|Q\|. \tag{4.51}$$

By L'Hospital's rule one gets

$$\begin{aligned}
\lim_{t\to\infty} g_{\|Q\|}(t) &= \lim_{t\to\infty} \frac{\frac{e^t a(t)}{a(t)+\|Q\|}}{\frac{e^t a(t)}{a(t)+\|Q\|} + \frac{e^t \dot{a}(t)\|Q\|}{(a(t)+\|Q\|)^2}} \\
&= \lim_{t\to\infty} \frac{1}{1 + \frac{\dot{a}(t)\|Q\|}{a(t)(a(t)+\|Q\|)}} = 1. \tag{4.52}
\end{aligned}$$

It follows from (4.51) and (4.52) that for an arbitrary small $\epsilon > 0$ there exists $t_\epsilon > 0$ such that

$$1 - \epsilon < g_\lambda(t) < 1 + \epsilon, \qquad \forall t \geq t_\epsilon, \quad 0 \leq \lambda \leq \|Q\|. \tag{4.53}$$

From (4.47) (4.48), (4.50), (4.51), and (4.53) one gets

$$(1 - \epsilon)^2 \leq \frac{\left\| e^{-t} \int_0^t e^s a(s) Q_{a(s)}^{-1} f_\delta \, ds \right\|^2}{a^2(t) \|Q_{a(t)}^{-1}\|^2} \leq (1 + \epsilon)^2, \qquad \forall t \geq t_\epsilon. \tag{4.54}$$

This implies that relation (4.46) holds.

From (4.38) one gets

$$\lim_{t \to \infty} \frac{e^{-t} \|A u_0 - f_\delta\|}{a(t) \|Q_{a(t)}^{-1} f_\delta\|} = \lim_{t \to \infty} \frac{\|A u_0 - f_\delta\|}{e^t a(t) \|Q_{a(t)}^{-1} f_\delta\|} = 0. \tag{4.55}$$

This, (4.45), and (4.46) imply

$$\lim_{t \to \infty} \frac{\|A v(t) - f_\delta\|}{a(t) \|Q_{a(t)}^{-1} f_\delta\|} = \lim_{t \to \infty} \frac{\left\| e^{-t} \int_0^t e^s a(s) Q_{a(s)}^{-1} f_\delta \, ds \right\|}{a(t) \|Q_{a(t)}^{-1} f_\delta\|} = 1. \tag{4.56}$$

It follows from (4.56) that $\lim_{t \to \infty} \|A v(t) - f_\delta\| \leq \delta$ since $\lim_{t \to \infty} h(t) \leq \delta$ as we have proved earlier. This implies the existence of t_δ.

Let us prove that

$$\lim_{\delta \to 0} t_\delta = \infty. \tag{4.57}$$

Arguing by contradiction, assume that (4.57) does not hold. Then there exists a sequence $0 < (\delta_n)_{n=1}^\infty \searrow 0$ such that the sequence $(t_{\delta_n})_{n=1}^\infty$ is bounded, along with a subsequence of $(\delta_n)_{n=1}^\infty$ denoted again by $(\delta_n)_{n=1}^\infty$ such that $t_{\delta_n} \to T$ as $n \to \infty$. Let $n \to \infty$. Then $\delta_n \to 0$ and $u(t_{\delta_n}) \to u(T)$. This and (4.42) imply the following relation

$$0 = \lim_{n \to \infty} \|A v(t_{\delta_n}) - f_{\delta_n}\| = \|A u(T) - f\|. \tag{4.58}$$

We have used the relation $u(t) = v(t)|_{\delta=0}, \forall t$. When $\delta = 0$, one has $A u_0 = A T_{a(0)}^{-1} A^* f = Q Q_{a(0)}^{-1} f$. So $A u_0 - f = -a(0) Q_{a(0)}^{-1} f$. Thus,

$$\begin{aligned} A u(t) - f &= e^{-t}(A u_0 - f) - \int_0^t e^{-(t-s)} a(s) Q_{a(s)}^{-1} f \, ds \\ &= -\left[e^{-t} a(0) Q_{a(0)}^{-1} + \int_0^t e^{-(t-s)} a(s) Q_{a(s)}^{-1} \, ds \right] f. \end{aligned} \tag{4.59}$$

This, (4.58) and the spectral theorem imply

$$
\begin{aligned}
0 &= -\langle f, Au(T) - f \rangle \\
&= -\left\langle f, -\left[e^{-T}a(0)Q_{a(0)}^{-1} + \int_0^T e^{-(T-s)}a(s)Q_{a(s)}^{-1}\, ds \right] f \right\rangle \\
&= \left\langle f, e^{-T} \int_0^{\|Q\|} \left[\frac{a(0)}{\lambda + a(0)} + \int_0^T e^s \frac{a(s)}{a(s) + \lambda}\, ds \right] dF_\lambda f \right\rangle \\
&\geq \|f\|^2 e^{-T} \left(\frac{a(0)}{\|Q\| + a(0)} + \int_0^T e^s \frac{a(s)}{a(s) + \|Q\|}\, ds \right) \\
&> 0.
\end{aligned}
\tag{4.60}
$$

This contradiction implies that (4.57) holds.

It follows from (4.56) that the first equation in (4.42) can be written as

$$
a(t)\|Q_{a(t)}^{-1} f_\delta\| = c_1 \delta,
\tag{4.61}
$$

where

$$
c_1 = c[1 + o(1)], \quad t \to \infty.
\tag{4.62}
$$

This and (4.57) imply

$$
\lim_{\delta \to 0} \|w_\delta(t_\delta) - y\| = 0, \qquad w_\delta(t) := T_{a(t)}^{-1} A^* f_\delta.
\tag{4.63}
$$

Thus, there exists $C = const > 0$ independent of δ such that

$$
\|w_\delta(t_\delta)\| \leq C \quad \text{as} \quad \delta \to 0.
\tag{4.64}
$$

Let us prove that

$$
\lim_{t \to \infty} \frac{\|v(t) - w_\delta(t)\|}{\|w_\delta(t)\|} = 0.
\tag{4.65}
$$

One has

$$
\|A^* f_\delta\| \leq \|T_{a(t)}\| \|T_{a(t)}^{-1} A^* f_\delta\| \leq (\|T\| + a(0))\|T_{a(t)}^{-1} A^* f_\delta\|.
$$

Thus,

$$
\|w_\delta(t)\| = \|T_{a(t)}^{-1} A^* f_\delta\| \geq \frac{\|A^* f_\delta\|}{\|T\| + a(0)}.
$$

This implies

$$
0 \leq \lim_{t \to \infty} \frac{e^{-t}\|u_0\|}{\|w_\delta(t)\|} \leq \lim_{t \to \infty} \frac{e^{-t}\|u_0\|(\|T\| + a(0))}{\|A^* f_\delta\|} = 0.
\tag{4.66}
$$

Thus, to prove (4.65) it suffices to show that

$$
J := \lim_{t \to \infty} \frac{\left\| e^{-t} \int_0^t e^s T_{a(s)}^{-1} A^* f_\delta\, ds - (1 - e^{-t}) T_{a(t)}^{-1} A^* f_\delta \right\|^2}{\|T_{a(t)}^{-1} A^* f_\delta\|^2} = 0.
\tag{4.67}
$$

Using the commutation formula $T_{a(s)}^{-1}A^* = A^*Q_{a(s)}^{-1}$ and the spectral theorem, one obtains

$$J = \lim_{t \to \infty} \frac{\int_0^{\|Q\|} \left(e^{-t} \int_0^t e^s \frac{\sqrt{\lambda}}{a(s)+\lambda} \, ds - \frac{\sqrt{\lambda}}{a(t)+\lambda} \right)^2 d\langle F_\lambda f_\delta, f_\delta \rangle}{\int_0^{\|Q\|} \frac{\lambda}{(a(t)+\lambda)^2} \, d\langle F_\lambda f_\delta, f_\delta \rangle}. \tag{4.68}$$

Using (4.49) and the arguments similar to the ones that yielded relations (4.49)–(4.53), one can show that for an arbitrary small $\epsilon > 0$, there exists t_ϵ such that

$$-\epsilon \leq \frac{e^{-t} \int_0^t e^s \frac{\sqrt{\lambda}}{a(s)+\lambda} \, ds - \frac{\sqrt{\lambda}}{a(t)+\lambda}}{\frac{\sqrt{\lambda}}{a(t)+\lambda}} \leq \epsilon, \qquad t \geq t_\epsilon, \, 0 \leq \lambda \leq \|Q\|. \tag{4.69}$$

This and (4.68) imply that (4.67) holds. Therefore, (4.65) holds. From our arguments it follows that (4.65) holds uniformly with respect to δ. Consequently,

$$\|v(t) - w_\delta(t)\| = \|w_\delta(t)\| o(1) \quad \text{as} \quad t \to \infty, \quad \forall \delta > 0. \tag{4.70}$$

This, relation (4.64), and (4.57) imply

$$0 \leq \lim_{\delta \to 0} \|v(t_\delta) - w_\delta(t_\delta)\| = 0. \tag{4.71}$$

This, (4.63) and the triangle inequality imply (4.43). Theorem 4.1.11 is proved. ∎

4.2 ANOTHER APPROACH

In this section we develop another approach to solving ill-posed equation (4.1). The operator A in Theorems 4.1.1–4.1.6 may be unbounded.

Let us first explain the main idea: Under suitable assumptions the inverse of an operator can be calculated by solving a Cauchy problem. Let B be a bounded linear operator which has a bounded inverse, and assume that

$$\lim_{t \to \infty} \|e^{Bt}\| = 0. \tag{4.72}$$

For example, a sufficient condition for (4.72) to hold is $B = B_1 + iB_2$, B_1 and B_2 are selfadjoint operators and $B_1 \leq -cI$, $c = const > 0$. The operator e^{Bt} is well defined not only for bounded operators B but for generators of C_0-semigroups (see [89]). One has

$$\int_0^t e^{Bs} ds = B^{-1}(e^{Bt} - I); \tag{4.73}$$

and if (4.72) holds, then

$$- \lim_{t \to \infty} \int_0^t e^{Bs} ds = B^{-1}. \tag{4.74}$$

The Cauchy problem

$$\dot{W} = BW + I, \quad W(0) = 0 \tag{4.75}$$

has a unique solution

$$W(t) = \int_0^t e^{Bs} ds. \tag{4.76}$$

Therefore the problem

$$\dot{u} = Bu + f, \quad u(0) = 0, \tag{4.77}$$

has a unique solution

$$u = W(t) f = \int_0^t e^{Bs} ds f, \tag{4.78}$$

so

$$- \lim_{t \to \infty} u(t) = B^{-1} f. \tag{4.79}$$

Therefore, if A in (4.1) is boundedly invertible operator, such that

$$\lim_{t \to \infty} \| e^{At} \| = 0,$$

then one can solve equation (4.1) by solving the Cauchy problem

$$\dot{u} = Au - f, \quad u(0) = 0, \tag{4.80}$$

which has a unique global solution

$$u(t) = - \int_0^t e^{A(t-s)} ds \, f, \tag{4.81}$$

and then calculating the solution $y = A^{-1} f$ by the formula

$$\lim_{t \to \infty} u(t) = A^{-1} f = y. \tag{4.82}$$

If A is not boundedly invertible, the above idea is also useful as we are going to show.

Let us assume that A is a linear self-adjoint densely defined operator, $\mathcal{N}$ is its null space, and E_s is its resolution of the identity, and consider the problem

$$\dot{u}_a = i A_{ia} u_a - if, \quad u(0) = 0, \quad a = const > 0, \quad A_{ia} := A + iaI. \tag{4.83}$$

For any $a > 0$ the inverse A_{ia}^{-1} is a bounded operator,

$$\| A_{ia}^{-1} \| \leq \frac{1}{a}.$$

Moreover,

$$\lim_{t \to \infty} \|e^{iA_{ia}t}\| = \lim_{t \to \infty} e^{-at} = 0, \tag{4.84}$$

so that (4.72) holds with $B = iA_{ia}$. Problem (4.83) has a unique global solution and, by (4.79), we have

$$\lim_{t \to \infty} u_a(t) = i(iA_{ia})^{-1}f = i(iA_{ia})^{-1}Ay, \quad y \perp \mathcal{N}. \tag{4.85}$$

Moreover,

$$\lim_{a \to 0} \|i(iA_{ia})^{-1}Ay - y\| = 0. \tag{4.86}$$

Indeed,

$$\lim_{a \to 0} \eta^2(a) := \lim_{a \to 0} \|A_{ia}^{-1}Ay - y\|^2$$

$$= \lim_{a \to 0} \int_{-\infty}^{\infty} \frac{a^2 d\langle E_s y, y \rangle}{s^2 + a^2} = \|P_{\mathcal{N}}y\|^2 = 0. \tag{4.87}$$

We have proved the following theorem.

Theorem 4.2.1 *Assume that A is a linear, densely defined self-adjoint operator, equation $Au = f$ is solvable (possibly nonuniquely), and y is its minimal-norm solution, $y \perp \mathcal{N} = \mathcal{N}(A)$. Then*

$$y = \lim_{a \to 0} \lim_{t \to \infty} u_a(t), \tag{4.88}$$

where $u_a(t) = u_a(t; f)$ is the unique solution to (4.83).

Remark 4.2.2 Practically one integrates (4.83) on a finite interval $[0, \tau]$ and, for a chosen $\tau > 0$ one takes $a = a(\tau) > 0$ such that

$$\lim_{\tau \to \infty} u_{a(\tau)}(\tau) = y. \tag{4.89}$$

Relation (4.89) holds if the following conditions are valid:

$$\lim_{\tau \to \infty} a(\tau) = 0, \quad \lim_{\tau \to \infty} \frac{e^{-a(\tau)\tau}}{a(\tau)} = 0. \tag{4.90}$$

For example, one may take $a(\tau) = \tau^{-\gamma}$, $\gamma \in (0, 1)$. To check that (4.90) implies (4.89), we note that

$$\|u_a(t) - u_a(\infty)\| \leq \frac{e^{-at}}{a}, \tag{4.91}$$

as follows from the proof of Theorem 4.2.1. Therefore

$$\|u_{a(\tau)}(\tau) - y\| \leq \|u_{a(\tau)}(\tau) - u_{a(\tau)}(\infty)\| + \|u_{a(\tau)}(\infty) - y\| \to 0 \tag{4.92}$$

as $\tau \to \infty$, provided that conditions (4.90) hold.

Remark 4.2.3 We set $u(0) = u_0$ in (4.83), but similarly one can treat the case $u(0) = u_0$ with an arbitrary u_0.

Let us consider the case of noisy data f_δ, $\|f_\delta - f\| \leq \delta$. We have

$$\|u_a(t; f) - u_a(t; f_\delta)\| \leq \left\| \int_0^t e^{i(A+ia)(t-s)} ds(-i)(f - f_\delta) \right\|$$
$$\leq \frac{\delta}{a} \|e^{i(A+ia)t} - I\| \leq \frac{2\delta}{a}, \tag{4.93}$$

where $u_a(t, f_\delta)$ solves (4.83) with f_δ replacing f. Thus,

$$\|u_a(t; f_\delta) - y\| \leq \frac{\delta}{a} + \|u_a(t; f) - y\|. \tag{4.94}$$

We estimate:

$$\|u_a(t; f) - y\| \leq \frac{e^{at}}{a} \|f\| + \eta(a), \quad \lim_{a \to 0} \eta(a) = 0. \tag{4.95}$$

From (4.94)–(4.95) one concludes that

$$\lim_{\delta \to 0} \|u_{a(\delta)}(t_\delta; f_\delta) - y\| = 0, \tag{4.96}$$

provided that

$$\lim_{\delta \to 0} a(\delta) = 0, \quad \lim_{\delta \to 0} t_\delta = \infty, \quad \lim_{\delta \to 0} \frac{\delta}{a(\delta)} = 0, \quad \lim_{\delta \to 0} \frac{e^{-t_\delta a(\delta)}}{a(\delta)} = 0. \tag{4.97}$$

Let us formulate the result.

Theorem 4.2.4 *Assume that equation $Au = f$ is solvable, $Ay = f$, $y \perp \mathcal{N}$, A is a linear selfadjoint densely defined operator in a Hilbert space H, $\|f_\delta - f\| \leq \delta$, and $u_a(t; f_\delta)$ solves (4.83) with f_δ replacing f. If (4.97) holds, then (4.96) holds.*

So far we have assumed that $a = const$. Let us take $a = a(t)$ and assume

$$0 < a(t) \searrow 0, \quad \int_0^\infty a(s) \, ds = \infty, \quad a' + a^2 \in L^1[0, \infty). \tag{4.98}$$

Consider the following DSM problem:

$$\dot{u} = i[A + ia(t)]u - if, \quad u(0) = 0. \tag{4.99}$$

Problem (4.99) has a unique solution

$$u(t) = \int_0^t e^{iA(t-s) - \int_s^t a(p)dp} ds(-if). \tag{4.100}$$

For exact data f we prove the following result.

Theorem 4.2.5 *If (4.98) holds, then*

$$\lim_{t\to\infty} \|u(t) - y\| = 0, \tag{4.101}$$

where $u(t)$ solves (4.99).

Proof: Substitute $f = Ay$ in (4.100) and integrate by parts to get

$$u(t) = e^{iA(t-s) - \int_s^t a\,dp} y \Big|_0^t - \int_0^t e^{iA(t-s)} a(s) e^{-\int_s^t a\,dp}\,ds\; y.$$

Thus

$$u(t) = y - e^{iAt - \int_0^t a\,dp} y - \int_0^t e^{iA(t-s)} a(s) e^{-\int_s^t a\,dp}\,ds\; y \tag{4.102}$$

and

$$\|u(t) - y\| \le e^{-\int_0^t a\,dp}\|y\| + \Big\|\int_0^t e^{iA(t-s)} a(s) e^{-\int_s^t a\,dp}\,ds\; y\Big\| := J_1 + J_2. \tag{4.103}$$

Assumption (4.98) implies $\lim_{t\to\infty} J_1 = 0$. Let us prove that

$$\lim_{t\to\infty} J_2 = 0.$$

Using the spectral theorem and denoting by E_λ the resolution of the identity corresponding to A, we get

$$J_2^2 = \int_{\infty}^\infty d\langle E_\lambda y, y\rangle \left|\int_0^t e^{i\lambda(t-s)} a(s) e^{-\int_s^t a\,dp}\,ds\right|^2. \tag{4.104}$$

We *claim* that assumptions (4.98) imply:

$$\lim_{t\to\infty} \int_0^t e^{i\lambda(t-s)} a(s) e^{-\int_s^t a\,dp}\,ds = 0, \qquad \forall \lambda \neq 0, \tag{4.105}$$

while for $\lambda = 0$ we have

$$\lim_{t\to\infty} \int_0^t a(s) e^{-\int_s^t a\,dp}\,ds = \lim_{t\to\infty} (1 - e^{-\int_0^t a\,dp}) = 1. \tag{4.106}$$

To verify (4.105) denote the integral in (4.105) by J_3, integrate by parts and get

$$J_3 = \frac{e^{i\lambda(t-s)}}{-i\lambda} a(s) e^{-\int_s^t a\,dp} \Big|_0^t + \frac{1}{i\lambda}\int_0^t e^{i\lambda(t-s)}[a'(s) + a^2(s)] e^{-\int_s^t a\,dp}\,ds. \tag{4.107}$$

Thus

$$J_3 = \frac{a(t)}{-i\lambda} + \frac{e^{i\lambda t - \int_0^t a\,dp}a(0)}{i\lambda} + J_4, \quad \lambda \neq 0, \tag{4.108}$$

where J_4 is the last integral in (4.107). The first two terms in (4.108) tend to zero as $t \to \infty$, due to assumptions (4.108), and these assumptions also imply $\lim_{t\to\infty} J_4 = 0$ if $\lambda \neq 0$.

Therefore, passing to the limit $t \to \infty$ in (4.104), one gets

$$\lim_{t\to\infty} J_2^2 = \|(E_0 - E_{-0})y\|^2 = \|P_\mathcal{N}y\|^2 = 0. \tag{4.109}$$

Theorem 4.2.5 is proved.   ■

Consider now the case of noisy data f_δ, $\|f_\delta - f\| \leq \delta$. Problem (4.100), with f_δ in place of f, has a unique solution given by formula (4.100) with f_δ in place of f. Let us denote this solution by $u_\delta(t)$. Then

$$\|u_\delta(t) - y\| \leq \|u_\delta(t) - u(t)\| + \|u(t) - y\| := I_1 + I_2. \tag{4.110}$$

We have

$$\int_0^t e^{-\int_s^t a\,dp}ds \leq \int_0^s \frac{a(s)}{a(t)}e^{-\int_s^t a\,dp}ds \leq \frac{1}{a(t)}.$$

Thus

$$I_1 \leq \delta \int_0^t e^{-\int_s^t a\,dp}ds \leq \frac{\delta}{a(t)}. \tag{4.111}$$

In Theorem 4.2.5 we have proved that

$$\lim_{t\to\infty} I_2 = 0. \tag{4.112}$$

From (4.110)–(4.112) we obtain the following result.

Theorem 4.2.6 *Assume (4.98) and*

$$\lim_{\delta\to 0} t_\delta = \infty, \quad \lim_{\delta\to 0} \frac{\delta}{a(t_\delta)} = 0. \tag{4.113}$$

Define $u_\delta := u_\delta(t_\delta)$, where $u_\delta(t)$ solves (4.99) with f_δ replacing f. Then

$$\lim_{\delta\to 0} \|u_\delta - y\| = 0. \tag{4.114}$$

Remark 4.2.7 It is of interest to find an optimal in some sense choice of the stopping time t_δ. Conditions (4.113) can be satisfied by many choices of t_δ.

Remark 4.2.8 In this section we assume that A is self-adjoint. If A is not self-adjoint, closed, and densely defined in H operator, equation (4.1) is solvable, and $Ay = f$, $y \perp \mathcal{N}$, then every solution to equation (4.1) generates a solution to the equation

$$A^*Au = A^*f \tag{4.115}$$

in the following sense. For a densely defined closed linear operator A the operator A^* is also densely defined and closed. However, the element f may not belong to the domain $D(A^*)$ of A^*, so equation (4.115) has to be interpreted for $f \notin D(A^*)$. We define the solution to (4.115) for any $f \in \mathcal{R}(A)$, that is, for $f = Ay$, $y \perp \mathcal{N}$, by the formula

$$u = \lim_{a \to 0} T_a^{-1} A^* f = y, \quad T = A^* A, \quad T_a := T + aI. \tag{4.116}$$

The existence of the limit and the formula

$$\lim_{a \to 0} T_a^{-1} A^* f = y,$$

valid for $y \perp \mathcal{N}$, have been established in Theorem 2.2.1. With the definition (4.116) of the solution to (4.115) for any $f \in \mathcal{R}(A)$, one has the same equivalence result for the solutions to (4.1) and (4.115) as in the case of bounded A.

Recall, that if A is bounded and $Au = f$, then $A^* Au = A^* f$, so u which solves (4.1) solves (4.115) as well. Conversely, if u solves (4.115) and $f = Ay$, then $A^* Au = A^* Ay$, so $\langle A^* A(u - y), u - y \rangle = 0$, and $\|Au - Ay\| = 0$, so $Au = f$.

If A is unbounded, then we modify the above equivalence claim: If $Ay = f$, $y \perp \mathcal{N}$, then y solves (4.115) and vice versa.

With this understanding of the notion of the solution to (4.115) we can use Theorems 4.2.1–4.2.6 of this section. Indeed, if equation (4.1) is solvable, $Ay = f$, $y \perp \mathcal{N}$, then we replace equation (4.1) by equation (4.115) with the self-adjoint operator $T = A^* A \geq 0$ and apply Theorems 4.2.1–4.2.6 of this section to equation (4.115). As we have explained, finding the minimal-norm solution to equation (4.1) is equivalent to finding the solution to equation (4.115) defined by formula (4.116).

4.3 EQUATIONS WITH UNBOUNDED OPERATORS

The results of Section 4.1 remain valid for unbounded, densely defined, closed linear operator A. In Theorem 2.2.3, we have proved that for such operator A the operator $T_a^{-1} A^*$, $a > 0$, can be considered as defined on all of H bounded operators whose norm is bounded by $\frac{1}{2\sqrt{a}}$ (see (2.76)).

Therefore the formulations and proofs of Theorems 4.1.1–4.1.6 remain valid.

The results of Section 4.2 have been formulated for unbounded, densely defined, closed linear operators.

In all the results we could assume that the initial condition $u(0)$ is an arbitrary element $u_0 \in H$, not necessarily zero. If u_0 is chosen suitably, for example, in a neighborhood of the solution y, this may accelerate the rate of convergence. *But our results are valid for any choice of u_0 because*

$$\lim_{t \to \infty} \|e^{i(A+ia)t} u_0\| = 0, \quad a > 0,$$

or, in the case $a = a(t)$,

$$\lim_{t \to \infty} \|e^{iAt - \int_0^t a(p)dp} u_0\| = 0.$$

Here the element

$$u_0(t) = e^{iAt - \int_0^t a(p)dp} u_0$$

solves the problem

$$\dot{u} = i[A + ia(t)]u, \quad u(0) = u_0,$$

so that this $u_0(t)$ is a contribution to the solution of (4.99) from the non zero initial condition u_0. If $\int_0^\infty a(t)\, dt = \infty$, then $\lim_{t \to \infty} u_0(t) = 0$.

4.4   ITERATIVE METHODS

The main results concerning iterative methods for solving linear ill-posed equation (4.1) have been formulated and proved in Section 2.4, Theorems 2.4.1–2.4.6.

In this section we show how to use these results for constructing a stable approximation to the minimal-norm solution y of equation (4.1), given noisy data f_δ, $\|f_\delta - f\| \leq \delta$.

Our approach is based on a general principle formulated in [104]:

Proposition 4.4.1 *If equation $Au = f$ has a solution and there is a convergent iterative process $u_{n+1} = T(u_n; f)$, $u_0 = u_0$, for solving this equation: $\lim_{n \to \infty} \|u_n - y\| = 0$, where $Ay = f$, $y \perp \mathcal{N}$, and T is a continuous operator, then there exists an $n(\delta)$, $\lim_{\delta \to 0} n(\delta) = \infty$, such that the same iterative process with f_δ replacing f produces a sequence $u_n(f_\delta)$ such that $\lim_{\delta \to 0} \|u_{n(\delta)}(f_\delta) - y\| = 0$.*

Proof: We have

$$\varepsilon(\delta) := \|u_{n(\delta)}(f_\delta) - y\| \leq \|u_{n(\delta)}(f_\delta) - u_{n(\delta)}(f)\| + \|u_{n(\delta)}(f) - y\|. \quad (4.117)$$

By one of the assumptions of Proposition 4.4.1, we have

$$\lim_{n \to \infty} \|u_n(f) - y\| = 0.$$

For any fixed n, by the continuity of T, we have

$$\lim_{\delta \to 0} \|u_n(f_\delta) - u_n(f)\| := \lim_{\delta \to 0} \eta_n(\delta) = 0. \quad (4.118)$$

Thus with

$$w(n) := \|u_n(f) - y\|, \quad \lim_{n \to \infty} w(n) = 0,$$

one gets

$$\varepsilon(\delta) \leq \eta_{n(\delta)}(\delta) + w(n(\delta)) \to 0 \quad \text{as } \delta \to 0, \quad (4.119)$$

where $n(\delta)$ is the minimizer with respect to n of $\eta_n(\delta) + w(n)$ for a small fixed $\delta > 0$.

Proposition 4.4.1 is proved. ∎

Let us consider iterative process (2.127) with f_δ replacing f. Then we have

$$\|u_{n+1}(f_\delta) - y\| \leq \|u_{n+1}(f_\delta) - u_{n+1}(f)\| + \|u_{n+1}(f) - y\|. \qquad (4.120)$$

It is proved in Theorem 2.4.1 that

$$\lim_{n\to\infty} \|u_n(f) - y\| = 0, \qquad (4.121)$$

where $Ay = f$, $y \perp \mathcal{N}$. Furthermore,

$$\varepsilon(n,\delta) := \|u_{n+1}(f_\delta) - u_{n+1}(f)\| \leq \|B[u_n(f_\delta) - u_n(f)]\| + \frac{\delta}{2\sqrt{a}}. \qquad (4.122)$$

Thus

$$\varepsilon(n,\delta) \leq \sum_{j=0}^{n} \|B\|^j \frac{\delta}{2\sqrt{a}} + \|B^{n+1}\| \, \|u_0(f_\delta) - u_0(f)\|$$

$$= \frac{\|B\|^{n+1} - 1}{\|B\| - 1} \frac{\delta}{2\sqrt{a}} := \gamma(n)\delta.$$

If one denotes

$$w(n) := \|u_n(f) - y\|,$$

then

$$\|u_{n+1}(f_\delta) - y\| \leq \gamma(n)\delta + w(n) \to 0 \text{ as } \delta \to 0, \quad \text{if } n = n(\delta), \qquad (4.123)$$

where $n(\delta)$ is obtained, for example, by minimizing the function

$$\gamma(n)\delta + w(n)$$

with respect to $n = 1, 2, ...$, for a fixed small $\delta > 0$. Note that if $\|B\| \geq 1$, then

$$\lim_{n\to\infty} \gamma(n) = \infty.$$

If $\|B\| < 1$, then

$$\lim_{n\to\infty} \gamma(n) = \frac{1}{(1 - \|B\|)2\sqrt{a}} = const.$$

In both cases one can take $n(\delta)$ such that $\lim_{\delta\to 0} n(\delta) = \infty$ and

$$\lim_{\delta\to 0} \varepsilon(n(\delta), \delta) = 0. \qquad (4.124)$$

We have proved the following result.

Theorem 4.4.2 *Let the assumptions of Theorem 2.4.1 hold, and*

$$\|f_\delta - f\| \leq \delta.$$

Then, if one uses the iterative process (2.127) with f_δ in place of f and chooses $n(\delta)$ such that (4.124) holds, then

$$\lim_{\delta \to 0} \|u_\delta - y\| = 0,$$

where $u_\delta := u_{n(\delta)}(f_\delta)$ is calculated by the above iterative process.

Consider now the iterative process (2.134) with f_δ in place of f. An argument, similar to the one given in the proof of Theorem 4.4.2, yields the following result.

Theorem 4.4.3 *Let the assumptions of Theorem 2.4.5 hold, $\|f_\delta - f\| \leq \delta$, and $n(\delta)$ is suitably chosen. Then $\lim_{\delta \to 0} \|u_{n(\delta)} - y\| = 0$, where $u_\delta := u_{n(\delta)}$ is calculated by the iterative process (2.134) with f_δ in place of f.*

The choice of $n(\delta)$ is quite similar to the choice which was made in the proof of Theorem 4.4.2 above. The role of the operator B is now played by the operator $B = iaA_{ia}^{-1}$ and the role of $\frac{\delta}{2\sqrt{a}}$ is played by $\frac{\delta}{a}$ because

$$\|A_{ia}^{-1}\| \leq \frac{1}{a}$$

if $A = A^*$.

Also, using the arguments similar to the ones given in the proof of Theorem 4.4.2 of this section, one can obtain an analog of Theorem 2.4.6 in the case of noisy data f_δ, $\|f_\delta - f\| \leq \delta$.

We leave this to the reader.

4.5   STABLE CALCULATION OF VALUES OF UNBOUNDED OPERATORS

Let us assume that A is a linear, closed, densely defined operator in H. If $u \in D(A)$ then $Au = f$.

The problem, we study in this section is:

How does one calculate f stably given u_δ, $\|u_\delta - u\| \leq \delta$?

The element u_δ may not belong to $D(A)$. Therefore this problem is ill-posed.

It was studied in [83] by the variational regularization method.

We propose a new method, an iterative method, for solving the above problem. The equation $Au = f$ is equivalent to the following one:

$$Bf = Fu, \tag{4.125}$$

where

$$B := (I + Q)^{-1}, \qquad Q = AA^*,$$
$$F := (I + Q)^{-1}A = A(I + T)^{-1}, \qquad T = A^*A. \tag{4.126}$$

It follows (see (2.76)) that F is a bounded operator,

$$\|F\| \leq \frac{1}{2}, \tag{4.127}$$

while B is a self-adjoint operator $0 < B \leq I$, whose eigenspace, corresponding to the eigenvalue $\lambda = 1$, is identical to the null space $\mathcal{N}(Q)$ of the operator Q, $\mathcal{N}(Q) = \mathcal{N}(A^*) := \mathcal{N}^*$.

If u_δ is given and $\|u_\delta - u\| \leq \delta$, then $\|Fu_\delta - Fu\| \leq \frac{\delta}{2}$. To solve equation (4.125) for f, given the noisy data Fu_δ, we propose to use the following iterative process:

$$f_{n+1} = (I - B)f_n + Fu_\delta := Vf_n + Fu_\delta, \qquad V := I - B, \tag{4.128}$$

and $f_0 \in H$ is arbitrary. Let g be the minimal-norm solution of equation (4.125), that is, $Bg = Fu$, $g = Vg + Fu$.

Theorem 4.5.1 *Let* $n = n(\delta)$ *be an integer such that*

$$\lim_{\delta \to 0} n(\delta) = \infty, \qquad \lim_{\delta \to 0} \delta n(\delta) = 0, \tag{4.129}$$

and $f_\delta := f_{n(\delta)}$, *where* f_n *is defined by (4.128). Then*

$$\lim_{\delta \to 0} \|f_\delta - g\| = 0. \tag{4.130}$$

Proof: Let $w_n := f_n - g$. Then

$$w_{n+1} = Vw_n + F(u_\delta - u), \qquad w_0 = f_0 - g. \tag{4.131}$$

From (4.131) we derive

$$w_n = \sum_{j=0}^{n-1} V^j Fv_\delta + V^n w_0, \qquad v_\delta := u_\delta - u, \qquad \|v_\delta\| \leq \delta. \tag{4.132}$$

Since $\|V\| \leq 1$ and $\|F\| \leq \frac{1}{2}$, the above equation implies

$$\|w_n\| \leq \frac{n\delta}{2} + \left[\int_0^1 (1 - s)^{2n} d\langle E_s w_0, w_0 \rangle \right]^{\frac{1}{2}}, \tag{4.133}$$

where E_s is the resolution of the identity corresponding to the self-adjoint operator B, $0 < B \leq I$. One has

$$\lim_{n \to \infty} \int_0^1 (1 - s)^{2n} d\langle E_s w_0, w_0 \rangle = \|Pw_0\|^2 = 0, \tag{4.134}$$

where P is the orthoprojector onto the subspace $\mathcal{N}(B)$, but $Pw_0 = 0$ because $\mathcal{N}(B) = \{0\}$. The conclusion (4.130) can now be established. Given an arbitrary small $\varepsilon > 0$, find $n = n(\delta)$, sufficiently large so that

$$\int_0^1 (1 - s)^{2n} d\langle E_s w_0, w_0 \rangle < \frac{\varepsilon^2}{4}.$$

This is possible as we have already proved (see (4.134)).

Simultaneously, we choose $n(\delta)$ so that $\delta n(\delta) < \varepsilon$. This is possible because $\lim_{\delta \to 0} \delta n(\delta) = 0$. Thus, if δ is sufficiently small, then (4.130) holds if $n(\delta)$ satisfies (4.129).

Theorem 4.5.1 is proved. ∎

CHAPTER 5

SOME INEQUALITIES

In this chapter some nonlinear inequalities are derived, which are used in other chapters.

5.1 BASIC NONLINEAR DIFFERENTIAL INEQUALITY

We will use much in what follows the following result.

Theorem 5.1.1 *Let $\alpha(t)$, $\beta(t)$, $\gamma(t)$ be continuous nonnegative functions on $[t_0, \infty)$, $t_0 \geq 0$ is a fixed number. If there exists a function*

$$\mu \in C^1[t_0, \infty), \quad \mu > 0, \quad \lim_{t \to \infty} \mu(t) = \infty, \tag{5.1}$$

such that

$$0 \leq \alpha(t) \leq \frac{\mu(t)}{2}\left[\gamma(t) - \frac{\dot{\mu}(t)}{\mu(t)}\right], \quad \dot{\mu}(t) = \frac{d\mu}{dt}, \tag{5.2}$$

$$\beta(t) \leq \frac{1}{2\mu(t)}\left[\gamma(t) - \frac{\dot{\mu}(t)}{\mu(t)}\right], \tag{5.3}$$

$$\mu(t_0)g(t_0) < 1, \tag{5.4}$$

Dynamical Systems Method and Applications: Theoretical Developments and Numerical Examples, First Edition. A. G. Ramm and N. S. Hoang.
Copyright © 2012 John Wiley & Sons, Inc.

and $g(t) \geq 0$ satisfies the inequality

$$\dot{g}(t) \leq -\gamma(t)g(t) + \alpha(t)g^2(t) + \beta(t), \quad t \geq t_0, \tag{5.5}$$

then $g(t)$ exists on $[t_0, \infty)$ and

$$0 \leq g(t) < \frac{1}{\mu(t)} \to 0 \quad as \quad t \to \infty. \tag{5.6}$$

Remark 5.1.2 There are several novel features in this result. Usually a nonlinear differential inequality is proved by integrating the corresponding differential equation, in our case the following equation:

$$\dot{g}(t) = -\gamma(t)g(t) + \alpha(t)g^2(t) + \beta(t), \tag{5.7}$$

and then using a comparison lemma. Equation (5.7) is the full Riccati equation, the solution to which, in general, may blow up on a finite interval. Assumptions (5.2)–(5.4) imply that the solutions to (5.7) and (5.5) exist globally. Furthermore, we do not assume that the coefficient $\alpha(t)$ in front of the senior nonlinear term in (5.5) is subordinate to other coefficients.

In fact, in our applications of Theorem 5.1.1 of this section (see Chapters 6, 7, and 10) the coefficient $\alpha(t)$ will tend to infinity as $t \to \infty$.

Remark 5.1.3 Let us give examples of the choices of α, β and γ satisfying assumption (5.2)–(5.4).

Let

$$\gamma(t) = c_1(1+t)^{\nu_1}, \quad \alpha(t) = c_2(1+t)^{\nu_2}, \quad \beta(t) = c_3(1+t)^{\nu_3},$$

where $c_j > 0$ are constants, and let $\mu = c(1+t)^\nu$, $c = const > 0$, $\nu > 0$. If

$$c_2(1+t)^{\nu_2} \leq \frac{c(1+t)^\nu}{2}\left(c_1(1+t)^{\nu_1} - \frac{1}{1+t}\right),$$

$$c_3(1+t)^{\nu_3} \leq \frac{1}{2c(1+t)^\nu}\left(c_1(1+t)^{\nu_1} - \frac{1}{1+t}\right), \quad g(0)c < 1,$$

then assumptions (5.2)–(5.4) hold. The above inequalities hold if, for example,

$$\nu + \nu_1 \geq \nu_2, \quad c_2 \leq \frac{c}{2}(c_1 - 1); \quad \nu_3 \leq \nu_1 - \nu, \quad c_3 \leq \frac{1}{2c}(c_1 - 1); \quad g(0)c < 1.$$

Let

$$\gamma = \gamma_0 = const > 0, \quad \alpha(t) = \alpha_0 e^{\nu t}, \quad \beta(t) = \beta_0 e^{-\nu t}, \quad \mu(t) = \mu_0 e^{\nu t},$$

where α_0, β_0, μ_0, and ν are positive constants. If

$$\alpha_0 \leq \frac{\mu_0}{2}(\gamma_0 - \nu), \quad \beta_0 \leq \beta_0 \leq \frac{1}{2\mu_0}(\gamma_0 - \nu), \quad g(0)\mu_0 < 1,$$

then assumptions (5.2)–(5.4) hold.

Let

$$\gamma(t) = [\ln(t + t_1)]^{-\frac{1}{2}}, \quad \mu(t) = c \ln(t + t_1),$$

$$0 \le \alpha(t) \le \frac{c}{2}\left[\sqrt{\ln(t + t_1)} - \frac{1}{t + t_1}\right],$$

and

$$0 \le \beta(t) \le \frac{1}{2c\ln(t + t_1)}\left[\frac{1}{\sqrt{\ln(t + t_1)}} - \frac{1}{(t + t_1)\ln(t + t_1)}\right],$$

where $t_1 = const > 1$. Then assumptions (5.2)–(5.4) hold.

Let us recall a known comparison lemma (see, e.g., [40]), whose proof we include for convenience of the reader.

Lemma 5.1.4 *Let $f(t, w)$ and $g(t, u)$ be continuous functions in the region $\Omega = [0, T) \times D$, $T \le \infty$, $D \subset \mathbb{R}$ is an interval, and $f(t, u) \le g(t, u)$ in Ω. Assume that the problem*

$$\dot{u} = g(t, u), \quad u(0) = u_0 \in D$$

has a unique solution defined on some interval $[0, \tau_u)$, where $\tau_u > 0$. If

$$\dot{w} \le f(t, w), \quad w(0) = w_0 \le u_0, \quad f(0, w_0) \le g(0, u_0), \quad w_0 \in D,$$

then $u(t) \ge w(t)$ for all t for which u and w are defined.

Proof of Lemma 5.1.4. Suppose first that $f(t, u) < g(t, u)$ in Ω and $f(0, w_0) < g(0, u_0)$. Since $w_0 \le u_0$ and $\dot{w}(0) \le f(0, w_0) < g(0, u_0) = \dot{u}(0)$, there exists a $\delta > 0$ such that $u(t) > w(t)$ on $(0, \delta]$. If $u(t_1) \le w(t_1)$ for some $t_1 > \delta$, then there is a $t_2 < t_1$ such that $u(t_2) = w(t_2)$ and $w(t) < u(t)$ for $t \in (0, t_2)$. Therefore

$$\dot{w}(t_2) \ge \dot{u}(t_2) = g(t_2, u(t_2)) > f(t_2, u(t_2)) = f(t_2, w(t_2)) \ge \dot{w}(t_2).$$

This contradiction proves that there is no point $t > 0$ at which $w(t) \ge u(t)$. Thus, Lemma 5.1.4 is proved under the additional assumption $f(t, u) < g(t, u)$ in Ω and $f(t, w_0) < g(t, u_0)$.

Let us drop this additional assumption and assume that $f(t, u) \le g(t, u)$ in Ω and $f(0, w_0) \le f(0, u_0)$.

Define $u_n(t)$ as the solution to the problem

$$\dot{u}_n = g(t, u_n) + \frac{1}{n}, \quad u_n(0) = u_0.$$

Then

$$f(0, w_0) \le g(0, u_0) < g(0, u_0) + \frac{1}{n},$$

$$f(t, u) \le g(t, u) < g(t, u) + \frac{1}{n}, \quad u \in D.$$

By what we have already proved, it follows that

$$w(t) < u_n(t), \qquad t > 0.$$

Passing to the limit $n \to \infty$, we obtain the conclusion of Lemma 5.1.4. Passing to the limit can be based on the following (also known) result: If $f_j(t, u)$ is a sequence of continuous functions in the region $\mathbb{R}_0 := [t_0, t_0 + a] \times |u - u_0| \leq \varepsilon$, such that $\lim_{j \to \infty} f_j(t, u) = f(t, u)$ uniformly in $\mathbb{R}_0$, and

$$\dot{u}_j = f_j(t, u_j), \qquad u_j(t_0) = u_0,$$

then

$$\lim_{k \to \infty} u_{j_k}(t) = u(t) \quad \text{uniformly on} \quad [t_0, t - 0 + a],$$

where $\dot{u} = f(t, u)$, $u(t_0) = u_0$; if this problem has a unique solution, then $u_j(t) \to u(t)$ uniformly on $[t_0, t_0 + a]$.

Lemma 5.1.4 is proved. ∎

Proof of Theorem 5.1.1. Denote $w(t) := g(t)e^{\int_{t_0}^{t} \gamma(s)ds}$. Then inequality (5.5) takes the form

$$\dot{w} \leq a(t)w^2 + b(t), \qquad w(t_0) = g(0) := g_0, \tag{5.8}$$

where

$$a(t) := \alpha(t)e^{-\int_{t_0}^{t} \gamma(s)ds}, \qquad b(t) := \beta(t)e^{\int_{t_0}^{t} \gamma(s)ds}. \tag{5.9}$$

Consider the following Riccati equation

$$\dot{u} = \frac{\dot{f}}{g}u^2(t) - \frac{\dot{g}(t)}{f(t)}. \tag{5.10}$$

Its solution can be written analytically:

$$u(t) = -\frac{g(t)}{f(t)} + \cfrac{1}{f^2(t)\left[c - \int_{t_0}^{t} \frac{\dot{f}(s)}{g(s)f^2(s)}ds\right]}, \qquad c = const, \tag{5.11}$$

(see [68]), and one can check that (5.11) solves (5.10) by direct calculation.

Define

$$f := \mu^{\frac{1}{2}}(t)e^{-\frac{1}{2}\int_{t_0}^{t} \gamma ds}, \qquad g(t) := -\frac{1}{\mu^{\frac{1}{2}}(t)}e^{\frac{1}{2}\int_{t_0}^{t} \gamma ds}, \tag{5.12}$$

and solve the Cauchy problem for equation (5.10) with the initial condition

$$u(t_0) = g(t_0). \tag{5.13}$$

The solution is given by formula (5.11) with

$$c = \frac{1}{\mu(t_0)g(t_0) - 1}. \tag{5.14}$$

Let us check assumptions (5.2)–(5.4). Note that

$$f(t)g(t) = -1, \quad \frac{f}{g} = -\mu(t)e^{-\int_{t_0}^t \gamma ds},$$ (5.15)

$$\frac{\dot{f}}{g} = \frac{\mu}{2}\left(\gamma - \frac{\dot{\mu}}{\mu}\right)e^{-\int_{t_0}^t \gamma ds} \geq a(t) \geq 0,$$ (5.16)

$$-\frac{\dot{g}}{f} = \frac{\left(\gamma - \frac{\dot{\mu}}{\mu}\right)}{2\mu}e^{\int_{t_0}^t \gamma ds} \geq b(t) \geq 0,$$ (5.17)

where conditions (5.2)–(5.3) imply (5.16) and (5.17). Condition (5.4) implies

$$c < 0.$$ (5.18)

Let us apply Lemma 5.1.4 to problems (5.8) and (5.10). The assumptions of this lemma are satisfied due to (5.13), (5.16), and (5.17). Thus, Lemma 5.1.4 implies

$$w(t) \leq u(t), \quad t \geq t_0,$$ (5.19)

that is

$$g(t)e^{\int_{t_0}^t \gamma ds} \leq \frac{e^{\int_{t_0}^t \gamma ds}}{\mu(t)}\left[1 - \frac{1}{\frac{1}{1-\mu(t_0)g(t_0)} + \frac{1}{2}\int_{t_0}^t \left(\gamma - \frac{\dot{\mu}}{\mu}\right)ds}\right].$$ (5.20)

This inequality implies

$$g(t) \leq \frac{1}{\mu(t)},$$ (5.21)

because $\gamma - \frac{\dot{\mu}}{\mu} \geq 0$.

Theorem 5.1.1 is proved. ∎

5.2 AN OPERATOR INEQUALITY

In this section we state and prove an operator version of the widely used Gronwall inequality for scalar functions. This inequality can be formulated as follows:

If $a(t), g(t), b(t) \geq 0$ are continuous functions, and

$$g(t) \leq b(t) + \int_0^t a(s)g(s)\, ds, \quad 0 \leq t \leq T,$$

then

$$g(t) \leq b(t) + \int_0^t e^{\int_s^t a(p)dp}b(s)\, ds, \quad 0 \leq t \leq T.$$ (5.22)

Its proof is simple and can be found in many books and in Section 5.4 below.

A nonlinear version of this result is:

If $u(t, g)$ is a continuous function, nondecreasing with respect to g in the region $R := [t_0, t_0 + a] \times \mathbb{R}$, and v solves the inequality

$$v(t) \leq v_0 + \int_{t_0}^{t} u(s, v(s)) \, ds, \quad v_0 \leq g_0,$$

then

$$v(t) \leq g(t), \quad t \in [t_0, t_0 + a],$$

where $g(t)$ is the maximal solution of the problem

$$\dot{g} = u(t, g), \quad g(t_0) = g_0,$$

and we assume that v and g exist on $[t_0, t_0 + a]$.

Several Gronwall-type inequalities for scalar functions are proved in Section 5.4. In this section we prove an operator version of the Gronwall inequality, which we will use in Chapter 10. The result is stated in the following Theorem.

Theorem 5.2.1 *Let $Q(t)$, $T(t)$ and $G(t)$ be bounded linear operator functions in a Hilbert space defined for $t \geq 0$, and assume that*

$$\dot{Q} = -T(t)Q(t) + G(t), \quad Q(0) = Q_0, \quad \dot{Q} := \frac{dQ}{dt}, \tag{5.23}$$

where the derivative can be understood in a weak sense. Assume that there exists a positive integrable function $\varepsilon(t)$ such that

$$\langle T(t)h, h \rangle \geq \varepsilon(t) \|h\|^2, \quad \forall h \in H. \tag{5.24}$$

Then

$$\|Q(t)\| \leq e^{-\int_0^t \varepsilon(s)ds} \left[\|Q_0\| + \int_0^t \|G(s)\| e^{-\int_0^s \varepsilon(p)dp} ds \right], \tag{5.25}$$

where $Q_0 := Q(0)$.

Proof: Denote

$$g(t) = \|Q(t)h\|,$$

where $h \in H$ is arbitrary. Equation (5.23) implies

$$g\dot{g} = \mathrm{Re}\langle \dot{Q}h, Qh \rangle = -\langle TQh, Qh \rangle + \mathrm{Re}\langle Gh, Qh \rangle \leq -\varepsilon(t)g^2 + \|Gh\|g.$$

Since $g \geq 0$, we get

$$\dot{g} \leq -\varepsilon(t)g + \|G(t)h\|. \tag{5.26}$$

This inequality implies

$$g(t) \leq g(0)e^{-\int_0^t \varepsilon(s)ds} + \int_0^t \|G(s)h\| e^{-\int_s^t \varepsilon(p)dp} ds. \tag{5.27}$$

Taking supremum in (5.27) with respect to all h on the unit sphere, $\|h\| = 1$, we obtain inequality (5.25).

Theorem 5.2.1 is proved. ∎

5.3   A NONLINEAR INEQUALITY

Let
$$\dot{u} \le -a(t)f(u(t)) + b(t), \quad u(0) = u_0, \quad u \ge 0. \tag{5.28}$$

Assume that $f(u) \in \text{Lip}_{loc}[0, \infty)$, $a(t)$, $b(t) \ge 0$ are continuous functions on $[0, \infty)$,

$$\int_0^\infty a(s)\,ds = \infty, \quad \lim_{t \to \infty} \frac{b(t)}{a(t)} = 0, \quad a \ge 0, \quad b \ge 0, \tag{5.29}$$

$$f(0) = 0; \quad f(u) > 0, \forall u > 0; \quad f(u) \ge c > 0, \forall u \ge 1, \tag{5.30}$$

where c is a positive constant.

Theorem 5.3.1 *These assumptions imply global existence of $u(t)$ and its decay at infinity:*

$$\lim_{t \to \infty} u(t) = 0. \tag{5.31}$$

Proof: Assumptions (5.29) about $a(t)$ allow one to introduce the new variable

$$s = s(t) := \int_0^t a(p)\,dp,$$

and claim that the map $t \to s$ maps $[0, \infty)$ onto $[0, \infty)$. In the new variable inequality (5.28) takes the form

$$w' \le -f(w) + \beta(s), \quad w(0) = u_0, \quad w \ge 0, \tag{5.32}$$

where

$$w = w(s) = u(t(s)), \quad w' = \frac{dw}{ds}, \quad \beta = \frac{b(t(s))}{a(t(s))}, \quad \lim_{s \to \infty} \beta(s) = 0.$$

Since
$$f(u) \in \text{Lip}_{loc}[0, \infty),$$

there exists a unique local solution to the Cauchy problem

$$\dot{v} = -f(v) + \beta(s), \quad v(0) = u_0.$$

Let us prove that (5.32) and assumptions (5.30) about f imply global existence of v, and, therefore, of w, and its decay at infinity:

$$\lim_{s \to \infty} w(s) = 0. \tag{5.33}$$

The global existence of v follows from a bound $\|v(s)\| \le c$ with c independent of s (see Lemma 2.6.1). This bound is proved as a similar bound for w, which follows from (5.33). If (5.33) is established, then Theorem 5.3.1 is proved.

To prove (5.33), define the set $E \subset \mathbb{R}_+ := [0, \infty)$:

$$E := \{s : f(w(s)) \le \beta(s) + \frac{1}{1+s}\}, \tag{5.34}$$

and let

$$F := \mathbb{R}_+ \backslash E.$$

Let us prove that

$$\sup E = +\infty. \tag{5.35}$$

If $\sup E := k < \infty$, then

$$f(w(s)) - \beta > \frac{1}{1+s}, \quad s > k.$$

This and inequality (5.32) imply

$$w' \le -\frac{1}{1+s}, \quad s > k. \tag{5.36}$$

Integrating (5.36) from k to ∞, one gets

$$\lim_{s \to \infty} w(s) = -\infty,$$

which contradicts the assumption $w \ge 0$. Thus (5.35) is established.

Let us derive (5.33) from (5.35). Let $s_1 \in E$ be such a point that $(s_1, s_2) \in F$, $s_2 > s_1$. Then

$$f(w(s)) > \beta(s) + \frac{1}{1+s}, \quad s \in (s_1, s_2), \tag{5.37}$$

so

$$w' \le -\frac{1}{1+s}, \quad s \in (s_1, s_2). \tag{5.38}$$

Integrating, we get

$$w(s) - w(s_1) \le -\ln\left(\frac{1+s}{1+s_1}\right) < 0.$$

Thus

$$w(s) < w(s_1), \quad s \in (s_1, s_2). \tag{5.39}$$

However, $s_1 \in E$, so

$$f(w(s_1)) \le \frac{1}{s_1 + 1} + \beta(s_1) \to 0 \quad \text{as} \quad s_1 \to \infty. \tag{5.40}$$

From (5.35), (5.39) and (5.40) the relation (5.33) follows, and the conclusion (5.31) of Theorem 5.3.1 is established.

Theorem 5.3.1 is proved.   ■

Let us now prove the following result.

Theorem 5.3.2 *Assume that $y(t)$ and $h(t)$ are nonnegative continuous functions, defined on $[0,\infty)$, and*

$$\int_0^\infty [h(t) + y(t)]\, dt < \infty,$$

and suppose that the following inequality holds:

$$y(t) - y(s) \leq \int_s^t f(y(p))\, dp + \int_s^t h(p)\, dp, \tag{5.41}$$

where $f > 0$ is a nondecreasing continuous function on $[0,\infty)$. Then

$$\lim_{t\to\infty} y(t) = 0. \tag{5.42}$$

Proof: Assume the contrary: There exists $t_n \to \infty$ such that $y(t_n) \geq a > 0$. Choose t_{n+1} such that

$$t_{n+1} - t_n \geq \frac{a}{2f(a)} := c. \tag{5.43}$$

Let

$$m_n := \min_{t_n - c \leq p \leq t_n} y(p) := y(p_n), \quad p_n \in [t_n - c, t_n]. \tag{5.44}$$

Denote

$$\lambda_n := \sup\{t : p_n < t < t_n, y(t) < a\}. \tag{5.45}$$

By the continuity of $y(t)$, we have $y(\lambda_n) = a$. Therefore

$$y(\lambda_n) - y(p_n) = a - m_n \leq \int_{p_n}^{\lambda_n} f(y(s))\, ds + \int_{p_n}^{\lambda_n} h(s)\, ds \leq cf(a) + \delta_n, \tag{5.46}$$

where $\delta_n := \int_{p_n}^{\lambda_n} h(s)\, ds$, and we took into account that $p_n \in [t_n - c, t_n]$ and $\lambda_n \in [p_n, t_n]$, so that $\lambda_n - p_n \leq c = \frac{a}{2f(a)}$. From (5.46) we derive

$$\frac{a}{2} \leq m_n + \delta_n \leq y(t) + \delta_n, \quad t \in [t_n - c, t_n]. \tag{5.47}$$

Integrate (5.47) over $[t_n - c, t_n]$ and get

$$\frac{ac}{2} \leq \int_{t_n - c}^{t_n} y(s)\, ds + c\delta_n. \tag{5.48}$$

Sum up (5.48) from $n = 1$ to $n = N$ to get

$$\frac{ac}{2} N \leq \sum_{j=1}^N \int_{t_j - c}^{t_j} y(s)\, ds + c \sum_{j=1}^N \delta_j. \tag{5.49}$$

Let $N \to \infty$. By assumption, $\int_0^\infty [y(s) + h(s)] \, ds < \infty$, so that the right-hand side of (5.49) is bounded as $N \to \infty$, while its left-hand side is not.

This contradiction proves Theorem 5.3.2.   ∎

Remark 5.3.3 Theorem 5.3.2 is proved in [197, p. 227] by a different argument. In [197] several applications of such result are given to nonlinear partial differential equations.

5.4   THE GRONWALL-TYPE INEQUALITIES

Let

$$\dot{v} \leq a(t)v + b(t), \quad t \geq t_0, \tag{5.50}$$

where $a(t)$ and $b(t)$ are integrable functions. Then the Gronwall inequality is

$$v(t) \leq v(t_0)e^{\int_{t_0}^t a(s)ds} + \int_{t_0}^t b(s)e^{\int_s^t a(p)dp}ds, \quad t \geq t_0. \tag{5.51}$$

To prove it, multiply (5.50) by $e^{-\int_{t_0}^t a(s)ds}$. Then (5.50) yields

$$\frac{d}{dt}[e^{-\int_{t_0}^t a(s)ds}v(t)] \leq b(t)e^{-\int_{t_0}^t a(s)ds}. \tag{5.52}$$

Thus

$$e^{-\int_{t_0}^t a(s)ds}v(t) \leq v(t_0) + \int_{t_0}^t b(s)e^{-\int_{t_0}^s a(p)dp}ds. \tag{5.53}$$

This is equivalent to (5.51).

Lemma 5.4.1 *Assume now that*

$$v, a, b \geq 0, \tag{5.54}$$

$$\int_t^{t+r} a(s)\,ds \leq a_1, \quad \int_t^{t+r} b(s)\,ds \leq a_2, \quad \int_t^{t+r} v(s)\,ds \leq a_3, \tag{5.55}$$

where $r \geq 0$. *Then* (5.50) *implies*

$$v(t + r) \leq (a_2 + \frac{a_3}{r})e^{a_1}, \quad t \geq t_0, \quad r \geq 0. \tag{5.56}$$

Proof: To prove (5.56), use (5.52) and (5.54) to get

$$\frac{d}{dt}[e^{-\int_{t_0}^t a\,dp}v(t)] \leq b(t). \tag{5.57}$$

From (5.57) we obtain

$$e^{-\int_{t_0}^{t+r} a\,dp}v(t+r) \leq e^{-\int_{t_0}^{t_1} a\,dp}v(t_1) + \int_{t_1}^{t+r} b(p)\,dp, \quad t_0 \leq t_1 < t + r, \tag{5.58}$$

and
$$v(t + r) \leq (v(t_1) + a_2)e^{a_1}. \tag{5.59}$$

Integrating (5.59) with respect to t_1 over the interval $[t, t + r]$ yields (5.56). Lemma 5.4.1 is proved. ∎

Inequality (5.56) is proved, for example, in [184].

Lemma 5.4.2 *Suppose c and M are positive constants,*

$$u_j \leq c + M \sum_{m=0}^{j-1} u_m, \quad u_0 \leq c, \quad u_m \in \mathbb{R}. \tag{5.60}$$

Then
$$u_p \leq c(1 + M)^p, \quad p = 1, 2, \dots . \tag{5.61}$$

Proof: Inequality (5.61) is easy to prove by induction: For $p = 0$ inequality (5.61) holds due to (5.60). If (5.61) holds for $p \leq n$, then it holds for $p_m = n+1$ due to (5.60).

$$u_{n+1} \leq M \sum_{m=0}^{n} c(1+M)^m \leq c(1 + M\frac{(1 + M)^{n+1} - 1}{1 + M - 1}) = c(1+M)^{n+1}. \tag{5.62}$$

Thus, (5.61) is proved.
Lemma 5.4.2 is proved. ∎

5.5 ANOTHER OPERATOR INEQUALITY

In this section we prove an inequality which holds in Banach spaces with a cone. The results of this type can be found in [21] and [177].

A closed subspace $\mathbb{K}$ in a Banach space X is called a cone if the following assumptions hold:

   i) If $u \in \mathbb{K}$, then $\lambda u \in \mathbb{K}$ for every $\lambda \geq 0$,

   ii) If $u, v \in \mathbb{K}$, then $u + v \in \mathbb{K}$,

   iii) If $u \in \mathbb{K}$ and $-u \in \mathbb{K}$, then $u = 0$.

For example, if $X = C(D)$, then the subspace of $u \in X$, $u \geq 0$ forms a cone.

If $u, v \in \mathbb{K}$, then we write $u \geq v$ if $u - v \in \mathbb{K}$.

Theorem 5.5.1 *If an operator $A : \mathbb{K} \to \mathbb{K}$ is monotone in the sense $u \geq v$ implies $A(u) \geq A(v)$, and is a contraction on $\mathbb{K}$, then inequality*

$$u \leq A(u) + f, \quad f \in \mathbb{K} \tag{5.63}$$

implies $u \leq w$, where $w \in \mathbb{K}$ is the (unique) solution to the equation $w = A(w) + f$.

Proof: Equation $w = A(w) + f$ has a unique solution in $\mathbb{K}$ by the contraction mapping principle. This solution can be obtained by iterations: $w_{n+1} = A(w_n) + f$, $w_1 = f$, $\lim_{n \to \infty} \|w_n - w\| = 0$. Iterating inequality (5.63), one obtains $u \leq w_n$. Passing to the limit in this inequality, yields the desired result $u \leq w$. Theorem 5.5.1 is proved. ∎

Remark 5.5.2 The result of Theorem 5.5.1 is often useful when A is a linear operator which leaves the cone of nonnegative functions invariant and has spectral radius $r(A) < 1$, or, less precise, $\|A\| < 1$.

Example 5.1

Let $Au := \int_D A(x, y)u(y) \, dy$, where $D \in \mathbb{R}^n$ is a bounded (or unbounded) domain, $A(x, y) \geq 0$ is a continuous function on $D \times D$, $X = C(D)$, $\mathbb{K}$ is a cone of nonnegative continuous functions in D, $\|A\| < 1$, and $f \in \mathbb{K}$. Then inequality $u(x) \leq \int_D A(x, y)u(y) \, dy + f(x)$, $x \in D$ implies $u(x) \leq w(x)$, where w solves the equation $w = Aw + f$.

Example 5.2

(cf. [21]). Let $a, b, c > 0$, $2b < c$, and

$$u(t) \leq ae^{-ct} + b \int_0^\infty e^{-c|t-s|} u(s) \, ds. \tag{5.64}$$

Then

$$u(t) \leq \frac{ac}{b} e^{-(c^2 - 2bc)^{1/2} t}. \tag{5.65}$$

Let us verify (5.65). If $2b < c$, then $\|A\| < 1$, where

$$Au = b \int_0^\infty e^{-c|t-s|} u(s) \, ds$$

is the operator in (5.64). This operator is considered in the Banach space $C(0, \infty)$. The kernel $be^{-c|t-s|}$ is nonnegative, and Theorem 5.5.1 is applicable. Equation $w = Aw + ae^{-ct}$ can be solved analytically in $C(0, \infty)$, and its solution is $w = \frac{ac}{b} e^{-(c^2 - 2bc)^{1/2} t}$, so inequality (5.65) is obtained.

5.6 A GENERALIZED VERSION OF THE BASIC NONLINEAR INEQUALITY

In this section we study the following nonlinear differential inequality

$$\dot{g}(t) \leq -\gamma(t)g(t) + \alpha(t, g(t)) + \beta(t), \qquad t \geq t_0, \quad \dot{g} = \frac{dg}{dt}, \quad g \geq 0. \tag{5.66}$$

In equation (5.66), $\beta(t)$ and $\gamma(t)$ are continuous functions, defined on $[t_0, \infty)$, where $t_0 \geq 0$ is a fixed number.

Inequality (5.66) was studied in Section 5.1 with $\alpha(t, y) = \tilde{\alpha}(t)y^2$, where $0 \leq \tilde{\alpha}(t)$ is a continuous function on $[t_0, \infty)$. This inequality arises in the study of the Dynamical Systems Method (DSM) for solving nonlinear operator equations. Sufficient conditions on β, α, and γ which yields an estimate for the rate of growth/decay of $g(t)$ were given in Section 5.1. A discrete analog of (5.66) was studied in [47]. An application to the study of a discrete version of the DSM for solving nonlinear equation was demonstrated in [47].

In [52] inequality (5.66) is studied in the case $\alpha(t, y) = \tilde{\alpha}(t)y^p$, where $p > 1$ and $0 \leq \tilde{\alpha}(t)$ is a continuous function on $[t_0, \infty)$. This equality allows one to study the DSM under weaker smoothness assumption on F than in the cited works. It allows one to study the convergence of the DSM under the assumption that F' is locally Hölder continuous. An application to the study of large time behavior of solutions to some partial differential equations was outlined in [52].

In this section we assume that $0 \leq \alpha(t, y)$ is a nondecreasing function of y on $[0, \infty]$ and continuous with respect to t on $[t_0, \infty)$. Under this weak assumption on α and some assumptions on β and γ, we give an estimate for the rate of growth/decay of $g(t)$ as $t \to \infty$ in Theorem 5.6.1.

A discrete version of (5.66) is studied and the result is stated in Theorem 5.6.6. In Section 5.6.2 an application of inequality (5.66) to the study of large time behavior of solutions to some partial equations is sketched.

5.6.1   Formulations and results

Throughout this section let us assume that the function $0 \leq \alpha(t, y)$ is locally Lipchitz-continuous, nondecreasing with respect to y, and continuous with respect to t on $[t_0, \infty)$.

Theorem 5.6.1 *Let $\beta(t)$ and $\gamma(t)$ be continuous functions on $[t_0, \infty)$. Suppose there exists a function $\mu(t) > 0$, $\mu \in C^1[t_0, \infty)$, such that*

$$\alpha\left(t, \frac{1}{\mu(t)}\right) + \beta(t) \leq \frac{1}{\mu(t)}\left[\gamma - \frac{\dot{\mu}(t)}{\mu(t)}\right], \qquad t \geq t_0. \tag{5.67}$$

Let $g(t) \geq 0$ be a solution to inequality (5.66) such that

$$\mu(t_0)g(t_0) < 1. \tag{5.68}$$

Then $g(t)$ exists globally and the following estimate holds:

$$0 \leq g(t) < \frac{1}{\mu(t)}, \qquad \forall t \geq t_0. \tag{5.69}$$

Consequently, if $\lim_{t \to \infty} \mu(t) = \infty$, then

$$\lim_{t \to \infty} g(t) = 0. \tag{5.70}$$

Proof: Denote

$$v(t) := g(t)e^{\int_{t_0}^{t} \gamma(s)ds}. \tag{5.71}$$

Then inequality (5.66) takes the form

$$\dot{v}(t) \le a(t)\alpha\big(t, v(t)e^{-\int_{t_0}^{t} \gamma(s)ds}\big) + b(t), \qquad v(t_0) = g(t_0) := g_0, \tag{5.72}$$

where

$$a(t) = e^{\int_{t_0}^{t} \gamma(s)ds}, \qquad b(t) := \beta(t)e^{\int_{t_0}^{t} \gamma(s)ds}. \tag{5.73}$$

Denote

$$\eta(t) = \frac{e^{\int_{t_0}^{t} \gamma(s)ds}}{\mu(t)}. \tag{5.74}$$

From inequality (5.68) and relation (5.74) one gets

$$v(t_0) = g(t_0) < \frac{1}{\mu(t_0)} = \eta(t_0). \tag{5.75}$$

It follows from the inequalities (5.67), (5.72), and (5.75) that

$$\begin{aligned}
\dot{v}(t_0) &\le \alpha\big(t_0, \frac{1}{\mu(t_0)}\big) + \beta(t_0)\\
&\le \frac{1}{\mu(t_0)}\Big[\gamma - \frac{\dot{\mu}(t_0)}{\mu(t_0)}\Big] = \frac{d}{dt}\left.\frac{e^{\int_{t_0}^{t} \gamma(s)ds}}{\mu(t)}\right|_{t=t_0} = \dot{\eta}(t_0).
\end{aligned} \tag{5.76}$$

From the inequalities (5.75) and (5.76) it follows that there exists $\delta > 0$ such that

$$v(t) < \eta(t), \qquad t_0 \le t \le t_0 + \delta. \tag{5.77}$$

To continue the proof we need two claims.

   Claim 1. If

$$v(t) \le \eta(t), \qquad \forall t \in [t_0, T], \quad T > t_0, \tag{5.78}$$

then

$$\dot{v}(t) \le \dot{\eta}(t), \qquad \forall t \in [t_0, T]. \tag{5.79}$$

Proof of Claim 1.

   It follows from inequalities (5.67) and (5.72) and the inequality $v(T) \le \eta(T)$ that

$$\begin{aligned}
\dot{v}(t) &\le e^{\int_{t_0}^{t} \gamma(s)ds}\alpha\big(t, \frac{1}{\mu(t)}\big) + \beta(t)e^{\int_{t_0}^{t} \gamma(s)ds}\\
&\le \frac{e^{\int_{t_0}^{t} \gamma(s)ds}}{\mu(t)}\Big[\gamma - \frac{\dot{\mu}(t)}{\mu(t)}\Big]\\
&= \frac{d}{dt}\left.\frac{e^{\int_{t_0}^{t} \gamma(s)ds}}{\mu(t)}\right|_{t=t} = \dot{\eta}(t), \qquad \forall t \in [t_0, T].
\end{aligned} \tag{5.80}$$

Claim 1 is proved. □

Denote

$$T := \sup\{\delta \in \mathbb{R}^+ : v(t) < \eta(t), \forall t \in [t_0, t_0 + \delta]\}. \tag{5.81}$$

Claim 2. One has $T = \infty$.

Claim 2 says that every nonnegative solution $g(t)$ to inequality (5.66), satisfying assumption (5.67), is defined globally.

Proof of Claim 2.

Assume the contrary, that is, $T < \infty$. The solution $v(t)$ to (5.72) is continuous at every point t at which it is bounded. From the definition of T and the continuity of v and η one gets

$$v(T) \leq \eta(T). \tag{5.82}$$

From inequalities (5.81) and (5.82) and *Claim 1*, it follows that

$$\dot{v}(t) \leq \dot{\eta}(t), \qquad \forall t \in [t_0, T]. \tag{5.83}$$

This implies

$$v(T) - v(t_0) = \int_{t_0}^{T} \dot{v}(s)\,ds \leq \int_{t_0}^{T} \dot{\eta}(s)\,ds = \eta(T) - \eta(t_0). \tag{5.84}$$

Since $v(t_0) < \eta(t_0)$ by assumption (5.68), it follows from inequality (5.84) that

$$v(T) < \eta(T). \tag{5.85}$$

Inequality (5.85) and inequality (5.83) with $t = T$ imply that there exists a $\delta > 0$ such that

$$v(t) < \eta(t), \qquad T \leq t \leq T + \delta. \tag{5.86}$$

This contradicts the definition of T in (5.81), and the contradiction proves the desired conclusion $T = \infty$.

Claim 2 is proved. □

From the definitions of $\eta(t)$, T, and $v(t)$ and from the relation $T = \infty$, it follows that

$$g(t) = e^{-\int_{t_0}^{t} \gamma(s)ds} v(t) < e^{-\int_{t_0}^{t} \gamma(s)ds} \eta(t) = \frac{1}{\mu(t)}, \qquad \forall t > t_0. \tag{5.87}$$

Theorem 5.6.1 is proved. ■

Theorem 5.6.2 *Let $\beta(t)$ and $\gamma(t)$ be as in Theorem 5.6.1. Assume that $0 < \alpha(t, y)$ is continuous with respect to t on $[t_0, \infty)$ and is locally Lipschitz-continuous and nondecreasing with respect to y on $[0, \infty)$. Let $0 \leq g(t)$ satisfy (5.66) and $0 < \mu(t)$ satisfy (5.67) and (5.68). Then*

$$g(t) \leq \frac{1}{\mu(t)}, \qquad \forall t \geq t_0. \tag{5.88}$$

Proof: Let $v(t)$ be defined in (5.71). Then inequality (5.72) holds. Let $w_n(t)$ solve the following differential equation:

$$\dot{w}_n(t) = a(t)\alpha\big(t, w_n(t)e^{-\int_{t_0}^{t} \gamma(s)ds}\big) + b(t), \qquad w_n(t_0) = g(t_0) - \frac{1}{n}, \quad (5.89)$$

where $n \geq n_0$, n_0 is sufficiently large and $g(t_0) > \frac{1}{n_0}$. Since $\alpha(t, y)$ is continuous with respect to t and locally Lipschitz-continuous with respect to y, there exists a unique local solution to (5.89).

From the proof of Theorem 5.6.1 one gets

$$w_n(t) < \frac{e^{\int_{t_0}^{t} \gamma(s)ds}}{\mu(t)}, \qquad \forall t \geq t_0, \forall n \geq n_0. \quad (5.90)$$

Let $t_0 < \tau < \infty$ be an arbitrary constant and

$$w(t) = \lim_{n \to \infty} w_n(t), \qquad \forall t \in [t_0, \tau]. \quad (5.91)$$

This and the fact that $w_n(t)$ is uniformly continuous on $[0, \tau]$ imply that $w(t)$ solves the following equation:

$$\dot{w}(t) = a(t)\alpha\big(t, w(t)e^{-\int_{t_0}^{t} \gamma(s)ds}\big) + b(t), \qquad w(t_0) = g(t_0), \quad (5.92)$$

for all $t \in [0, \tau]$. Note that the solution $w(t)$ to (5.92) is unique since $\alpha(t, y)$ is continuous with respect to t and locally Lipschitz-continuous with respect to y. From (5.72), (5.92), Lemma 5.1.4, and the continuity of $w_n(t)$ with respect to $w_0(t_0)$ on $[0, \tau]$, and (5.90), one gets

$$v(t) \leq w(t) \leq \frac{e^{\int_{t_0}^{t} \gamma(s)ds}}{\mu(t)}, \qquad \forall t \in [t_0, \tau], \ \forall n \geq n_0. \quad (5.93)$$

Since $\tau > t_0$ is arbitrary, inequality (5.88) follows from (5.93).

Theorem 5.6.2 is proved. ∎

Remark 5.6.3 Inequality (5.67) can be written as

$$-\gamma(t)u + \alpha\big(t, \frac{1}{\mu(t)}\big) + \beta(t) \leq \frac{i}{\mu(t)},$$

while inequality (5.66) is

$$\dot{g}(t) \leq -\gamma(t)g(t) + \alpha(t, g(t)) + \beta(t).$$

If $\frac{1}{\mu(t_0)} \geq g(t_0)$ and the Cauchy problem

$$\dot{w}(t) = -\gamma(t)w(t) + \alpha(t, w(t)) + \beta(t), \qquad w(t_0) = w_0$$

has a unique solution, then, from a comparison Lemma 5.1.4, one concludes that $g(t)$ exists globally, and inequality (5.21) holds.

Remark 5.6.4 The results of Theorems 5.6.1 and 5.6.2 are closely related to the known comparison-type results in differential equations. The comparison lemmas, or lemmas about differential inequalities, are described in many books and papers—for example, in [40], [73], [79], [177], and [182], to mention a few.

In this remark we give an alternative proof of Theorem 5.6.1, which uses comparison results for differential inequalities. In this proof we have to assume that the Cauchy problem, corresponding to a differential equation, obtained from differential inequality (5.66) (see problem (5.94) below) has a unique solution. This assumption was not used in the proof of Theorem 5.6.1, given above. Uniqueness of the solution to this Cauchy problem holds, for example, if the function $a(t, g)$ satisfies Lipschitz condition with respect to g. Less restrictive conditions (e.g., one-sided inequalities) sufficient for the uniqueness of the solution to the Cauchy problem (5.94) are known (e.g., see [73]).

Let $\phi(t)$ solve the following Cauchy problem:

$$\dot{\phi}(t) = -\gamma(t)\phi(t) + \alpha(t, \phi(t)) + \beta(t), \quad t \geq t_0, \quad \phi(t_0) = \phi_0. \tag{5.94}$$

Inequality (5.67) can be written as

$$-\gamma(t)\mu^{-1}(t) + \alpha(t, \mu^{-1}(t)) + \beta(t) \leq \frac{d\mu^{-1}(t)}{dt}. \tag{5.95}$$

From the known comparison result (see, e.g., [40], Theorem III.4.1) it follows that

$$\phi(t) \leq \mu^{-1}(t) \quad \forall t \geq t_0, \tag{5.96}$$

provided that $\phi(t_0) \leq \mu^{-1}(t_0)$, where $\phi(t)$ is the *minimal* solution to problem (5.94).

Inequality (5.66) implies that

$$g(t) \leq \phi(t) \quad \forall t \geq t_0, \tag{5.97}$$

provided that $g(t_0) \leq \phi(t_0)$, where $\phi(t)$ is the *maximal* solution to problem (5.94).

Therefore, if problem (5.94) has at most one local solution, and

$$g(t_0) \leq \mu^{-1}(t_0), \tag{5.98}$$

then

$$g(t) \leq \mu^{-1}(t) \quad \forall t \geq t_0. \tag{5.99}$$

Since $\mu(t)$ is defined for all $t \geq t_0$, it follows that the solution to problem (5.94) with $\phi(t_0) = \mu^{-1}(t_0)$ is also defined for all $t \geq t_0$. Consequently, $g(t)$ is defined for all $t \geq t_0$.   □

From Theorem 5.6.2 one has the following corollary:

Corollary 5.6.5 *Let $\alpha(t)$, $\beta(t)$, and $\gamma(t)$ be continuous functions on $[t_0, \infty)$ and $\alpha(t) \geq 0$, $\forall t \geq t_0$. Suppose there exists a function $\mu(t) > 0$, $\mu \in C^1[t_0, \infty)$, such that*

$$\frac{\alpha(t)}{\mu^p(t)} + \beta(t) \leq \frac{1}{\mu(t)}\left[\gamma(t) - \frac{\dot{u}(t)}{\mu(t)}\right].$$

Let $g(t) \geq 0$ be a solution to the following inequality

$$\dot{g}(t) \leq -\gamma(t)g(t) + \alpha(t)g^p(t) + \beta(t), \qquad \mu(t_0)g(t_0) \leq 1, \quad t \geq t_0, \quad p > 1.$$

Then $g(t)$ exists globally and the following estimate holds:

$$0 \leq g(t) \leq \frac{1}{\mu(t)}, \qquad \forall t \geq t_0.$$

Consequently, if $\lim_{t \to \infty} \mu(t) = \infty$, then

$$\lim_{t \to \infty} g(t) = 0.$$

Let us consider a *discrete analog* of Theorem 5.6.1.
Let

$$\frac{g_{n+1} - g_n}{h_n} \leq -\gamma_n g_n + \alpha(n, g_n) + \beta_n, \qquad h_n > 0, \quad 0 < h_n\gamma_n < 1, \quad p > 1,$$

and the inequality

$$g_{n+1} \leq (1 - \gamma_n)g_n + \alpha(n, g_n) + \beta_n, \quad n \geq 0, \qquad 0 < \gamma_n < 1, \quad p > 1,$$

where g_n, β_n, and γ_n are positive sequences of real numbers.

Under suitable assumptions on β_n and γ_n, we obtain an upper bound for g_n as $n \to \infty$. In particular, we give sufficient conditions for the validity of the relation $\lim_{n \to \infty} g_n = 0$, and we estimate the rate of growth/decay of g_n as $n \to \infty$. This result can be used in a study of evolution problems, in a study of iterative processes, and in a study of nonlinear PDE.

Theorem 5.6.6 *Let β_n, and g_n be nonnegative sequences of numbers. Assume that*

$$\frac{g_{n+1} - g_n}{h_n} \leq -\gamma_n g_n + \alpha(n, g_n) + \beta_n, \qquad 0 < h_n < \frac{1}{\gamma_n}, \tag{5.100}$$

or, equivalently,

$$g_{n+1} \leq g_n(1 - h_n\gamma_n) + h_n\alpha(n, g_n) + h_n\beta_n, \qquad 0 < h_n < \frac{1}{\gamma_n}. \tag{5.101}$$

If there is a sequence of positive numbers $(\mu_n)_{n=1}^{\infty}$ such that the conditions

$$\alpha(n, \frac{1}{\mu_n}) + \beta_n \le \frac{1}{\mu_n}\left(\gamma_n - \frac{\mu_{n+1} - \mu_n}{\mu_n h_n}\right), \tag{5.102}$$

$$g_0 \le \frac{1}{\mu_0} \tag{5.103}$$

hold, then

$$0 \le g_n \le \frac{1}{\mu_n} \qquad \forall n \ge 0. \tag{5.104}$$

Therefore, if $\lim_{n\to\infty} \mu_n = \infty$, then $\lim_{n\to\infty} g_n = 0$.

Proof: Let us prove (5.104) by induction. Inequality (5.104) holds for $n = 0$ by assumption (5.103). Suppose that (5.104) holds for all $n \le m$. From inequalities (5.100) and (5.102) and from the induction hypothesis $g_n \le \frac{1}{\mu_n}$, $n \le m$, one gets

$$\begin{aligned}
g_{m+1} &\le g_m(1 - h_m\gamma_m) + h_m\alpha(m, g_m) + h_m\beta_m \\
&\le \frac{1}{\mu_m}(1 - h_m\gamma_m) + h_m\alpha(m, \frac{1}{\mu_m}) + h_m\beta_m \\
&\le \frac{1}{\mu_m}(1 - h_m\gamma_m) + \frac{h_m}{\mu_m}\left(\gamma_m - \frac{\mu_{m+1} - \mu_m}{\mu_m h_m}\right) \\
&= \frac{1}{\mu_{m+1}} - \frac{\mu_{m+1}^2 - 2\mu_{m+1}\mu_m + \mu_m^2}{\mu_n^2 \mu_{m+1}} \le \frac{1}{\mu_{m+1}}.
\end{aligned} \tag{5.105}$$

Therefore, inequality (5.104) holds for $n = m + 1$. Thus, inequality (5.104) holds for all $n \ge 0$ by induction. Theorem 5.6.6 is proved. ∎

Setting $h_n = 1$ in Theorem 5.6.6, one obtains the following result:

Theorem 5.6.7 *Let β_n, γ_n, and g_n be sequences of nonnegative numbers, and*

$$g_{n+1} \le g_n(1 - \gamma_n) + \alpha(n, g_n) + \beta_n, \qquad 0 < \gamma_n < 1. \tag{5.106}$$

If there is sequence $(\mu_n)_{n=1}^{\infty} > 0$ such that the conditions

$$g_0 \le \frac{1}{\mu_0}, \qquad \alpha(n, \frac{1}{\mu_n}) + \beta_n \le \frac{1}{\mu_n}\left(\gamma_n - \frac{\mu_{n+1} - \mu_n}{\mu_n h_n}\right), \qquad \forall n \ge 0, \tag{5.107}$$

hold, then

$$g_n \le \frac{1}{\mu_n}, \qquad \forall n \ge 0. \tag{5.108}$$

5.6.2 Applications

Here we sketch an idea for possible applications of our inequalities in a study of dynamical systems in a Hilbert space H.

In this section we assume without loss of generality that $t_0 = 0$. Let

$$\dot{u} + Au = h(t, u) + f(t), \qquad u(0) = u_0, \qquad \dot{u} := \frac{du}{dt}, \qquad t \geq 0. \qquad (5.109)$$

To explain the ideas, let us make simplifying assumptions: $A > 0$ is a self-adjoint time-independent operator in a real Hilbert space H, $h(t, u)$ is a non-linear operator in H, locally Lipschitz with respect to u and continuous with respect to $t \in \mathbb{R}_+ := [0, \infty)$, and f is a continuous function on $\mathbb{R}_+$ with values in H, $\sup_{t \geq 0} \|f(t)\| < \infty$. The scalar product in H is denoted $\langle u, v \rangle$. Assume that

$$\langle Au, u \rangle \geq \gamma \langle u, u \rangle, \quad \|h(t, u)\| \leq \alpha(t, \|u\|), \qquad \forall u \in D(A), \qquad (5.110)$$

where $\gamma = const > 0$, $\alpha(t, y) \leq c|y|^p$, $p > 1$ and $c > 0$ are constants, and $\alpha(t, y)$ is a nondecreasing $C^1([0, \infty))$ function of y. Our approach works when $\gamma = \gamma(t)$ and $c = c(t)$; see Examples 5.3 and 5.4 below. The problem is to estimate the behavior of the solution to (5.109) as $t \to \infty$ and to give sufficient conditions for a global existence of the unique solution to (5.109). Our approach consists of a reduction of this problem to the inequality (5.66) and an application of Theorem 5.6.1. A different approach, studied in the literature (see, e.g., [70] and [89]), is based on the semigroup theory.

Let $g(t) := \|u(t)\|$. Problem (5.109) has a unique local solution under our assumptions. This local solution exists globally if $\sup_{t \geq 0} \|u(t)\| < \infty$. Multiply (5.109) by u and use (5.110) to get

$$\dot{g}g \leq -\gamma(t)g^2 + \alpha(t, g)g + \beta(t)g, \qquad \beta(t) := \|f(t)\|. \qquad (5.111)$$

Since $g \geq 0$, one gets

$$\dot{g} \leq -\gamma(t)g + \alpha(t, g(t)) + \beta(t). \qquad (5.112)$$

Now Theorem 5.6.1 is applicable and yields sufficient conditions (5.67) and (5.68) for the global existence of the solution to (5.109) and estimate (5.69) for the behavior of $\|u(t)\|$ as $t \to \infty$. The choice of $\mu(t)$ in Theorem 5.6.1 is often straightforward. For example, if $\alpha(t, g(t)) = \frac{c_0}{a(t)}g^2$, where $\lim_{t \to \infty} a(t) = 0$, $\dot{a}(t) < 0$, then one can often choose $\mu(t) = \frac{\lambda}{a(t)}$, $\lambda = const > 0$; see Theorem 6.2.1 in Section 6.2, and [46, p. 487], for examples of applications of this approach.

The outlined approach is applicable to stability of the solutions to non-linear differential equations, to semilinear parabolic problems, to hyperbolic problems, and to other problems. There is a large literature on the stability of the solutions to differential equations (see, e.g., [21] and [22] and references therein). Our approach yields some novel results. If the self-adjoint operator A depends on t, $A = A(t)$, and $\gamma = \gamma(t) > 0$, $\lim_{t \to \infty} \gamma(t) = 0$, one can treat problems with degenerate, as $t \to \infty$, elliptic operators A.

For instance, if the operator A is a second-order elliptic operator with matrix $a_{ij}(x,t)$, and the minimal eigenvalue $\lambda(x,t)$ of this matrix satisfies the condition $\min_x \lambda(x,t) := \gamma(t) \to 0$ as $t \to \infty$, then Theorem 5.6.1 is applicable under suitable assumptions on $\gamma(t)$, $h(t,u)$, and $f(t)$.

Example 5.3

Consider

$$\dot{u} = -\gamma(t)u + a(t)u(t)|u(t)|^p + \frac{1}{(1+t)^q}, \qquad u(0) = 0, \qquad (5.113)$$

where $\gamma(t) = \frac{c}{(1+t)^b}$, $a(t) = \frac{1}{(1+t)^m}$, p, q, b, c, and m are positive constants. Our goal is to give sufficient conditions for the solution to the above problem to converge to zero as $t \to \infty$. Multiply (5.113) by u, denote $g := u^2$, and get the following inequality:

$$\dot{g} \leq -2\frac{c}{(1+t)^b}g + 2\frac{1}{(1+t)^m}g(t)^{1+0.5p} + 2\frac{1}{(1+t)^q}g^{0.5}, \qquad g = u^2. \quad (5.114)$$

Choose $\mu(t) = \lambda(1+t)^\nu$, where $\lambda > 0$ and $\nu > 0$ are constants.

Inequality (5.67) takes the form

$$\frac{2}{(1+t)^m}[\lambda(1+t)^\nu]^{-1-0.5p} + \frac{2}{(1+t)^q}[\lambda(1+t)^\nu]^{-0.5}$$

$$\leq [\lambda(1+t)^\nu]^{-1}\left(2\frac{c}{(1+t)^b} - \frac{\nu}{1+t}\right). \quad (5.115)$$

Choose p, q, m, c, λ and ν so that inequality (5.115) is valid and $\lambda u(0)^2 < 1$, so that condition (5.68) with $t_0 = 0$ holds. If this is done, then $u^2(t) \leq \frac{1}{\lambda(1+t)^\nu}$, so $\lim_{t\to\infty} u(t) = 0$. For example, choose $b = 1$, $\nu = 1$, $q = 1.5$, $m = 1$, $\lambda = 4$, $c = 4$, $p \geq 1$. Then inequality (5.115) is valid, and if $u(0)^2 < 1/4$, then (5.68) with $t_0 = 0$ holds, so $\lim_{t\to\infty} u(t) = 0$. The choice of the parameters can be varied. In particular, the nonlinearity growth, governed by p, can be arbitrary in power scale. If $b = 1$, then three inequalities, $m + 0.5p\nu \geq 1$, $q - 0.5\nu \geq 1$, and $\lambda^{1/2} + \lambda^{-0.5p} \leq c - 0.5\nu$, together with $u(0)^2 < \lambda^{-1}$, are sufficient for (5.68) and (5.115) to hold, so they imply $\lim_{t\to\infty} u(t) = 0$.

Example 5.4

Consider problem (5.109) with A, h, and f satisfying (5.110) with $\gamma \equiv 0$. So one gets inequality (5.112) with $\gamma(t) \equiv 0$. Choose

$$\mu(t) := c + \lambda(1+t)^{-b}, \qquad c > 0, \, b > 0, \, \lambda > 0, \quad (5.116)$$

where c, λ, and b are constants. Inequality (5.67) takes the form

$$\alpha(t, \frac{1}{\mu(t)}) + \beta(t) \leq \frac{1}{\mu(t)}\frac{b\lambda}{(1+t)[\lambda + c(1+t)^b]}. \quad (5.117)$$

Let $\theta \in (0,1)$, $p > 0$, and $C > 0$ be constants. Assume that

$$\alpha(t, |y|) \leq \frac{\theta C|y|^p b\lambda}{(\lambda + c)(1+t)^{1+b}}, \qquad \beta(t) \leq \frac{(1-\theta)b\lambda}{(c+\lambda)^2(1+t)^{1+b}}, \quad (5.118)$$

for all $t \geq 0$, and

$$C = \begin{cases} c^{p-1} & \text{if} \quad p > 1, \\ (\lambda + c)^{p-1} & \text{if} \quad p \leq 1. \end{cases} \tag{5.119}$$

Let us verify that inequality (5.117) holds given that (5.118) and (5.119) hold.

It follows from (5.116) that $c < \mu(t) \leq c + \lambda$, $\forall t \geq 0$. This and (5.118) imply

$$\beta(t) \leq (1-\theta)\frac{1}{(c+\lambda)^2(1+t)^{1+b}} \leq (1-\theta)\frac{1}{\mu(t)}\frac{1}{(1+t)(c+\lambda(1+t)^b)}. \tag{5.120}$$

From (5.119) and (5.116) one gets

$$\frac{C}{\mu^{p-1}(t)} \leq C\max(c^{1-p}, (c+\lambda)^{1-p}) \leq 1, \qquad \forall t \geq 0. \tag{5.121}$$

From (5.118) and (5.121) one obtains

$$\alpha(t, \frac{1}{\mu(t)}) \leq \theta C \frac{1}{\mu(t)}\frac{1}{\mu^{p-1}(t)}\frac{b\lambda}{(1+t)[\lambda(1+t)^b + c(1+t)^b]}$$
$$\leq \theta\frac{1}{\mu(t)}\frac{b\lambda}{(1+t)[\lambda + c(1+t)^b]}. \tag{5.122}$$

Inequality (5.117) follows from (5.120) and (5.122). From (5.117) and Theorem 5.6.1 one obtains

$$g(t) \leq \frac{1}{\mu(t)} < \frac{1}{c}, \qquad \forall t > 0, \tag{5.123}$$

provided that $g(0) < (c+\lambda)^{-1}$. From (5.112) with $\gamma(t) = 0$ and (5.118)–(5.123), one gets $\dot{g}(t) = O(\frac{1}{(1+t)^{1+b}})$. Thus, there exists finite limit

$$\lim_{t \to \infty} g(t) = g(\infty) \leq c^{-1}.$$

5.7   SOME NONLINEAR INEQUALITIES AND APPLICATIONS

The stability study of many evolution equations is a study of large time behavior of the solutions to these equations. In this section we reduce such a study to a study of the behavior of a solution $y(t)$ to some nonlinear inequalities. Assume that a nonnegative continuous function $y(t)$ satisfies the following conditions:

$$\int_0^\infty \omega(y(t))\frac{1}{(1+t)^\alpha}\,dt < \infty, \qquad 0 \leq \alpha \leq 1, \tag{5.124}$$

and

$$y(t) - y(s) \leq \int_s^t f(x, y(x))\,dx, \qquad 0 \leq s \leq t, \tag{5.125}$$

where $f(x, y)$ is a nonnegative continuous function on $[0, \infty) \times [0, \infty)$, $0 \leq w(t)$ is a nondecreasing continuous function, and $w(t) = 0$ implies $t = 0$.

The question arises:

Under what condition on $f(y, t)$ does it follow that

$$\lim_{t \to \infty} y(t) = 0? \tag{5.126}$$

There is a very large literature on inequalities (see, e.g., [11], [15], and references therein). The Barbalat's lemma is an integral inequality used in applied nonlinear control [181]. The inequalities, derived in this section, are new and are useful in many applications. In [197, p. 227] inequality (5.124) is studied for $w(t) = t$ and $\alpha = 0$. In this case, condition (5.124) becomes $y \in L^1[0, \infty)$. In [197] it is proved that (5.126) holds if $y \in L^1[0, \infty)$ and the following two conditions hold:

$$y(t) - y(s) \leq \int_s^t f(y(x)) \, dx + \int_s^t h(x) \, dx, \quad \int_0^\infty h(x) \, dx < \infty. \tag{5.127}$$

Here f is continuous, nondecreasing and nonnegative function, and h is nonnegative function. Proofs of this result can be found in [197] and in [151]. Applications of this result to the stability study of evolution equations can be found in [197] and references therein. This result is not applicable if $y(t) = O(\frac{1}{t^\beta})$ as $t \to \infty$, where $\beta \in (0, 1)$, because then $y(t)$ is not in $L^1[0, \infty)$. Also, this result is not applicable if (5.125) holds instead of (5.127) and f depends on x.

The second nonlinear inequality we study is the following one:

$$\dot{g}(t) \leq -a(t)f(g(t)) + b(t), \quad t \geq 0, \tag{5.128}$$

where a, b, and g are nonnegative functions on $[0, \infty)$, $g \in C^1([0, \infty))$, $a \in C([0, \infty))$, and $b \in L^1_{loc}([0, \infty))$. In [151] a sufficient condition for the relation $\lim_{t \to \infty} g(t) = 0$ to hold is proposed and justified. In this section inequality (5.128) is studied by a different method and some new sufficient conditions for (5.126) to hold are proposed and justified.

The section is organized as follows. In Theorems 5.7.1, 5.7.4, and 5.7.7 and their corollaries, sufficient conditions for (5.127) to hold are formulated and justified. In Theorems 5.7.11, 5.7.13, and 5.7.14, sufficient conditions for the relation $\lim_{t \to \infty} g(t) = 0$ to hold are proposed and justified under the assumption that $f(t)$ is a continuous and nondecreasing function on $[0, \infty)$. In Section 5.7.2, applications of the new results to the stability study of evolution equations are given.

5.7.1 Formulations and results

Throughout the section we assume that

$\quad w(t) \geq 0$ *is a nondecreasing continuous function; and if $w(t) = 0$, then* $t = 0$.

This assumption is standing and is not repeated.

Theorem 5.7.1 *Let $y(t) \geq 0$ be a continuous function on $[0, \infty)$,*

$$\int_0^\infty \omega(y(t)) \, dt < \infty, \tag{5.129}$$

and

$$y(t) - y(s) \leq \int_s^t f(\xi, y(\xi)) \, d\xi, \qquad 0 \leq s \leq t, \tag{5.130}$$

where $f(t, y)$ is a nonnegative continuous function on $[0, \infty) \times [0, \infty)$. Define

$$F(t, v) := \int_0^t \max_{0 \leq \zeta \leq v} f(\xi, \zeta) \, d\xi, \qquad v, t \geq 0. \tag{5.131}$$

If there exists a constant $a > 0$ such that the function $F(t, a)$ is uniformly continuous with respect to t on $[0, \infty)$, then

$$\lim_{t \to \infty} y(t) = 0. \tag{5.132}$$

Proof: If (5.132) does not hold, then there exists an $\epsilon > 0$ and a sequence $(t_n)_{n=1}^\infty$ such that

$$0 < t_n \nearrow \infty, \qquad y(t_n) \geq \epsilon, \qquad \forall n \geq 1. \tag{5.133}$$

Without loss of generality we assume that $\epsilon < a$.

Since $F(t, a)$ is uniformly continuous with respect to t, there exists $\delta > 0$ such that

$$\int_t^{t+\delta} \max_{0 \leq y \leq a} f(\xi, y) d\xi = F(t + \delta, a) - F(t, a) < \frac{\epsilon}{2}, \qquad \forall t \geq 0. \tag{5.134}$$

Let us prove that

$$y(t) \geq \frac{\epsilon}{2}, \qquad \forall t \in [t_n - \delta, t_n], \qquad \forall n \geq 1. \tag{5.135}$$

Assume that (5.135) does not hold. Then there exists $\tilde{n} > 0$ and $\xi \in [t_{\tilde{n}} - \delta, t_{\tilde{n}})$ such that

$$y(\xi) < \frac{\epsilon}{2}. \tag{5.136}$$

Let

$$\nu = \min\{x : \xi < x \leq t_{\tilde{n}}, \, y(x) \geq \epsilon\}. \tag{5.137}$$

From the continuity of y, (5.133), and (5.136)–(5.137) one obtains

$$t_{\tilde{n}} - \delta \leq \xi < \nu \leq t_{\tilde{n}}, \qquad y(\nu) = \epsilon, \tag{5.138}$$

and

$$0 \leq y(x) \leq y(\nu) = \epsilon, \qquad \xi \leq x \leq \nu. \tag{5.139}$$

It follows from (5.130), (5.136), and (5.138)–(5.139) that

$$
\begin{aligned}
\frac{\epsilon}{2} &< y(\nu) - y(\xi) \leq \int_{\xi}^{\nu} f(x, y(x))\, dx \leq \int_{\xi}^{\nu} \sup_{0 \leq \zeta \leq \epsilon} f(x, \zeta)\, dx \\
&\leq \int_{t_n - \delta}^{t_n} \sup_{0 \leq \zeta \leq a} f(x, \zeta)\, dx < \frac{\epsilon}{2}.
\end{aligned}
\tag{5.140}
$$

This contradiction proves that (5.135) holds.

From (5.135) one gets

$$
\int_{t_n - \delta}^{t_n} \omega(y(x))\, dx \geq \delta \omega\left(\frac{\epsilon}{2}\right) > 0, \qquad \forall n \geq 1.
\tag{5.141}
$$

This contradicts the Cauchy criterion for the convergence of the integral (5.129). Thus, (5.132) holds.

Theorem 5.7.1 is proved. ∎

Remark 5.7.2 If $F(t, a)$ is uniformly continuous with respect to t on $[0, \infty)$, then $F(t, v)$ is uniformly continuous with respect to t on $[0, \infty)$ for all $v \in [0, a]$. However, $F(t, v)$ may be not uniformly continuous with respect to t on $[0, \infty)$ for some $v > a$. Here is an example:

Let

$$
f(x, y) := \begin{cases} 1 & \text{if} \quad 0 \leq y \leq 1, \\ 1 + (y - 1)t & \text{if} \quad y \geq 1. \end{cases}
\tag{5.142}
$$

By a simple calculation, one gets

$$
F(t, u) = \begin{cases} t & \text{if} \quad 0 \leq u \leq 1 \\ t + (u - 1)\frac{t^2}{2} & \text{if} \quad u \geq 1. \end{cases}
\tag{5.143}
$$

It follows from (5.143) that $F(t, u)$ is uniformly continuous with respect to t on $[0, \infty)$ if and only if $u \in [0, 1]$.

From Theorem 5.7.1 one derives the following corollary.

Corollary 5.7.3 *Assume that $y(t) \geq 0$ is a continuous function satisfying inequality (5.129),*

$$
y(t) - y(s) \leq \int_{s}^{t} [g(\xi)\varphi(y(\xi)) + h(\xi)]\, d\xi, \qquad 0 \leq s \leq t,
\tag{5.144}
$$

where g and h are nonnegative locally integrable functions on $[0, \infty)$, $\varphi \geq 0$ is a continuous function on $[0, \infty)$, and the functions $\int_0^t g(x)\, dx$ and $\int_0^t h(x)\, dx$ are uniformly continuous with respect to t on $[0, \infty)$. Then (5.132) holds.

Proof: Let

$$
f(x, y) := g(x)\varphi(y) + h(x), \qquad x \geq 0, \qquad y \geq 0.
$$

It follows from the uniform continuity of $\int_0^t g(x)\,dx$ and $\int_0^t h(x)\,dx$ that the function F, defined in (5.131), is uniformly continuous. Thus, (5.132) follows from Theorem 5.7.1. ∎

Theorem 5.7.4 *Assume that $y(t) \geq 0$ is a continuous function on $[0, \infty)$,*

$$\int_0^\infty \omega(y(t))\varphi(t)\,dt < \infty, \tag{5.145}$$

where $\varphi(t) \geq 0$ is a continuous function on $[0, \infty)$, and there exists a constant $C > 0$ such that

$$\lim_{t \to \infty}\left(t - \frac{C}{\varphi(t)}\right) = \infty, \qquad M := \limsup_{t \to \infty} \frac{\max_{\xi \in [t - \frac{C}{\varphi(t)}, t]} \varphi(\xi)}{\min_{\xi \in [t - \frac{C}{\varphi(t)}, t]} \varphi(\xi)} < \infty, \tag{5.146}$$

where $f(x, y) \geq 0$ is a continuous on $[0, \infty) \times [0, \infty)$ function, which satisfies condition (5.130). If there exist constants $a > 0$ and $\theta > 0$ such that the following condition holds:

$$\int_s^t \sup_{0 \leq \zeta \leq a} f(x, \zeta)\,dx \leq (t - s)\theta a \max_{\xi \in [s,t]} \varphi(\xi), \qquad t > s \gg 1, \tag{5.147}$$

then (5.132) holds.

Remark 5.7.5 In (5.147) and below, the notation $s \gg 1$ means "for all sufficiently large $s > 0$".

Proof: Let us consider first Case 1, namely, $0 < \theta < 1$. Later we reduce Case 2, namely, $\theta \geq 1$, to Case 1.

Assume that (5.132) does not hold. Then there exists an $\epsilon > 0$, a sequence $(t_n)_{n=1}^\infty$ such that

$$0 < t_n \nearrow \infty, \qquad y(t_n) \geq \epsilon, \qquad \forall n \geq 1, \tag{5.148}$$

and without loss of generality one assumes that

$$\epsilon \leq 2aMC. \tag{5.149}$$

Let us prove that

$$y(t) \geq (1 - \theta)\epsilon, \qquad \forall t \in [\tilde{t}_n, t_n], \qquad \forall n \gg 1, \tag{5.150}$$

where

$$\tilde{t}_n := t_n - \frac{\epsilon}{2aM\varphi(t_n)} < t_n. \tag{5.151}$$

Assume that (5.150) does not hold. Then there exists a sufficiently large $\tilde{n} > 0$ and a $\xi \in [\tilde{t}_{\tilde{n}}, t_{\tilde{n}})$ such that

$$y(\xi) < (1 - \theta)\epsilon. \tag{5.152}$$

Let
$$\nu = \min\{x : \xi \leq x \leq t_{\tilde{n}},\, y(x) \geq \epsilon\}. \tag{5.153}$$

Then
$$\tilde{t}_{\tilde{n}} \leq \xi < \nu \leq t_{\tilde{n}} \tag{5.154}$$

and
$$0 \leq y(x) \leq y(\nu) = \epsilon, \qquad \xi \leq x \leq \nu. \tag{5.155}$$

It follows from (5.130), (5.148), (5.152), and (5.154)–(5.155) that

$$
\begin{aligned}
\theta\epsilon < y(\nu) - y(\xi) &\leq \int_{\xi}^{\nu} f(x, y(x))\, dx \leq \int_{\xi}^{\nu} \sup_{0 \leq \zeta \leq \epsilon} f(x, \zeta)\, dx \\
&\leq \int_{\tilde{t}_{\tilde{n}}}^{t_{\tilde{n}}} \sup_{0 \leq \zeta \leq a} f(x, \zeta)\, dx \leq (t_{\tilde{n}} - \tilde{t}_{\tilde{n}})\theta a \max_{\xi \in [\tilde{t}_{\tilde{n}}, t_{\tilde{n}}]} \varphi(\xi) \quad (5.156) \\
&= \theta a \frac{\epsilon \max_{\xi \in [\tilde{t}_{\tilde{n}}, t_{\tilde{n}}]} \varphi(\xi)}{2Ma\varphi(t_{\tilde{n}})} \leq \theta\epsilon.
\end{aligned}
$$

This contradiction proves (5.150). In the derivation of (5.156) we have used the following inequality:

$$\frac{\max_{\xi \in [\tilde{t}_{\tilde{n}}, t_{\tilde{n}}]} \varphi(\xi)}{\varphi(t_{\tilde{n}})} \leq \frac{\max_{\xi \in [t_{\tilde{n}} - C\varphi^{-1}(t_{\tilde{n}}), t_{\tilde{n}}]} \varphi(\xi)}{\min_{\xi \in [t_{\tilde{n}} - C\varphi^{-1}(t_{\tilde{n}}), t_{\tilde{n}}]} \varphi(\xi)} < 2M, \qquad \tilde{n} \gg 1, \quad (5.157)$$

which follows from (5.146) for sufficiently large $t_{\tilde{n}}$, and the factor 2 in (5.157) can be replaced by any fixed factor $1 + q$, where $q > 0$ can be arbitrarily small if $t_{\tilde{n}}$ is sufficiently large.

Since $\omega(t)$ is nondecreasing, it follows from (5.150) that

$$
\begin{aligned}
\int_{\tilde{t}_n}^{t_n} \omega(y(x))\varphi(x)\, dx &\geq (t_n - \tilde{t}_n)\omega\big((1 - \theta)\epsilon\big) \min_{\tilde{t}_n \leq \xi \leq t_n} \varphi(\xi) \\
&\geq \omega\big((1 - \theta)\epsilon\big) \frac{\epsilon}{2aM} \frac{\min_{\tilde{t}_n \leq \xi \leq t_n} \varphi(\xi)}{\max_{\tilde{t}_n \leq \xi \leq t_n} \varphi(\xi)} \qquad (5.158) \\
&\geq \omega\big((1 - \theta)\epsilon\big) \frac{\epsilon}{2aM(M + q)} > 0,
\end{aligned}
$$

where $q > 0$ is arbitrarily small for all sufficiently large n. From (5.149), (5.146), and (5.148), one gets

$$\lim_{n \to \infty} \left(t_n - \frac{\epsilon}{2aM\varphi(t_n)} \right) \geq \lim_{n \to \infty} \left(t_n - \frac{C}{\varphi(t_n)} \right) = \infty. \tag{5.159}$$

Inequalities (5.158) and (5.159) contradict the Cauchy criterion for the convergence of integral (5.145). Thus, (5.132) holds. b

Consider Case 2, namely $\theta \geq 1$. In this case one replaces θ by $\theta_1 = \frac{1}{2}$, C by $C_1 = 2\theta C$, M by $M_1 = M$, defined in (5.146) with the C_1 in place of C, and, therefore, one reduces the problem to Case 1 with $\theta = \frac{1}{2} < 1$.

Let us give a more detailed argument. Let $\varphi_1(t) := 2\theta\varphi(t)$ and $C_1 := 2\theta C$. Then

$$\frac{C_1}{\varphi_1(t)} = \frac{C}{\varphi(t)}, \qquad \forall t \geq 0. \tag{5.160}$$

This, (5.145), (5.146), and (5.147) imply

$$\int_0^\infty \omega(y(t))\varphi_1(t)\,dt < \infty, \tag{5.161}$$

$$\lim_{t\to\infty}\left(t - \frac{C_1}{\varphi_1(t)}\right) = \infty, \qquad \limsup_{t\to\infty}\frac{\max_{\xi\in[t-\frac{C_1}{\varphi_1(t)},t]}\varphi_1(\xi)}{\min_{\xi\in[t-\frac{C_1}{\varphi_1(t)},t]}\varphi_1(\xi)} = M < \infty \tag{5.162}$$

$$\int_s^t \sup_{0\leq\zeta\leq a} f(x,\zeta)dx \leq (t-s)\frac{a}{2}\max_{\xi\in[s,t]}\varphi_1(\xi), \qquad t > s \gg 1. \tag{5.163}$$

Theorem 5.7.4 is proved. ∎

Remark 5.7.6   (i) Conditions (5.146) hold if

$$\liminf_{t\to\infty} t\varphi(t) > 0, \qquad \limsup_{t\to\infty}\frac{\max_{\xi\in[(1-\epsilon)t,t]}\varphi(\xi)}{\min_{\xi\in[(1-\epsilon)t,t]}\varphi(\xi)} < \infty, \tag{5.164}$$

for a sufficiently small $\epsilon > 0$.

(ii) If $y(t)$ is differentiable, then (5.130) is equivalent to

$$y'(t) \leq f(t, y(t)), \qquad t \geq 0. \tag{5.165}$$

(iii) Theorem 5.7.4 holds if in place of (5.147) one assumes that

$$\sup_{0\leq\zeta\leq a} f(t,\zeta) \leq \tilde{C}\varphi(t), \qquad t \gg 1, \qquad \tilde{C} = const > 0. \tag{5.166}$$

Indeed, if (5.166) holds, then

$$\int_s^t \sup_{0\leq\zeta\leq a} f(\xi,\zeta)\,d\xi \leq \int_s^t \tilde{C}\varphi(\xi)\,d\xi \leq \tilde{C}(t-s)\max_{s\leq\xi\leq t}\varphi(\xi).$$

(iv) If $\varphi(t)$ is nonincreasing, then the second relation in (5.146) becomes

$$M := \limsup_{t\to\infty}\frac{\varphi(t - \frac{C}{\varphi(t)})}{\varphi(t)} < \infty. \tag{5.167}$$

From Theorem 5.7.4 we derive the following theorem.

Theorem 5.7.7 *Assume that* $y(t) \geq 0$ *is a continuous on* $[0,\infty)$ *function,*

$$\int_0^\infty \omega(y(t))\frac{1}{(1+t)^\alpha}\,dt < \infty, \qquad 0 < \alpha \leq 1, \tag{5.168}$$

$$y(t) - y(s) \leq \int_s^t f(x, y(x)) \, dx, \qquad 0 \leq s \leq t, \tag{5.169}$$

and there exist constants $a > 0$ and $\kappa > 0$ such that

$$\int_s^t \sup_{0 \leq \zeta \leq a} f(x, \zeta) \, dx \leq \kappa a \frac{t - s}{s^\alpha}, \qquad \kappa > 0, \quad t > s \gg 1. \tag{5.170}$$

Then,

$$\lim_{t \to \infty} y(t) = 0. \tag{5.171}$$

Proof: Let $\varphi(t) := \frac{1}{(1+t)^\alpha}$, $\alpha \in (0, 1]$. Then one can easily verify that conditions (5.146) hold with $C = 1/2$. Condition (5.147) also holds for this choice of φ and $\theta = 2\kappa$. Thus, Theorem 5.7.7 follows from Theorem 5.7.4. ∎

Remark 5.7.8 The assumption $\alpha \in (0, 1]$ in (5.168) is essential: If $\alpha > 1$, then inequality (5.146) does not hold for $\varphi(t) = \frac{1}{(1+t)^\alpha}$ whatever fixed $C > 0$ is.

Corollary 5.7.9 *Let $y(t) \geq 0$ be a continuous function on $[0, \infty)$ and*

$$\int_0^\infty \omega\big(y(t)\big)\varphi(t) \, dt < \infty, \tag{5.172}$$

where $\varphi(t) > 0$ is a continuous function on $[0, \infty)$. Assume that there exists a constant $C > 0$ such that

$$\lim_{t \to \infty} \left(t - \frac{C}{\varphi(t)} \right) = \infty, \qquad M := \limsup_{t \to \infty} \frac{\max_{\xi \in [t - \frac{C}{\varphi(t)}, t]} \varphi(\xi)}{\min_{\xi \in [t - \frac{C}{\varphi(t)}, t]} \varphi(\xi)} < \infty, \tag{5.173}$$

$$y(t) - y(s) \leq \int_s^t h(\xi) \, d\xi, \qquad 0 \leq s \leq t, \tag{5.174}$$

where $h(t) \geq 0$, $\forall t \in [0, \infty)$, and

$$A := \limsup_{t \to \infty} \frac{h(t)}{\varphi(t)} < \infty. \tag{5.175}$$

Then,

$$\lim_{t \to \infty} y(t) = 0. \tag{5.176}$$

Proof: Let

$$f(t, y) := h(t), \qquad t \geq 0, \qquad y \geq 0. \tag{5.177}$$

Let us check that condition (5.147) holds for this $f(t, y)$ and $a = 2A$. From (5.177) one gets

$$f(t, y) \leq 2A\varphi(t), \qquad t \gg 1, \qquad \forall y \geq 0. \tag{5.178}$$

This implies

$$\int_s^t \max_{0 \le \zeta \le 2A} f(\xi, \zeta) \, d\xi \le \int_s^t 2A\varphi(\xi) \, d\xi \le 2A(t-s)\varphi(s), \qquad t > s \gg 1.$$
(5.179)

This and Theorem 5.7.4 imply (5.176). ∎

Corollary 5.7.10 *Let* $y(t) \ge 0$ *be a continuous function on* $[0, \infty)$,

$$\int_0^\infty \omega\big(y(t)\big) \frac{1}{(1+t)^\alpha} \, dt < \infty, \qquad 0 < \alpha \le 1,$$
(5.180)

and

$$y(t) - y(s) \le \int_s^t h(\xi) \, d\xi, \qquad 0 \le s \le t,$$
(5.181)

where $h(t) \ge 0$, $\forall t \in [0, \infty)$. *If*

$$A := \limsup_{t \to \infty} h(t) t^\alpha < \infty,$$
(5.182)

then

$$\lim_{t \to \infty} y(t) = 0.$$
(5.183)

Proof: Let $\varphi(t) = \frac{1}{(t+1)^\alpha}$, $\alpha \in (0, 1]$. Then conditions (5.173) hold with $C = \frac{1}{2}$ and $M = 1$, and condition (5.175) also holds. Thus, (5.183) follows from Corollary 5.7.9. ∎

Theorem 5.7.11 *Assume that* $g \ge 0$ *is a continuously differentiable function on* $[0, \infty)$,

$$\dot{g}(t) \le -a(t) f(g(t)) + b(t), \qquad g(0) = g_0,$$
(5.184)

where $f(t)$ *is a nonnegative continuous function on* $[0, \infty)$, $f(0) = 0$, $f(t) > 0$ *for* $t > 0$, *and*

$$m(\epsilon) := \inf_{x \ge \epsilon} f(x) > 0, \qquad \forall \epsilon > 0.$$
(5.185)

If $a(t) > 0$, $b(t) \ge 0$ *are continuous on* $[0, \infty)$ *functions, and*

$$\int_0^\infty a(s) \, ds = \infty, \qquad \lim_{t \to \infty} \frac{b(t)}{a(t)} = 0,$$
(5.186)

then

$$\lim_{t \to \infty} g(t) = 0.$$
(5.187)

Proof: Let

$$s := s(t) := \int_0^t a(\xi) \, d\xi.$$
(5.188)

It follows from (5.186) that the map $t \to s$ maps $[0, \infty)$ onto $[0, \infty)$. Let $t(s)$ be the inverse map and define $w(s) = g(t(s))$. Then (5.184) takes the form

$$w'(s) \leq -f(w) + \beta(s), \qquad w(0) = g_0, \tag{5.189}$$

where

$$w' = \frac{dw}{ds}, \qquad \beta(s) = \frac{b(t(s))}{a(t(s))}, \qquad \lim_{s \to \infty} \beta(s) = 0. \tag{5.190}$$

Assume that (5.187) does not hold. Then there exist $\epsilon > 0$ and $(s_n)_{n=1}^{\infty}$ such that

$$0 < s_n \nearrow \infty, \qquad w(s_n) > \epsilon, \qquad \forall n. \tag{5.191}$$

From the last relation in (5.190) it follows that there exists $T > 0$ such that

$$\beta(s) < \frac{m(\epsilon)}{2}, \qquad \forall s \geq T. \tag{5.192}$$

Since $s_n \nearrow \infty$, there exists $N > 0$ such that $s_n > T$, $\forall n \geq N$. Thus,

$$w'(s_n) \leq -f(w(s_n)) + \beta(s_n) \leq -m(\epsilon) + \frac{m(\epsilon)}{2} < 0, \qquad \forall n \geq N. \tag{5.193}$$

Since $w(s)$ is continuously differentiable on the interval (s_{n-1}, s_n) and $w'(s_n) < 0$, $\forall n \geq N$, there are two possibilities:

Case 1: $w'(s) < 0$, $n \geq N$, for all $s \in (s_{n-1}, s_n)$.

Case 2: There exists a point $t_n \in (s_{n-1}, s_n)$ such that $w'(s) < 0$, $\forall s \in (t_n, s_n)$ and $w'(t_n) = 0$ where $n \geq N$.

We claim that Case 2 cannot happen if $n \geq N$ is sufficiently large, namely so large that $\beta(t_n) < m(\epsilon)$. Indeed, if Case 2 holds for such n, then

$$w'(t_n) = 0, \qquad w(t_n) > w(s_n) > \epsilon. \tag{5.194}$$

This and (5.189) imply

$$0 = w'(t_n) \leq -f(w(t_n)) + \beta(t_n) < -m(\epsilon) + \beta(t_n), \tag{5.195}$$

that is, $0 < m(\epsilon) < \beta(t_n)$. This contradicts the assumption $\lim_{t \to \infty} \beta(t) = 0$ because if n is sufficiently large then t_n is so large that $\beta(t_n) < m(\epsilon)$.

Since Case 2 cannot happen for all sufficiently large n, there exists $N_1 > 0$ sufficiently large so that

$$w'(s) < 0, \qquad \forall s \in (s_{n-1}, s_n), \qquad n \geq N_1. \tag{5.196}$$

Thus,

$$w'(t) \leq 0, \qquad \forall t \geq s_{N_1}. \tag{5.197}$$

Therefore $w(t)$ decays monotonically for all sufficiently large t. Since $w(t) \geq 0$, one concludes that the following limit $W \geq 0$ exists and is finite:

$$\lim_{t \to \infty} w(t) = W < \infty. \tag{5.198}$$

This and (5.189) imply

$$\limsup_{t\to\infty} w'(t) \le \lim_{t\to\infty} [-f(w(t)) + \beta(t)] \le -m(W). \qquad (5.199)$$

If $W \ne 0$, then $m(W) > 0$ and

$$\limsup_{t\to\infty} w'(t) \le -m(W) < 0. \qquad (5.200)$$

This is impossible since $w(t) \ge 0$, $\forall t$. This contradiction implies that $W = 0$, so (5.187) holds.

Theorem 5.7.11 is proved. $\blacksquare$

Remark 5.7.12 Theorem 5.7.11 is proved in [151] under the assumption that $f \in Lip_{loc}[0, \infty)$ and

$$f(0) = 0, \qquad f(u) > 0 \quad \text{for} \quad t > 0, \quad f(u) \ge c > 0 \quad \text{for} \quad u \ge 1, \quad (5.201)$$

where $c = const$. The assumption $f \in Lip_{loc}[0, \infty)$ was used in [151] in order to prove the global existence of $g(t)$. In this section we assume the global existence of $g(t)$ and give a new simple proof of Theorem 5.7.11.

Theorem 5.7.13 *Assume that $g \ge 0$ is a $C^1([0, \infty))$ function,*

$$\dot{g}(t) \le -a(t)f(g(t)) + b(t), \qquad g(0) = g_0, \qquad (5.202)$$

where $f(t) \ge 0$ is a nondecreasing function on $[0, \infty)$, $f(0) = 0$, $f(t) > 0$ if $t > 0$. If $a(t) > 0$, $b(t) \ge 0$ are continuous on $[0, \infty)$ functions, and

$$\int_0^\infty a(s)\, ds = \infty, \qquad \int_0^\infty \beta(s)\, ds < \infty, \qquad \beta(t) := \frac{b(t)}{a(t)}, \qquad (5.203)$$

then

$$\lim_{t\to\infty} g(t) = 0. \qquad (5.204)$$

Proof: Let s be defined in (5.188) and $w(s) = g(t(s))$. From (5.189) one gets

$$w(t) - w(0) + \int_0^t f(w(s))\, ds \le \int_0^t \beta(s)\, ds \le \int_0^\infty \beta(s)\, ds < \infty, \qquad (5.205)$$

for all $t \ge 0$. This and the assumption that $w \ge 0$ imply

$$\int_0^\infty f(w(s))\, ds < \infty. \qquad (5.206)$$

From (5.189) one obtains

$$w'(s) \le \beta(s), \qquad \forall s \ge 0. \qquad (5.207)$$

Since $\int_0^\infty \beta(s)\,ds < \infty$, the function $\psi(t) := \int_0^t \beta(s)\,ds < \infty$ is uniformly continuous with respect to t on $[0,\infty)$. This, relation (5.206), inequality (5.207), and Theorem 5.7.1 imply

$$\lim_{s\to\infty} w(s) = 0. \tag{5.208}$$

This and the relation $w(s) = g(t(s))$ imply (5.204).

Theorem 5.7.13 is proved. ∎

Theorem 5.7.14 *Assume that $g \geq 0$ is a $C^1([0,\infty))$ function,*

$$\dot{g}(t) \leq -a(t)f(g(t)) + b(t), \qquad g(0) = g_0, \tag{5.209}$$

where $f(t) \geq 0$ is a nondecreasing continuous function on $[0,\infty)$, $f(0) = 0$, $f(t) > 0$ if $t > 0$, $a(t) > 0$, and $b(t) \geq 0$ are continuous functions on $[0,\infty)$, and there exists a constant $C > 0$ such that

$$\lim_{t\to\infty}\left(t - \frac{C}{a(t)}\right) = \infty, \qquad \limsup_{t\to\infty}\frac{\max_{\xi\in[t-\frac{C}{a(t)},t]}a(\xi)}{\min_{\xi\in[t-\frac{C}{a(t)},t]}a(\xi)} < \infty. \tag{5.210}$$

If

$$K := \limsup_{t\to\infty}\frac{b(t)}{a(t)} < \infty, \qquad \int_0^\infty b(s)\,ds < \infty, \tag{5.211}$$

then

$$\lim_{t\to\infty} g(t) = 0. \tag{5.212}$$

Proof: From (5.209) one gets for all $t \geq 0$ the following inequalities:

$$g(t) - g(0) + \int_0^t a(s)f(g(s))\,ds \leq \int_0^t b(s)\,ds \leq \int_0^\infty b(s)\,ds < \infty. \tag{5.213}$$

Thus

$$\int_0^\infty a(s)f(g(s))\,ds < \infty. \tag{5.214}$$

This relation, (5.211), (5.210), and Corollary 5.7.9 imply (5.212). Theorem 5.7.14 is proved. ∎

Remark 5.7.15 If $a(t) = O(\frac{1}{(1+t)^\gamma})$, $\gamma \in [0,1)$, then conditions (5.210) hold for any $C > 0$. If $a(t) = O(\frac{1}{1+t})$, then conditions (5.210) hold if $C > 0$ is sufficiently small.

5.7.2 Applications

Let H be a real Hilbert space. Consider the following problem

$$\dot{u} = A(t,u) + f(t), \qquad u(0) = u_0; \qquad f \in C([0,\infty);H), \tag{5.215}$$

where $u_0 \in H$, $A(t, u) : [0, \infty) \times H \to H$ is continuous with respect to t and u. Assume that

$$A(t, 0) = 0, \qquad \forall t \geq 0, \tag{5.216}$$

$$\langle A(t, u) - A(t, v), u - v \rangle \leq -\gamma(t) \|u - v\| \omega(\|u - v\|), \qquad u, v \in H, \tag{5.217}$$

where $\gamma(t) > 0$ for all $t \geq 0$ is a continuous function and $\omega(t) \geq 0$ is continuous and strictly increasing function on $[0, \infty)$, $\omega(0) = 0$.

The above assumptions are standing and are not repeated. Assumption (5.217) means that A is a dissipative operator. Existence of the solution to problem (5.215) with such operators was discussed in the literature ([82], [178], and [151]).

Let

$$\beta(t) := \|f(t)\|.$$

Consider the following three assumptions:

- Assumption A

$$\int_0^\infty \gamma(t)\, dt = \infty, \qquad \lim_{t \to \infty} \frac{\beta(t)}{\gamma(t)} = 0. \tag{5.218}$$

- Assumption B

$$\int_0^\infty \gamma(t)\, dt = \infty, \qquad \int_0^\infty \frac{\beta(t)}{\gamma(t)}\, dt < \infty. \tag{5.219}$$

- Assumption C

$$\int_0^\infty \beta(t)\, dt < \infty, \quad \gamma(t) = O\left(\frac{1}{(1+t)^\alpha}\right), \quad \limsup_{t \to \infty} \beta(t) t^\alpha < \infty, \tag{5.220}$$

where $\alpha = const \in (0, 1]$.

Remark 5.7.16 Assumption (5.216) is not an essential restriction: If it does not hold, define $f_1(t) := f(t) + A(t, 0)$, and $A_1(t, u) := A(t, u) - A(t, 0)$. Then $A_1(t, u)$ satisfies assumptions (5.216) and (5.217) and $f_1(t)$ plays the role of $f(t)$.

Lemma 5.7.17 *If assumptions* (5.216)–(5.217) *hold, then there exists a unique global solution* $u(t)$ *to* (5.215).

Proof: *Let us first prove the uniqueness of solution to* (5.215).
Assume that u and v are two solutions to (5.215). Then one gets

$$\dot{u} - \dot{v} = A(t, u) - A(t, v), \qquad t \geq 0. \tag{5.221}$$

Multiply (5.221) by $u - v$ and use (5.217) to obtain

$$\frac{1}{2}\frac{d}{dt}\|u - v\|^2 = \langle A(t, u) - A(t, v), u - v \rangle \leq 0. \tag{5.222}$$

Integrating (5.222), one gets

$$\frac{1}{2}\left(\|u(t) - v(t)\|^2 - \|u(0) - v(0)\|^2\right) \leq 0, \qquad \forall t \geq 0. \tag{5.223}$$

This implies $u(t) = v(t)$, $\forall t \geq 0$, since $u(0) = v(0)$.

Let us prove the local existence of a solution to (5.215).

In this proof an argument similar to the one in [22] or [151] is used. Let $u_n(t)$, called Peano's approximation of u, solve the following equation:

$$u_n(t) = u_0 + \int_0^t \left[A\left(s, u_n\left(s - \frac{1}{n}\right)\right) + f(s)\right] ds, \qquad t \geq 0, \tag{5.224}$$

and $u_n(t) = u_0$, $\forall t \leq 0$.

Fix some positive numbers $r > 0$ and $t_1 > 0$. Let

$$B(u_0, r) := \{u : \|u - u_0\| \leq r\},$$

and

$$c := \sup_{(t,u) \in [0,t_1] \times B(u_0,r)} (\|A(t, u)\| + \|f(t)\|) < \infty. \tag{5.225}$$

From (5.224) one gets

$$\|u_n(t) - u_0\| \leq ct \leq r, \qquad 0 \leq t \leq T_1, \qquad T_1 := \min\left(t_1, \frac{r}{c}\right) > 0, \tag{5.226}$$

Define

$$w_{nm} := u_n(t) - u_m(t), \qquad g_{mn} := \|w_{mn}(t)\|, \qquad t \geq 0. \tag{5.227}$$

From (5.224) one obtains

$$g_{mn}(t)\dot{g}_{mn}(t) = \left\langle A\big(t, u_n(t_n)\big) - A\big(t, u_m(t_m)\big), u_n(t) - u_m(t) \right\rangle$$

$$= \left\langle A\big(t, u_n(t_n)\big) - A\big(t, u_m(t_m)\big), u_n(t_n) - u_m(t_m) \right\rangle$$

$$+ \left\langle A\big(t, u_n(t_n)\big) - A\big(t, u_m(t_m)\big), u_n(t) - u_n(t_n) \right\rangle$$

$$+ \left\langle A\big(t, u_n(t_n)\big) - A\big(t, u_m(t_m)\big), u_m(t_m) - u_m(t) \right\rangle, \tag{5.228}$$

where $t_n = t - \frac{1}{n}$ and $t_m = t - \frac{1}{m}$. From (5.228), (5.225), and (5.217), one obtains

$$\frac{1}{2}\frac{d}{dt}g_{nm}^2(t) \le 4c^2\left(\frac{1}{n} + \frac{1}{m}\right), \qquad m, n \ge 0, \quad t \in [0, T_1]. \qquad (5.229)$$

Integrating (5.229), using the relation $g_{mn}(0) = 0$, and taking the limit as $m, n \to \infty$, one obtains

$$0 \le \lim_{n,m\to\infty} g_{nm}^2(t) \le \lim_{n,m\to\infty} t4c^2(\frac{1}{n} + \frac{1}{m}) = 0, \qquad \forall t \in [0, T_1]. \qquad (5.230)$$

It follows from (5.230) and the Cauchy criterion for convergence of a sequence that the following limit exists:

$$u(t) := \lim_{n\to\infty} u_n(t), \qquad 0 \le t \le T_1. \qquad (5.231)$$

Passing to the limit $n \to \infty$ in equation (5.224) and using the continuity of $A(t, u)$ on $[0, \infty) \times H$ and (5.231), one concludes that $u(t)$ solves the equation

$$u(t) = u_0 + \int_0^t [A(s, u(s)) + f(s)]\,ds, \qquad \forall t \in [0, T_1]. \qquad (5.232)$$

Thus, the local existence of the solution $u(t)$ to equation (5.215) is proved.

Let us prove the global existence of $u(t)$.

Assume that $u(t)$ does not exist globally. Let $[0, T]$ be the maximal existence interval of $u(t)$. Then, $0 < T < \infty$. It follows from relation (5.242) that

$$\|u(t)\| = g(t) < c = const < \infty, \qquad \forall t \in [0, T). \qquad (5.233)$$

Let us prove the existence of the finite limit

$$\lim_{t\to T_-} u(t) = u_T. \qquad (5.234)$$

Let $z_h(t) := u(t + h) - u(t)$, $0 < t \le t + h < T$. From (5.195) one gets

$$\dot{z}_h(t) = A(t, u(t + h)) - A(t, u(t)) + f(t + h) - f(t), \qquad (5.235)$$

where $0 < t \le t + h < T$. Multiply (5.235) by $z_h(t)$ and get

$$\frac{1}{2}\frac{d}{dt}\|z_h(t)\|^2 \le -\gamma(\|z_h(t)\|)\|z_h(t)\|\omega(\|z_h(t)\|)$$
$$+ \|z_h(t)\|\|f(t + h) - f(t)\|. \qquad (5.236)$$

This implies

$$\frac{d}{dt}\|z_h(t)\| \le \|f(t + h) - f(t)\|, \qquad 0 < t \le t + h < T. \qquad (5.237)$$

Integrating (5.237), one gets

$$\|z_h(t)\| \le \|z_h(0)\| + \int_0^t \|f(x+h) - f(x)\|\, dx$$
$$\le \|z_h(0)\| + T \max_{0 \le t \le T-h} \|f(t+h) - f(t)\|. \tag{5.238}$$

Since $\lim_{h \to +0} \|u(h) - u(0)\| = 0$ and $\lim_{h \to +0} \max_{0 \le t \le T-h} \|f(t+h) - f(t)\| = 0$, one concludes that

$$\lim_{h \to 0} \|u(t+h) - u(t)\| = 0, \tag{5.239}$$

and this relation holds uniformly with respect to t and $t+h$ such that $t < t+h < T$. Relation (5.239) and the Cauchy criterion for convergence imply the existence of the finite limit in (5.234).

Consider the following Cauchy problem:

$$\dot{u} = A(t,u) + f(t), \qquad u(T) = u_T. \tag{5.240}$$

By the arguments similar to those given above, one derives that there exists a unique solution $u(t)$ to (5.240) on $[T, T+\delta]$, where $\delta > 0$ is a sufficiently small number. From the continuity of $f(t)$ and $u(t)$ in t and $A(t,u)$ in both t and u, one gets

$$\lim_{t \to T_-} \dot{u}(t) = \lim_{t \to T_-} A(t,u) + f(t) = \lim_{t \to T_+} A(t,u) + f(t) = \lim_{t \to T_+} \dot{u}(t), \tag{5.241}$$

and the above limits are finite. Thus, the solution to (5.215) can be extended to the interval $[0, T+\delta]$. This contradicts the definition of T. Thus, $T = \infty$, that is, $u(t)$ exists globally.

Lemma 5.7.17 is proved. ∎

The main result of this section is the following theorem.

Theorem 5.7.18 *If $u(t)$ solves problem (5.215) globally and at least one of the assumptions A, B, or C holds, then*

$$\lim_{t \to \infty} \|u(t)\| = 0. \tag{5.242}$$

Proof: The global existence of $u(t)$ follows from Lemma 5.7.17 under the assumptions of this lemma, or from the results in [82] and [178].

Let us prove relation (5.242).

Multiplying (5.215) by u, one obtains

$$\frac{1}{2}\frac{d}{dt}\|u(t)\|^2 = \langle A(t,u), u \rangle + \langle f(t), u \rangle \le -\gamma(t)\|u\|\omega(\|u\|) + \|f(t)\|\|u\|. \tag{5.243}$$

Since $u(t)$ is continuously differentiable, so is $\|u(t)\|$ at the points t at which $\|u(t)\| > 0$. At these points, inequality (5.243) implies

$$\frac{d}{dt}\|u(t)\| \le -\gamma(t)\omega(\|u(t)\|) + \|f(t)\|, \qquad t \ge 0. \tag{5.244}$$

If $\|u(t)\| = 0$ on an open interval $(a, b) \subset [0, \infty)$, then $\frac{d}{dt}\|u(t)\| = 0$ on (a, b), and inequality (5.244) holds trivially, because $\omega(0) = 0$ and $\|f(t)\| \geq 0$. If $\|u(s)\| = 0$ at an isolated point $s > 0$, then the right-sided derivative of $\|u(t)\|$ at the point s exists, and

$$\frac{d}{dt}\|u(t)\| = \lim_{\tau \to +0} \frac{\|u(s+\tau)\|}{\tau} = \|\frac{d}{dt}u(s)\|,$$

and inequality (5.244) holds for this derivative. In what follows we understand by $\frac{d}{dt}\|u(t)\|$ the right-sided derivative at the points s at which $\|u(s)\| = 0$. The left-sided derivative of $\|u(t)\|$ also exists at such points and is equal to $-\|\frac{d}{dt}u(s)\|$, but we will not need this left-sided derivative.

Let $g(t) := \|u(t)\|$ and $\beta(t) := \|f(t)\|$. From (5.244) one gets

$$\dot{g}(t) \leq -\gamma(t)\omega(g(t)) + \beta(t), \qquad t \geq 0. \tag{5.245}$$

Let $a(t) := \gamma(t)$ and $b(t) := \beta(t)$.

If Assumption A holds, then (5.242) follows from this assumption and Theorem 5.7.11.

If Assumption B holds, then (5.242) follows from this assumption and Theorem 5.7.13.

If Assumption C holds, then (5.242) follows from this assumption and Theorem 5.7.14.

Thus, (5.242) holds.

Theorem 5.7.18 is proved.   ■

Example 5.5

Let $D \subset \mathbb{R}^3$ be a bounded domain with a smooth boundary. Consider problem (5.215) with $\gamma(t) = (1+t)^{-\alpha}$, $\alpha = const \in (0, 1]$, $f(t) = O(\frac{1}{(1+t)^k})$, $k > 1$, and $A(t, u) := \gamma(t)[Lu - u^3]$, where L is a second-order negative-definite self-adjoint Dirichlet elliptic operator in D, for example, $L = \Delta$, where Δ is the Dirichlet Laplacian in D. Then

$$\beta(t) = \|f(t)\| \leq \frac{c}{(1+t)^k}, \qquad k > 1,$$

conditions (5.216) and (5.217) are satisfied for $u, v \in D(A)$, where $D(A)$ is the domain of definition of the operator A, $A(u) := \Delta u - u^3$, $D(A) = H^2(D) \cap H_0^1(D) \subset C(D)$, H^ℓ, $\ell = 1, 2$, are the usual Sobolev spaces, and the inclusion holds by the Sobolev embedding theorem in R^3. The function $\omega(r)$ in (5.217) in this example is $\omega(r) = cr$, where $c > 0$ is a constant. This follows from the known inequality

$$-\langle Lu, u \rangle = \|\nabla u\| \geq c(D)\|u\|,$$

valid for $u \in H_0^1(D)$; $c(D) = const$ does not depend on $u \in H_0^1(D)$. In this example the operator A is not continuous in H, but the global solution to problem (5.215) exists and is unique (see, e.g., [82], [178], and [197]). One checks that Assumption C is satisfied; and one concludes, using Theorem (5.7.18) that (5.242) holds for the solution to (5.215) in this example.

Theorem (5.7.18) can be applied regardless of the method by which the global existence of the unique solution to problem (5.215) is established and inequality (5.245) is derived for this solution.

Let $\langle \cdot, \cdot \rangle$ denote the inner product and $\| \cdot \|$ denote the norm in $L^2(D)$. Then the usual ellipticity constant $c_1 = \gamma(t)c(D)$ in the inequality

$$c_1 \|u\|^2 \le -\gamma(t)\langle Lu, u \rangle$$

tends to zero as $t \to \infty$, so one deals with a degenerate elliptic operator as $t \to \infty$ in problem (5.215) in this example.

One can extend the result in this example to much more general nonlinearities. For instance, if $A(t, u) = \gamma(t)[Lu - h(u)]$, where $uh(u) \ge 0$ for all $u \in R$, and h satisfies a local Lipschitz condition, then one can derive an a priori bound for the solution $u(t)$ of (5.215) $\sup_{t \ge 0} \|u(t)\| \le c$, and prove the global existence and uniqueness of the solution $u(t)$ to problem (5.215) using, for instance, the method from [149]. The assumption $uh(u) \ge 0$ for all $u \in R$ makes it possible to consider nonlinearities $h(u)$ with an arbitrary large speed of growth at infinity. Let us outline the derivation of the above bound. Multiplying (5.215) by u and using the estimate $\langle Lu, u \rangle \le -c\|u\|^2$, the assumption $uh(u) \ge 0$, and denoting $g := \|u\|^2$, one gets the following inequality:

$$\dot{g} \le -2c\gamma(t)g + 2\|f\|g^{1/2}, \qquad g(0) = \|u_0\|^2.$$

For simplicity and without loss of generality, assume that $u_0 = 0$. Then it is not difficult to derive the following inequality:

$$\|u(t)\| \le \int_0^t \|f(s)\| e^{-c \int_s^t \gamma(\tau)d\tau} ds.$$

Using the assumption $k > 1$, one obtains from this inequality the following estimate:

$$\sup_{t \ge 0} \|u(t)\| \le \frac{1}{k-1}.$$

CHAPTER 6

DSM FOR MONOTONE OPERATORS

In this chapter the DSM method is developed for solving equations with monotone operators. Convergence of the DSM is proved for any initial approximation. The DSM yields the unique minimal-norm solution.

6.1 AUXILIARY RESULTS

In this chapter we study the equation

$$F(u) = f, \tag{6.1}$$

assuming that F is monotone in the sense

$$\langle F(u) - F(v), u - v \rangle \geq 0, \quad \forall u, v \in H, \tag{6.2}$$

where H is a Hilbert space. We assume also that $F \in C^2_{\text{loc}}$, that is, assumptions (1.10) hold, but we do not assume that assumption (1.9) hold. Therefore, the problem of solving (6.1) cannot be solved, in general, by Newton's method. We assume that equation (6.1) has a solution, possibly nonunique.

Dynamical Systems Method and Applications: Theoretical Developments and Numerical Examples, First Edition. A. G. Ramm and N. S. Hoang.

For $F \in C^1_{\text{loc}}$, condition (6.2) is equivalent to

$$F'(u) \geq 0. \tag{6.3}$$

We will use the notations

$$A := F'(u), \quad A_a := A + aI. \tag{6.4}$$

In Section 6.2 we formulate a result which contains a justification of a DSM for solving equation (6.1) with exact data f, and then we show how to use this DSM for a stable approximation of the solution y given noisy data f_δ, $\|f_\delta - f\| \leq \delta$. In this section we prove some auxiliary results used in the next section.

Let us recall some (known) properties of monotone operators. For convenience of the reader we prove the results we use in this chapter. These results are auxiliary for the rest of the chapter.

Lemma 6.1.1 *Equation (6.1) holds if and only if*

$$\langle F(v) - f, v - u \rangle \geq 0, \quad \forall v \in H. \tag{6.5}$$

Proof: If u solves (6.1), then (6.5) holds by the monotonicity (6.2). Conversely, if (6.5) holds, take $v = u + \lambda w$, $\lambda = const > 0$, and $w \in H$ is arbitrary; then (6.5) yields

$$\langle F(u + \lambda w) - f, w \rangle \geq 0. \tag{6.6}$$

Let $\lambda \to 0$ and use the continuity of F to get

$$\langle F(u) - f, w \rangle \geq 0. \tag{6.7}$$

Since w is arbitrary, this implies (6.1). Lemma 6.1.1 is proved. ∎

Lemma 6.1.2 *The set $\mathcal{N} := \{u : F(u) - f = 0\}$ is closed and convex.*

Proof: The set $\mathcal{N}$ is closed if F is continuous. Let us prove that $\mathcal{N}$ is convex. Assume that $u, w \in \mathcal{N}$. We want to prove that $\lambda u + (1 - \lambda)w \in \mathcal{N}$ for any $\lambda \in (0, 1)$. By Lemma 6.1.1 we have to prove

$$0 \leq \langle F(v) - f, v - \lambda u - (1 - \lambda)w \rangle$$
$$= \lambda \langle F(v) - f, v - u \rangle + (1 - \lambda)\langle F(v) - f, v - w \rangle,$$

so that the desired inequality follows from the assumption $u, w \in \mathcal{N}$ and from Lemma 6.1.1.

Lemma 6.1.2 is proved. ∎

Lemma 6.1.3 *A closed and convex set $\mathcal{N}$ in a Hilbert space has a unique element of minimal norm.*

Proof: Assume that

$$\|u\| \leq \|v\|, \quad \forall v \in \mathcal{N}$$

and

$$\|w\| \le \|v\|, \quad \forall v \in \mathcal{N}.$$

Since $\mathcal{N}$ is convex, we have

$$\|u\| \le \|\frac{u+w}{2}\|, \quad \|w\| \le \|\frac{u+w}{2}\|, \quad \|u\| = \|w\|.$$

Thus $\langle u, w \rangle = \|u\| \|w\|$, so $u = v$. Lemma 6.1.3 is proved. ∎

Remark 6.1.4 A Banach space X is called strictly convex if the condition

$$\|u\| = \|w\| = \|\frac{u+w}{2}\|$$

implies $u = w$. Hilbert spaces are strictly convex, $L^p(a, b)$ space for $p > 1$ is strictly convex, but $C([a, b])$ and $L^1([a, b])$ are not strictly convex. If X is strictly convex, then $\|u + w\| = \|u\| + \|w\|$ implies $w = \lambda u$, where λ is a real number.

Lemma 6.1.5 (*w-closedness*). *If (6.2) holds and F is continuous, then the assumptions*

$$u_n \rightharpoonup u, \quad F(u_n) \to f \tag{6.8}$$

imply (6.1). Here $\rightharpoonup$ denotes weak convergence.

Proof: Using (6.2), we have

$$\langle F(v) - F(u_n), v - u_n \rangle \ge 0. \tag{6.9}$$

Passing to the limit $n \to \infty$ in (6.9) and using (6.8), we obtain

$$\langle F(v) - f, u - v \rangle \ge 0, \quad \forall v \in H. \tag{6.10}$$

By Lemma 6.1.1, this implies $F(u) = f$.

Lemma 6.1.5 is proved. ∎

Lemma 6.1.6 *If (6.2) holds and $F \in C^1_{loc}$, then*

$$\|A_a^{-1}\| \le \frac{1}{a}, \quad a = const > 0; \quad A := F'(u). \tag{6.11}$$

Proof: By (6.3), one has $A \ge 0$. Thus,

$$\langle A_a v, v \rangle \ge a \|v\|^2, \quad \forall v \in H.$$

Since $\langle Av, v \rangle \le \|Av\| \|v\|$, one gets $\|A_a v\| \ge a \|v\|$. Thus, with $A_a v := w$, one has

$$a^{-1} \|w\| \ge \|A_a^{-1} w\|.$$

This implies (6.11).

Lemma 6.1.6 is proved.   ∎

Lemma 6.1.7 *Equation*

$$F(u) + au = f, \quad a = const > 0 \tag{6.12}$$

has a unique solution for every $f \in H$ *if* $F \in C_{loc}^2$ *satisfies (6.2).*

Proof: Consider the problem

$$\dot{u} = -A_a^{-1}[F(u) + au - f], \quad u(0) = u_0, \tag{6.13}$$

where $u_0 \in H$ is arbitrary and A_a is defined in (6.4). Since $F \in C_{loc}^2$, the operator in the right-hand side of (6.13) satisfies locally a Lipschitz condition, so problem (6.13) has a unique local solution $u(t)$. Define

$$g(t) := \|F(u(t)) + au(t) - f\|.$$

Then, by equation (6.13), one gets

$$\dot{g} = -g, \quad g(0) = \|F(u_0) + au_0 - f\| := g_0; \quad g(t) \le g_0 e^{-t}. \tag{6.14}$$

Using (6.13) again and applying (6.11), we get

$$\|\dot{u}\| \le \frac{g_0}{a} e^{-t},$$

so

$$\int_0^\infty \|\dot{u}\| \, dt \le \frac{g_0}{a},$$

there exists $u(\infty)$, and

$$\|u(t) - u_0\| \le \frac{g_0}{a}, \quad \|u(\infty) - u(t)\| \le \frac{g_0}{a} e^{-t}. \tag{6.15}$$

From (6.11) and (6.15) we conclude that $u(\infty)$ solves equation (6.12). The solution to this equation is unique: If u and v solve (6.12), then

$$F(u) - F(v) + a(u - v) = 0.$$

Multiplying this equation by $u - v$ and using (6.2), we derive $u = v$.
    Lemma 6.1.7 is proved.   ∎

Remark 6.1.8 The assumption $F \in C_{loc}^2$ in Lemma 6.1.7 can be relaxed: F is hemicontinuous, defined on all of H, suffices for the conclusion of Lemma 6.1.7 to hold. ([190]).

Lemma 6.1.9 *Assume that (6.11) is solvable, y is its minimal-norm solution, and assumption (6.2) holds. Then*

$$\lim_{a \to 0} \|u_a - y\| = 0, \tag{6.16}$$

where u_a solves (6.12).

Proof of Lemma 6.1.9. We note that $F(y) = f$ and write equation (6.12) as

$$F(u_a) - F(y) + au_a = 0, \quad a > 0. \tag{6.17}$$

Multiply this equation by $u_a - y$, use (6.12), and get

$$\langle u_a, u_a - y \rangle \le 0. \tag{6.18}$$

Thus

$$\|u_a\| \le \|y\|. \tag{6.19}$$

Inequality (6.19) implies the existence of a sequence $u_n := u_{a_n}$, $a_n \to 0$ as $n \to \infty$, such that

$$u_n \rightharpoonup u. \tag{6.20}$$

From equation (6.12) it follows that

$$F(u_n) \to f. \tag{6.21}$$

From (6.20), (6.21), and Lemma 6.1.5 it follows that

$$F(u) = f. \tag{6.22}$$

Let us prove that $u = y$. Indeed, (6.19) implies

$$\overline{\lim_{n \to \infty}} \|u_n\| \le \|y\|, \tag{6.23}$$

while (6.20) implies

$$\|u\| \le \varliminf_{n \to \infty} \|u_n\|. \tag{6.24}$$

Combine (6.23) and (6.24) to get

$$\|u\| \le \|y\|. \tag{6.25}$$

Since u and y are solutions to equation (6.1) and y is the minimal-norm solution, which is unique by Lemmas 6.1.1 and 6.1.2, it follows that $u = y$. Therefore (6.20) implies

$$u \rightharpoonup y, \tag{6.26}$$

while (6.23) and (6.24) imply

$$\lim_{n \to \infty} \|u_n\| = \|y\|. \tag{6.27}$$

From (6.26) and (6.27) it follows that

$$\lim_{n \to \infty} \|u_n - y\| = 0. \tag{6.28}$$

Indeed,

$$\|u_n - y\|^2 = \|u_n\|^2 + \|y\|^2 - 2\mathrm{Re}\langle u_n, y\rangle \xrightarrow[n\to\infty]{} 0. \qquad (6.29)$$

Therefore every convergent sequence $u_n = u_{a_n}$ converges strongly to y. This implies (6.16).

Lemma 6.1.9 is proved. ∎

Remark 6.1.10 One can estimate the rate $\|\dot{V}\|$ of convergence of $V(t)$ to y when $V(t)$ solves equation (6.12) with $a = a(t)$. We assume

$$0 < a(t) \searrow 0, \quad \lim_{t\to\infty} \frac{\dot{a}(t)}{a(t)} = 0, \quad \frac{|\dot{a}(t)|}{a(t)} \le \frac{1}{2}. \qquad (6.30)$$

For example, one may take

$$a(t) = \frac{c_1}{(c_0 + t)^b},$$

where c_0, c_1, and b are positive constants, and $2b \le c_0$.

Differentiating equation (6.12) with respect to t, one gets

$$A_{a(t)}\dot{V} = -\dot{a}V. \qquad (6.31)$$

Thus

$$\|\dot{V}\| \le |\dot{a}|\, \|A_{a(t)}^{-1}V\| \le \frac{|\dot{a}(t)|}{a(t)}\|V\| \le \frac{|\dot{a}(t)|}{a(t)}\|y\|. \qquad (6.32)$$

Lemma 6.1.11 *If*

$$\lim_{t\to\infty} a(t) = 0, \quad a(t) > 0,$$

then

$$\lim_{t\to\infty} \|V(t) - y\| = 0. \qquad (6.33)$$

Proof: The conclusion (6.33) follows immediately from (6.16) and the assumption $\lim_{t\to\infty} a(t) = 0$.

If one assumes more about $a(t)$, then one still cannot estimate the rate of convergence in (6.33), see also Remark 6.1.12 below.

Lemma 6.1.11 is proved. ∎

Remark 6.1.12 It is not possible to use estimate (6.32) in order to estimate $\|V(t) - y\|$ because such an estimate requires the integral $\int_t^\infty \frac{|\dot{a}(s)|}{a(s)}\,ds$ to converge. However, this integral diverges for any $a(t)$ satisfying (6.30). Indeed $|\dot{a}| = -\dot{a}$, so

$$\int_t^\infty \frac{|\dot{a}(s)|}{a(s)}\,ds = -\lim_{N\to\infty}\int_t^N \frac{\dot{a}}{a}\,ds = -\lim_{N\to\infty}\ln\frac{a(N)}{a(t)} = \lim_{N\to\infty}\ln\frac{a(t)}{a(N)} = \infty,$$

because $\lim_{N\to\infty} a(N) = 0$.

Therefore, without extra assumptions about f in equation (6.1), it is not possible to estimate the rate of convergence in (6.33).

6.2 FORMULATION OF THE RESULTS AND PROOFS

Theorem 6.2.1 *Consider the DSM*

$$\dot{u} = -A_{a(t)}^{-1}(u)[F(u) + a(t)u - f], \quad u(0) = u_0, \tag{6.34}$$

where $u_0 \in H$ is arbitrary, F satisfies (6.2) and (1.10), and $a(t)$ satisfies (6.30). Assume that equation (6.2) has a solution. Then problem (6.34) has a unique global solution $u(t)$, there exists $u(\infty)$, and $u(\infty)$ solves equation (6.1).

Proof: Denote
$$w := u(t) - V(t),$$
where $V(t)$ solves equation (6.12) with $a = a(t)$. Then

$$\|u(t) - y\| \leq \|w\| + \|V(t) - y\|. \tag{6.35}$$

We have proved (6.33). We want to prove

$$\lim_{t \to \infty} \|w(t)\| = 0. \tag{6.36}$$

We derive a differential inequality (5.5) for $g(t) = \|w(t)\|$ and then use Theorem 5.1.1.

Let us proceed according to this plan. Write equation (6.34) as

$$\dot{w} = -\dot{V} - A_{a(t)}^{-1}[F(u) - F(V) + a(t)w]. \tag{6.37}$$

We use Taylor's formula and get

$$F(u) - F(V) + aw = A_a(u)w + \varepsilon, \quad \|\varepsilon\| \leq \frac{M_2}{2}\|w\|^2, \tag{6.38}$$

where M_2 is the constant from the estimate

$$\sup_{u \in B(u_0, R)} \|F''(u)\| \leq M_2(R) := M_2.$$

Denote
$$g(t) := \|w(t)\|.$$

Multiply (6.37) by w, and using (6.38), get

$$g\dot{g} \leq -g^2 + \frac{M_2}{2}\|A_{a(t)}^{-1}\|g^3 + \|\dot{V}\|g. \tag{6.39}$$

Since $g \geq 0$, we get the inequality

$$\dot{g} \leq -g(t) + \frac{c_0}{a(t)}g^2 + \frac{|\dot{a}|}{a(t)}c_1, \quad c_0 := \frac{M_2}{2}, \quad c_1 := \|y\|, \tag{6.40}$$

where we have used the estimates (6.11) and (6.32). Inequality (6.40) is of the type (5.5) with

$$\gamma(t) := 1, \quad \alpha(t) := \frac{c_0}{a(t)}, \quad \beta(t) := c_1 \frac{|\dot{a}|}{a(t)}. \tag{6.41}$$

Let us check assumptions (5.2)–(5.4). We take

$$\mu(t) = \frac{\lambda}{a(t)}, \quad \lambda = const.$$

Assumption (5.2) is

$$\frac{c_0}{a(t)} \le \frac{\lambda}{2a(t)} \left[1 - \frac{|\dot{a}|}{a(t)} \right], \tag{6.42}$$

where we took into account that $|\dot{a}| = -\dot{a}$. Since

$$\frac{|\dot{a}|}{a} \le \frac{1}{2},$$

assumption (6.42) will be satisfied if the following inequality holds:

$$c_0 \le \frac{\lambda}{4}. \tag{6.43}$$

Take $\lambda \ge 4c_0$ and (6.43) is satisfied. Assumption (5.3) is

$$c_1 \frac{|\dot{a}|}{a(t)} \le \frac{a(t)}{2\lambda} \left[1 - \frac{|\dot{a}|}{a} \right]. \tag{6.44}$$

This assumption holds if

$$4\lambda c_1 \frac{|\dot{a}(t)|}{a^2(t)} \le 1. \tag{6.45}$$

The scaling transformation

$$a(t) \to \nu a(t),$$

where $\nu = const > 0$, does not change assumptions (6.30), but makes the ratio $\frac{|\dot{a}|}{a^2}$ as small as one wishes if ν is sufficiently large. So taking $\nu a(t)$ in place of $a(t)$, one can satisfy inequality (6.45). Finally, assumption (5.4) is

$$\frac{\lambda}{a(0)} g(0) < 1. \tag{6.46}$$

This assumption is satisfied for any fixed $g(0)$ and λ if $a(0)$ is sufficiently large. Again, taking $\nu a(t)$ in place of $a(t)$, one can obtain arbitrarily large $\nu a(0)$ and satisfy inequality (6.46) (with $a(0)$ replaced by $\nu a(0)$ with a sufficiently large constant $\nu > 0$).

Thus, Theorem 5.1.1 yields

$$g(t) < \frac{a(t)}{\lambda} \to 0 \quad \text{as } t \to \infty. \tag{6.47}$$

This implies uniform with respect to $t \in [0, \infty)$ boundedness of the norm $\|u(t)\|$ of the unique local solution to (6.34) and therefore existence of its unique global solution, existence of the limit $u(\infty)$, and the relation $u(\infty) = y$, where y is the (unique) minimal-norm solution to equation (6.1).

Theorem 6.2.1 is proved.   ∎

Theorem 6.2.2 *Assume that $a = const > 0$, and*

$$\dot{u} = -A_a^{-1}[F(u) + au - f], \quad u(0) = u_0, \tag{6.48}$$

where $u_0 \in H$ is arbitrary, and (6.2) and (1.10) hold. Assume that equation (6.1) is solvable and y is its minimal-norm solution. Then problem (6.48) has a unique global solution $u_a(t)$ and

$$\lim_{a \to 0} \lim_{t \to \infty} \|u_a(t) - y\| = 0. \tag{6.49}$$

Proof: Local existence of the unique solution to (6.48) follows from the local Lipschitz condition satisfied by the operator in the right-hand side of (6.48). Global existence of this solution follows from the uniform boundedness with respect to $t \to \infty$ of the norm $\|u_a(t)\|$ of the solution to (6.48). This boundedness is established by the same argument as in the proof of Lemma 6.1.7. Also, in this proof one establishes the existence of $u_a(\infty) = \lim_{t \to \infty} u_a(t)$ and the relation

$$F(u_a(\infty)) + au(\infty) = f.$$

Finally, this and Lemma 6.1.9 from Section 6.1 imply (6.49).

Theorem 6.2.2 is proved.   ∎

Remark 6.2.3 If we integrate problem (6.48) over an interval of length τ, then we can choose $a(\tau)$ such that

$$\lim_{\tau \to \infty} a(\tau) = 0$$

and

$$\lim_{\tau \to \infty} \|u_{a(\tau)}(\tau) - y\| = 0. \tag{6.50}$$

This observation allows us to choose $a = a(\tau)$ if the length τ of the interval of integration is chosen a priori and then increased. Consequently, one calculates the solution y using one limiting process in equation (6.50) rather than two limiting processes in equation (6.49).

6.3   THE CASE OF NOISY DATA

Let us use Theorem 6.2.1 in the case when f is replaced by f_δ in equation (6.34). Denote by u_δ the solution of (6.34) with f_δ replacing f. Then

$$\|u_\delta(t) - y\| \leq \|u_\delta(t) - V_\delta(t)\| + \|V_\delta(t) - V(t)\| + \|V(t) - y\|, \qquad (6.51)$$

where $V(t)$ solves equation (6.12) and V_δ solves equation (6.2) with f_δ in place of f.

We have already proved (6.33). We want to prove that if $t = t_\delta$ in (6.51), where t_δ, the stopping time, is properly chosen, then

$$\lim_{\delta \to 0} \|u_\delta(t_\delta) - V_\delta(t_\delta)\| = 0, \quad \lim_{\delta \to 0} \|V_\delta(t_\delta) - V(t_\delta)\| = 0. \qquad (6.52)$$

Lemma 6.3.1 *One has*

$$\|V_\delta(t) - V(t)\| \leq \frac{\delta}{a(t)}. \qquad (6.53)$$

Thus,

$$\lim_{\delta \to 0} \|V_\delta(t_\delta) - V(t_\delta)\| = 0,$$

if

$$\lim_{\delta \to 0} \frac{\delta}{a(t_\delta)} = 0. \qquad (6.54)$$

Proof: From equation (6.12) we derive

$$F(V_\delta) - F(V) + a(t)(V_\delta - V) - (f_\delta - f) = 0. \qquad (6.55)$$

Multiplying (6.55) by $V_\delta - V$, using (6.2) and the inequality $\|f_\delta - f\| \leq \delta$, we obtain

$$a(t)\|V_\delta - V\|^2 \leq \delta\|V_\delta - V\|.$$

This implies (6.53). Lemma 6.3.1 is proved.   ∎

Let us estimate $\|u_\delta(t) - V_\delta(t)\| := g(t)$. We use the ideas of the proof of Theorem 6.2.1. Let $w := u_\delta - V_\delta(t)$. Then

$$\dot{w} = -\dot{V}_\delta - A_{a(t)}^{-1}[F(u_\delta) - F(V_\delta) + a(t)w],$$

which is an equation similar to (6.37). As in the proof of Theorem 6.2.1, we obtain

$$\dot{g} \leq -g(t) + \frac{c_0}{a(t)}g^2 + \frac{|\dot{a}|}{a(t)}c_1 \qquad (6.56)$$

and conclude that the estimate similar to (6.47) holds:

$$g(t) \leq \frac{a(t)}{\lambda}. \qquad (6.57)$$

Therefore, if t_δ is such that

$$\lim_{\delta \to 0} t_\delta = \infty, \quad \lim_{\delta \to 0} \frac{\delta}{a(t_\delta)} = 0, \tag{6.58}$$

then (6.52) holds and

$$\lim_{\delta \to 0} \| u_\delta(t_\delta) - y \| = 0. \tag{6.59}$$

We have proved the following result.

Theorem 6.3.2 *Assume that* $\| f_\delta - f \| \le \delta$, t_δ *satisfies (6.58), and* $u_\delta(t)$ *solves problem (6.34) with* f_δ *in place of* f. *Then (6.59) holds.*

Remark 6.3.3 The results of this chapter can be generalized: The condition

$$\text{Re}\langle F(u) - F(v), u - v \rangle \ge 0 \tag{6.60}$$

can be used in place of (6.2).

CHAPTER 7

DSM FOR GENERAL NONLINEAR OPERATOR EQUATIONS

In this chapter we construct a convergent DSM for solving any solvable non-linear equation $F(u) = 0$, under the assumptions $F \in C^2_{\text{loc}}$, $F'(y) \neq 0$, where $F(y) = 0$.

7.1 FORMULATION OF THE PROBLEM. THE RESULTS AND PROOFS

Consider the equation

$$F(u) = 0, \tag{7.1}$$

where $F : H \to H$ satisfies assumption (1.10), equation (7.1) has a solution y, possibly nonunique, and

$$\tilde{A} := F'(y) \neq 0. \tag{7.2}$$

Under these assumptions we want to construct a DSM method for solving equation (7.1).

Consider the following DSM method:

$$\dot{u} = -T^{-1}_{a(t)}[A^* F(u) + a(t)(u - z)], \quad u(0) = u_0, \tag{7.3}$$

Dynamical Systems Method and Applications: Theoretical Developments and Numerical Examples, First Edition. A. G. Ramm and N. S. Hoang.

where $u_0 \in H$, $z \in H$. We use the following notations:

$$A := F'(u), \quad T := A^*A, \quad T_a := T + aI, \quad \tilde{T} := \tilde{A}^*\tilde{A}, \quad \tilde{A} := F'(y). \quad (7.4)$$

If assumption (7.2) holds, then one can find an element $v \neq 0$ such that

$$\tilde{T}v \neq 0.$$

Thus, there is an element z such that

$$y - z = \tilde{T}v,$$

and one can choose z such that $\|v\| \ll 1$; that is, $\|v\| > 0$ is as small as one wishes:

$$y - z = \tilde{T}v, \quad \|v\| \ll 1. \quad (7.5)$$

Since y is unknown, we do not give an algorithm for finding such a z, but only prove its existence.

Let us denote

$$w := u(t) - y.$$

Since $F(y) = 0$, one has, using the Taylor's formula,

$$F(u) = F(u) - F(y) = Aw + \varepsilon, \quad \|\varepsilon\| \leq \frac{M_2}{2}\|w\|^2.$$

Here and below $M_j, j = 1, 2$, are constants from (1.10). Equation (7.3) can be written as

$$\dot{w} = -T_{a(t)}^{-1}\left[\left(A^*A + a(t)\right)w + A^*\varepsilon + a(t)\tilde{T}v\right].$$

Let

$$g(t) := \|w(t)\|.$$

Multiplying the above equation by w, we get

$$g\dot{g} \leq -g^2 + \frac{M_2 g^3}{4\sqrt{a(t)}} + a(t)\|T_{a(t)}^{-1}\tilde{T}\|\|v\|g, \quad (7.6)$$

where we have used estimate (7.4) and the inequality (2.13):

$$\|T_a^{-1}A^*\| \leq \frac{1}{2\sqrt{a(t)}}. \quad (7.7)$$

Since $g \geq 0$, we get

$$\dot{g} \leq -g + \frac{c_0 g^2}{\sqrt{a(t)}} + a(t)\|v\|\|T_{a(t)}^{-1}\tilde{T}\|, \quad c_0 := \frac{M_2}{4}. \quad (7.8)$$

Let us transform the last factor:

$$\|T_{a(t)}^{-1}\tilde{T}\| \leq \left\|\left(T_{a(t)}^{-1} - \tilde{T}_{a(t)}^{-1}\right)\tilde{T}\right\| + \|\tilde{T}_{a(t)}^{-1}\tilde{T}\|. \tag{7.9}$$

We have

$$\|\tilde{T}_{a(t)}^{-1}\tilde{T}\| \leq 1, \quad \|aT_a^{-1}\| \leq 1,$$

and

$$T_a^{-1} - \tilde{T}_a^{-1} = T_a^{-1}\left(\tilde{T} - T\right)\tilde{T}_a^{-1}. \tag{7.10}$$

Moreover,

$$\|\tilde{T} - T\| = \|\tilde{A}^*\tilde{A} - A^*A\| \leq 2M_1 M_2\|u - y\| = 2M_1 M_2 g. \tag{7.11}$$

Therefore, choosing $\|v\|$ so that

$$2M_1 M_2\|v\| \leq \frac{1}{2}, \tag{7.12}$$

one can rewrite (7.8) as

$$\dot{g} \leq -\frac{1}{2}g(t) + \frac{c_0 g^2(t)}{\sqrt{a(t)}} + a(t)\|v\|, \quad t \geq 0, \quad c_0 := \frac{M_2}{4}. \tag{7.13}$$

Let

$$\mu = \frac{\lambda}{\sqrt{a(t)}}. \tag{7.14}$$

Let us apply Theorem 5.1.1 to inequality (7.13). We have

$$\gamma(t) = \frac{1}{2}, \quad \alpha(t) = \frac{c_0}{\sqrt{a(t)}}, \quad \beta(t) = a(t)\|v\|.$$

Condition (5.2) is

$$\frac{c_0}{\sqrt{a(t)}} \leq \frac{\lambda}{2\sqrt{a(t)}}\left(\frac{1}{2} + \frac{\dot{a}}{2a}\right) = \frac{\lambda}{4\sqrt{a(t)}}\left(1 - \frac{|\dot{a}|}{a}\right). \tag{7.15}$$

Let us assume that

$$0 < a(t) \searrow 0, \quad \frac{|\dot{a}(t)|}{a} \leq \frac{1}{2}. \tag{7.16}$$

For example, one can take

$$a(t) = \frac{c_1}{(c_2 + t)^b},$$

where c_1, c_2, and b are positive constants, $2b \leq c_2$. Inequality (7.16) is satisfied for rapidly decaying $a(t)$ as well, for example,

$$a(t) = e^{-ct}, \quad c \leq \frac{1}{2}.$$

Inequality (7.15) holds if

$$\lambda \geq 8c_0. \tag{7.17}$$

Condition (5.3) is

$$a(t)\|v\| \leq \frac{\sqrt{a(t)}}{2\lambda}\left(1 - \frac{|\dot{a}|}{a}\right). \tag{7.18}$$

Because of inequality (7.16), this inequality holds if

$$4\lambda\sqrt{a(0)}\|v\| \leq 1. \tag{7.19}$$

If

$$\|v\| \leq \frac{1}{4\lambda\sqrt{a(0)}}, \tag{7.20}$$

then (7.19) holds.

Finally, condition (5.4) is

$$\frac{\lambda}{\sqrt{a(0)}}g(0) < 1. \tag{7.21}$$

This condition holds if

$$\|u_0 - y\| < \frac{\sqrt{a(0)}}{\lambda}. \tag{7.22}$$

Inequality (7.22) is valid for any initial approximation u_0, provided that λ is sufficiently large. For any fixed λ, inequality (7.20) holds if $\|v\|$ is sufficiently small. Therefore, if conditions (7.16), (7.17), (7.20), and (7.22) hold, then (5.6) implies

$$\|u(t) - y\| < \frac{\sqrt{a(t)}}{\lambda} \to 0 \text{ as } t \to \infty. \tag{7.23}$$

Let us formulate the result we have just proved.

Theorem 7.1.1 *Assume that equation (7.1) has a solution y, $F(y) = 0$, possibly nonunique, and the following conditions are satisfied: (1.10), (7.2), (7.5), (7.16), (7.17), (7.20), and (7.22). Then problem (7.3) has a unique solution $u(t)$, there exists $u(\infty)$, $u(\infty) = y$, and the rate of convergence of $u(t)$ to the solution y is given by (7.23).*

7.2   NOISY DATA

Assume now that the noisy data f_δ are given,

$$\|f_\delta - f\| \leq \delta,$$

equation (7.1) is of the form

$$F(u) = f, \tag{7.24}$$

and this equation has a solution y, $F(y) = f$.

Consider the DSM similar to (7.3):

$$\dot{u}_\delta = -T_{a(t)}^{-1}[A^*(F(u_\delta) - f_\delta) + a(t)(u_\delta - z)], \quad u_\delta(0) = u_0. \tag{7.25}$$

As in Section 7.1, denote

$$w = w_\delta := u_\delta - y$$

and

$$g := g_\delta := \|w_\delta(t)\|.$$

We want to prove that for a suitable stopping time t_δ, $\lim_{\delta \to 0} t_\delta = \infty$, we have the quantity $w_\delta(t_\delta) \to 0$ as $\delta \to 0$; in other words,

$$\lim_{\delta \to 0} \|u_\delta - y\| = 0, \quad u_\delta := u_\delta(t_\delta). \tag{7.26}$$

Let us argue as in Section 7.1. We have now

$$f_\delta = f + \eta_\delta, \quad \|\eta_\delta\| \le \delta.$$

Therefore, inequality (7.13) will be replaced by

$$\dot{g} \le -\frac{1}{2}g(t) + \frac{c_0 g^2(t)}{\sqrt{a(t)}} + a(t)\|v\| + \frac{\delta}{2\sqrt{a(t)}}, \quad c_0 := \frac{M_2}{4}, \tag{7.27}$$

where (7.12) holds. Let us choose

$$\mu(t) = \frac{\lambda}{\sqrt{a(t)}}$$

as in (7.14), and assume (7.16), (7.17), and (7.20). Then conditions (5.2) and (5.4) are satisfied. Condition (5.3) is satisfied if

$$a(t)\|v\| + \frac{\delta}{2\sqrt{a(t)}} \le \frac{\sqrt{a(t)}}{4\lambda}. \tag{7.28}$$

Let us choose v small enough, so that

$$a(t)\|v\| \le \frac{1}{2}\frac{\sqrt{a(t)}}{4\lambda},$$

that is,

$$4\lambda\sqrt{a(0)}\|v\| \le \frac{1}{2}, \tag{7.29}$$

and t_δ such that

$$\frac{\delta}{2\sqrt{a(t)}} \le \frac{1}{2}\frac{\sqrt{a(t)}}{4\lambda},$$

that is,

$$\frac{2\lambda\delta}{a(t)} \leq \frac{1}{2}, \qquad t \leq t_\delta. \tag{7.30}$$

Then (7.28) holds for $t \leq t_\delta$, and estimate (5.6) yields

$$\|u_\delta - y\| \leq \frac{\sqrt{a(t)}}{\lambda}, \quad t \leq t_\delta. \tag{7.31}$$

Therefore, if t_δ is chosen as the solution to the equation

$$a(t) = 4\lambda\delta, \tag{7.32}$$

then $u_\delta := u_\delta(t_\delta)$ satisfies the estimate

$$\|u_\delta - y\| \leq 2\sqrt{\frac{\delta}{\lambda}}. \tag{7.33}$$

We have proved the following result.

Theorem 7.2.1 *Assume equation (7.24) has a solution y, possibly nonunique, $\|f_\delta - f\| \leq \delta$, and the following conditions are satisfied: (1.10), (7.2), (7.3), (7.16), (7.17), (7.22), (7.29); also t_δ satisfies (7.32). Then problem (7.25) has a unique solution $u_\delta(t)$ on the interval $[0, t_\delta]$, and $u_\delta := u_\delta(t_\delta)$ satisfies inequality (7.33).*

Remark 7.2.2 Theorem 7.2.1 shows that, in principle, DSM (7.25) can be used for solving stably any solvable operator equation (7.24) for which $F \in C^2_{\text{loc}}$, that is, assumption (1.10) holds, and F satisfies assumption (7.2), where y is a solution to equation (7.24).

These are rather weak assumptions. However, as it was mentioned already, we do not give an algorithmic choice of z in (7.25). Under an additional assumption, e.g., if one assumes $y = \tilde{T}v$, $\|v\| \ll 1$, it is possible to take $z = 0$ and, using the arguments given in the proofs of Theorems 7.1.1 and 7.2.1, establish the conclusions of these theorems.

7.3 ITERATIVE SOLUTION

Let us prove the following result.

Theorem 7.3.1 *Under the assumptions of Theorem 7.1.1, the iterative process*

$$u_{n+1} = u_n - h_n T_{a_n}^{-1}[A^*(u_n)F(u_n) + a_n(u_n - z)], \qquad u_0 = u_0, \tag{7.34}$$

where $h_n > 0$ and $a_n > 0$ are suitably chosen, generates the sequence u_n converging to y.

Remark 7.3.2 The suitable choices of a_n and h_n are discussed in the proof of Theorem 7.3.1.

Lemma 7.3.3 *Let*

$$g_{n+1} \leq \gamma g_n + pg_n^2, \quad g_0 := m > 0, \quad 0 < \gamma < 1, \quad p > 0.$$

If

$$m < \frac{q - \gamma}{p}, \quad \gamma < q < 1,$$

then

$$\lim_{n \to \infty} g_n = 0, \quad and \quad g_n \leq g_0 q^n.$$

Proof: The estimate

$$g_1 \leq \gamma m + pm^2 \leq qm$$

holds if

$$m \leq \frac{q - \gamma}{p}, \quad \gamma < q < 1.$$

Assume that $g_n \leq g_0 q^n$. Then

$$g_{n+1} \leq \gamma g_0 q^n + p(g_0 q^n)^2 = g_0 q^n (\gamma + pg_0 q^n) < g_0 q^{n+1},$$

because

$$\gamma + pg_0 q^n < \gamma + pg_0 q \leq q.$$

Lemma 7.3.3 is proved. ∎

Proof of Theorem 7.3.1.
    Let

$$w_n := u_n - y, \quad g_n := \|w_n\|.$$

As in the proof of Theorem 7.1.1, we assume

$$2M_1 M_2 \|v\| \leq \frac{1}{2}$$

and rewrite (7.34) as

$$w_{n+1} = w_n - h_n T_{a_n}^{-1} \left[A^*(u_n)(F(u_n) - F(y)) + a_n w_n + a_n(y - z) \right],$$
$$w_0 = \|u_0 - y\|.$$

Using the Taylor formula

$$F(u_n) - F(y) = A(u_n)w_n + K(w_n), \quad \|K\| \leq \frac{M_2 g_n^2}{2},$$

the estimate

$$\|T_{a_n}^{-1} A^*(u_n)\| \leq \frac{1}{2\sqrt{a_n}},$$

and the formula

$$y - z = \tilde{T}v,$$

we get

$$w_{n+1} = (1 - h_n)w_n - h_n T_{a_n}^{-1} A^*(u_n) K(w_n) - h_n a_n T_{a_n}^{-1} \tilde{T}v. \qquad (7.35)$$

Taking into account that

$$\|\tilde{T}_a^{-1}\tilde{T}\| \le 1,$$

and

$$a\|T_a^{-1}\| \le 1 \quad \text{if} \quad a > 0,$$

we obtain

$$\|T_{a_n}^{-1}\tilde{T}v\| \le \|(T_{a_n}^{-1} - \tilde{T}_{a_n}^{-1})\tilde{T}\|\|v\| + \|v\|$$

and

$$\|(T_{a_n}^{-1} - \tilde{T}_{a_n}^{-1})\tilde{T}\| = \|T_{a_n}^{-1}(\tilde{T}_{a_n} - T_{a_n})\tilde{T}_{a_n}^{-1}\tilde{T}\| \le \frac{2M_1 M_2 g_n}{a_n} := \frac{c_1 g_n}{a_n}.$$

Let

$$c_0 := \frac{M_2}{4}.$$

Then we obtain from (7.35) the following inequality:

$$g_{n+1} \le (1 - h_n)g_n + \frac{c_0 h_n g_n^2}{\sqrt{a_n}} + c_1 h_n \|v\| g_n + h_n a_n \|v\|.$$

We have assumed in the proof of Theorem 7.3.1 that

$$c_1 \|v\| \le \frac{1}{2}.$$

Thus

$$g_{n+1} \le (1 - \frac{h_n}{2})g_n + \frac{c_0 h_n}{\sqrt{a_n}} g_n^2 + h_n a_n \|v\|.$$

Choose

$$a_n = 16 c_0^2 g_n^2.$$

Then $\frac{c_0 g_n}{\sqrt{a_n}} = \frac{1}{4}$, and

$$g_{n+1} \le (1 - \frac{h_n}{4})g_n + 16 c_0^2 h_n \|v\| g_n^2, \qquad g_0 = \|u_0 - y\| \le R, \qquad (7.36)$$

where $R > 0$ is defined in (1.10). Take $h_n = h \in (0,1)$ and choose $g_0 := m$ such that

$$m < \frac{q + \frac{h}{4} - 1}{16 c_0^2 h \|v\|},$$

where $q \in (0,1)$ and $q + h > 1$.

Then Lemma 7.3.3 implies

$$\|u_n - y\| \le g_0 q^n \to 0 \quad \text{as} \quad n \to \infty.$$

Theorem 7.3.1 is proved.   ∎

7.4 STABILITY OF THE ITERATIVE SOLUTION

Assume that the equation we want to solve is $F(u) = f$, f is unknown, f_δ is known, and $\|f_\delta - f\| \leq \delta$. Consider the iterative process similar to (7.34):

$$v_{n+1} = v_n - h_n T_{a_n}^{-1}\big[A^*(v_n)\big(F(v_n) - f_\delta\big) + a_n(v_n - z)\big], \quad v_0 = u_0, \quad (7.37)$$

Let

$$w_n := v_n - y, \quad \|w_n\| := \psi_n,$$

and choose $h_n = h$ independent of n, $h \in (0,1)$. Later we impose a positive lower bound on h; see formula (7.42) below. The inequality similar to (7.36) is

$$\psi_{n+1} \leq \gamma\psi_n + p\psi_n^2 + \frac{h\delta}{2\sqrt{a_n}}, \quad \psi_0 = \|u_0 - y\|, \quad (7.38)$$

where

$$\gamma := 1 - \frac{h}{4}, \quad p := 16c_0^2\|v\|, \quad a_n = 16c_0^2\psi_n^2. \quad (7.39)$$

We stop iterations in (7.38) when $n = n(\delta)$, where $n(\delta)$ is the largest integer for which the inequality

$$\frac{h\delta}{2\sqrt{a_n}} \leq \kappa\gamma\psi_n, \quad \kappa \in \left(0, \frac{1}{3}\right)$$

holds. One can use formula (7.39) for a_n and rewrite this inequality in the form

$$\frac{h\delta}{8c_0\kappa\gamma} \leq \psi_n^2, \quad \kappa \in \left(0, \frac{1}{3}\right). \quad (7.40)$$

If (7.40) holds, then (7.38) implies

$$\psi_{n+1} \leq (1 + \kappa)\gamma\psi_n + p\psi_n^2, \quad (1 + \kappa)\gamma < 1, \quad (7.41)$$

where the conditions

$$\gamma = 1 - \frac{h}{4}, \quad 0 < \kappa < \frac{1}{3}, \quad h \in \left(\frac{4\kappa}{1+\kappa}, 1\right) \quad (7.42)$$

imply that $(1 + \kappa)\gamma < 1$. If

$$\psi_0 < \frac{q - (1+\kappa)\gamma}{p}, \quad \text{where} \quad (1+\kappa)\gamma < q < 1, \quad \gamma = 1 - \frac{h}{4}, \quad (7.43)$$

then (7.41) and Lemma 7.3.3 imply

$$\psi_n \leq \psi_0 q^n, \quad \text{for} \quad n < n(\delta), \quad (1+\kappa)\gamma < q < 1, \quad 0 < \kappa < \frac{1}{3}, \quad (7.44)$$

where $n(\delta)$ is the largest integer for which inequality (7.40) holds. We have

$$\lim_{\delta \to 0} n(\delta) = \infty.$$

Thus

$$\lim_{\delta \to 0} \psi_{n(\delta)} = 0. \tag{7.45}$$

We have proved the following result.

Theorem 7.4.1 *Let the assumptions of Theorem 7.1.1 and conditions (7.40), (7.42), and (7.43) hold, and let ψ_n be defined by (7.37). Then relations (7.44) and (7.45) hold, so $\lim_{\delta \to 0} \|v_{n(\delta)} - y\| = 0$.*

CHAPTER 8

DSM FOR OPERATORS SATISFYING A SPECTRAL ASSUMPTION

In this chapter we introduce a spectral assumption (8.1) and obtain some results based on this assumption.

8.1 SPECTRAL ASSUMPTION

In this chapter we assume that the operator equation (7.24) is solvable, y is its solution, possibly nonunique, $F \in C^2_{\text{loc}}$, and the linear operator $A = F'(u)$ has the set $\{z : |\arg z - \pi| \le \varphi_0, \, 0 < |z| < r_0\}$ consisting of regular points of F', where $\varphi_0 > 0$ and $r_0 > 0$ are arbitrary small fixed numbers, and $u \in H$ is arbitrary. This is a *spectral assumption* on F. If this condition is satisfied then

$$\|(A + \varepsilon I)^{-1}\| \le \frac{c_0}{\varepsilon}, \quad 0 < \varepsilon \le r_0, \tag{8.1}$$

where $c_0 = \frac{1}{\sin \varphi_0}$.

Condition (8.1) is much weaker than the assumption of monotonicity of F. If F is monotone, then $A = F'(u) \ge 0$ and $\|A_\varepsilon^{-1}\| \le \frac{1}{\varepsilon}$ for all $\varepsilon > 0$, where $A_\varepsilon = A + \varepsilon I$. If $c_0 = 1$ and $r_0 = \infty$ in (8.1), then A is a generator of a semigroup of contractions [89].

Dynamical Systems Method and Applications: Theoretical Developments and Numerical Examples, First Edition. A. G. Ramm and N. S. Hoang.
Copyright © 2012 John Wiley & Sons, Inc.

It is known that if F is monotone, hemicontinuous, and $\varepsilon > 0$, then the equation

$$F(u) + \varepsilon u = f, \quad \varepsilon = const > 0, \tag{8.2}$$

is uniquely solvable for any $f \in H$. We want to prove a similar result assuming (8.1).

Our first result is the following.

Theorem 8.1.1 *Assume that F satisfies conditions (1.10) and (8.1). Then equation (8.2) has a solution.*

Proof: From our assumptions it follows that the problem

$$\dot{u} = -A_\varepsilon^{-1}[F(u) + \varepsilon u - f], \quad u(0) = u_0, \tag{8.3}$$

is locally solvable. We will prove the uniform bound

$$\sup_{t \geq 0} \|u(t)\| \leq c, \tag{8.4}$$

where $c = const > 0$ does not depend on t and the supremum is over all t for which $u(t)$, the local solution to (8.3), does exist. If (8.4) holds, then, as we have proved in Section 2.6, Lemma 2.6.1, the local solution is a global one, and it exists on $[0, \infty)$. We also prove that there exists $u(\infty)$ and that $u(\infty)$ solves (8.2).

Let us prove estimate (8.4). Consider the function

$$g(t) := \|F(u(t)) + \varepsilon u(t) - f\|. \tag{8.5}$$

We have

$$g\dot{g} = \text{Re}\langle [F'(u(t)) + \varepsilon]\dot{u}, F(u) + \varepsilon u - f \rangle = -g^2. \tag{8.6}$$

Thus

$$g(t) = g_0 e^{-t}, \quad g_0 := g(0). \tag{8.7}$$

From (8.1), (8.3), and (8.7) we get

$$\|\dot{u}\| \leq \frac{c_0 g_0 e^{-t}}{\varepsilon}. \tag{8.8}$$

This implies the existence of $u(\infty)$ and the estimates

$$\|u(t) - u(0)\| \leq \frac{c_0 g_0}{\varepsilon}, \quad \|u(t) - u(\infty)\| \leq \frac{c_0 g_0 e^{-t}}{\varepsilon}. \tag{8.9}$$

For any fixed u_0 and $\varepsilon > 0$, one can take $R > 0$ such that

$$\frac{c_0 g_0}{\varepsilon} \leq R,$$

so that the trajectory $u(t)$ stays in the ball $B(u_0, R)$. Passing to the limit $t \to \infty$ in (8.7) yields

$$F(u(\infty)) + \varepsilon u(\infty) - f = 0. \tag{8.10}$$

Therefore $u(\infty)$ solves equation (8.2), so this equation is solvable. Moreover, the DSM (8.3) converges to a solution $u(\infty)$ at an exponential rate; see (8.9).

Theorem 8.1.1 is proved. ∎

Remark 8.1.2 Estimate (8.1) follows from the spectral assumption because of the known estimate of the norm of the resolvent of a linear operator. Namely, if A is a linear operator and $(A - zI)^{-1}$ is its resolvent, where z is a complex number in the set of regular points of the operator A. A point z is called a regular point of A if the operator $A - zI$ has a bounded inverse defined on the whole space. Otherwise, z is called a point of spectrum σ of A.

The estimate we have mentioned is

$$\|(A - zI)^{-1}\| < \frac{1}{\rho(z, \sigma)}, \tag{8.11}$$

where $\rho(z, \sigma)$ is the distance (on the complex plane $\mathbb{C}$) from the point z to the set σ of points of spectrum of A. To check this, consider the function

$$(A - zI - \mu I)^{-1} = (A - zI)^{-1}[I - \mu(A - zI)^{-1}]^{-1}.$$

If

$$|\mu| \, \|(A - zI)^{-1}\| < 1,$$

then the above function is an analytic function of μ. Therefore, if μ is smaller than the distance from z to the nearest point of spectrum of A, then the point $z + \mu$ is a regular point of A.

Thus (8.11) follows.

Remark 8.1.3 We have proved in Theorem 8.1.1 that for $\varepsilon \in (0, r_0)$ equation (8.2) is solvable. This does not imply that the limiting equation

$$F(u) = f \tag{8.12}$$

is solvable.

For example, the equation $e^u + \varepsilon u = 0$ is solvable in $\mathbb{R}$ for any $\varepsilon > 0$, but the limiting equation $e^u = 0$ has no solutions.

Therefore it is of interest to study the following problem:

If the limiting equation (8.12) is solvable, then what are the conditions under which the solution to equation (8.2), or a more general equation

$$F(u_\varepsilon) + \varepsilon(u_\varepsilon - z) = f, \tag{8.13}$$

where $z \in H$ is some element, converges to a solution to (8.12) as $\varepsilon \to 0$?

This question is discussed in Chapter 9.

8.2   EXISTENCE OF A SOLUTION TO A NONLINEAR EQUATION

Let $D \subset \mathbb{R}^3$ be a bounded domain with Lipschitz boundary S, $k = const > 0$, and let $f : \mathbb{R} \to \mathbb{R}$ be a function such that

$$uf(u) \geq 0, \quad \text{for} \quad |u| \geq a \geq 0, \tag{8.14}$$

where a is an arbitrary nonnegative fixed number. We assume that f is continuous in the region $|u| \geq a$, and bounded and piecewise-continuous with at most finitely many discontinuity points u_j, such that $f(u_j + 0)$ and $f(u_j - 0)$ exist, in the region $|u| \leq a$.

Consider the problem

$$(-\Delta + k^2)u + f(u) = 0 \quad \text{in} \quad D, \tag{8.15}$$

$$u = 0 \quad \text{on} \quad S. \tag{8.16}$$

There is a large literature on problems of this type. Usually it is assumed that f does not grow too fast or f is monotone (see, e.g., [13] and references therein).

The novel point in this section is the absence of the monotonicity restrictions on f and of the growth restrictions on f as $|u| \to \infty$, except for the assumption (8.14).

This assumption allows an arbitrary behavior of f inside the region $|u| \leq a$, where $a \geq 0$ can be arbitrary large, and an arbitrarily rapid growth of f to $+\infty$ as $u \to +\infty$, or arbitrarily rapid decay of f to $-\infty$ as $u \to -\infty$.

Our result is:

Theorem 8.2.1 *Under the above assumptions, problem* (8.15)–(8.16) *has a solution* $u \in H^2(D) \cap \overset{\circ}{H}{}^1(D) := H_0^2(D)$.

Here $H^\ell(D)$ is the usual Sobolev space, and $\overset{\circ}{H}{}^1(D)$ is the closure of $C_0^\infty(D)$ in the norm $H^1(D)$. Uniqueness of the solution does not hold without extra assumptions.

The ideas of our proof are: First, we prove that if

$$\sup_{u \in \mathbb{R}} |f(u)| \leq \mu,$$

then a solution to (8.15)–(8.16) exists by the Schauder's fixed-point theorem (see Appendix A, Theorem A.8.1). Here μ is a constant. Secondly, we prove an a priori bound

$$\|u\|_\infty \leq a.$$

If this bound is proved, then problem (8.15)–(8.16) with the nonlinearity f replaced by

$$F(u) := \begin{cases} f(u), & |u| \leq a, \\ f(a), & u \geq a, \\ f(-a), & u \leq -a, \end{cases} \tag{8.17}$$

has a solution, and this solution solves the original problem (8.15)–(8.16). The bound

$$\|u\|_\infty \le a$$

is proved by using some integral inequalities. An alternative proof of this bound is also given. This proof is based on the maximum principle for elliptic equation (8.15).

We use some ideas from [103]. Our presentation follows [149].

Proof of Theorem 8.2.1. If $u \in L^\infty := L^\infty(D)$, then problem (8.15)–(8.16) is equivalent to the integral equation

$$u = -\int_D G(x,y)f(u(y))\,dy := T(u), \tag{8.18}$$

where

$$(-\Delta + k^2)G = -\delta(x-y) \quad \text{in} \quad D, \qquad G\,|_{x\in S} = 0. \tag{8.19}$$

By the maximum principle,

$$0 \le G(x,y) < g(x,y) := \frac{e^{-k|x-y|}}{4\pi|x-y|}, \qquad x,y \in D. \tag{8.20}$$

The map T is a continuous and compact map in the space $C(D) := X$, and

$$\|u\|_{C(D)} := \|u\| \le \mu \sup_x \int_D \frac{e^{-k|x-y|}}{4\pi|x-y|}\,dy \le \mu \int_{\mathbb{R}^3} \frac{e^{-k|y|}}{4\pi|y|}\,dy \le \frac{\mu}{k^2}. \tag{8.21}$$

This is an a priori estimate of any bounded solution to (8.15)–(8.16) for a bounded nonlinearity f such that

$$\sup_{u\in\mathbb{R}} |f(u)| \le \mu.$$

Thus, Schauder's fixed-point theorem yields the existence of a solution to (8.18), and consequently to problem (8.15)–(8.16), for bounded f. Indeed, if B is a closed ball of radius $\frac{\mu}{k^2}$, then the map T maps this ball into itself by (8.21), and since T is compact, the Schauder principle is applicable. Thus, the following lemma is proved.

Lemma 8.2.2 *If* $\sup_{u\in\mathbb{R}} |f(u)| \le \mu$, *then problems* (8.18) *and* (8.15)–(8.16) *have a solution in* $C(D)$, *and this solution satisfies estimate* (8.21).

Let us now prove an a priori bound for any solution $u \in C(D)$ of the problem (8.15)–(8.16) without assuming that $\sup_{u\in\mathbb{R}} |f(u)| < \infty$.

Let

$$u_+ := \max(u,0).$$

Multiply (8.15) by $(u-a)_+$, and integrate over D and then by parts to get

$$0 = \int_D [\nabla u \cdot \nabla(u-a)_+ + k^2 u(u-a)_+ + f(u)(u-a)_+]\,dx, \tag{8.22}$$

where the boundary integral vanishes because

$$(u - a)_+ = 0 \quad \text{on} \quad S \quad \text{for} \quad a \geq 0.$$

Each of the terms in (8.22) is nonnegative, the last one due to (8.14). Thus (8.22) implies

$$u \leq a. \tag{8.23}$$

Similarly, using (8.14) again, and multiplying (8.15) by $(-u - a)_+$, one gets

$$-a \leq u. \tag{8.24}$$

We have proved:

Lemma 8.2.3 *If* (8.14) *holds, then any solution* $u \in H_0^2(D)$ *to* (8.15)–(8.16) *satisfies the inequality*

$$|u(x)| \leq a. \tag{8.25}$$

Consider now equation (8.18) in $C(D)$ with an arbitrary continuous f satisfying (8.14). Any $u \in C(D)$ which solves (8.18) will also solve (8.15)–(8.16) and therefore satisfies (8.25) and belongs to $H_0^2(D)$. This u solves problem (8.15)–(8.16) with f replaced by F, defined in (8.17), and vice versa. Since F is a bounded nonlinearity, equation (8.18) and problem (8.15)–(8.16) (with f replaced by F) have a solution by Lemma 8.2.2.

Theorem 8.2.1 is proved. ∎

Let us sketch an alternative derivation of the inequality (8.25) using the maximum principle. Let us derive (8.23). The derivation of (8.24) is similar.

Assume that (8.23) fails. Then $u > a$ at some point in D. Therefore at a point y, at which u attains its maximum value, one has $u(y) \geq u(x)$ for all $x \in D$ and $u(y) > a$. The function u attains its maximum value, which is positive, at some point in D, because u is continuous, vanishes at the boundary of D, and is positive at some point of D by the assumption $u > a$. At the point y, where the function u attains its maximum, one has $-\Delta u \geq 0$ and $k^2 u(y) > 0$. Moreover, $f(u(y)) > 0$ by the assumption (8.14), since $u(y) > a$. Therefore the left-hand side of equation (8.15) is positive, while its left-hand side is zero. Thus, we have a contradiction, and estimate (8.23) is proved. Similarly, one proves estimate (8.24). Thus, (8.25) is proved. ∎

CHAPTER 9

DSM IN BANACH SPACES

In this chapter we generalize some of our results to operator equations in Banach spaces.

9.1 WELL-POSED PROBLEMS

Consider the equation

$$F(u) = f, \tag{9.1}$$

where $F : X \to Y$ is a map from a Banach space X into a Banach space Y.

Consider first the case when both assumptions (1.9) and (1.10) hold:

$$\sup_{u \in B(u_0, R)} \|[F'(u)]^{-1}\| \leq m(R), \tag{9.2}$$

$$\sup_{u \in B(u_0, R)} \|F^{(j)}(u)\| \leq M_j(R), \quad 0 \leq j \leq 2. \tag{9.3}$$

In Section 9.3 we use assumption (9.1.3) with $j \leq 3$.

Dynamical Systems Method and Applications: Theoretical Developments and Numerical Examples, First Edition. A. G. Ramm and N. S. Hoang.
Copyright © 2012 John Wiley & Sons, Inc.

We construct the Newton-type DSM:

$$\dot{u} = -[F'(u)]^{-1}[F(u) - f], \quad u(0) = u_0, \tag{9.4}$$

which makes sense due to (9.2). As in Sections 3.1–3.2, the right-hand side of (9.4) satisfies a Lipschitz condition due to assumptions (9.2)–(9.3). Therefore problem (9.4) has a unique local solution. As in Section 2.6, Lemma 2.6.1, this local solution is a global one, provided that bound (2.161) is established with the constant c independent of time. Denote

$$g(t) := < F(u(t)) - f, h >,$$

where $< w, h >$ is the value of a linear functional $h \in Y^*$ at the element $F(u(t)) - f$.

We have, using equation (9.4),

$$\dot{g} = < F'(u)\dot{u}, h >= -g. \tag{9.5}$$

Therefore

$$g(t) = g(0)e^{-t}.$$

Taking the supremum over $h \in Y^*$, $\|h\| = 1$, one gets

$$G(t) = G(0)e^{-t}, \quad G(t) := \|F(u(t)) - f\|. \tag{9.6}$$

Using equation (9.4), estimate (9.2), and formula (9.6), one obtains

$$\|\dot{u}\| \le m(R)G(0)e^{-t}. \tag{9.7}$$

Therefore

$$\|u(t) - u(0)\| \le m(R)G(0). \tag{9.8}$$

We assume that

$$m(R)\|F(u_0) - f\| \le R. \tag{9.9}$$

This assumption ensures that the trajectory $u(t)$ stays inside the ball $B(u_0, R)$ for all $t \ge 0$, so that the local solution to problem (9.4) is the global one.

Moreover, (9.7) and (9.8) imply the existence of $u(\infty)$ and the estimate

$$\|u(t) - u(\infty)\| \le m(R)\|F(u_0) - f\|e^{-t}. \tag{9.10}$$

This estimate shows exponential rate of convergence of $u(t)$ to $u(\infty)$. Finally, passing to the limit $t \to \infty$ in (9.6), one checks that $u(\infty)$ solves equation (9.1). Let us summarize the result.

Theorem 9.1.1 *Assume that $F : X \to Y$ is a map from a Banach space X into a Banach space Y, and (9.2), (9.3), and (9.9) hold. Then equation (9.1) has a solution, problem (9.4) has a unique global solution $u(t)$, there exists $u(\infty)$, $u(\infty)$ solves equation (9.1), and estimates (9.8) and (9.10) hold.*

Remark 9.1.2 Theorem 9.1.1 gives a sufficient condition (condition(9.9)) for the existence of a solution to equation (9.1).

If one knows a priori that equation (9.1) has a solution, then condition (9.9) is always satisfied if one chooses the initial approximation u_0 sufficiently close to this solution, because then the quantity $\|F(u_0) - f\|$ can be made as small as one wishes.

The results in Theorem 9.1.1 are similar to the results obtained in Section 3.2 for equation (9.1) with the operator F which acted in a Hilbert space.

9.2   ILL-POSED PROBLEMS

Consider equation (9.1) assuming that condition (9.3) holds, but condition (9.2) does not hold, and that equation (9.1) has a solution y, possibly non-unique.

Equation (9.4) cannot be used now because the operator $[F'(u)]^{-1}$ is not defined. Therefore we need some additional assumptions on F to treat ill-posed problems.

Let us denote

$$A := F'(u),$$

and make the spectral assumption from Section 8.1:

$$\|A_\varepsilon^{-1}\| \leq \frac{c_0}{\varepsilon}, \quad 0 < \varepsilon < r_0, \tag{9.11}$$

where c_0 and r_0 are some positive constants.

Theorem 8.1.1 from Section 8.1 claims that assumptions (9.11) and (9.3) imply existence of a solution to equation (8.2). The proof of Theorem 8.1.1 is valid, after suitable modifications, in the case when $F : X \to X$ is an operator from a Banach space X into X. Let us point out these modifications.

Definition (8.5) should be replaced by

$$g(t) :=< F(u(t)) + \varepsilon u(t) - f, h >, \tag{9.12}$$

where $h \in X^*$ is arbitrary, $\|h\| = 1$. We have

$$G(t) := \|F(u(t)) + \varepsilon u(t) - f\| = \sup_{\|h\|=1, \ h \in X^*} |g(t)|. \tag{9.13}$$

As in Section 9.1, we prove that the problem

$$\dot{u} = -A_\varepsilon^{-1}[F(u) + \varepsilon u(t) - f], \quad u(0) = u_0, \tag{9.14}$$

has a unique local solution $u(t)$ and

$$g(t) \leq g(0)e^{-t}.$$

Formula (9.13) yields

$$G(t) \leq G(0)e^{-t}, \quad G(0) = \|F(u_0) + \varepsilon u_0 - f\|. \tag{9.15}$$

Equation (9.14) and estimates (9.11) and (9.15) imply

$$\|\dot{u}\| \leq \frac{c_0 G(0)}{\varepsilon} e^{-t}. \tag{9.16}$$

For any fixed u_0 the inequality

$$\frac{c_0 G(0)}{\varepsilon} \leq R \tag{9.17}$$

holds if $\varepsilon > 0$ is sufficiently large or $G(0)$ is sufficiently small.

If (9.17) holds, then

$$\|u(t) - u(0)\| \leq \frac{c_0 G(0)}{\varepsilon} \leq R, \tag{9.18}$$

so that $u(t) \in B(u_0, R)$ for $t \geq 0$. This implies, as in Section 2.6 (see Lemma 2.6.1), that the local solution $u(t)$ to (9.14) is the global one. Moreover, (9.14) implies the existence of $u(\infty)$ and the estimate

$$\|u(t) - u(\infty)\| \leq \frac{c_0 G(0)}{\varepsilon} e^{-t} \leq R e^{-t}. \tag{9.19}$$

Passing to the limit $t \to \infty$ in (9.15), we see that $u(\infty)$ solves the equation

$$F(u(\infty)) + \varepsilon u(\infty) = f. \tag{9.20}$$

Let us formulate the result we have just proved.

Theorem 9.2.1 *Assume that $F : X \to X$ and that (9.11), (9.3), and (9.17) hold. Then problem (9.14) has a unique global solution $u(t)$, there exists $u(\infty)$, $u(\infty)$ solves equation (9.20), and estimates (9.18), (9.19) hold.*

The problem is:

Under what assumptions can we establish that $u(t) := u_\varepsilon(t)$ converges, as $\varepsilon \to 0$, to a solution of the limiting equation?

This question is discussed in the next section.

9.3   SINGULAR PERTURBATION PROBLEM

In Section 9.2 we have proved that equation

$$F(u) + \epsilon u = f, \quad 0 < \epsilon < r_0, \tag{9.21}$$

has a solution $u = u_\epsilon$. This does not imply, in general, the existence of the solution to the limiting equation

$$F(y) = f. \tag{9.22}$$

A simple example was given in Section 8.1 below formula (8.12).

Therefore we have to assume that equation (9.22) has a solution.

If equation (9.22) is solvable, then our aim is to give sufficient conditions for a solution $w = w_\epsilon$ to the equation

$$F(w) + \epsilon(w - z) = f \tag{9.23}$$

to converge, as $\epsilon \to 0$, to a solution of the limiting equation (9.22).

Let us formulate our result.

Theorem 9.3.1 *Assume that equation (9.22) is solvable, conditions (9.11) and (9.3) with $j \leq 3$ hold, and z in (9.23) is chosen so that*

$$y - z = \tilde{A}v, \quad \|v\| < \frac{1}{2M_2 c_0(1 + c_0)},$$

where c_0 is the constant from (9.11), and $\tilde{A} = F'(y) \neq 0$. Then

$$\lim_{\epsilon \to 0} \|w_\epsilon - y\| = 0, \tag{9.24}$$

where y is a solution to (9.22).

Proof: Using Taylor's formula, let us write equation (9.23) as

$$0 = F(w) - F(y) + \epsilon(w - y) + \epsilon(y - z) = \tilde{A}_\epsilon \psi + \eta + \epsilon \tilde{A}v, \tag{9.25}$$

where $\tilde{A}_\epsilon := \tilde{A} + \epsilon I$ and

$$\psi = w - y, \quad \|\eta\| \leq \frac{M_2 \|\psi\|^2}{2}, \quad \eta = \eta(w). \tag{9.26}$$

Equation (9.25) can be written as

$$\psi = T\psi, \tag{9.27}$$

where

$$T\psi := -\tilde{A}_\epsilon^{-1}\eta - \epsilon \tilde{A}_\epsilon^{-1}\tilde{A}v. \tag{9.28}$$

We *claim* that the operator T maps a ball

$$B_R := \{\psi : \|\psi\| \leq R\}$$

into itself and is a contraction mapping in this ball. If this claim is established, then the contraction mapping principle yields existence and uniqueness of the solution to (9.27) in the ball B_R. We choose $R = R(\epsilon)$ so that

$$\lim_{\epsilon \to 0} R(\epsilon) = 0.$$

Thus

$$\|w_\epsilon - y\| \leq R(\epsilon) \to 0 \quad \text{as} \quad \epsilon \to 0, \tag{9.29}$$

and Theorem 9.3.1 is proved.

Let us verify the *claim.* Note that

$$\|\tilde{A}_\epsilon^{-1}\tilde{A}\| = \|\tilde{A}_\epsilon^{-1}(\tilde{A} + \epsilon - \epsilon)\| \leq 1 + c_0.$$

Thus we have $TB_R \subseteq B_R$, provided that

$$\|T\psi\| \leq \frac{c_0}{\epsilon}\frac{M_2}{2}R^2 + \epsilon(1 + c_0)\|v\| \leq R. \tag{9.30}$$

This inequality holds if

$$\frac{\epsilon}{c_0 M_2}(1 - \rho) \leq R \leq \frac{\epsilon}{c_0 M_2}(1 + \rho), \tag{9.31}$$

where

$$\rho = \sqrt{1 - 2c_0(1 + c_0)M_2\|v\|}, \qquad \|v\| < \frac{1}{2c_0(1 + c_0)M_2}. \tag{9.32}$$

Let us now verify that T is a contraction mapping in B_R, where

$$R = \frac{\epsilon}{c_0 M_2}(1 - \rho). \tag{9.33}$$

Let $p, q \in B_R$. Then

$$Tp - Tq = -\tilde{A}_\epsilon^{-1}[\eta(p) - \eta(q)], \tag{9.34}$$

where

$$\eta(p) = \int_0^1 (1 - s)F''(y + sp)pp\, ds \tag{9.35}$$

is the remainder in the Taylor formula

$$F(p) - F(y) = \tilde{A}(p - y) + \eta(p). \tag{9.36}$$

If $p, q \in B_R$, then

$$\begin{aligned}
\|\eta(p) - \eta(q)\| &= \left\| \int_0^1 (1 - s)\big[F''(y + sp)pp - F''(y + sq)qq\big]\, ds \right\| \\
&\leq \int_0^1 (1 - s)\bigg(\big\|\big[F''(y + sp) - F''(y + sq)\big]pp\big\| \\
&\qquad + \|F''(y + sq)(pp - qq)\| \bigg)\, ds \\
&\leq \int_0^1 (1 - s)(M_3 sR^2 + 2M_2 R)\|p - q\|\, ds.
\end{aligned} \tag{9.37}$$

Thus,

$$\|\eta(p) - \eta(q)\| \leq \left(M_2 R + \frac{M_3 R^2}{6} \right)\|p - q\|. \tag{9.38}$$

From (9.34), (9.38), and (9.11) we get

$$\|Tp - Tq\| \leq c_0 \frac{M_2 R}{\epsilon} \left(1 + \frac{M_3 R}{6 M_2}\right) \|p - q\|. \tag{9.39}$$

This and (9.33) imply

$$\|Tp - Tq\| \leq (1 - \rho + O(\epsilon)) \|p - q\|, \quad \epsilon \to 0, \quad p, q \in B_R. \tag{9.40}$$

Therefore, for sufficiently small ϵ, the mapping T is a contraction mapping in B_R, where R is given by (9.33).

Theorem 9.3.1 is proved. ∎

Remark 9.3.2 If $\tilde{A} := F(y) \neq 0$, then one can always choose z so that

$$y - z = \tilde{A}v, \quad \|v\| < b, \tag{9.41}$$

where $b > 0$ is an arbitrary small fixed number. Indeed, if $\tilde{A} \neq 0$, then there exists an element $p \neq 0$ such that $\tilde{A}p \neq 0$. Choose z such that $y - z = \lambda \tilde{A}p$, where $\lambda = const$ and $|\lambda| \|p\| < b$. Then $v = \lambda p$, $\|v\| < b$.

Remark 9.3.3 If one can choose $z = 0$ in Theorem 9.3.1, that is, if $y = \tilde{A}v$ and $\|v\|$ is sufficiently small, then the solution to equation (9.21) tends to y as $\epsilon \to 0$.

In this case the DSM (9.14) yields a solution to equation (9.22) by the formula

$$y = \lim_{\epsilon \to 0} \lim_{t \to \infty} u_\epsilon(t), \tag{9.42}$$

where $u_\epsilon(t)$ is the solution to problem (9.14).

CHAPTER 10

DSM AND NEWTON-TYPE METHODS WITHOUT INVERSION OF THE DERIVATIVE

In this chapter we construct the DSM so that there is no need to invert $F'(u)$.

10.1 WELL-POSED PROBLEMS

A Newton-type DSM requires calculation of $[F'(u)]^{-1}$; see (3.28).

This is a difficult and time-consuming step. Can one avoid such a step? In this section we assume that conditions (9.2) and (9.3) hold, and equation

$$F(u) = f \tag{10.1}$$

has a solution y.

Consider the following DSM:

$$\dot{u} = -Q[F(u) - f], \qquad u(0) = u_0, \qquad t \geq 0, \tag{10.2}$$

$$\dot{Q} = -TQ + A^*, \qquad Q(0) = Q_0 \qquad t \geq 0, \tag{10.3}$$

where

$$A := F'(u), \quad T := A^*A, \tag{10.4}$$

Dynamical Systems Method and Applications: Theoretical Developments and Numerical Examples, First Edition. A. G. Ramm and N. S. Hoang.
Copyright © 2012 John Wiley & Sons, Inc.

$u_0 \in H$, and $Q(t)$ is an operator-valued function.

Our result is the following theorem.

Theorem 10.1.1 *Assume (9.2), (9.3). Suppose that equation (10.1) has a solution y, u_0 is sufficiently close to y, and Q_0 is sufficiently close to $\tilde{A}^{-1}$, where $\tilde{A} := F'(y)$. Then problem (10.2)–(10.3) has a unique global solution, there exists $u(\infty)$, $u(\infty) = y$, that is,*

$$\lim_{t \to \infty} \|u(t) - y\| = 0, \tag{10.5}$$

and

$$\lim_{t \to \infty} \|Q(t) - \tilde{A}^{-1}\| = 0. \tag{10.6}$$

Proof of Theorem 10.1.1. From our assumptions the existence and uniqueness of the local solution to problem (10.2)–(10.3) follows. If we derive a uniform with respect to t bound on the norm $\|u(t)\| + \|Q(t)\| \leq c$, then, as in Section 2.6, we prove that the local solution to (10.2)–(10.3) is a global one. First, let us estimate the norm $\|Q(t)\|$ uniformly with respect to t. We use Theorem 5.2.1 and estimate (5.25). Since A is invertible, there is a constant $\epsilon = const > 0$ such that

$$\langle Th, h \rangle \geq \epsilon \|h\|^2. \tag{10.7}$$

Therefore estimate (5.25) yields

$$\|Q(t)\| \leq e^{-\epsilon t} \left(\|Q_0\| + \int_0^t M_1 e^{\epsilon s} ds \right) \leq \|Q_0\| + \frac{M_1}{\epsilon} := c_1. \tag{10.8}$$

Thus, $\|Q(t)\|$ is bounded uniformly with respect to t.

Let us now estimate $\|u(t)\|$. We denote

$$w := u(t) - y, \quad \|w(t)\| := g(t),$$

and use Taylor's formula:

$$F(u) - f = F(u) - F(y) = \tilde{A}w + K(w), \tag{10.9}$$

where

$$\|K(w)\| \leq \frac{M_2}{2} \|w\|^2,$$

and M_2 is the constant from (9.3). Thus equation (10.2) can be written as

$$\dot{w} = -Q\tilde{A}w - QK(w). \tag{10.10}$$

Define

$$\Lambda := I - Q\tilde{A}. \tag{10.11}$$

Then

$$\dot{w} = -w + \Lambda w - QK(w). \tag{10.12}$$

Multiply this equation by w and get

$$g\dot{g} \le -g^2 + \lambda g^2 + \frac{c_1 M_2}{2} g^3$$

and

$$\dot{g} \le -\gamma g + c_2 g^2, \quad \gamma := 1 - \lambda, \quad \gamma \in (0,1). \tag{10.13}$$

Here we have used the estimate

$$\langle \Lambda w, w \rangle \le \lambda \|w\|^2, \quad \lambda \in (0,1), \tag{10.14}$$

which will be proved later.

From (10.13) one derives

$$g(t) \le re^{-\gamma t}, \quad r := \frac{g(0)}{1 - g(0)c_2}, \tag{10.15}$$

and we assume that

$$g(0)c_2 < 1. \tag{10.16}$$

Assumption (10.16) is always justified if $g(0)$ is sufficiently small, that is, if $\|u_0 - y\|$ is sufficiently small.

Let us verify inequality (10.14). Using definition (10.11) and differentiating it with respect to time, one gets

$$\dot{\Lambda} = -\dot{Q}\tilde{A}. \tag{10.17}$$

This equation, (10.11), and (10.3) yield

$$\dot{\Lambda} = (TQ - A^*)\tilde{A} = T(I - \Lambda) - A^*\tilde{A} = -T\Lambda + A^*(A - \tilde{A}). \tag{10.18}$$

We have

$$\|A^*(A - \tilde{A})\| \le M_1 M_2 g(t) \le M_1 M_2 re^{-\gamma t}, \tag{10.19}$$

where r is given by (10.15).

Let us apply estimate (5.25) and inequality (10.7) to equation (10.18). This yields

$$\|\Lambda(t)\| \le e^{-\epsilon t}\left(\|\Lambda(0)\| + \int_0^t M_1 M_2 re^{-\gamma s}e^{\epsilon s}\, ds \right). \tag{10.20}$$

Thus

$$\|\Lambda(t)\| \le \|\Lambda(0)\| + cg(0) := \lambda, \tag{10.21}$$

where

$$c := \frac{M_1 M_2}{1 - g(0)c_2} \sup_{t \ge 0} \frac{e^{-\gamma t} - e^{-\epsilon t}}{\epsilon - \gamma}. \tag{10.22}$$

Estimate (10.21) shows that $\lambda \in (0,1)$ if $\|\Lambda(0)\|$ and $\|u_0 - y\|$ are sufficiently small.

To complete the proof of Theorem 10.1.1, we need to verify formula (10.16). Estimate (10.20) implies

$$\lim_{t\to\infty} \|\Lambda(t)\| \le O(e^{-\min(\epsilon,\gamma)t}) \to 0 \quad \text{as} \quad t \to \infty. \tag{10.23}$$

This formula and (10.11) imply

$$\lim_{t\to\infty} \|Q(t)\tilde{A} - I\| = 0,$$

and since $\tilde{A}^{-1}$ is a bounded operator, we get

$$\lim_{t\to\infty} \|Q(t) - \tilde{A}^{-1}\| = 0. \tag{10.24}$$

Theorem 10.1.1 is proved.   ∎

10.2 ILL-POSED PROBLEMS

In this section we assume (9.3), but not (9.2). Let

$$F'(u) := A, \quad T := A^*A, \quad \tilde{T} = \tilde{A}^*\tilde{A}, \quad \tilde{A} := F'(y), \tag{10.25}$$

where y is a solution to the equation

$$F(u) = f. \tag{10.26}$$

Consider the following DSM:

$$\dot{u} = -Q[A^*(F(u) - f) + \epsilon(t)(u - z)], \quad u(0) = u_0, \tag{10.27}$$

$$\dot{Q} = -T_{\epsilon(t)}Q + I, \quad Q(0) = Q_0, \tag{10.28}$$

where z is an element which we choose later.

Assume that

$$0 < \epsilon(t) \searrow 0, \quad \frac{|\dot{\epsilon}|}{\epsilon} \le \frac{\gamma}{2}, \quad \frac{|\dot{\epsilon}(t)|}{\epsilon^2(t)} \le b, \quad \lim_{t\to\infty} \frac{|\dot{\epsilon}(t)|}{\epsilon^2(t)} = 0,$$

$$\int_0^\infty \epsilon(s)\, ds = \infty, \tag{10.29}$$

where γ is defined in formula (10.37) below, and

$$\|w(t)\| = g(t), \quad b = const > 0.$$

Let

$$w := u(t) - y,$$

and assume that z is chosen so that

$$y - z = \tilde{T}v, \quad \|v\| \le \frac{\lambda\gamma}{4c_2\epsilon^2(0)}; \tag{10.30}$$

see inequality (10.44) below.

Define

$$\Lambda := I - Q\tilde{T}_{\epsilon(t)}. \tag{10.31}$$

From equation (10.28) and estimate (5.25) we get

$$\|Q(t)\| \le e^{-\int_0^t \epsilon(s)ds}\left(\|Q(0)\| + \int_0^t e^{-\int_0^s \epsilon(p)dp}ds\right)$$

$$\le \|Q(0)\| + \frac{1}{\epsilon(t)} \le \frac{c_1}{\epsilon(t)}. \tag{10.32}$$

Let us write equation (10.27) as

$$\dot{w} = -w + \Lambda w + Q(\tilde{A}^* - A^*)\tilde{A}w + Q(\tilde{A}^* - A^*)K - \epsilon Q\tilde{T}v, \tag{10.33}$$

where v is defined (10.30) and K is defined in (10.9). We have

$$\|\tilde{A}\| \le M_1, \quad \|\tilde{A}^* - A^*\| \le M_2 g, \quad \|Q\| \le \frac{c_1}{\epsilon(t)}, \quad \|w\| := g,$$

so

$$\|Q(\tilde{A}^* - A^*)\tilde{A}w\| \le \frac{c_1}{\epsilon(t)}M_1 M_2 g^2(t)$$

and

$$\|Q(\tilde{A}^* - A^*)K\| \le \frac{c_1}{\epsilon(t)}2M_1\frac{M_2}{2}g^2(t) = \frac{c_1 M_1 M_2 g^2}{\epsilon(t)}. \tag{10.34}$$

We will prove below that

$$\|Q\tilde{T}\| \le c_2 \tag{10.35}$$

and

$$\|\Lambda\| \le \lambda_0 < 1, \tag{10.36}$$

where λ_0 is a positive constant independent of t.

We define constant γ by the formula

$$\gamma := 1 - \lambda_0, \quad \gamma \in (0, 1). \tag{10.37}$$

Multiply (10.33) by w and use estimates (10.34)–(10.37) to get

$$\dot{g}g \le -\gamma g^2 + \frac{c_0}{\epsilon(t)}g^3 + c_2\epsilon(t)\|v\|g, \tag{10.38}$$

where

$$c_0 := 2c_1 M_1 M_2 \tag{10.39}$$

and c_2 is the constant from (10.35).

Since $g(t) \ge 0$, we obtain from (10.38) the inequality

$$\dot{g} \le -\gamma g(t) + \frac{c_0}{\epsilon(t)}g^2 + c_2\epsilon(t)\|v\|, \quad \gamma \in (0, 1). \tag{10.40}$$

We apply to this differential inequality Theorem 5.1.1. Let us check conditions (5.2)–(5.4), which are the assumptions in this theorem. We take

$$\mu = \frac{\lambda}{\epsilon(t)}, \quad \text{where} \quad \lambda = const > 0.$$

Condition (5.2) for inequality (10.40) takes the form

$$\frac{c_0}{\epsilon(t)} \leq \frac{\lambda}{2\epsilon(t)} \left(\gamma - \frac{|\dot{\epsilon}(t)|}{\epsilon(t)} \right). \tag{10.41}$$

Using (10.29), one concludes that this inequality is satisfied if

$$\frac{4c_0}{\gamma} \leq \lambda. \tag{10.42}$$

Condition (5.3) for inequality (10.40) is

$$c_0 \epsilon(t) \|v\| \leq \frac{\lambda}{2\epsilon(t)} \left(\gamma - \frac{|\dot{\epsilon}(t)|}{\epsilon(t)} \right). \tag{10.43}$$

This inequality is satisfied if

$$\frac{4c_2}{\lambda\gamma} \epsilon^2(0) \|v\| \leq 1. \tag{10.44}$$

Inequality (10.44) is satisfied if $\|v\|$ is sufficiently small. Finally, condition (5.4) is

$$\|u_0 - y\| \frac{\lambda}{\epsilon(0)} < 1. \tag{10.45}$$

Inequality (10.45) holds if $\|u_0 - y\|$ is sufficiently small.
  Thus, if

$$\lambda \geq \frac{4c_0}{\gamma},$$

then condition (10.42) holds; if

$$\|v\| \leq \frac{\lambda\gamma}{4c_2\epsilon^2(0)},$$

then condition (10.44) holds; and, finally, if

$$\|u_0 - y\| < \frac{\epsilon(0)}{\lambda},$$

then condition (10.45) holds. If these conditions hold, then the inequality (5.6) yields

$$\|u(t) - y\| < \frac{\epsilon(t)}{\lambda}. \tag{10.46}$$

This estimate yields a uniform bound on the norm $\|u(t)\|$ and therefore implies the global existence of the solution to problem (10.27)–(10.28), the existence of $u(\infty)$, and the relation $u(\infty) = y$, where y solves equation (10.26).

To complete the argument, we have to prove estimates (10.35) and (10.36). Estimate (10.35) follows from (10.36) and from the definition (10.31). Let us prove (10.36). Using definition (10.31), we derive

$$\dot{\Lambda} = -\dot{Q}\tilde{T}_{\epsilon(t)} - Q\dot{\epsilon}. \tag{10.47}$$

Using equations (10.28) and (10.47), we obtain

$$\dot{\Lambda} = -T_{\epsilon(t)}\Lambda + T_{\epsilon(t)} - \tilde{T}_{\epsilon(t)} - Q\dot{\epsilon}. \tag{10.48}$$

This equation and Theorem 5.2.1 yield

$$\begin{aligned}
\|\Lambda(t)\| \leq & e^{-\int_0^t \epsilon(s)ds}\left[\|\Lambda(0)\| + \int_0^t e^{-\int_0^s \epsilon(p)dp}\|T_{\epsilon(t)} - \tilde{T}_{\epsilon(t)}\|\,ds\right] \\
& + e^{-\int_0^t \epsilon(s)ds}\int_0^t e^{-\int_0^s \epsilon(p)dp}\|Q(s)\|\,|\dot{\epsilon}(s)|\,ds.
\end{aligned} \tag{10.49}$$

Using inequality (10.46), we derive the following estimate:

$$\begin{aligned}
\|T_{\epsilon(t)} - \tilde{T}_{\epsilon(t)}\| = \|A^*A - \tilde{A}^*\tilde{A}\| &\leq 2M_1 M_2\|u(t) - y\| \\
&\leq \frac{2M_1 M_2 \epsilon(t)}{\lambda}.
\end{aligned} \tag{10.50}$$

Using (10.32), we get

$$\|Q(s)\|\,|\dot{\epsilon}(s)| \leq c_1 \frac{|\dot{\epsilon}(s)|}{\epsilon(s)}. \tag{10.51}$$

From (10.49)–(10.51) we obtain

$$\|\Lambda(t)\| \leq \|\Lambda(0)\| + \frac{2M_1 M_2}{\lambda} + b, \tag{10.52}$$

where the following estimate

$$e^{-\int_0^t \epsilon(s)ds}\int_0^t \epsilon(s)e^{-\int_0^s \epsilon(p)dp}ds = 1 - e^{-\int_0^t \epsilon(s)ds} \leq 1 \tag{10.53}$$

was used.

Assume that

$$\|\Lambda(0)\| + \frac{2M_1 M_2}{\lambda} + b \leq \lambda_0 < 1. \tag{10.54}$$

Then condition (10.36) holds and, therefore, inequality (10.35) holds. Condition (10.54) holds if

$$\|I - Q(0)\tilde{T}_{\epsilon(0)}\| < 1.$$

Indeed, one can choose $\epsilon(t)$ such that $b > 0$ is as small as one wishes. For example, if

$$\epsilon(t) = \frac{C}{(a+t)^q},$$

then

$$\frac{|\dot{\epsilon}|}{\epsilon^2} = \frac{q}{C(a+t)^{1-q}} \leq \frac{q}{Ca^{1-q}}, \quad 0 < q < 1, \tag{10.55}$$

and this number can be made as small as one wishes by taking C sufficiently large. The term $\frac{2M_1 M_2}{\lambda}$ can be made as small as one wishes by choosing λ sufficiently large.

Let us formulate the result we have proved.

Theorem 10.2.1 *Assume that equation (10.26) has a solution y, possibly nonunique, that $\epsilon(t)$ satisfies (10.29), and assume that $\|u_0 - y\|$ and $\|I - Q(0)\tilde{T}_{\epsilon(0)}\|$ are sufficiently small. Then problem (10.27)–(10.28) has a unique global solution and estimate (10.46) holds.*

Remark 10.2.2 Estimate (10.46) implies that $u(\infty) = y$ solves equation (10.26).

Remark 10.2.3 In contrast to the well-posed case (see Theorem 10.1.1 from Section 10.1), we do not prove in the case of ill-posed problems the convergence of $Q(t)$ as $t \to \infty$. The reason is: The inverse of the limiting operator T^{-1} does not exist, in general.

Remark 10.2.4 If $\|\tilde{T}\| < \infty$, then the condition

$$\|I - Q(0)\tilde{T}_{\epsilon(0)}\| < 1$$

can be satisfied algorithmically in spite of the fact that we do not know y and, consequently, we do not know $\tilde{T}_{\epsilon(0)}$. For example, if one takes

$$Q(0) = cI,$$

where $c > 0$ is constant, and denotes $\epsilon(0) := \nu$, then

$$\|I - Q(0)\tilde{T}_{\epsilon(0)}\| = \sup_{0 \leq s \leq \|\tilde{T}\|} \left|1 - \frac{c}{\nu + s}\right| = 1 - \frac{c}{\nu + \|\tilde{T}_{\epsilon(0)}\|} < 1,$$

provided that

$$\frac{c}{\nu + \|\tilde{T}_{\epsilon(0)}\|} < 1.$$

CHAPTER 11

DSM AND UNBOUNDED OPERATORS

In this chapter we consider the case when the operator F in the equation $F(u) = 0$ is unbounded.

11.1 STATEMENT OF THE PROBLEM

Consider the equation

$$G(u) := Lu + g(u) = 0, \tag{11.1}$$

where L is a linear closed, unbounded, densely defined operator in a Hilbert space H, which has a bounded inverse,

$$\|L^{-1}\| \le m, \tag{11.2}$$

while g is a nonlinear operator which satisfies assumptions (1.10). The operator G in (11.1) is not continuously Fréchet differentiable in the standard sense if L is unbounded. Recall that the standard definition of the Fréchet derivative at a point u requires the existence of a bounded linear operator $A(u)$ such that

$$F(u + h) = F(u) + A(u)h + o(\|h\|) \quad \text{as} \quad \|h\| \to 0. \tag{11.3}$$

Dynamical Systems Method and Applications: Theoretical Developments and Numerical Examples, First Edition. A. G. Ramm and N. S. Hoang.

If G is defined by formula (11.1), then formally

$$G'(u) := A := A(u) = L + g'(u),$$

but in the standard definition the element $h \in H$ in (11.3) is arbitrary, subject to the condition $\|h\| \to 0$. In the case of unbounded G, defined in (11.1), the element h cannot be arbitrary: It has to belong to the domain $D(L)$ of L. The operator $A = L + g'(u)$ has the domain $D(A) = D(L)$ dense in H, and $D(L)$ is a linear manifold in H. If assumption (11.2) holds, then equation (11.1) is equivalent to

$$F(u) := u + L^{-1}g(u) = 0. \tag{11.4}$$

To this equation we can apply the theory developed for equations which satisfy assumption (1.10). We have

$$F'(u) = I + L^{-1}g(u), \qquad \sup_{u \in B(u_0, R)} \|F'(u)\| \leq M_1(R), \tag{11.5}$$

and

$$\sup_{u \in B(u_0, R)} \|F''(u)\| \leq M_2(R).$$

If equation (11.1) has a solution, then this solution solves equation (11.4), so we may apply the results of Chapters 3 and 6–10 to equation (11.4). For example, Theorem 3.2.1 from Section 3.2 yields the following result.

Theorem 11.1.1 *Assume that equation (11.1) is solvable, L is a densely defined, closed linear operator, (11.2) holds, and the operator F, defined in (11.4), satisfies assumptions (1.9), (1.10), and (3.32). Then the problem*

$$\dot{u} = -[F'(u)]^{-1}F(u), \quad u(0) = u_0, \tag{11.6}$$

has a unique global solution $u(t)$, there exists $u(\infty)$, and $F(u(\infty)) = 0$.

If condition (1.9) is not satisfied, then problem (11.4) can be treated by the methods developed in Chapters 6–10.

An example of such treatment is given in Section 11.2.

Example 11.1

Consider a semilinear elliptic problem:

$$-\Delta u + g(u) = f \text{ in } D, \quad u|_S = 0, \tag{11.7}$$

where $D \subset \mathbb{R}^3$ is a bounded domain with a smooth boundary S, $f \in L^2(D)$ is a given function, $g(u)$ is a smooth nonlinear function, and we assume that

$$g(u) \geq 0, \quad g'(u) \geq 0. \tag{11.8}$$

Then one can check easily that problem (11.7) has no more than one solution. If the solution u of problem (11.7) exists, then it solves the problem

$$F(u) := u + L^{-1}g(u) = h, \quad h := L^{-1}f, \tag{11.9}$$

where $L^{-1} := (-\Delta)^{-1}$, and $-\Delta$ is the Dirichlet Laplacian. It is well known that estimate (11.2) holds for $L^{-1} = (-\Delta)^{-1}$ in a bounded domain. The operator F in (11.9) satisfies conditions (1.9) and (1.10).

Thus, one can solve equation (11.9) by the DSM (11.6).

11.2 ILL-POSED PROBLEMS

In this section we consider equation (11.1) under the assumption (11.2) and also consider the equivalent equation (11.4), and we assume that the operator F satisfies condition (1.10) but not (1.9). We assume that equations (11.1) and, therefore, equation (1.4) have a solution y, and

$$\tilde{A} := F'(y) \neq 0.$$

Under these assumptions we may apply Theorem 7.1.1.

Consider the equation

$$F(u) := u + B(u) = f, \quad B(u) := L^{-1}g(u), \tag{11.10}$$

which is equation (11.4), and the DSM for solving the equation

$$\dot{u} = -T_{a(t)}^{-1}\left[A^*(F(u) - f) + a(t)(u - z)\right], \quad u(0) = u_0, \tag{11.11}$$

where the notations are the same as in Section 7.1; see (7.4).

Theorem 7.1.1 from Section 7.1 gives sufficient conditions for the existence and uniqueness of the global solution $u(t)$ to problem (11.11), for the existence of $u(\infty) = y$, where $F(y) = 0$.

Theorem 7.2.1 from Section 7.2, applied to equation (11.10) with noisy data f_δ, $\|f_\delta - f\| \leq \delta$, given in place of the exact data f, allows us to use DSM (11.11) with f_δ in place of f. In particular, this theorem yields the existence of the stopping time t_δ such that the element $u_\delta(t_\delta)$ converges to y as $\delta \to 0$. Here $u_\delta(t)$ is the solution to problem (11.11) with f_δ in place of f and y is a solution to equation (11.10).

CHAPTER 12

DSM AND NONSMOOTH OPERATORS

In this chapter we study the equation $F(u) = 0$ with monotone operator F without assuming that $F \in C^2_{\text{loc}}$.

12.1 FORMULATION OF THE RESULTS

Consider the equation

$$F(u) = 0, \tag{12.1}$$

where F *is a monotone operator in a Hilbert space* H:

$$\langle F(u) - F(v), u - v \rangle \geq 0, \quad \forall u, v \in H. \tag{12.2}$$

We do not assume in this chapter that $F \in C^2_{\text{loc}}$. We assume that F *is defined on all of* H *and is hemicontinuous.*

Definition 12.1.1 F *is called hemicontinuous if* $s_n \to +0$ *implies*

$$F(u + s_n v) \rightharpoonup F(u), \quad \forall u, v \in H,$$

Dynamical Systems Method and Applications: Theoretical Developments and Numerical Examples, First Edition. A. G. Ramm and N. S. Hoang.
Copyright © 2012 John Wiley & Sons, Inc.

and is called demicontinuous if $u_n \to u$ implies $F(u_n) \rightharpoonup F(u)$. Here $\rightharpoonup$ denotes weak convergence in H.

We also assume that equation (12.1) has a solution, and we denote by y the unique minimal-norm solution to equation (12.1). This solution is well-defined due to Lemmas 6.1.2 and 6.1.3 from Section 6.1. In Lemma 6.1.2 we assumed that F was continuous, but the conclusion of this lemma remain valid if F is hemicontinuous and monotone.

Lemma 12.1.1 *If F is hemicontinuous and monotone, then the set*

$$\mathcal{N}_f := \{u : F(u) = f\}$$

is closed and convex.

Proof: Let $u_n \to u$ and $F(u_n) = f$. We want to prove that $F(u) = f$, that is, the set $\mathcal{N}_f$ is closed. Since F is monotone, we have

$$\langle F(u_n) - F(u - tz), u_n - u + tz \rangle \geq 0, \quad t = const > 0, \quad \forall z \in H. \quad (12.3)$$

Taking $n \to \infty$ in (12.3), we get

$$\langle f - F(u - tz), tz \rangle \geq 0. \quad (12.4)$$

Let us take $t \to +0$ and use the hemicontinuity of F. The result is

$$\langle f - F(u), z \rangle \leq 0. \quad (12.5)$$

Since z is arbitrary, it follows that $F(u) = f$. To check that $\mathcal{N}_f$ is convex, we use the argument used in the proof of Lemma 6.1.2.

Lemma 12.1.1 is proved. ∎

Lemma 12.1.2 *If F is hemicontinuous and monotone, then F is demicontinuous.*

Proof: Let $u_n \to u$. We want to prove that $F(u_n) \rightharpoonup F(u)$. Since F is monotone and $D(F) = H$, it is locally bounded in H (see, e.g., [25, p. 97]). Thus,

$$\|F(u_n)\| \leq c,$$

where c does not depend on n. Bounded sets in H are weakly precompact. Consequently, we may assume that

$$F(u_n) \rightharpoonup f.$$

Let us prove that $f = F(u)$. This will complete the proof of Lemma 12.1.2. By the monotonicity of F, we have inequality (12.3). Let $n \to \infty$ in (12.3) and get

$$\langle f - F(u - tz), z \rangle \geq 0, \quad \forall z \in H. \quad (12.6)$$

Taking $t \to +0$ and using hemicontinuity of F, we obtain

$$\langle f - F(u), z \rangle \geq 0. \tag{12.7}$$

Since z is arbitrary, inequality (12.7) implies

$$F(u) = f.$$

Lemma 12.1.2 is proved.   ∎

Consider the following DSM for solving equation (12.1):

$$\dot{u} = -F(u) - au, \quad u(0) = u_0, \quad a = const > 0. \tag{12.8}$$

Lemma 12.1.3 *Assume that F is monotone, hemicontinuous, and $D(F) = H$. Then problem (12.8) has a unique global solution for any $u_0 \in H$ and any $a \geq 0$.*

Proof: In this proof we use the known argument based on Peano approximations (see, e.g, [25, p. 99] and [131]). Uniqueness of the solution to (12.8) can be proved easily: If $u(t)$ and $v(t)$ solve (12.8), then $w := u - v$ solves the problem

$$\dot{w} = -[F(u) - F(v)] - aw, \quad w(0) = 0. \tag{12.9}$$

Multiply (12.9) by w and use the monotonicity of F to get

$$\langle \dot{w}, w \rangle \leq -a\langle w, w \rangle, \quad w(0) = 0, \quad a = const > 0. \tag{12.10}$$

Integrating (12.10) yields $w(t) = 0$, $\forall t > 0$, that is, $u = v$. To prove the existence of the solution to (12.8), we define the solution to (12.8) as the solution to the equation

$$u(t) = u_0 - \int_0^t [F(u(s)) + au(s)]\, ds. \tag{12.11}$$

We prove the existence of the solution to (12.11) with $a = 0$. The proof for $a > 0$ is even simpler. Consider the solution to the following equation

$$u_n(t) = u_0 - \int_0^t F(u_n(s - \tfrac{1}{n}))\, ds, \quad t \geq 0, \quad u_n(t) = u_0, \quad \forall t \leq 0. \tag{12.12}$$

From (12.12) it follows that

$$\|u_n(t) - u_0\| \leq ct, \quad 0 \leq t \leq \frac{r}{c}, \tag{12.13}$$

where

$$c = \sup_{u \in B(u_0, r)} \|F(u)\|, \quad B(u_0, r) := \{u : \|u - u_0\| \leq r\},$$

and $c < \infty$ because of the local boundedness of monotone operators defined on all of H. From equation (12.8) (with $a = 0$) it follows that $\|\dot{u}\| \leq c$. Define

$$z_{nm} := u_n(t) - u_m(t), \quad \|z_{nm}\| = g_{nm}(t). \tag{12.14}$$

Equation (12.8) with $a = 0$ implies

$$J := g_{nm}(t)\dot{g}_{nm}(t)$$
$$= -\left\langle F\left(u_n\left(t - \frac{1}{n}\right)\right) - F\left(u_m\left(t - \frac{1}{m}\right)\right), u_n(t) - u_m(t) \right\rangle. \tag{12.15}$$

One has

$$J = -\left\langle F\left(u_n\left(t - \frac{1}{n}\right)\right) - F\left(u_m\left(t - \frac{1}{m}\right)\right), u_n\left(t - \frac{1}{n}\right) - u_m\left(t - \frac{1}{m}\right) \right\rangle$$
$$- \left\langle F\left(u_n\left(t - \frac{1}{n}\right)\right) - F\left(u_m\left(t - \frac{1}{m}\right)\right), u_n(t) - u_n\left(t - \frac{1}{n}\right) \right\rangle$$
$$- \left\langle F\left(u_n\left(t - \frac{1}{n}\right)\right) - F\left(u_m\left(t - \frac{1}{m}\right)\right), u_m(t) - u_m\left(t - \frac{1}{m}\right) \right\rangle.$$

Using the monotonicity of F and the estimates

$$\|F(u)\| \leq c \quad \text{and} \quad \|\dot{u}\| \leq c,$$

we obtain

$$J \leq \left(\left\|F\left(u_n\left(t - \frac{1}{n}\right)\right)\right\| + \left\|F\left(u_m\left(t - \frac{1}{m}\right)\right)\right\|\right)c\left(\frac{1}{n} + \frac{1}{m}\right)$$
$$\leq 2c^2\left(\frac{1}{n} + \frac{1}{m}\right).$$

Thus (12.15) implies

$$\frac{d}{dt}g_{nm}^2(t) \leq 4c^2\left(\frac{1}{n} + \frac{1}{m}\right) \to 0 \quad \text{as} \quad n, m \to \infty; \quad g_{nm}(0) = 0. \tag{12.16}$$

Consequently we obtain

$$\lim_{n,m \to \infty} g_{nm}(t) = 0, \quad 0 \leq t \leq \frac{r}{c}. \tag{12.17}$$

From (12.17) we conclude that the following limit exists:

$$\lim_{n \to \infty} u_n(t) = u(t), \quad 0 \leq t \leq \frac{r}{c}. \tag{12.18}$$

Since F is demicontinuous, this implies

$$F(u_n(t)) \rightharpoonup F(u(t)). \tag{12.19}$$

Using (12.18) and (12.19), let us pass to the limit $n \to \infty$ in (12.12) and obtain

$$u(t) = u_0 - \int_0^t F(u(s)) \, ds, \qquad t \le \frac{r}{c}. \tag{12.20}$$

Here we have used the relation

$$F\left(u_n\left(t - \frac{1}{n}\right)\right) \rightharpoonup F(u(t)) \quad \text{as} \quad n \to \infty. \tag{12.21}$$

This relation follows from the formula

$$u_n\left(t - \frac{1}{n}\right) \to u(t) \quad \text{as} \quad n \to \infty. \tag{12.22}$$

Indeed

$$\left\| u_n\left(t - \frac{1}{n}\right) - u(t) \right\| \le \left\| u_n\left(t - \frac{1}{n}\right) - u_n(t) \right\| + \| u_n(t) - u(t) \|.$$

The second term on the right converges to zero due to (12.18), and the first term is estimated as follows:

$$\left\| u_n\left(t - \frac{1}{n}\right) - u_n(t) \right\| \le c\frac{1}{n} \to 0 \quad \text{as} \quad n \to \infty, \tag{12.23}$$

where c is the constant from (12.13). We can differentiate the solution $u(t)$ of the equation (12.20) in the weak sense because convergence in (12.19) is the weak convergence. Therefore (12.20) implies that

$$\dot{u} = -F(u), \quad u(0) = u_0, \tag{12.24}$$

where the derivative $\dot{u}$ is understood in the weak sense, that is,

$$\frac{d}{dt}\langle u(t), \eta \rangle = -\langle F(u(t)), \eta \rangle, \qquad \forall \eta \in H.$$

If we would assume that F is continuous rather than hemicontinuous, then the derivative $\dot{u}$ could be understood in the strong sense.

Let us prove that the solution $u(t)$ of the equation (12.20) exists for all $t \ge 0$. Assume the contrary: The solution $u(t)$ to (12.24) exists on the interval $[0, T]$ but does not exist on $[0, T+d]$, where $d > 0$ is arbitrarily small. This assumption leads to a contradiction if we prove that the limit

$$\lim_{t \to T-0} u(t) = u(T)$$

exists and is finite, because in this case one can take $u(T)$ as the initial data at $t = T$ and construct the solution to equation (12.24) on the interval $[T, T+d]$, $d > 0$, so that the solution $u(t)$ will exist on $[0, T+d]$, contrary to our assumption.

To prove that the finite limit $u(T)$ exists, let

$$u(t + h) - u(t) := z(t), \quad \|z(t)\| = g(t), \quad t + h < T. \tag{12.25}$$

We have

$$\dot{z} = -[F(u(t + h)) - F(u(t))], \quad z(0) = u(h) - u(0). \tag{12.26}$$

Multiply (12.26) by z and use the monotonicity of F, to get

$$\langle \dot{z}, z \rangle \le 0, \quad z(0) = u(h) - u(0). \tag{12.27}$$

Thus,

$$\|u(t + h) - u(t)\| \le \|u(h) - u(0)\|. \tag{12.28}$$

We have

$$\lim_{h \to 0} \|u(h) - u(0)\| = 0.$$

Therefore,

$$\lim_{h \to 0} \|u(t + h) - u(t)\| = 0. \tag{12.29}$$

This relation holds uniformly with respect to t and $t + h$ such that $t + h < T$ and $t < T$.

By the Cauchy criterion for convergence, relation (12.29) implies the existence of a finite limit

$$\lim_{t \to T - 0} u(t) := u(T). \tag{12.30}$$

Lemma 12.1.3 is proved.   ∎

Lemma 12.1.4 *If we denote by $u_a(t)$ the unique solution to (12.8), then*

$$\lim_{a \to 0} \|u_a(t) - u(t)\| = 0 \tag{12.31}$$

uniformly with respect to $t \in [0, T]$. Here $T > 0$ is an arbitrary large fixed number, and $u(t)$ solves (12.24).

Proof: First, we check that

$$\sup_{t \ge 0} \|u_a(t)\| \le c, \tag{12.32}$$

where the constant c does not depend on t. To prove (12.32), we start with the equation

$$\dot{u}_a = -[F(u_a) - F(0)] - F(0) - au_a,$$

multiply it by u_a and denote

$$g(t) := \|u_a(t)\|.$$

Then, using the monotonicity of F, we get

$$g\dot{g} \leq c_0 g - ag^2, \quad c_0 = \|F(0)\|.$$

Thus

$$\|u_a(t)\| = g(t) \leq g(0)e^{-at} + \frac{c_0(1 - e^{-at})}{a} \leq g(0) + \frac{c_0}{a}. \tag{12.33}$$

To prove (12.31), denote

$$w(t) := u_a(t) - u(t).$$

Then

$$\dot{w} = -[F(u_a) - F(u)] - au_a, \quad w(0) = 0. \tag{12.34}$$

Multiply this equation by w, and use the monotonicity of F, to get

$$p\dot{p} \leq -a(u_a, w) \leq a\|u_a(t)\|p(t),$$
$$p := \|w(t)\| = \|u_a(t) - u(t)\|, \quad p(0) = 0. \tag{12.35}$$

Thus

$$p(t) \leq a \int_0^t \|u_a(s)\| \, ds. \tag{12.36}$$

If $t \in [0, T]$, then the first inequality (12.33) implies

$$\max_{0 \leq t \leq T} a\|u_a(t)\| \leq ag(0) + c_0 \max_{0 \leq t \leq T} |1 - e^{-at}|. \tag{12.37}$$

Passing to the limit $a \to 0$ in (12.37) we obtain

$$\lim_{a \to 0} \max_{0 \leq t \leq T} a\|u_a(t)\| = 0. \tag{12.38}$$

From (12.38) and (12.36) the relation (12.31) follows.
Lemma 12.1.4 is proved. ∎

Let us now state our main results.

Theorem 12.1.5 *Assume that equation $F(u) = 0$ has a solution, possibly nonunique, that $F : H \to H$ is monotone and continuous, $D(F) = H$, and $a = \text{const} > 0$. Then problem (12.8) has a unique global solution $u_a(t)$ and*

$$\lim_{a \to 0} \lim_{t \to \infty} \|u_a(t) - y\| = 0, \tag{12.39}$$

where y is the unique minimal-norm solution to equation $F(u) = 0$.

Theorem 12.1.6 *Under the assumption of Theorem 12.1.5, let $a = a(t)$. Let us assume that*

$$0 < a(t) \searrow 0, \quad \dot{a} \leq 0, \quad \ddot{a} \geq 0; \quad \lim_{t \to \infty} ta(t) = \infty;$$

$$\lim_{t \to \infty} \frac{\dot{a}}{a} = 0; \quad \int_0^\infty a(s)\, ds = \infty, \tag{12.40}$$

$$\int_0^t e^{-\int_s^t a(\tau)d\tau} |\dot{a}(s)|\, ds = O\left(\frac{1}{t}\right) \quad as \quad t \to \infty.$$

Then the problem

$$\dot{u} = -F(u) - a(t)u, \quad u(0) = u_0 \tag{12.41}$$

has a unique global solution and

$$\lim_{t \to \infty} \|u(t) - y\| = 0, \tag{12.42}$$

where $u(t)$ solves (12.41).

In Section 12.2, proofs of Theorems 12.1.5 and 12.1.6 are given.

Remark 12.1.7 Conditions (12.40) are satisfied, for example, by the function

$$a(t) = \frac{c_1}{(c_0 + t)^b}, \tag{12.43}$$

where c_1, c_0, and b are positive constants, $b \in (0, 1)$. The slow decay of $a(t)$ is important only for large t. For small t the function $a(t)$ may decay arbitrarily fast.

12.2 PROOFS

Proof of Theorem 12.1.5. The global existence of the solution $u_a(t)$ to problem (12.8) has been proved in Lemma 12.1.3. Let us prove the existence of the limit $u_a(\infty)$. Denote

$$w := u_a(t + h) - u_a(t), \quad \|w(t)\| = g(t), \quad h = const > 0.$$

Then (12.8) implies

$$\dot{w} = -F(u_a(t + h)) - F(u_a(t)) - aw, \quad w(0) = u(h) - u(0). \tag{12.44}$$

Multiply (12.44) by w and use the monotonicity of F to get

$$g\dot{g} \leq -ag^2, \quad g(0) = \|u(h) - u(0)\|. \tag{12.45}$$

Thus

$$\|u_a(t + h) - u_a(t)\| \leq e^{-at} \|u_a(h) - u_a(0)\|. \tag{12.46}$$

From (12.46), (12.32), and the Cauchy criterion for the existence of the limit as $t \to \infty$, we conclude that the limit

$$\lim_{t \to \infty} u_a(t) = u_a \tag{12.47}$$

exists. Let us denote

$$z_h(t) := \frac{u_a(t+h) - u_a(t)}{h}, \qquad \psi_h(t) := \|z_h(t)\|.$$

Then

$$\dot{z}_h = -\frac{1}{h}[F(u_a(t+h)) - F(u_a(t))] - az_h. \tag{12.48}$$

Multiply (12.48) by z_h and use the monotonicity of F to get

$$\dot{\psi}_h \psi_h \leq -a\psi_h^2. \tag{12.49}$$

Thus

$$\psi_h(t) \leq \psi_h(0)e^{-at}. \tag{12.50}$$

Let $h \to 0$ in (12.50). Since we assumed F continuous, the derivative $\dot{u}(t)$ exists in the strong sense, as the limit

$$\dot{u}(t) = \lim_{h \to 0} z_h(t),$$

and one has

$$\lim_{h \to 0} \psi_h(t) = \|\dot{u}(t)\|.$$

Thus, as $h \to 0$, formula (12.50) yields

$$\|\dot{u}_a(t)\| \leq \|\dot{u}_a(0)\|e^{-at}. \tag{12.51}$$

This inequality implies the existence of $u(\infty)$ and the following estimates:

$$\|u_a(t) - u_a(0)\| \leq \|\dot{u}_a(0)\|\frac{1 - e^{-at}}{a} \tag{12.52}$$

and

$$\|u_a(t) - u_a(\infty)\| \leq \|\dot{u}_a(0)\|\frac{e^{-at}}{a}. \tag{12.53}$$

Estimate (12.51), the existence of $u(\infty)$, and the continuity of F allow us to pass to the limit $t \to \infty$ in equation (12.8) and get

$$F(u_a(\infty)) + au_a(\infty) = 0. \tag{12.54}$$

Let us prove that

$$\lim_{a \to 0} u_a(\infty) = y, \tag{12.55}$$

where y is the unique minimal-norm solution to the equation

$$F(y) = 0. \tag{12.56}$$

The proof of (12.55) is the same as the proof of Lemma 6.1.9: This proof is valid for a hemicontinuous monotone operator F.

Theorem 12.1.5 is proved. ∎

Proof of Theorem 12.1.6. The global existence of the unique solution to problem (12.41) follows from Lemma 12.1.3. Let us prove relation (12.41). Equation (12.41) and the existence of a solution to (12.56) imply

$$\sup_{t \geq 0} \|u(t)\| < c. \tag{12.57}$$

Indeed, let

$$p(t) := u(t) - y, \quad \|p(t)\| = q(t).$$

Equation (12.41) implies

$$\dot{p} = -[F(u) - F(y)] - a(t)p - a(t)y. \tag{12.58}$$

Multiply this equation by p and use the monotonicity of F to get

$$q\dot{q} \leq -a(t)q^2 + a(t)\|y\|q$$

and

$$\dot{q} \leq -a(t)q(t) + a(t)\|y\|. \tag{12.59}$$

This implies

$$q(t) \leq e^{-\int_0^t a(s)ds} \left[q(0) + \int_0^t e^{\int_0^s a(\tau)d\tau} a(s)\,ds\|y\| \right]$$
$$\leq \|q(0)\| + \|y\|, \tag{12.60}$$

because

$$\int_0^t e^{-\int_s^t a(\tau)d\tau} a(s)ds = 1 - e^{-\int_0^t a(\tau)d\tau} \leq 1.$$

Therefore

$$\|u(t)\| \leq q(t) + \|y\| \leq \|q(0)\| + 2\|y\| := c, \tag{12.61}$$

so (12.57) is established.

Let us now prove that $q(\infty) = 0$, that is, the existence of $u(\infty)$ and the relation $u(\infty) = y$.

Using (12.41), we obtain

$$\dot{w} = -[F(u(t+h)) - F(u(t))] - [a(t+h)u(t+h) - a(t)u(t)],$$

where

$$w := u(t+h) - u(t).$$

Multiplying this equation by w and using the monotonicity of F, we derive

$$g\dot{g} \le |a(t+h) - a(t)| \|u(t+h)\| g - a(t)g^2, \qquad (12.62)$$

where

$$g(t) := \|w(t)\|$$

and

$$\dot{g} \le -a(t)g(t) + ch|\dot{a}(t)|. \qquad (12.63)$$

Here we have used estimate (12.57) and the assumptions $\dot{a} \le 0$ and $\ddot{a} > 0$, which imply

$$|a(t+h) - a(t)| \le h|\dot{a}(t)|.$$

Inequality (12.63) implies

$$g(t) \le e^{-\int_0^t a(s)ds} \left[g(0) + ch \int_0^t e^{\int_0^s a(\tau)d\tau} |\dot{a}(s)| \, ds \right]. \qquad (12.64)$$

Let us derive from estimate (12.64) and assumptions (12.40) that

$$\lim_{t\to\infty} g(t) = 0 \qquad (12.65)$$

uniformly with respect to h running through any fixed compact subset of $\mathbb{R}_+ := [0, \infty)$.

Indeed, the last assumption (12.40) implies

$$\lim_{t\to\infty} g(0)e^{-\int_0^t a(s)ds} = 0, \qquad (12.66)$$

and an application of the L'Hospital rule yields

$$\lim_{t\to\infty} \frac{\int_0^t e^{\int_0^s a(\tau)d\tau} |\dot{a}(s)|ds}{e^{\int_0^t a(s)ds}} = \lim_{t\to\infty} \frac{|\dot{a}(t)|}{a(t)} = 0, \qquad (12.67)$$

because of (12.40). Thus, the relation (12.65) is established.

From (12.57) it follows that there exists a sequence $t_n \to \infty$ such that

$$u(t_n) \rightharpoonup v. \qquad (12.68)$$

We want to pass to the limit $t_n \to \infty$ in equation (12.41) and obtain $F(v) = 0$. To do this, we note that estimate (12.57) and the property $\lim_{t\to\infty} a(t) = 0$ imply

$$\lim_{t\to\infty} a(t)u(t) = 0. \qquad (12.69)$$

Dividing inequality (12.64) by h and letting $h \to 0$, we obtain

$$\|\dot{u}(t)\| \le e^{-\int_0^t a(s)ds} \left[\|\dot{u}(0)\| + c \int_0^t e^{\int_0^s a(\tau)d\tau} |\dot{a}(s)| \, ds \right]. \qquad (12.70)$$

This inequality together with (12.66) and (12.67) implies

$$\lim_{t\to\infty} \|\dot{u}(t)\| = 0. \tag{12.71}$$

From (12.69), (12.71), and (12.41) it follows that

$$\lim_{t\to\infty} F(u(t)) = 0. \tag{12.72}$$

From (12.68), (12.72), and Lemma 6.1.5 it follows that

$$F(v) = 0. \tag{12.73}$$

Let us prove now that

$$\lim_{n\to\infty} u(t_n) = v \quad \text{and} \quad v = y. \tag{12.74}$$

From (12.68) we conclude that

$$\|v\| \le \liminf_{n\to\infty} \|u(t_n)\|. \tag{12.75}$$

Let us prove that

$$\limsup_{n\to\infty} \|u(t_n)\| \le \|v\|. \tag{12.76}$$

If (12.76) is proved, then (12.75) and (12.76) imply

$$\lim_{n\to\infty} \|u(t_n)\| = \|v\|. \tag{12.77}$$

This relation together with weak convergence (12.68) imply strong convergence (12.74).

So, *let us verify (12.76)*. By the last assumption (12.40) we have

$$e^{-\int_0^t a(s)ds} \int_0^t e^{\int_0^s a(\tau)d\tau} |\dot{a}(s)|\, ds = O\left(\frac{1}{t}\right) \quad \text{as} \quad t\to\infty. \tag{12.78}$$

Thus, estimate (12.70) for $a(t)$, defined in (12.43), implies

$$\|\dot{u}(t)\| = O\left(\frac{1}{t}\right) \quad \text{as} \quad t\to\infty. \tag{12.79}$$

Equation (12.73) and (12.41) imply

$$\langle F(u(t_n)) - F(v), u(t_n) - v\rangle + a(t_n)\langle u(t_n), u(t_n) - v\rangle$$
$$= -\langle \dot{u}(t_n), u(t_n) - v\rangle. \tag{12.80}$$

This relation holds for any solution to equation (12.73). Equation (12.80) and the monotonicity of F imply

$$\langle u(t_n), u(t_n) - v\rangle \le -\frac{1}{a(t_n)}\langle \dot{u}(t_n), u(t_n) - v\rangle. \tag{12.81}$$

From (12.57), (12.79), and (12.81) we get

$$\langle u(t_n), u(t_n) - v \rangle \leq \frac{c}{t_n a(t_n)} \to 0 \quad \text{as} \quad n \to \infty, \tag{12.82}$$

because

$$\lim_{t \to \infty} ta(t) = \infty$$

by the assumption (12.40). From (12.82) we obtain (12.76). Therefore, the first relation (12.74) is verified.

Let us prove that $v = y$. In formula (12.80) we can replace v by any solution to equation (12.73). Let us replace v by y. Then (12.82) is replaced by

$$\|u(t_n)\|^2 \leq \|u(t_n)\| \|y\| + \frac{c}{t_n a(t_n)}. \tag{12.83}$$

Passing to the limit $n \to \infty$ yields

$$\limsup_{n \to \infty} \|u(t_n)\| \leq \|y\|. \tag{12.84}$$

We have already proved that

$$\limsup_{n \to \infty} \|u(t_n)\| = \lim_{n \to \infty} \|u(t_n)\| = \|v\|. \tag{12.85}$$

Thus

$$\|v\| = \|y\|. \tag{12.86}$$

Since y is the unique minimum-norm solution of equation (12.73) and v solves this equation and has a norm greater than $\|y\|$, it follows that $v = y$.

Theorem 12.1.6 is proved. ∎

CHAPTER 13

DSM AS A THEORETICAL TOOL

In this chapter we give a sufficient condition for a nonlinear map to be surjective and to be a global homeomorphism.

13.1 SURJECTIVITY OF NONLINEAR MAPS

In this section we prove the following result.

Theorem 13.1.1 *Assume that $F : X \to X$ is a C^2_{loc} map in a Banach space X and assumptions (1.9) and (1.10) hold. Then F is surjective if*

$$\limsup_{R \to \infty} \frac{R}{m(R)} = \infty. \tag{13.1}$$

Proof: Consider the DSM:

$$\dot{u} = -[F'(u)]^{-1}(F(u) - f), \quad u(0) = u_0, \tag{13.2}$$

where f and u_0 are arbitrary. Arguing as in the proof of Theorem 3.2.1, we establish the existence and uniqueness of the global solution to (1.10),

Dynamical Systems Method and Applications: Theoretical Developments and Numerical Examples, First Edition. A. G. Ramm and N. S. Hoang.

provided that condition analogous to (3.32) holds:

$$\|F(u_0) - f\|m(R) \le R. \tag{13.3}$$

If this condition holds, then $u(\infty)$ exists and

$$F(u(\infty)) = f. \tag{13.4}$$

Since f is arbitrary, the map F is surjective. Our assumption (13.1) implies that for sufficiently large R and an arbitrary fixed u_0 condition (13.3) is satisfied.

Theorem 13.1.1 is proved.   ∎

13.2   WHEN IS A LOCAL HOMEOMORPHISM A GLOBAL ONE?

Let $F : X \to X$ satisfy conditions (1.9) and (1.10). Condition (1.9) implies that F is a local homeomorphism; that is, F maps a sufficiently small neighborhood of an arbitrary point u_0 onto a small neighborhood of the point $F(u_0)$ bijectively and bicontinuously.

It is well known that a local homeomorphism may be not a global one. See an example in Section 2.6. Here is another one.

■ **EXAMPLE 13.1**

Let

$$F : \mathbb{R}^2 \to \mathbb{R}^2, \qquad F(u) = \begin{pmatrix} \arctan u_1 \\ u_2(1 + u_1^2) \end{pmatrix}.$$

Then $\det F'(u) = 1$, but F is not surjective. Indeed

$$F'(u) = \begin{pmatrix} \frac{1}{1+u_1^2} & 0 \\ 2u_1 u_2 & 1 + u_1^2 \end{pmatrix}, \qquad \det F'(u) = 1,$$

so

$$[F'(u)]^{-1} = \begin{pmatrix} 1 + u_1^2 & 0 \\ -2u_1 u_2 & \frac{1}{1+u_1^2} \end{pmatrix}.$$

There is no point $u = \begin{pmatrix} u_1 \\ u_2 \end{pmatrix}$ such that $F(u) = \begin{pmatrix} \frac{\pi}{2} \\ 1 \end{pmatrix}$.

This example can be found in [87].

J. Hadamard has proved in 1906 (see [38]) the following result:

Proposition 13.2.1 *If*

$$F : \mathbb{R}^n \to \mathbb{R}^n, \quad F \in C^1_{loc}(\mathbb{R}^n), \quad and \quad \sup_{u \in \mathbb{R}^n} \|[F'(u)]^{-1}\| \le m, \tag{13.5}$$

then F is a global homeomorphism of $\mathbb{R}^n$ onto $\mathbb{R}^n$.

This result was generalized later to Hilbert and Banach spaces (see [87] and references therein).

Our aim is to give the following generalization of the Hadamard's result.

Theorem 13.2.2 *Assume that $F : H \to H$ satisfies assumption (1.10) and*

$$\|[F'(u)]^{-1}\| \leq \psi(\|u\|), \tag{13.6}$$

where H is a Hilbert space and $\psi(s) > 0$ is a continuous function on $[0, \infty)$ such that

$$\int_0^\infty \frac{ds}{\psi(s)} = \infty. \tag{13.7}$$

Then F is a global homeomorphism of H onto H.

Remark 13.2.3 One can take, for example, $\psi(s) = m = const$, and then the assumption (13.6) reduces to (13.5). One can take $\psi(s) = as + b$, where a and b are positive constants.

Proof of Theorem 13.2.2. It follows from (13.6) that F is a local homeomorphism. We want to prove that F is injective and surjective.

Let us first prove that (13.6) implies surjectivity of F.

Consider problem (13.2). Denote

$$\|F(u(t)) - f\| := g(t). \tag{13.8}$$

Under the assumptions of Theorem 13.2.2, problem (13.2) has a unique local solution. We will establish a uniform with respect to t bound

$$\sup_t \|u(t)\| \leq c, \tag{13.9}$$

and then the local solution to (13.2) is a global one (see Lemma 2.6.1). Then we prove the existence of $u(\infty)$ and that $u(\infty)$ solves equation (13.4). This will prove the surjectivity of F.

We have, using (13.2),

$$g\dot{g} = \mathrm{Re}\langle F'(u)\dot{u}, F(u) - f\rangle = -g^2.$$

Thus

$$g(t) = g(0)e^{-t}. \tag{13.10}$$

This and (13.2) imply

$$\|\dot{u}\| \leq \psi(\|u\|)g(0)e^{-t}. \tag{13.11}$$

Note that

$$\frac{d}{dt}\|u\| \leq \|\dot{u}\|. \tag{13.12}$$

Indeed, differentiate $\langle u, u\rangle = |u\|^2$ with respect to t and get

$$2\|u\|\frac{d}{dt}\|u\| = 2\mathrm{Re}\langle \dot{u}, u\rangle \leq 2\|\dot{u}\|\|u\|.$$

If $\|u(t)\| > 0$, then we get (13.12). If the closure of the set of points t at which $u(t) = 0$ contains an open set, then on this set $\frac{d}{dt}\|u\| = \|\dot{u}\| = 0$, so (13.12) holds. If inequality (13.12) holds everywhere except on a nowhere dense set, then by continuity of $\|\dot{u}(t)\|$ it holds everywhere.

Thus, denoting $\|u(t)\| := s$, we can derive from (13.11) the inequality

$$\int_{\|u_0\|}^{\|u(t)\|} \frac{ds}{\psi(s)} \leq g(0)(1 - e^{-t}). \tag{13.13}$$

Thus inequality and assumption (13.7) imply estimate (13.9). From (13.9) and (13.11) it follows that

$$\|\dot{u}(t)\| \leq c_1 e^{-t}, \tag{13.14}$$

where

$$c_1 := g(0) \max_{0 \leq s \leq c} \psi(s).$$

Estimate (13.14) implies the existence of $u(\infty)$ and the inequalities

$$\|u(t) - u(\infty)\| \leq c_1 e^{-t}, \quad \|u(t) - u_0\| \leq c_1. \tag{13.15}$$

Passing to the limit $t \to \infty$ in equation (13.2) and taking into account (13.14), we derive equation

$$F(v) - f = 0, \quad v := u(\infty). \tag{13.16}$$

Since f is arbitrary, the map F is surjective.

Let us now prove that F is injective, that is, $F(v) = F(w)$ implies $w = v$. The idea of our proof is simple. We started with the initial approximation u_0 in problem (13.2) and constructed $v = u(\infty; u_0)$. We wish to consider the straight line

$$u_0(s) = u_0 + s(w - u_0), \quad u_0(0) = u_0, \quad u_0(1) = w, \tag{13.17}$$

and for each $u_0(s)$ we construct

$$u(t, s) := u(t, u_0(s)).$$

We will show that

$$u(\infty, s) = w, \quad \forall s \in [0, 1]. \tag{13.18}$$

This implies $w = v$.

To verify (13.18), we show that if $\|u(\infty, s) - u(\infty, s + \sigma)\|$ is sufficiently small, then the equation

$$F(u(\infty, s)) = F(u(\infty, s + \sigma)) = f$$

implies

$$u(\infty, s) = u(\infty, s + \sigma), \tag{13.19}$$

because F is a local homeomorphism.

We prove that

$$\|u(\infty, s) - u(\infty, s + \sigma)\| \le c\|u(0, s) - u(0, s + \sigma)\| \le c_2\sigma, \tag{13.20}$$

where

$$c_2 := c\|w - u_0\|,$$

so that $\|u(\infty, s) - u(\infty, s + \sigma)\|$ is as small as we wish if σ is sufficiently small. Thus, in finitely many steps we can reach the point $s = 1$, and at each step we will have equation (13.18). So, our proof will be complete if we verify estimate (13.20).

Let us do this. Denote

$$x(t) := u(t, s + \sigma) - u(t, s), \quad \eta(t) := \|x(t)\|, \tag{13.21}$$
$$z := u(t, s + \sigma), \quad \zeta := u(t, s), \quad z - \zeta = x = x(t). \tag{13.22}$$

Using (13.14), (13.2) and (1.10), we get

$$\begin{aligned}
\eta\dot\eta &= - \left\langle [F'(z)]^{-1}(F(z) - f) - [F'(\zeta)]^{-1}(F(\zeta) - f), x(t) \right\rangle \\
&= \left\langle ([F'(z)]^{-1} - [F'(\zeta)]^{-1})(F(z) - f), x(t) \right\rangle \\
&\quad - \left\langle [F'(z)]^{-1}(F(z) - F(\zeta)), x(t) \right\rangle \\
&\le ce^{-t}\eta^2 - \eta^2 + c\eta^3, \quad \eta \ge 0.
\end{aligned} \tag{13.23}$$

Here we have used the following estimates:

$$\|[F'(z)]^{-1} - [F'(\zeta)]^{-1}\| \le c\|z - \zeta\|, \tag{13.24}$$

$$\|F(z) - f\| \le ce^{-t}, \tag{13.25}$$

$$F(z) - F(\zeta) = F'(\zeta)(z - \zeta) + K, \quad \|K\| \le \frac{M_2\eta^2(t)}{2}, \tag{13.26}$$

$$\|[F'(\zeta)]^{-1}\| \le c, \tag{13.27}$$

where $c > 0$ denotes different constants independent of time. From equation (13.23) we obtain the following inequality:

$$\dot\eta \le -\eta + ce^{-t}\eta + c\eta^2, \quad \eta(0) = \|u(0, s + \sigma) - u(0, s)\| := \delta. \tag{13.28}$$

Let us define $q = q(t)$ by the formula

$$\eta = q(t)e^{-t}. \tag{13.29}$$

Then (13.28) yields

$$\dot q \le ce^{-t}(q + q^2), \quad q(0) = \delta. \tag{13.30}$$

We integrate (13.30) and get

$$\int_\delta^{q(t)} \frac{dq}{q + q^2} \le c(1 - e^{-t}). \tag{13.31}$$

This implies

$$\ln \frac{q(\delta + 1)}{(q + 1)\delta} \le c, \qquad \frac{q}{q + 1} \le \frac{\delta}{\delta + 1} e^c := c_2 \delta, \qquad (13.32)$$

where $c_2 = \frac{e^c}{\delta + 1}$. We assume that

$$c_2 \delta < 1. \qquad (13.33)$$

Then (13.32) yields

$$q(t) \le c_3 \delta. \qquad (13.34)$$

Therefore

$$\|u(t, s + \sigma) - u(t, s)\| = \eta(t)$$
$$\le c_3 \delta e^{-t} \|u(0, s + \sigma) - u(0, s)\| \le c_4 e^{-t} \sigma. \qquad (13.35)$$

We have verified estimate (13.20). This completes the proof of injectivity of F.

Theorem 13.2.2 is proved.   ∎

CHAPTER 14

DSM AND ITERATIVE METHODS

14.1 INTRODUCTION

In this chapter a general approach to constructing convergent iterative schemes is developed. This approach is based on the DSM. The idea is simple. Suppose we want to solve an operator equation $F(u) = 0$ which has a solution, and we have justified a DSM method

$$\dot{u} = \Phi(t, u), \quad u(0) = u_0 \tag{14.1}$$

for solving equation $F(u) = 0$. Suppose also that a dicretization scheme

$$u_{n+1} = u_n + h_n \Phi(t_n, u_n), \quad u_0 = u_0, \quad t_{n+1} = t_n + h \tag{14.2}$$

converges to the solution of (14.1) and one can choose h_n so that

$$\sum_{n=1}^{\infty} h_n = t,$$

where $t \in \mathbb{R}_+$ is arbitrary; then (14.2) is a convergent iterative process for solving equation $F(u) = 0$.

Dynamical Systems Method and Applications: Theoretical Developments and Numerical Examples, First Edition. A. G. Ramm and N. S. Hoang.
Copyright © 2012 John Wiley & Sons, Inc.

Iterative methods for solving linear equations have been discussed in Sections 2.4 and 4.4. The basic result we have proved in these sections can be described as follows:

Every solvable linear equation $Au = f$ with a closed densely defined operator in a Hilbert space H can be solved by a convergent iterative process.

If the data f_δ are given, such that $\|f_\delta - f\| \leq \delta$, then the iterative process gives a stable approximation to the minimal-norm solution of the equation $Au = f$, provided that iterations are stopped at $n = n(\delta)$, where $n(\delta)$ is suitably chosen.

In Section 14.2 we construct an iterative method for solving well-posed problems. We prove that every solvable well-posed operator equation can be solved by an iterative method which converges at an exponential rate. In Section 14.3 we construct an iterative process for solving ill-posed problems with monotone operators. We prove that any solvable nonlinear operator equation $F(u) = f$ with C^2_{loc} monotone operator F can be solved by an iterative process which converges to the unique minimal-norm solution y of this equation for any choice of the initial approximation. In Section 14.4 we deal with iterative processes for general nonlinear equations.

14.2 ITERATIVE SOLUTION OF WELL-POSED PROBLEMS

Consider the equation

$$F(u) = 0, \tag{14.3}$$

where $F : H \to H$ satisfies conditions (1.9) and (1.10) and H is a Hilbert space. Let a DSM for solving equation (14.3) be of the form

$$\dot{u} = \Phi(u), \quad u(0) = u_0, \tag{14.4}$$

and assume that conditions (3.21) and (3.22) hold. Note that in Sections 3.1 to 3.6 the Φ did not depend on t. That is why we take $\Phi(u)$ rather than $\Phi(t, u)$ in (14.4). We assume that Φ is locally Lipschitz.

Consider the following iterative process:

$$u_{n+1} = u_n + h\Phi(u_n), \quad u_0 = u_0, \tag{14.5}$$

where the initial approximation u_0 is chosen so that condition (3.23) hold.

Theorem 14.2.1 *If F satisfies conditions (1.9), (1.10), and (3.23) and Φ satisfies conditions (3.21) and (3.22), and if $h > 0$ in (14.5) is sufficiently small, then iterative process (14.5) produces u_n such that*

$$\|u_n - y\| \leq Re^{-chn}, \quad \|F(u_n)\| \leq \|F_0\|e^{-chn}, \tag{14.6}$$

where

$$F_0 := F(u_0), \quad c = const > 0, \quad c < c_1,$$

c_1 is the constant from condition (3.21), $R > 0$ is a constant from conditions (1.9) and (1.10), and y solves equation (14.3).

Proof: For $n = 0$ the second estimate (14.6) is obvious, and the first one follows from (3.25) and (3.23). Assuming that estimates (14.6) hold for $n \leq m$, let us prove that they hold for $n = m + 1$. Then, by induction, they hold for any n. Denote by $w_{n+1}(t)$ the solution to the problem

$$\dot{w}(t) = \Phi(w), \quad w(nh) = u_n, \quad t_n := nh \leq t \leq (n+1)h =: t_{n+1}. \quad (14.7)$$

Estimate (3.24) yields

$$\|w(t) - y\| \leq \frac{c_2}{c_1}\|F_n\|e^{-c_1 t} \leq Re^{-chn - c_1(t - nh)}, \quad t \geq t_n. \quad (14.8)$$

We have

$$\|u_{n+1} - y\| \leq \|u_{n+1} - w(t_{n+1})\| + \|w(t_{n+1}) - y\| \quad (14.9)$$

and

$$\|u_{n+1} - w(t_{n+1})\| \leq \int_{t_n}^{t_{n+1}} \|\Phi(u_n) - \Phi(w(s))\| \, ds$$

$$\leq L_1 \int_{t_n}^{t_{n+1}} \|u_n - w(s)\| \, ds$$

$$\leq L_1 \int_{t_n}^{t_{n+1}} ds \| \int_{t_n}^{s} \Phi(w(s)) \, ds \|$$

$$\leq L_1 h c_2 \int_{t_n}^{t_{n+1}} \|F(w(s))\| \, ds$$

$$\leq L_1 c_2 h^2 \|F(u_n)\|$$

$$\leq L_1 c_2 h^2 \|F_0\|e^{-chn}. \quad (14.10)$$

From (14.9), (14.10), and (14.8) with $t = (n+1)h$ we obtain

$$\|u_{n+1} - y\| \leq Re^{-chn}\left(e^{-c_1 h} + L_1 c_1 h^2\right) \leq Re^{-ch(n+1)}, \quad (14.11)$$

provided that

$$e^{-c_1 h} + L_1 c_1 h^2 \leq e^{-ch}. \quad (14.12)$$

Inequality (14.12) holds if $c_1 > c$ and h is sufficiently small.

Let us estimate $\|F(u_{n+1})\|$. We have

$$\|F(u_{n+1})\| \leq \|F(u_{n+1}) - F(w(t_{n+1}))\| + \|F(w(t_{n+1}))\|. \quad (14.13)$$

Furthermore, using (14.10) and (1.10), we get

$$\|F(u_{n+1}) - F(w(t_{n+1}))\| \leq M_1\|u_{n+1} - w(t_{n+1})\|$$

$$\leq M_1 L_1 c_2 h^2 \|F_0\|e^{-chn} \quad (14.14)$$

and using (3.26) with $u_0 = u_n$ and $t = t_{n+1} - t_n = h$, we get

$$\|F(w(t_{n+1}))\| \le \|F(u_n)\|e^{-c_1 h} \le \|F_0\|e^{-chn - c_1 h}. \tag{14.15}$$

From (14.13)–(14.15) we obtain

$$\|F(u_{n+1})\| \le \|F_0\|e^{-chn}\left(e^{-c_1 h} + M_1 L_1 c_2 h^2\right) \le \|F_0\|e^{-ch(n+1)}, \tag{14.16}$$

provided that

$$e^{-c_1 h} + M_1 L_1 c_2 h^2 \le e^{-ch}. \tag{14.17}$$

This inequality holds if $c_1 > c$ and h is sufficiently small.

Finally, the assumption (3.23) implies the existence of a solution y to equation (14.3).

Theorem 14.2.1 is proved.   ∎

Remark 14.2.2 If condition (3.23) is dropped but we assume that equation (14.3) has a solution y and $\|u_0 - y\|$ is sufficiently small, then our proof remains valid and yields the conclusion (14.6) of Theorem 14.2.1.

14.3 ITERATIVE SOLUTION OF ILL-POSED EQUATIONS WITH MONOTONE OPERATOR

Assume now that equation (14.3) has a solution y, the operator F in this equation satisfies assumption (1.10) but not (1.9), and F is monotone:

$$\langle F(u) - F(v), u - v \rangle \ge 0, \quad \forall u, v \in H, \quad F : H \to H, \tag{14.18}$$

where H is a Hilbert space. Let y be the (unique) minimal-norm solution to equation (14.3).

Theorem 14.3.1 *Under the above assumptions one can choose $h_n > 0$ and $a_n > 0$ such that the iterative process*

$$u_{n+1} = u_n - h_n A_n^{-1}[F(u_n) + a_n u_n], \quad u_0 = u_0, \tag{14.19}$$

where $A_n := F'(u_n) + a_n I$, and $u_0 \in H$ is arbitrary, converges to y:

$$\lim_{n \to \infty} \|u_n - y\| = 0. \tag{14.20}$$

Proof: Denote by V_n the solution of the equation

$$F(V_n) + a_n V_n = 0. \tag{14.21}$$

This equation has a unique solution as we have proved in Lemma 6.1.7. Let

$$z_n := u_n - V_n, \quad \|z_n\| := g_n. \tag{14.22}$$

We have

$$\|u_n - y\| \le g_n + \|V_n - y\|. \tag{14.23}$$

In Lemma 6.1.9 we have proved that

$$\lim_{n \to \infty} \|V_n - y\| = 0, \tag{14.24}$$

provided that

$$\lim_{n \to \infty} a_n = 0. \tag{14.25}$$

Let us prove that

$$\lim_{n \to \infty} \|u_n - V_n\| = 0. \tag{14.26}$$

Denote

$$b_n := \|V_{n+1} - V_n\|.$$

Then

$$\lim_{n \to \infty} b_n = 0.$$

Let us write (14.19) as

$$z_{n+1} = (1 - h_n)z_n - h_n A_n^{-1} K(z_n) - (V_{n+1} - V_n). \tag{14.27}$$

Here we have used Taylor formula:

$$F(u_n) + a_n u_n = F(v_n) - F(V_n) + a_n(u_n - V_n) = A_n z_n + K(z_n), \tag{14.28}$$

where

$$\|K(z_n)\| \le \frac{M_2}{2}\|z_n\|^2 := cg_n^2. \tag{14.29}$$

From (14.27) we obtain

$$g_{n+1} \le (1 - h_n)g_n + \frac{ch_n}{a_n}g_n^2 + b_n, \qquad 0 < h_n \le 1, \tag{14.30}$$

where the estimate

$$\|A_n^{-1}\| \le \frac{1}{a_n}$$

is used. Choose

$$a_n = 2cg_n. \tag{14.31}$$

Then

$$\begin{aligned} g_{n+1} &\le (1 - h_n)g_n + \frac{h_n}{2}g_n + b_n \\ &= \left(1 - \frac{h_n}{2}\right)g_n + b_n := (1 - \gamma_n)g_n + b_n. \end{aligned}$$

Therefore

$$g_{n+1} \le (1 - \gamma_n)g_n + b_n, \tag{14.32}$$

where
$$\gamma_n := \frac{h_n}{2}, \quad 0 < \gamma_n \le \frac{1}{2},$$

and $g_1 \ge 0$ is arbitrary.

Assume that
$$\sum_{n=1}^{\infty} h_n = \infty. \tag{14.33}$$

Then (14.32) implies
$$\lim_{n \to \infty} g_n = 0. \tag{14.34}$$

Indeed, (14.32) implies
$$g_{n+1} \le b_n + \sum_{k=1}^{n-1} b_k \prod_{j=k+1}^{n} (1 - \gamma_j) + g_1 \prod_{j=1}^{n} (1 - \gamma_j). \tag{14.35}$$

Assumption (14.33) implies
$$\lim_{n \to \infty} \prod_{j=1}^{n} (1 - \gamma_j) = 0. \tag{14.36}$$

Assume that (14.33) holds and
$$\lim_{n \to \infty} \frac{b_{n-1}}{h_n} = 0. \tag{14.37}$$

Then
$$\lim_{n \to \infty} \sum_{k=1}^{n-1} b_k \prod_{j=k+1}^{n} (1 - \gamma_j) = 0. \tag{14.38}$$

This follows from a discrete analog of L'Hospital's rule:

If $p_n, q_n > 0$, $\lim_{n \to \infty} q_n = \infty$, and $\lim_{n \to \infty} \frac{p_n - q_{n-1}}{q_n - q_{n-1}}$ exists, then
$$\lim_{n \to \infty} \frac{p_n}{q_n} = \lim_{n \to \infty} \frac{p_n - p_{n-1}}{q_n - q_{n-1}}.$$

In our case
$$p_n = \sum_{k=1}^{n-1} b_k \prod_{j=1}^{k} (1 - \gamma_j), \quad q_n = \prod_{j=1}^{n} (1 - \gamma_j)^n, \quad \lim_{n \to \infty} q_n = \infty,$$

and
$$\lim_{n \to \infty} \frac{p_n - p_{n-1}}{q_n - q_{n-1}} = \lim_{n \to \infty} \frac{b_{n-1} \prod_{j=1}^{n-1} (1 - \gamma_j)}{\prod_{j=1}^{n} (1 - \gamma_j) - \prod_{j=1}^{n-1} (1 - \gamma_j)}$$
$$= \lim_{n \to \infty} \frac{b_{n-1}}{-\gamma_n} = 0, \tag{14.39}$$

where we have used assumption (14.37).

Theorem 14.3.1 is proved.   ∎

Remark 14.3.2 If $h = const$, $h \in (0,1)$, then condition (14.33) holds, so one can use $h_n = h$, $h \in (0,1)$, in the iterative process (14.19) with a_n chosen as in (14.31). Then (14.32) takes the form

$$g_{n+1} \leq qg_n + b_n, \quad q := 1 - h \in (0,1), \quad \lim_{n \to \infty} b_n = 0, \tag{14.40}$$

g_1 is arbitrary, and (14.40) implies $\lim_{n \to \infty} g_n = 0$.

14.4   ITERATIVE METHODS FOR SOLVING NONLINEAR EQUATIONS

In this section we do not assume that $F : H \to H$ is monotone. We assume that condition (1.10) holds, that y solves equation (14.3), $F(y) = 0$, and that

$$\tilde{A} := F'(y) \neq 0. \tag{14.41}$$

We want to construct a convergent iterative process for solving equation (14.3).

The DSM for solving this equation has been given in Section 7.1, formula (7.3), and we use the notations and the results from this section. Consider the corresponding iterative method:

$$u_{n+1} = u_n - h_n T_{a_n}^{-1}[A^*(u_n)F(u_n) + a_n(u_n - z)], \quad u_0 = u_0, \tag{14.42}$$

where $a_n, h_n > 0$ are some sequences,

$$T := A^*(u_n)A(u_n), \quad T_{a_n} := T + a_n I.$$

Following the method developed in Section 7.1, let us denote

$$u_n - y := w_n, \quad \|w_n\| = g_n, \tag{14.43}$$

and rewrite (14.42) as follows:

$$w_{n+1} = w_n - h_n T_{a_n}^{-1}[A^*(u_n)(F(u_n) - F(y)) + a_n w_n + a_n(y - z)]. \tag{14.44}$$

As was proved in Section 7.1, we can choose z so that

$$y - z = \tilde{T}v, \quad 2M_1 M_2 \|v\| \leq \frac{1}{2}, \tag{14.45}$$

where M_1 and M_2 are constants from (1.10). This is possible if assumption (14.41) holds.

Let us use the formulas:

$$F(u_n) - F(y) = A(u_n)w_n + K(w_n), \quad \|K(w_n)\| \le \frac{M_2}{2}g_n^2, \qquad (14.46)$$

$$\|T_{a_n}^{-1}A^*(u_n)\| \le \frac{1}{2\sqrt{a_n}}, \qquad (14.47)$$

and rewrite (14.44) as follows:

$$w_{n+1} = (1 - h_n)w_n - h_nT_{a_n}^{-1}A^*(u_n)K(w_n) - h_na_nT_{a_n}^{-1}\tilde{T}v. \qquad (14.48)$$

We have

$$T_{a_n}^{-1}\tilde{T} = \left(T_{a_n}^{-1} - \tilde{T}_{a_n}^{-1}\right)\tilde{T} + \tilde{T}_{a_n}^{-1}\tilde{T}.$$

Therefore

$$\|T_{a_n}^{-1}\tilde{T}v\| \le \|\left(T_{a_n}^{-1} - \tilde{T}_{a_n}^{-1}\right)\tilde{T}\|\|v\| + \|v\|, \qquad (14.49)$$

where the inequality

$$\|\tilde{T}_a^{-1}\tilde{T}\| \le 1, \quad a \ge 0,$$

was used.

From (14.47)–(14.49) we obtain

$$g_{n+1} \le (1 - h_n)g_n + h_n\frac{M_2}{4\sqrt{a_n}}g_n^2 + h_na_n\|v\|$$

$$+ h_na_n\|v\|\|\left(T_{a_n}^{-1} - \tilde{T}_{a_n}^{-1}\right)\tilde{T}\|. \qquad (14.50)$$

Denote

$$C_0 := \frac{M_2}{2},$$

and use the following estimate:

$$\|\left(T_{a_n}^{-1} - \tilde{T}_{a_n}^{-1}\right)\tilde{T}\| = \|T_{a_n}^{-1}\left[A^*(u_n)A(u_n) - \tilde{A}^*(y)\tilde{A}(y)\right]\tilde{T}_{a_n}^{-1}\tilde{T}\|$$

$$\le \frac{2M_1M_2g_n}{a_n} := \frac{c_1g_n}{a_n}. \qquad (14.51)$$

From (14.50) and (14.51) we get

$$g_{n+1} \le \left(1 - \frac{h_n}{2}\right)g_n + \frac{c_0h_n}{\sqrt{a_n}}g_n^2 + h_na_n\|v\|, \qquad (14.52)$$

where we have used inequality (14.45), which implies

$$c_1\|v\| \le \frac{1}{2}.$$

Let us choose a_n so that

$$\frac{c_0}{\sqrt{a_n}}g_n = \frac{1}{4}, \quad \text{i.e.,} \quad a_n = 16c_0^2g_n^2. \qquad (14.53)$$

Then (14.52) takes the form

$$g_{n+1} \leq \left(1 - \frac{h_n}{4}\right) g_n + 16 h_n \|v\| c_0^2 g_n^2, \quad g_0 = \|u_0 - y\| \leq R, \qquad (14.54)$$

where $R > 0$ is the radius of the ball in condition (1.10).

We can choose $h_n = h \in (0,1)$ and $g_0 > 0$ such that (14.54) implies

$$\lim_{n \to \infty} g_n = 0. \qquad (14.55)$$

This possibility is seen from the following lemma.

Lemma 14.4.1 *Let*

$$g_{n+1} \leq \gamma g_n + p g_n^2, \quad g_0 = m > 0; \quad 0 < \gamma < 1, \quad p > 0, \qquad (14.56)$$

where p is an arbitrary fixed positive constant. If $m > 0$ is sufficiently small, namely, if

$$m < (q - \gamma)/p,$$

where $q \in (\gamma, 1)$ is an arbitrary number, then the sequence g_n, generated by (14.56), tends to zero:

$$\lim_{n \to \infty} g_n = 0 \qquad (14.57)$$

at a rate of a geometrical progression,

$$g_n \leq m q^n, \quad 0 < q < 1.$$

Proof: Assumption (14.56) implies

$$g_{n+1} < q^{n+1} m, \quad \gamma < q < 1, \quad n = 0, 1, 2, ..., \qquad (14.58)$$

provided that

$$m < \frac{q - \gamma}{p}, \quad \gamma < q < 1. \qquad (14.59)$$

Let us prove (14.58) by induction. We have

$$q_1 \leq \gamma m + p m^2 < q m, \quad m := g_0,$$

provided (14.59) holds. So (14.58) holds for $n = 0$. Assuming that it holds for some n, let us check that it holds for $n + 1$:

$$g_{n+2} \leq \gamma g_{n+1} + p g_{n+1}^2 \leq \gamma q^{n+1} m + p q^{2n+2} m^2$$
$$\leq q^{n+1} (\gamma m + p m^2) < q^{n+2} m. \qquad (14.60)$$

So Lemma 14.4.1 is proved. ∎

From Lemma 14.4.1 it follows that we can choose a constant $h_n = h$ in (14.54) such that by choosing g_0 sufficiently small we get (14.55).

Let us formulate the result we have proved.

Theorem 14.4.2 *Assume tht $F : F \to H$ satisfies conditions (1.10) and (14.41), where y solves equation (14.3). Then the iterative process (14.42) produces a sequence u_n such that*

$$\lim_{n \to \infty} \|u_n - y\| = 0, \tag{14.61}$$

provided that a_n is chosen as in (14.20) and $h_n = h$ is a constant, $h \in (0,1)$, $u_0 - y$ is sufficiently small, and z is chosen so that (14.45) holds. The convergence in (14.61) is at the rate of a geometrical series at least.

Remark 14.4.3 Theorem 14.4.2 guarantees that any solvable equation (14.3) can be solved by a convergent iterative process if conditions (1.10) and (14.41) hold. We do not give an algorithm for choosing z, but only prove the existence of such a z that (14.45) holds.

14.5 ILL-POSED PROBLEMS

Suppose that the assumptions of Theorem 14.3.1 or Theorem 14.4.2 hold, that equation

$$F(u) = f \tag{14.62}$$

is being considered, and that noisy data f_δ, $\|f_\delta - f\| \leq \delta$, are given in place of the exact data f for which equation (14.62) is solvable. We want to show that iterative processes (14.19) and (14.42) can be used for constructing a stable approximation to a solution y of equation (14.62). This is done by the method developed in Section 4.4. Namely, we stop iterations at the stopping number $n = n(\delta)$, $\lim_{\delta \to 0} n(\delta) = \infty$, which can be chosen so that $u_\delta := u_{n(\delta)}(f_\delta)$ is the desired stable approximation of y in the sense

$$\lim_{\delta \to 0} \|u_\delta - y\| = 0. \tag{14.63}$$

Here $u_{n(\delta)}(f_\delta)$ is the sequence, generated by the iterative process (14.19) (or (14.42)) with $F(u_n) - f_\delta$ replacing $F(u_n)$. Consider, for example, iterative process (14.19), where u_n is replaced by $u_{n,\delta} := u_n(f_\delta)$ and $F(u_n)$ is replaced by $F(u_n) - f_\delta$. We have proved in Theorem 14.3.1 that

$$\lim_{n \to \infty} \|u_n(f) - y\| = 0, \tag{14.64}$$

where $u_n := u_n(f)$ is generated by (14.19) with $F(u_n) - f_\delta$ replacing $F(u_n)$, f being the exact data. We have

$$\|u_n(f_\delta) - y\| \leq \|u_n(f_\delta) - u_n(f)\| + \|u_n(f) - y\|. \tag{14.65}$$

If we choose $n = n(\delta)$ such that

$$\lim_{\delta \to 0} n(\delta) = \infty \tag{14.66}$$

and

$$\lim_{\delta \to 0} \|u_{n(\delta)}(f_\delta) - u_{n(\delta)}(f)\| = 0, \tag{14.67}$$

then, due to (14.64), the element $u_\delta := u_{n(\delta)}(f_\delta)$ gives the desired stable approximation of the solution y. Due to the continuity of $u_n(f)$ with respect to f for any fixed n, we have

$$\|u_n(f_\delta) - u_n(f)\| \le \epsilon(n, \delta), \tag{14.68}$$

where

$$\lim_{\delta \to 0} \epsilon(n, \delta) = 0, \tag{14.69}$$

n being fixed. Therefore, denoting

$$\|u_n(f) - y\| := \omega(n), \quad \lim_{n \to \infty} \omega(n) = 0, \tag{14.70}$$

we rewrite inequality (14.65) as

$$\|u_n(f_\delta) - y\| \le \epsilon(n, \delta) + \omega(n). \tag{14.71}$$

Minimizing the right-hand side of (14.71) with respect to n for a fixed small $\delta > 0$, one finds $n = n(\delta)$, and

$$\lim_{\delta \to 0} [\epsilon(n(\delta), \delta) + \omega(n(\delta))] = 0, \quad \lim_{\delta \to 0} n(\delta) = \infty. \tag{14.72}$$

One can also find $n(\delta)$ by solving the equation

$$\epsilon(n, \delta) = \omega(n) \tag{14.73}$$

for n for a fixed small $\delta > 0$. For a fixed $\delta > 0$ the quantity $\epsilon(n, \delta)$ grows as $n \to \infty$, so that equation (14.73) has a solution $n(\delta)$ with the properties (14.72). This was proved in Section 2.4 for linear operators.

For nonlinear operators one may choose an arbitrary small number $\eta > 0$, find $n = n(\eta)$, such that

$$\omega(n(\eta)) < \frac{\eta}{2},$$

then for a fixed $n(\eta)$ find $\delta = \delta(\eta)$ so small that

$$\epsilon(n(\eta), \delta(\eta)) < \frac{\eta}{2}.$$

Then

$$\epsilon(n(\eta), \delta(\eta)) + \omega(n(\eta)) < \eta.$$

One can consider the functions $n = n(\eta)$ and $\delta = \delta(\eta)$ as a parametric representation of the dependence $n = n(\delta)$. The function $\omega(n)$ can be chosen without loss of generality monotonically decaying to zero as $n \to \infty$. Thus the equation $\omega(n) = \frac{\eta}{2}$ determines a unique monotone function $n = n(\eta)$, so that for a given n one can find a unique $\eta(n)$ such that $n(\eta(n)) = n$.

The function $\epsilon(n, \delta) = \delta p(n)$, where one may assume $\lim_{n \to \infty} p(n) = \infty$ and $p(n)$, grows monotonically when n grows.

Indeed

$$\epsilon(n, \delta) = \|u_n(f_\delta) - u_n(f)\| \leq \sup_{\|\xi\| \in B(f, \delta)} \|u_n'(\xi)\| \delta := p(n)\delta,$$

where $u_n'(\xi)$ is the Fréchet derivative of the operator $f \mapsto u_n(f)$ for fixed n. Since $p(n)$ is an upper bound on the norm of this operator, one may assume that $p(n)$ grows monotonically with n. One may minimize the right-hand side of (14.71) with respect to n for a fixed δ analytically if $\omega(n)$ and $p(n)$ are given analytically.

CHAPTER 15

NUMERICAL PROBLEMS ARISING IN APPLICATIONS

15.1 STABLE NUMERICAL DIFFERENTIATION

Let $f \in C^\mu(a,b)$, $\mu > 0$ be the space of μ times differentiable functions. If $0 < \mu < 1$, then the norm in $C^\mu(a,b)$ is defined as follows:

$$\|f\|_{C^\mu(a,b)} = \sup_{x \in (a,b)} |f(x)| + \sup_{x,y \in (a,b),\ x \neq y} \frac{|f(x) - f(y)|}{|x - y|^\mu}.$$

If $m < \mu < m + 1$, then

$$\|f\|_{C^\mu(a,b)} = \sup_{x \in (a,b)} \sum_{j=0}^{m} |f^{(j)}(x)| + \sup_{x,y \in (a,b),\ x \neq y} \frac{|f^{(m)}(x) - f^{(m)}(y)|}{|x - y|^{\mu - m}}.$$

If $\mu = m$, where $m \geq 0$ is an integer, then

$$\|f\|_{C^m(a,b)} = \sup_{x \in (a,b)} \sum_{j=0}^{m} |f^{(j)}(x)|.$$

Suppose that $\mu \geq 1$ and f_δ is given in place of f,

$$\|f_\delta - f\| \leq \delta.$$

Dynamical Systems Method and Applications: Theoretical Developments and Numerical Examples, First Edition. A. G. Ramm and N. S. Hoang.
Copyright © 2012 John Wiley & Sons, Inc.

We are going to discuss the problem of stable numerical differentiation of f given the noisy data f_δ. By a stable approximation of f' one usually means an expression $R_\delta f_\delta$ such that

$$\lim_{\delta \to 0} \|R_\delta(f_\delta) - f'\| = 0, \tag{15.1}$$

where the norm is C^0 norm, that is, the usual sup-norm. Thus

$$\|R_\delta(f_\delta) - f'\| := \eta(\delta) \to 0 \quad \text{as} \quad \delta \to 0, \tag{15.2}$$

where R_δ is some, not necessarily linear, operator acting on the noisy data f_δ. Since the data are $\{\delta, f_\delta\}$ and the exact data f are unknown, we think that it is natural to change the standard definition of the regularizer and to define a regularizer R_δ by the relation

$$\sup_{f \in B(f_\delta, \delta) \cap K} \|R_\delta(f_\delta) - f'\| := \eta(\delta) \to 0 \quad \text{as} \quad \delta \to 0, \tag{15.3}$$

where K is a compactum to which the exact data belong, and

$$B(f_\delta, \delta) := \{f : \|f - f_\delta\| \le \delta\}.$$

The above definition is the general Definition 2.1.3, applied to the problem of stable numerical differentiation.

In many applications K is defined as

$$K = K_\mu := \{f : \|f\|_\mu \le M_\mu\}, \quad \mu > 1, \tag{15.4}$$

where $\|\cdot\|_\mu$ in (15.4) denotes C^μ norm.

We will prove that

If $\mu \le 1$, then there does not exist R_δ, linear or nonlinear, such that (15.3) holds. If $\mu > 1$, then a linear operator R_δ exists such that (15.3) holds.

Moreover, R_δ will be given explicitly, analytically, and $\eta(\delta)$ will be specified.

If $\mu = 2$, we will prove that R_δ that we construct is the best possible operator among all linear and nonlinear operators in the following sense:

$$\sup_{f \in K_{2,\delta}} \|R_\delta f_\delta - f'\| = \inf_{T \in N} \sup_{f \in K_{2,\delta}} \|T f_\delta - f'\|, \tag{15.5}$$

where

$$K_{2,\delta} := K_2 \cap B(f_\delta, \delta),$$

and N is the set of all operators from $L^\infty(a, b)$ into $L^\infty(a, b)$.

Stable numerical differentiation is of practical interest in many applications. For example, if one measures the distance $s(t)$, traveled by a particle by the time t, and wants to find the velocity of this particle, one has to estimate $s'(t)$ stably, given $s_\delta(t)$, $\|s_\delta(t) - s(t)\| \le \delta$. The number of examples can be easily increased.

The first question to ask is:

When is it possible, in principle, to find R_δ satisfying (15.3)?
We split the second question into two questions:

1. *How does one construct R_δ?*

2. *What is the error estimate in formula (15.1.3)?*

Let us answer these questions.

Theorem 15.1.1 *If $\mu \leq 1$, then there is no operator R_δ such that (15.3) holds.*

Proof: Without loss of generality, let $(a, b) = (0, 1)$, and let

$$f_1(x) = -\frac{Mx(x - 2h)}{2}, \quad 0 \leq x \leq 2h, \quad f_2(x) = -f_1(x). \tag{15.6}$$

Let us extend f_1 from $[0, 2h]$ to $[2h, 1]$ by zero and denote again by f_1 the extended function. Thus $f_1 \in W^{1,\infty}(0, 1)$, where $W^{l,p}$ is the Sobolev space. We have

$$\|f_k\| := \sup_{x \in (0,1)} |f_k(x)| = \frac{Mh^2}{2}, \quad k = 1, 2. \tag{15.7}$$

Let

$$h = \sqrt{\frac{2\delta}{M}}. \tag{15.8}$$

Then

$$\frac{Mh^2}{2} = \delta.$$

For $f_\delta = 0$ we have

$$\|f_k - f_\delta\| = \|f_k\| = \delta. \tag{15.9}$$

Denote

$$T(0)|_{x=0} := b, \tag{15.10}$$

where $T = T_\delta$ is some operator,

$$T : L^\infty(0, 1) \to L^\infty(0, 1).$$

Let K_1 be given by (15.4) with a constant M_1. We have

$$f_1'(0) = Mh.$$

Thus

$$\gamma := \inf_T \sup_{f \in K_{1,\delta}} \|T(f_\delta) - f'\| \geq \inf_b \max\{|b - f_1'(0)|, |b + f_1'(0)|\}$$

$$= \inf_b \max\{|b - Mh|, |b + Mh|\}$$

$$= Mh = \sqrt{2\delta M} := \epsilon(\delta). \tag{15.11}$$

If the constant M_1 in the definition of K_1 is fixed, then

$$\sup_{x \in (0,1)} |f'_k(x)| = \sup_{x \in (0,2h)} M|h - x| = Mh = \sqrt{2\delta M} \le M_1. \qquad (15.12)$$

Since M in (15.6) is arbitrary, we can choose $M = M(\delta)$ so that $\sqrt{2\delta M} = M_1$, that is, $M = \frac{M_1^2}{2\delta}$. In this case $\gamma \ge M_1 > 0$, so that relation (15.3) does not hold no matter what the choice of R_δ is. If $\mu < 1$, then for no choice of R_δ the relation (15.3) holds because f' does not exist in $L^\infty(0,1)$.

Theorem 15.1.1 is proved. ∎

An argument similar to the above leads to a similar conclusion if the derivatives are understood in L^2 sense, $\| \cdot \|$ is the norm in the Sobolev space $H^\mu = H^\mu(0,1)$, that is, the space of real-valued functions with the norm defined for an integer $\mu \ge 0$ by the formula

$$\|u\|_\mu = \left(\sum_{j=0}^{\mu} \|u^{(j)}\|_0^2 \right)^{\frac{1}{2}},$$

where

$$\|u\|_0^2 := \int_0^1 |u|^2 dx,$$

and for arbitrary values $\mu > 0$ the norm is defined by interpolation; see, for instance, [12].

The main differences between C^μ and H^μ cases consists in the calculation of the norms $\|f_1\|_0$ and $\|f'_1\|_0$:

$$\|f_1\|_{L^2(0,1)} := \|f_1\|_0 = c_0 M h^{\frac{5}{2}}, \quad \|f'_1\| = c_1 M h^{\frac{3}{2}}, \qquad (15.13)$$

where $c_0, c_1 > 0$ are constants.

Indeed

$$\|f_1\|_0^2 = \int_0^{2h} \frac{M^2}{4} x^2 (x - 2h)^2 dx = \frac{M^2 h^5}{4} (2h)^2 \left(\frac{1}{5} - \frac{1}{4} + \frac{1}{3} \right)$$

$$= M^2 h^5 \frac{34}{15} := c_0^2 M^2 h^5,$$

and

$$\|f'_1\|_0^2 = \int_0^{2h} M^2 (x - h)^2 dx = \frac{2}{3} M^2 h^3 := c_1^2 M^2 h^3.$$

Choose h from the condition (15.9):

$$c_0 M h^{\frac{5}{2}} = \delta, \quad h = \left(\frac{\delta}{c_0 M} \right)^{\frac{2}{5}}. \qquad (15.14)$$

Let

$$f_\delta = 0, \quad K_{1,\delta}^{(2)} := \{ f : \|f\|_0 + \|f'\|_0 \le M_1 \}.$$

Then

$$\gamma_2 := \inf_{T} \sup_{f \in K_{1,\delta}^{(2)}} \|T(f_\delta) - f'\|$$

$$\geq \inf_{T} \max\{\|T(0) - f_1'\|_0, \|T(0) - f_1'\|_0\}. \tag{15.15}$$

Denote

$$\varphi := T(0) \in L^2(0,1),$$

and set

$$\varphi = cf_1' + \psi, \quad \langle \psi, f_1' \rangle = 0, \quad c = const. \tag{15.16}$$

Then (15.15) implies

$$\gamma_2 \geq \inf_{c \in \mathbb{R}, \psi \perp f_1'} \max\left\{ \sqrt{|1 - c|^2 \|f_1'\|_0^2 + \|\psi\|_0^2}, \sqrt{|1 + c|^2 \|f_1'\|_0^2 + \|\psi\|_0^2} \right\}$$

$$= \inf_{c \in \mathbb{R}} \max\{|1 - c|, |1 + c|\} \|f_1'\|_0 = \|f_1'\|_0 = c_1 M h^{\frac{3}{2}}. \tag{15.17}$$

From (15.14) and (15.17) we get

$$\gamma_2 \geq \frac{c_1}{c_0^{\frac{3}{5}}} M^{\frac{2}{5}} \delta^{\frac{3}{5}}. \tag{15.18}$$

We choose h by formula (15.14), and then condition (15.9) holds. We then choose M such that

$$c_1 M h^{\frac{3}{2}} = M_1. \tag{15.19}$$

Then (15.17) yields

$$\gamma_2 \geq M_1 > 0. \tag{15.20}$$

Thus we have an analog of Theorem 15.1.1 for L^2-norm.

Theorem 15.1.2 *If $\mu \leq 1$, then there is no operator R_δ such that (15.3) holds with $\| \cdot \| = \| \cdot \|_{L^2(0,1)}$.*

Proof: We have proved this theorem for $\mu = 1$. If $\mu < 1$, then f' does not exist in $L^2(0,1)$, so (15.3) does not hold with $\| \cdot \| = \| \cdot \|_{L^2(0,1)}$.

Let us now prove that if $\mu > 1$, then relation (15.3) holds. We obtain an estimate for $\eta_s(\delta)$. Define

$$R_\delta f_\delta := \begin{cases} \frac{f_\delta(x+h(\delta)) - f_\delta(x)}{h(\delta)}, & 0 \leq x \leq h(\delta), \\[2mm] \frac{f_\delta(x+h(\delta)) - f_\delta(x-h(\delta))}{2h(\delta)}, & h(\delta) \leq x \leq 1 - h(\delta), \\[2mm] \frac{f_\delta(x) - f_\delta(x-h(\delta))}{h(\delta)}, & 1 - h(\delta) \leq x \leq 1, \end{cases} \tag{15.21}$$

where $h(\delta) > 0$ will be specified below. Our argument is valid for L^p-norms with any $p \in [1, \infty]$. We have

$$\|R_\delta f_\delta - f'\| \le \|R_\delta f_\delta - R_\delta f\| + \|R_\delta f - f'\| \le \|R_\delta\|\delta + \|R_\delta f - f'\|. \quad (15.22)$$

For simplicity let us assume that f is periodic, $f(x+1) = f(x)$, and, therefore, f is defined on all of $\mathbb{R}$. Then we can use the middle line in (15.21) as the definition of $R_\delta f_\delta$, yielding

$$\|R_\delta\| \le \frac{1}{h(\delta)}, \quad (15.23)$$

and

$$\|R_\delta(f_\delta) - R_\delta(f)\| \le \frac{\delta}{h(\delta)}. \quad (15.24)$$

Let us estimate the last term in (15.22). Let $h(\delta) := h$, and assume $f \in C^2$. Then, for the L^∞-norm we have

$$\left\| \frac{f(x+h) - f(x-h)}{2h} - f' \right\|$$

$$= \left\| \frac{f(x) + hf'(x) + \frac{h^2}{2}f''(\xi_+) - f(x) + hf'(x) - \frac{h^2}{2}f''(\xi_-)}{2h} - f' \right\|$$

$$\le \frac{M_2 h}{2}, \quad (15.25)$$

where $\xi_\pm$ are the points in the remainder of the Taylor formula, and

$$M_m = \sup_{x \in (0,1)} |f^{(m)}(x)|.$$

Thus (15.22), (15.24), and (15.25) yield

$$\|R_\delta f_\delta - f'\| \le \frac{\delta}{h} + \frac{M_2 h}{2} := \eta(\delta, h). \quad (15.26)$$

Minimizing the right-hand side of (15.26) with respect to h, we get

$$h = \sqrt{\frac{2\delta}{M_2}}, \qquad \eta_s(\delta) := \eta(\delta, h(\delta)) = \sqrt{2M_2\delta}. \quad (15.27)$$

These formulas are obtained under the assumption $f \in C^2$.

If $f \in C^\mu$, $1 < \mu < 2$, then the estimate, analogous to (15.25), is

$$\left\| \frac{f(x+h) - f(x-h)}{2h} - f' \right\| = \left\| \frac{1}{2h} \int_{x-h}^{x+h} [f'(t) - f'(x)] \, dt \right\|$$

$$\le \frac{1}{2h} \left\| \int_{x-h}^{x+h} M_{\mu-1} |t - x|^{\mu-1} dt \right\|$$

$$= \frac{2M_{\mu-1}}{2h} \left\| \int_0^h s^{\mu-1} ds \right\|$$

$$= \frac{M_{\mu-1} h^{\mu-1}}{\mu}. \quad (15.28)$$

Therefore, estimates similar to (15.26) and (15.27) take the form

$$\|R_\delta f_\delta - f'\| \le \frac{\delta}{h} + \frac{M_{\mu-1}h^{\mu-1}}{\mu} = \eta(\delta, h), \tag{15.29}$$

$$h = h(\delta) = c_\mu \delta^{\frac{1}{\mu}}, \quad c_\mu := \left[\frac{\mu}{(\mu-1)M_{\mu-1}} \right]^{\frac{1}{\mu}};$$

$$\eta(\delta) = \eta(\delta, h(\delta)) = C_\mu \delta^{\frac{\mu-1}{\mu}}, \tag{15.30}$$

where

$$C_\mu = \frac{1}{\mu} + \frac{M_{\mu-1}}{\mu} c_\mu^{\mu-1}. \tag{15.31}$$

∎

We have proved the following result.

Theorem 15.1.3 *If $\mu \in (1, 2)$, then R_δ, defined in (15.21), satisfies estimate (15.3) if $h = h(\delta)$ is given in (15.30), and then the error $\eta(\delta)$ is given also in (15.30).*

If $\mu > 2$, then one can define

$$R_h^{(Q)} := \frac{1}{h} \sum_{k=-Q}^{Q} A_k^{(Q)} f\left(x + \frac{kh}{Q}\right). \tag{15.32}$$

Suppose

$$\sup_x |f^{(m)}(x)| \le M_m, \tag{15.33}$$

where $m = 2q$ or $m = 2q + 1$, and $q \ge 0$ is an integer. Let us take $Q = q$ in (15.32) and choose the coefficients $A_j^{(Q)}$ so that the difference $R_h^{(Q)} - f'$ has the highest order of smallness as $h \to 0$. Then these coefficients solve the following linear algebraic system:

$$\sum_{k=-Q}^{Q} \frac{1}{j!} \left(\frac{k}{Q}\right)^j A_k^{(Q)} = \delta_{1j}, \quad 0 \le j \le 2Q, \quad \delta_{kj} = \begin{cases} 1, & k = j, \\ 0, & k \ne j. \end{cases} \tag{15.34}$$

We set $A_0^{(Q)} = 0$. Then system (15.34) is a linear algebraic system for $2Q$ unknown coefficients $A_k^{(Q)}$, $k = \pm 1, ..., \pm Q$, with Vandermonde matrix whose determinant does not vanish. Thus, all the coefficients

$$A_k^{(Q)}, \quad 1 \le |k| \le Q,$$

are uniquely determined from the system (15.34). For example:

$$A_0^{(1)} = 0, \quad A_{\pm 1}^{(1)} = \pm\frac{1}{2}$$

$$A_0^{(2)} = 0, \quad A_{\pm 1}^{(2)} = \pm\frac{4}{3}, \quad A_{\pm 2}^{(2)} = \pm\frac{1}{6}$$

$$A_0^{(3)} = 0, \quad A_{\pm 1}^{(3)} = \pm\frac{9}{4}, \quad A_{\pm 2}^{(3)} = \pm\frac{9}{20}, \quad A_{\pm 3}^{(3)} = \pm\frac{1}{20},$$

etc.

Let $m = 2q + 1 > 1$, $q = Q$. We have

$$|R_h^{(Q)} f - f'| \le 2N_{2Q+1} h^{2Q}, \quad \|R_h^{(Q)}\|_{L^\infty} = \frac{c(Q)}{h}, \tag{15.35}$$

where

$$c(Q) = \sum_{k=-Q}^{Q} |A_k^{(Q)}|.$$

Therefore an inequality analogous to (15.26) with L^∞-norm will be of the form

$$\|R_h^{(Q)} f_\delta - f'\| \le \frac{c(Q)\delta}{h} + 2M_m h^{m-1} := \eta(\delta, h). \tag{15.36}$$

Minimizing the right-hand side of (15.36) with respect to h, we get

$$h = h(\delta) = \left[\frac{c(Q)\delta}{2M_m(m-1)} \right]^{\frac{1}{m}} \delta^{\frac{1}{m}} \tag{15.37}$$

and

$$\eta(\delta) = O\left(\delta^{1-\frac{1}{m}} \right) \quad \text{as} \quad \delta \to 0.$$

We have proved the following result.

Theorem 15.1.4 *If $f \in C^m$, $m > 2$, then the operator $R_\delta := R_{h(\delta)}^{(Q)}$, where $Q = \frac{m-1}{2}$ if m is odd, $Q = \frac{m}{2}$ if m is even, and $h(\delta)$ is given in (15.37), yields a stable approximation $R_\delta f_\delta$ of f' with the error $\eta(\delta) = O(\delta^{1-\frac{1}{m}})$ as $\delta \to 0$, and estimate (15.3) holds. A similar result holds for L^p-norm, $p \in [1, \infty]$.*

Finally, we state the following result, which follows from (15.11) and (15.30) with $\mu = 2$:

$$\|R_\delta f_\delta - f'\| \le \sqrt{2M_2 \delta}, \quad h(\delta) = \sqrt{\frac{2\delta}{M_2}}. \tag{15.38}$$

Let $\mu = 2$ and the norm be L^∞-norm. Then, among all operators T, linear and nonlinear, acting from the space of L^∞ periodic functions with period 1 into itself, the operator

$$T f_\delta = R_\delta f_\delta := \frac{f_\delta(x + h(\delta)) - f_\delta(x - h(\delta))}{2h(\delta)}$$

with $h(\delta) = \sqrt{\frac{2\delta}{M_2}}$ gives the best possible estimate of f', given the data f_δ, $\|f_\delta - f\| \le \delta$, on the class of all f such that $\sup_x |f''(x)| \le M_2$.

15.2   STABLE DIFFERENTIATION OF PIECEWISE-SMOOTH FUNCTIONS

Let f be a piecewise-$C^2([0,1])$ function,

$$0 < x_1 < x_2 < \cdots < x_J, \quad 1 \le j \le J,$$

be the discontinuity points of f. We do not assume their locations x_j and their number J known a priori. We assume that the limits $f(x_j \pm 0)$ exist, and

$$\sup_{x \ne x_j, 1 \le j \le J} |f^{(m)}(x)| \le M_m, \quad m = 0, 1, 2. \tag{15.39}$$

Assume that f_δ is given,

$$\|f - f_\delta\| := \sup_{x \ne x_j, 1 \le j \le J} |f - f_\delta| \le \delta,$$

where $f_\delta \in L^\infty(0,1)$ are the noisy data.

The problem is:

Given $\{f_\delta, \delta\}$, *where* $\delta \in (0, \delta_0)$ *and* $\delta_0 > 0$ *is a small number, estimate stably* f', *find the locations of discontinuity points* x_j *of* f *and their number* J, *and estimate the jumps*

$$p_j := f(x_j + 0) - f(x_j - 0)$$

of f *across* x_j, $1 \le j \le J$.

A stable estimate $R_\delta f_\delta$ of f' is an estimate satisfying the relation

$$\lim_{\delta \to 0} \|R_\delta f_\delta - f'\| = 0.$$

There is a large literature on stable differentiation of noisy smooth functions, but the problem stated above was not solved for piecewise-smooth functions by the method given below. A statistical estimation of the location of discontinuity points from noisy discrete data is given in [72]. In [71], [115], and [118], various approaches to finding discontinuities of functions from the measured values of these functions are developed.

The following formula (see Section 15.1) was proposed originally (in 1968; see [98]) for stable estimation of $f'(x)$, assuming $f \in C^2([0,1])$, $M_2 \ne 0$, and given noisy data f_δ:

$$R_\delta f_\delta := \frac{f_\delta(x + h(\delta)) - f_\delta(x - h(\delta))}{2h(\delta)},$$

$$h(\delta) := \left(\frac{2\delta}{M_2} \right)^{\frac{1}{2}}, \quad h(\delta) \le x \le 1 - h(\delta), \tag{15.40}$$

and

$$\|R_\delta f_\delta - f'\| \le \sqrt{2 M_2 \delta} := \varepsilon(\delta), \tag{15.41}$$

where the norm in (15.41) is $L^\infty(0,1)$ norm. Numerical efficiency and stability of the stable differentiation method proposed in [98] has been demonstrated in [123]. Moreover (cf. [107, p. 345] and Section 15.1),

$$\inf_{T} \sup_{f \in K(M_2,\delta)} \|Tf_\delta - f'\| \geq \varepsilon(\delta), \tag{15.42}$$

where $T : L^\infty(0,1) \to L^\infty(0,1)$ runs through the set of all bounded operators,

$$K(M_2,\delta) := \{f : \|f''\| \leq M_2, \ \|f - f_\delta\| \leq \delta\}.$$

Therefore, estimate (15.40) is the best possible estimate of f', given noisy data f_δ and assuming $f \in K(M_2,\delta)$.

In [137] this result was generalized to the case

$$f \in K(M_a,\delta), \quad \|f^{(a)}\| \leq M_a, \quad 1 < a \leq 2,$$

where

$$\|f^{(a)}\| := \|f\| + \|f'\| + \sup_{x,x'} \frac{|f'(x) - f'(x')|}{|x - x'|^{a-1}}, \quad 1 < a \leq 2,$$

and $f^{(a)}$ is the fractional-order derivative of f.

The aim of this section is to extend the above results to the case of piecewise-smooth functions.

Theorem 15.2.1 *Formula (15.40) gives stable estimate of f' on the set $S_\delta :=$ $[h(\delta), 1 - h(\delta)] \setminus \bigcup_{j=1}^{J}(x_j - h(\delta), x_j + h(\delta))$, and (15.41) holds with the norm $\|\cdot\|$ taken on the set S_δ. Assuming $M_2 > 0$ and computing the quantities*

$$f_j := \frac{f_\delta(jh + h) - f_\delta(jh - h)}{2h},$$

where

$$h := h(\delta) := \left(\frac{2\delta}{M_2}\right)^{\frac{1}{2}}, \quad 1 \leq j < [\frac{1}{h}],$$

for sufficiently small δ, one finds the location of discontinuity points of f with accuracy $2h$, along with their number J. Here $[\frac{1}{h}]$ is the integer smaller than $\frac{1}{h}$ and closest to $\frac{1}{h}$. The discontinuity points of f are located on the intervals $(jh - h, jh + h)$ such that $|f_j| \gg 1$ for sufficiently small δ, where $\varepsilon(\delta)$ is defined in (15.41). The size p_j of the jump of f across the discontinuity point x_j is estimated by the formula

$$p_j \approx f_\delta(jh + h) - f_\delta(jh - h),$$

and the error of this estimate is $O(\sqrt{\delta})$.

Let us assume that $\min_j |p_j| := p \gg h(\delta)$, where $\gg$ means "much greater than". Then x_j is located on the jth interval $[jh - h, jh + h]$, $h := h(\delta)$, such that

$$|f_j| := \left| \frac{f_\delta(jh + h) - f_\delta(jh - h)}{2h} \right| \gg 1, \qquad (15.43)$$

so that x_j is localized with the accuracy $2h(\delta)$. More precisely,

$$|f_j| \geq \frac{|f(jh + h) - f(jh - h)|}{2h} - \frac{\delta}{h}$$

and

$$\frac{\delta}{h} = 0.5\varepsilon(\delta),$$

where $\varepsilon(\delta)$ is defined in (15.41) . One has

$$|f(jh + h) - f(jh - h)| \geq, |p_j| - |f(jh + h) - f(x_j + 0)|$$
$$- |f(jh - h) - f(x_j - 0)|,$$
$$\geq |p_j| - 2M_1 h.$$

Thus,

$$|f_j| \geq \frac{|p_j|}{2h} - M_1 - 0.5\varepsilon(\delta) = c_1 \frac{|p_j|}{\sqrt{\delta}} - c_2 \gg 1,$$

where

$$c_1 := \frac{\sqrt{M_2}}{2\sqrt{2}} \quad \text{and} \quad c_2 := M_1 + 0.5\varepsilon(\delta).$$

The jump p_j is estimated by the formula

$$p_j \approx [f_\delta(jh + h) - f_\delta(jh - h)], \qquad (15.44)$$

and the error estimate of this formula can be given:

$$|p_j - [f_\delta(jh + h) - f_\delta(jh - h)]| \leq 2\delta + 2M_1 h$$
$$= 2\delta + 2M_1 \sqrt{\frac{2\delta}{M_2}} = O(\sqrt{\delta}). \qquad (15.45)$$

Thus, the error of the calculation of p_j by the formula

$$p_j \approx f_\delta(jh + h) - f_\delta(jh - h)$$

is $O(\delta^{\frac{1}{2}})$ as $\delta \to 0$.

Proof of Theorem 15.2.1. If $x \in S_\delta$, then, using Taylor's formula, one gets

$$|(R_\delta f_\delta)(x) - f'(x)| \leq \frac{\delta}{h} + \frac{M_2 h}{2}. \qquad (15.46)$$

Here we assume that $M_2 > 0$ and the interval $(x - h(\delta), x + h(\delta)) \subset S_\delta$, i.e., this interval does not contain discontinuity points of f. If for all sufficiently small h, not necessarily for $h = h(\delta)$, inequality (15.46) fails, i.e., if

$$|(R_\delta f_\delta)(x) - f'(x)| > \frac{\delta}{h} + \frac{M_2 h}{2}$$

for all sufficiently small $h > 0$, then the interval $(x - h, x + h)$ contains a point $x_j \notin S_\delta$, that is, a point of discontinuity of f or f'. This observation can be used for locating the position of an isolated discontinuity point x_j of f with any desired accuracy, provided that the size $|p_j|$ of the jump of f across x_j is greater than $k\delta$, where $k > 2$ is a constant, $|p_j| > k\delta$, and that h can be taken as small as desirable.

Indeed, if $x_j \in (x - h, x + h)$, then we have

$$|p_j| - 2\delta - 2hM_1 \leq |f_\delta(x + h) - f_\delta(x - h)| \leq |p_j| + 2hM_1 + 2\delta.$$

The above estimate follows from the relation

$$|f_\delta(x + h) - f_\delta(x - h)|$$
$$= |f(x + h) - f(x_j + 0) + p_j + f(x_j - 0) - f(x - h) \pm 2\delta|$$
$$= |p_j \pm (2hM_1 + 2\delta)|.$$

Here $|p \pm b|$, where $b > 0$, denotes a quantity such that

$$|p| - b \leq |p \pm b| \leq |p| + b.$$

Thus, if h is sufficiently small and $|p_j| > k\delta$, where $k > 2$, then the inequality

$$(k - 2)\delta - 2hM_1 < |f_\delta(x + h) - f_\delta(x - h)|$$

can be checked, and therefore the inclusion $x_j \in (x - h, x + h)$ can be checked. Since $h > 0$ is arbitrarily small in this argument, it follows that the location of the discontinuity point x_j of f, at which $|p_j| > k\delta$ with $k > 2$, can be established with arbitrary accuracy.

A discussion of the case when a discontinuity point x_j belongs to the interval $(x - h(\delta), x + h(\delta))$ will be given below.

Minimizing the right-hand side of (15.46) with respect to h yields formula (15.40) for the minimizer $h = h(\delta)$ defined in (15.40), along with estimate (15.41) for the minimum of the right-hand side of (15.46).

If $|p| \gg h(\delta)$, and (15.43) holds, then the discontinuity points are located with the accuracy $2h(\delta)$, as we prove now by an argument very similar to the one given above.

Consider the case when a discontinuity point x_j of f belongs to the interval $(jh - h, jh + h)$, where $h = h(\delta)$. Then estimate (15.44) can be obtained as follows. For $jh - h \leq x_j \leq jh + h$, one has

$$|f(x_j + 0) - f(x_j - 0) - f_\delta(jh + h) + f_\delta(jh - h)|$$
$$\leq 2\delta + |f(x_j + 0) - f(jh + h)|$$
$$+ |f(x_j - 0) - f(jh - h)| \leq 2\delta + 2hM_1, \quad h = h(\delta).$$

This yields formulas (15.44) and (15.45). Computing the quantities f_j for $1 \le j < [\frac{1}{h}]$, and finding the intervals on which (15.43) holds for sufficiently small δ, one finds the location of discontinuity points of f with accuracy $2h$, and the number J of these points. For a small fixed $\delta > 0$ the above method allows one to recover the discontinuity points of f at which

$$|f_j| \ge \frac{|p_j|}{2h} - \frac{\delta}{h} - M_1 \gg 1.$$

This is the inequality (15.43). If $h = h(\delta)$, then

$$\frac{\delta}{h} = 0.5\varepsilon(\delta) = O(\sqrt{\delta})$$

and

$$|2hf_j - p_j| = O(\sqrt{\delta}) \quad \text{as} \quad \delta \to 0,$$

provided that $M_2 > 0$. Theorem 15.2.1 is proved. ∎

Remark 15.2.2 Similar results can be derived if

$$\|f^{(\mu)}\|_{L^\infty(S_\delta)} := \|f^{(\mu)}\|_{S_\delta} \le M_\mu, \qquad 1 < \mu \le 2.$$

In this case

$$h = h(\delta) = c_\mu \delta^{\frac{1}{\mu}},$$

where $c_\mu = \left[\frac{2}{M_\mu(\mu-1)} \right]^{\frac{1}{\mu}}$, $R_\delta f_\delta$ is defined in (15.40), and the error of the estimate is

$$\|R_\delta f_\delta - f'\|_{S_\delta} \le \mu M_\mu^{\frac{1}{\mu}} \left(\frac{2}{\mu-1} \right)^{\frac{\mu-1}{\mu}} \delta^{\frac{\mu-1}{\mu}}.$$

The proof is similar to the given above. In Section 15.1 it is proved that for C^μ-functions given with noise, it is possible to construct stable differentiation formulas if $\mu > 1$ and it is impossible to construct such formulas if $\mu \le 1$. The obtained formulas are useful in applications. One can also use L^p-norm on S_δ in the estimate $\|f^{(\mu)}\|_{S_\delta} \le M_\mu$.

Remark 15.2.3 The case when $M_2 = 0$ requires a discussion. In this case the last term on the right-hand side of formula (15.46) vanishes and the minimization with respect to h becomes meaningless: It requires that h be as large as possible, but one cannot take h arbitrarily large because estimate (15.46) is valid only on the interval $(x - h, x + h)$ which does not contain discontinuity points of f, and these points are unknown. If $M_2 = 0$, then f is a piecewise-linear function. The discontinuity points of a piecewise-linear function can be found if the sizes $|p_j|$ of the jumps of f across these points satisfy the inequality

$$|p_j| >> 2\delta + 2hM_1$$

for some choice of h. This follows from the estimate

$$|f_\delta(jh + h) - f_\delta(jh - h)| \geq |p_j| - 2hM_1 - 2\delta,$$

if the discontinuity point x_j lies on the interval $(jh - h, jh + h)$. For instance, if $h = \frac{\delta}{M_1}$, then $2\delta + 2M_1 h = 4\delta$. So, if $|p_j| >> 4\delta$, then the location of discontinuity points of f can be found in the case when $M_2 = 0$. The discontinuity points x_j of f are located on the intervals for which

$$|f_\delta(jh + h) - f_\delta(jh - h)| >> 4\delta,$$

where $h = \frac{\delta}{M_1}$.

The size $|p_j|$ of the jump of f across a discontinuity point x_j can be estimated by formula (15.44) with $h = \frac{\delta}{M_1}$, and one assumes that $x_j \in (jh - h, jh + h)$ is the only discontinuity point on this interval. The error of the formula (15.44) is estimated as in the proof of Theorem 15.2.1. This error is not more than $2\delta + 2M_1 h = 4\delta$ for the above choice of $h = \frac{\delta}{M_1}$.

One can estimate the derivative of f at the point of smoothness of f assuming $M_2 = 0$, provided that this derivative is not too small. If $M_2 = 0$, then $f = a_j x + b_j$ on every interval Δ_j between the discontinuity points x_j, where a_j and b_j are some constants. If $(jh - h, jh + h) \subset \Delta_j$ and

$$f_j := \frac{f_\delta(jh + h) - f_\delta(jh - h)}{2h},$$

then

$$|f_j - a_j| \leq \frac{\delta}{h}.$$

Choose $h = \frac{t\delta}{M_1}$, where $t > 0$ is a parameter and $M_1 = \max_j |a_j|$. Then the relative error of the approximate formula

$$a_j \approx f_j$$

for the derivative $f' = a_j$ on Δ_j equals

$$\frac{|f_j - a_j|}{|a_j|} \leq \frac{M_1}{t|a_j|}.$$

Thus, if, for example, $|a_j| \geq \frac{M_1}{2}$ and $t = 20$, then the relative error of the above approximate formula is not more than 0.1.

Suppose now that $\xi \in (mh - h, mh + h)$, where $m > 0$ is an integer, and ξ is a point at which f is continuous but $f'(\xi)$ does not exist. Thus, the jump of f across ξ is zero, but ξ is not a point of smoothness of f. *How does one locate the point ξ?*

The algorithm we propose consists of the following. We assume that $M_2 > 0$ on S_δ. Calculate the numbers

$$f_j := \frac{f_\delta(jh + h) - f_\delta(jh - h)}{2h}$$

and

$$|f_{j+1} - f_j|, \quad j = 1, 2, \ldots, \quad h = h(\delta) = \sqrt{\frac{2\delta}{M_2}}.$$

Inequality (15.41) implies

$$f_j - \varepsilon(\delta) \le f'(jh) \le f_j + \varepsilon(\delta),$$

where $\varepsilon(\delta)$ is defined in (15.41).

Therefore, if

$$|f_j| > \varepsilon(\delta),$$

then

$$\text{sign } f_j = \text{sign } f'(jh).$$

One has

$$J - \frac{2\delta}{h} \le |f_{j+1} - f_j| \le J + \frac{2\delta}{h},$$

where

$$\frac{\delta}{h} = 0.5\varepsilon(\delta)$$

and

$$J := \left| \frac{f(jh + 2h) - f(jh) - f(jh + h) + f(jh - h)}{2h} \right|.$$

Using Taylor's formula, one derives the estimate:

$$0.5[J_1 - \varepsilon(\delta)] \le J \le 0.5[J_1 + \varepsilon(\delta)], \tag{15.47}$$

where

$$J_1 := |f'(jh + h) - f'(jh)|.$$

If the interval $(jh - h, jh + 2h)$ belongs to S_δ, then

$$J_1 = |f'(jh + h) - f'(jh)| \le M_2 h = \varepsilon(\delta).$$

In this case

$$J \le \varepsilon(\delta),$$

so

$$|f_{j+1} - f_j| \le 2\varepsilon(\delta) \quad \text{if} \quad (jh - h, jh + 2h) \subset S_\delta. \tag{15.48}$$

Conclusion: If

$$|f_{j+1} - f_j| > 2\varepsilon(\delta),$$

then the interval $(jh - h, jh + 2h)$ does not belong to S_δ; that is, there is a point $\xi \in (jh - h, jh + 2h)$ at which the function f is not twice continuously differentiable with $|f''| \le M_2$. Since we assume that either at a point ξ the function is twice differentiable or at this point f' does not exist, it follows that if $|f_{j+1} - f_j| > 2\varepsilon(\delta)$, then there is a point $\xi \in (jh - h, jh + 2h)$ at which f' does not exist.

If

$$f_j f_{j+1} < 0 \tag{15.49}$$

and

$$\min(|f_{j+1}|, |f_j|) > \varepsilon(\delta), \tag{15.50}$$

then (15.49) implies $f'(jh)f'(jh+h) < 0$, so the interval $(jh, jh+h)$ contains a critical point ξ of f, or a point ξ at which f' does not exist. To determine which one of these two cases holds, let us use the right inequality (15.47). If ξ is a critical point of f and $\xi \in (jh, jh+h) \subset S_\delta$, then $J_1 \leq \varepsilon(\delta)$, and in this case the right inequality (15.47) yields $J \leq \varepsilon(\delta)$. Thus

$$|f_{j+1} - f_j| \leq 2\varepsilon(\delta). \tag{15.51}$$

Conclusion: If (15.49)–(15.51) hold, then ξ is a critical point. If (15.49) and (15.50) hold and

$$|f_{j+1} - f_j| > 2\varepsilon(\delta),$$

then ξ is a point of discontinuity of f'.

If ξ is a point of discontinuity of f', we would like to estimate the jump

$$P := |f'(\xi + 0) - f'(\xi - 0)|.$$

Using Taylor's formula, one gets

$$f_{j+1} - f_j = \frac{P}{2} \pm 2.5\varepsilon(\delta). \tag{15.52}$$

The expression $A = B \pm b$, $b > 0$, means that $B - b \leq A \leq B + b$. Therefore,

$$P = 2(f_{j+1} - f_j) \pm 5\varepsilon(\delta). \tag{15.53}$$

We have proved the following theorem:

Theorem 15.2.4 *If $\xi \in (jh - h, jh + 2h)$ is a point of continuity of f and $|f_{j+1} - f_j| > 2\varepsilon(\delta)$, then ξ is a point of discontinuity of f'. If (15.49) and (15.50) hold, and $|f_{j+1} - f_j| \leq 2\varepsilon(\delta)$, then ξ is a critical point of f. If (15.49) and (15.50) hold and $|f_{j+1} - f_j| > 2\varepsilon(\delta)$, then $\xi \in (jh, jh + h)$ is a point of discontinuity of f'. The jump P of f' across ξ is estimated by formula (15.53).*

Let us give a method for finding *nonsmoothness points of piecewise-linear functions.*

Assume that f is a piecewise-linear function on the interval $[0, 1]$ and $0 < x_1 < \cdots < x_J < 1$ is its nonsmoothness points, that is, the discontinuity points of f or these of f'. Assume that f_δ is known at a grid mh, $m = 0, 1, 2, \ldots, M$,

$$h = \frac{1}{M}, \quad f_{\delta,m} = f_\delta(mh), \quad |f(mh) - f_{\delta,m}| \leq \delta \; \forall m,$$

where $f_m = f(mh)$. If mh is a discontinuity point, $mh = x_j$, then we define its value as $f(x_j - 0)$ or $f(x_j + 0)$, depending on which of these two numbers satisfy the inequalty $|f(mh) - f_{\delta,m}| \leq \delta$.

The problem is:

Given $f_{\delta,m}$ $\forall m$, estimate the location of the discontinuity points x_j and estimate their number J, find out which of these points are points of disconti- nuity of f and which are points of discontinuity of f' but points of continuity of f, and estimate the sizes of the jumps of f

$$|p_j| = |f(x_j + 0) - f(x_j - 0)|$$

and the sizes of the jumps of f'

$$q_j = |f'(x_j + 0) - f'(x_j - 0)|$$

at the continuity points of f which are discontinuity points of f'.

Let us solve this problem. Consider the quantities

$$G_m := \frac{f_{\delta,m+1} - 2f_{\delta,m} + f_{\delta,m-1}}{2h^2} := g_m + w_m$$

where

$$g_m := \frac{f_{m+1} - 2f_m + f_{m-1}}{2h^2}$$

and

$$w_m := \frac{f_{\delta,m+1} - f_{m+1} - 2(f_{\delta,m} - f_m) + f_{\delta,m-1} - f_m}{2h^2}.$$

We have

$$|w_m| \leq \frac{4\delta}{2h^2} = \frac{2\delta}{h^2}$$

and

$$g_m = 0 \text{ if } x_j \notin (mh - h, mh + h), \quad \forall j.$$

Therefore, the following claim holds:

Claim:

If $\min_j |x_{j+1} - x_j| > 2h$ and

$$|G_m| > \frac{2\delta}{h^2}, \tag{15.54}$$

then the interval $(mh - h, mh + h)$ must contain a discontinuity point of f.

Condition (15.54) is sufficient for the interval $(mh - h, mh + h)$ to contain a discontinuity point of f, but is not a necessary condition: It may happen that the interval $(mh - h, mh + h)$ contains more than one discontinuity point (this is only possible if the assumption $\min_j |x_{j+1} - x_j| > 2h$ does not hold) without changing g_m or G_m, so that one cannot detect these points by the above method. We have proved the following result.

Theorem 15.2.5 *Condition (15.54) is a sufficient condition for the interval* $(mh - h, mh + h)$ *to contain a nonsmoothness point of* f. *If one knows a priori that* $\min_j |x_{j+1} - x_j| > 2h$, *then condition (15.54) is a necessary and sufficient condition for the interval* $(mh - h, mh + h)$ *to contain exactly one point of nonsmoothness of* f.

Let us estimate the size of the jump $|p_j| = |f(x_j + 0) - f(x_j - 0)|$. Let us assume that (15.54) holds, $x_{j+1} - x_j > 2h$, so there is only one discontinuity point x_j of f on the interval $(mh - h, mh + h)$, and assume that $x_j \in (mh - h, mh)$. The case when $x_j \in (mh, mh + h)$ is treated similarly. Let

$$f(x) = a_j x + b_j \quad \text{when} \quad mh < x < x_j$$

and

$$f(x) = a_{j+1} x + b_{j+1} \quad \text{when} \quad x_j < x < (m + 1)h,$$

where a_j and b_j are constants. One has

$$g_m = \frac{-(a_{j+1} - a_j)(mh - h) - (b_{j+1} - b_j)}{2h^2},$$

and

$$|p_j| = |(a_{j+1} - a_j)x_j + b_{j+1} - b_j|.$$

Thus

$$|g_m| = \left| \frac{-(a_{j+1} - a_j)x_j - (b_{j+1} - b_j) - (a_{j+1} - a_j)(mh - h - x_j)}{2h^2} \right|$$

$$= \frac{|p_j|}{2h^2} \pm \frac{|a_{j+1} - a_j||x_j - (mh - h)|}{2h^2}, \tag{15.55}$$

where $A \pm B$ denotes a quantity such that $A - B \le A \pm B \le A + B$, $A, B > 0$.

Let

$$|a_{j+1} - a_j| = q_j.$$

Note that

$$|x_j - (mh - h)| \le h \quad \text{if} \quad mh - h < x_j < mh.$$

Thus,

$$|G_m| = \frac{|p_j|}{2h^2} \pm \left(\frac{q_j h}{2h^2} + \frac{2\delta}{h^2} \right).$$

Therefore,

$$|G_m| = \frac{|p_j|}{2h^2} \left[1 \pm \left(\frac{q_j h}{|p_j|} + \frac{4\delta}{|p_j|} \right) \right],$$

provided that $|p_j| > 0$.

If

$$\frac{q_j h + 4\delta}{|p_j|} \ll 1 \quad \text{and} \quad |p_j| > 0,$$

then

$$|p_j| \approx 2h^2 |G_m|.$$

If $p_j = 0$, then $x_j = \frac{b_j - b_{j+1}}{a_{j+1} - a_j}$ and

$$|G_m| = \frac{q_j(x_j - mh + h)}{2h^2} \pm \frac{2\delta}{h^2},$$

because $x_j > mh - h$ by the assumption. Thus,

$$q_j \approx \frac{2h^2 |G_m|}{x_j - mh + h},$$

provided that $p_j = 0$ and $\delta << q_j(x_j - mh + h)$.
    If

$$\min_j |x_{j+1} - x_j| > 2h,$$

then the number J of the nonsmoothness points of f can be determined as the number of intervals on which (15.54) holds.

15.3 SIMULTANEOUS APPROXIMATION OF A FUNCTION AND ITS DERIVATIVE BY INTERPOLATION POLYNOMIALS

In this section we present a result from [102]. We want to construct an interpolation polynomial which gives a stable approximation of a continuous function $f \in C(I)$, $I := [-1, 1]$, given noisy values $f_\delta(x_j)$, $|f_\delta(x_j) - f(x_j)| < \delta$, and approximation should have the property that its derivative gives a stable approximation of f'.
    The idea is to use Lagrange interpolating polynomial $L_n(x)$ with nodes at the roots

$$x_j = x_{j,n+1} := \cos\left(\frac{2j-1}{2n+2}\pi\right), \quad 1 \le j \le n+1,$$

of the Tchebyshev polynomial $T_{n+1}(x)$, $x \in I$,

$$T_n(x) := 2^{-(n-1)} \cos(n \arccos x), \quad L_n = L_n(x, f) = \sum_{j=1}^{n+1} f(x_j) l_j(x), \quad (15.56)$$

where

$$l_j(x) := \frac{T_{n+1}(x)}{T'_{n+1}(x_j)(x - x_j)}. \quad (15.57)$$

Let

$$\lambda_n := \max_{x \in I} \sum_{j=1}^{n+1} |l_j(x)|, \quad \lambda'_n(x) := \sum_{j=1}^{n+1} |l'_j(x)|, \quad \lambda'_n := \max_{x \in I} \lambda'_n(x). \quad (15.58)$$

It is known (see [5]) that

$$\frac{\ln(n+1)}{8\sqrt{\pi}} < \lambda_n < 8 + \frac{4}{\pi}\ln(n+1). \tag{15.59}$$

Let

$$E_n(f) := \min_{P \in \mathcal{P}_n} \max_{x \in I} |f(x) - P(x)|,$$

where $\mathcal{P}_n$ is the set of all polynomials of degree $\leq n$. The polynomial of the best approximation exists and is unique for any $f \in C(I)$.

If $C^r = C^r(I)$ is the set of r times continuously differentiable functions on I with the norm

$$\|f\|_r = \max_{x \in I}\{|f(x)| + |f^{(r)}(x)|\}$$

and $f \in C^1$, then (see [188])

$$\|f - L_n\| \leq (1 + \lambda_n)E_n(f), \tag{15.60}$$

where $\|\cdot\| := \|\cdot\|_0$.

Let us state our results.

Theorem 15.3.1 *Let $f \in C^1$. Then*

$$\|f' - L_n'\| \leq (1 + \lambda_n')E_n(f'), \tag{15.61}$$

$$|f'(x) - L_n'(x)| \leq \left(1 + \frac{n\lambda_n}{\sqrt{1-x^2}}\right)E_{n-1}(f'), \quad \lambda_n' \leq n^2\lambda_n, \tag{15.62}$$

$$\lambda_n'(x) \leq \frac{n\lambda_n}{\sqrt{1-x^2}}. \tag{15.63}$$

Let

$$L_{n,\delta}(x) := \sum_{j=1}^{n+1} f_\delta(x_j)l_j(x).$$

Our second result is the following theorem:

Theorem 15.3.2 *Let $f \in C^r$, $r > 3$. Then there exists a function $n(\delta)$ such that*

$$\|L_{n(\delta),\delta} - f\| = O(\delta|\ln\delta|), \quad \delta \to 0, \tag{15.64}$$

$$|\frac{d}{dx}L_{n(\delta),\delta}(x) - f'(x)| \leq E_{n-1}(f')[1 + \lambda_n'] + \delta\lambda_n'(x), \tag{15.65}$$

$$\|L_{n(\delta),\delta} - f\| \leq (1 + \lambda_n)E_n(f) + \delta\lambda_n. \tag{15.66}$$

The following lemma will be used in the proofs.

Lemma 15.3.3 *Let* $P_j \in \mathcal{P}_n$ *and*

$$\sum_{j=1}^{m} |P_j(x)| \leq M, \quad x \in I, \tag{15.67}$$

where $M = const$ *and* $m \geq 1$ *is an arbitrary integer. Then*

$$\sum_{j=1}^{m} |P_j'(x)| \leq \frac{Mn}{\sqrt{1-x^2}}, \tag{15.68}$$

$$\sum_{j=1}^{m} |P_j'(x)| \leq Mn^2. \tag{15.69}$$

This lemma is a generalization of the known for $m = 1$ inequalities of S. Bernstein.

Proof of Lemma 15.3.3. Let

$$x = \cos\theta, \quad P_j(x) = \mathcal{P}_j(\theta).$$

Then (see [188, p. 227]):

$$\mathcal{P}_j'(\theta) = \frac{1}{4n} \sum_{k=1}^{2n} (-1)^{k+1} \mathcal{P}_j(\theta + \theta_k) \frac{1}{\sin^2(\frac{\theta_k}{2})}, \quad \theta_k = \frac{(2k-1)\pi}{2n}, \tag{15.70}$$

and

$$\frac{1}{4n} \sum_{k=1}^{2n} \frac{1}{\sin^2\left(\frac{\theta_k}{2}\right)} = n. \tag{15.71}$$

From (15.67), (15.70), and (15.71) we get

$$\sum_{j=1}^{m} |\mathcal{P}_j'(\theta)| \leq \sum_{j=1}^{m} \left| \frac{1}{4n} \sum_{k=1}^{2n} (-1)^{k+1} \mathcal{P}_j(\theta + \theta_k) \frac{1}{\sin^2(\frac{\theta_k}{2})} \right|$$

$$\leq \frac{1}{4n} \sum_{k=1}^{2n} (-1)^{k+1} \frac{1}{\sin^2(\frac{\theta_k}{2})} \sum_{j=1}^{m} |\mathcal{P}_j(\theta + \theta_k)| \leq Mn. \tag{15.72}$$

Using (15.72), we get

$$\sum_{j=1}^{m} |P_j'(x)| = \sum_{j=1}^{m} |\mathcal{P}_j'(\theta)| \left| \frac{d\theta}{dx} \right| \leq \frac{Mn}{\sqrt{1-x^2}}, \tag{15.73}$$

so (15.68) is proved.

Let us prove (15.69).

If $|x| \leq \cos\left(\frac{\pi}{2n}\right)$, then, using the inequality

$$\sin x \geq \frac{2x}{\pi}, \quad 0 \leq x \leq \frac{\pi}{2},$$

we get

$$\sqrt{1 - x^2} \geq \sin\left(\frac{\pi}{2n}\right) > \frac{\pi}{2n}\frac{2}{\pi} = \frac{1}{n}, \tag{15.74}$$

so (15.69) follows from (15.73) if $|x| \leq \cos\left(\frac{\pi}{2n}\right)$.
  If

$$\cos\left(\frac{\pi}{2n}\right) \leq |x| \leq 1,$$

then we use the following known formula (see [90], problem VI.71):

$$P'_k(x) = \frac{2^{n-1}}{n} \sum_{j=1}^{n} (-1)^{j-1} \sqrt{1 - x_{j,n}^2}\, P'_k(x_{j,n}) \frac{T_n(x)}{x - x_{j,n}}, \tag{15.75}$$

and get

$$\sum_{k=1}^{m} |P'_k(x)| \leq \frac{2^{n-1}}{n} \sum_{j=1}^{n} \sum_{k=1}^{m} \sqrt{1 - x_{j,n}^2}\, |P'_k(x_{j,n})| \frac{|T_n(x)|}{|x - x_{j,n}|}$$

$$\leq Mn \frac{2^{n-1}}{n} \sum_{j=1}^{n} \frac{|T_n(x)|}{|x - x_{j,n}|} \leq Mn^2. \tag{15.76}$$

Here we have used inequality (15.73):

$$\sum_{k=1}^{m} \sqrt{1 - x_{j,n}^2}\, |P'_k(x_{j,n})| \leq Mn, \tag{15.77}$$

and took into account that (see [90], problem VI.80)

$$\frac{2^{n-1}}{n} \sum_{j=1}^{n} \frac{|T_n(x)|}{|x - x_{j,n}|} = \frac{2^{n-1}}{n} \sum_{j=1}^{n} \frac{T_n(x)}{x - x_{j,n}} = \frac{T'_n(x)}{n} \leq n, \tag{15.78}$$

if $\cos\left(\frac{\pi}{2n}\right) \leq |x| \leq 1$.
  Lemma 15.3.3 is proved. ∎

Proof of Theorem 15.3.1. Let $Q_n(x)$ be the polynomial of the best approximation for the function $f(x)$, that is,

$$\|f - Q_n\| = E_n(f).$$

Note that

$$L_n(x, Q_n) = Q_n(x).$$

Therefore

$$\begin{aligned}
\|f(x) - L_n(x, f)\| \leq &\|f - Q_n\| + \|Q_n - L_n(x, Q_n)\| \\
&+ \|L_n(x, Q_n) - L_n(x, f)\| \\
\leq &E_n(f) + \lambda_n E_n(f),
\end{aligned} \tag{15.79}$$

where λ_n is defined in (15.58)

Let $P_{n-1}(x)$ be the polynomial of the best approximation for f', that is,

$$\|f' - P_{n-1}\| = E_{n-1}(f').$$

Let

$$P_n(x) := f(0) + \int_0^x P_{n-1}(t)\,dt, \quad P_n' = P_{n-1}. \tag{15.80}$$

We have

$$L_n'(x, P_n) = P_n' = P_{n-1},$$

$$|L_n'(x, P_n) - L_n'(x, f)| \leq \|f - P_n\| L_n'(x, f), \tag{15.81}$$

and

$$\|f - P_n\| \leq \left\| \int_0^x |f'(t) - P_{n-1}(t)|dt \right\| \leq E_{n-1}(f'). \tag{15.82}$$

Using estimates (15.80)–(15.82), we obtain

$$\begin{aligned}
|f'(x) - L_n'(x, f)| \leq &|f'(x) - P_{n-1}(x)| + |P_{n-1}(x) - L_n'(x, P_n)| \\
&+ |L_n'(x, P_n) - L_n'(x, f)| \\
\leq &E_{n-1}(f') + \lambda_n'(x) E_{n-1}(f'),
\end{aligned} \tag{15.83}$$

which is inequality (15.58).

Lemma 15.3.3 and the definition (15.58) of λ_n imply

$$\lambda_n'(x) \leq \frac{n\lambda_n}{\sqrt{1 - x^2}}, \quad \lambda_n' \leq n^2 \lambda_n. \tag{15.84}$$

This argument allows one to estimate

$$\lambda_n^{(k)}(x) := \sum_{j=1}^{n+1} |l_j^{(k)}(x)|$$

and

$$\lambda_n^{(k)} := \|\lambda_n^{(k)}(x)\|_0.$$

For example,

$$\lambda_n^{(k)} \leq n^{2k} \lambda_n.$$

Theorem 15.3.1 is proved. ∎

Proof of Theorem 15.3.2. We have

$$\|f(x) - L_{n,\delta}(x)\| \le \|f(x) - L_n(x,f)\| - \|L_n(x, f - f_\delta)\|$$
$$\le (1 + \lambda_n)E_n(f) + \delta\lambda_n, \tag{15.85}$$

where we have used the estimate (15.79). We assume that n is large. If one minimizes the right-hand side of (15.85) with respect to n for a small fixed δ, then one gets the value of n, $n = n(\delta)$, for which $L_{n(\delta),\delta}(x)$ approximates $f(x)$ best.

It is known (see [5]) that if $f \in C^r(I)$, then

$$E_n(f) \le \frac{K_r M_r}{n^r}, \quad M_r = \|f^{(r)}\|, \quad 1 < K_r \le \frac{\pi}{2}, \tag{15.86}$$

where $r \ge 1$ is an integer. Let us denote

$$\mu_r := K_r M_r.$$

For large n minimization of the right-hand side of (15.85) can be done analytically. We have for large n

$$(1 + \lambda_n)E_n(f) + \delta\lambda_n \asymp \frac{\mu_r \ln n}{n^r} + \delta \ln n, \tag{15.87}$$

where the notation $a \asymp b$ means $c_1 a \le b \le c_2 a$ and $c_1, c_2 > 0$ are constants independent of n.

A necessary condition for the minimizer of the right-hand side of (15.87) is

$$\delta = \mu_r \frac{\ln n}{n^r}\left(1 + O\left(\frac{1}{\ln n}\right)\right), \quad n \to 0. \tag{15.88}$$

Thus

$$n = n(\delta) = \left(\frac{r\mu_r}{\delta}\ln\frac{\mu_r}{\delta}\right)^{\frac{1}{r}}, \quad \delta \to 0. \tag{15.89}$$

From (15.85) and (15.89) it follows that formula (15.64) holds. Similarly, we derive (15.65) and (15.66)

Theorem 15.3.2 is proved. ∎

If $f \in C^r(I)$, then

$$\|f^{(k)}(x) - L_n^{(k)}(x,f)\| \le c\omega\left(\frac{1}{n}, f^{(r)}\right)n^{k-r}(1 + n^k \ln n), \quad k \le r, \tag{15.90}$$

where $c > 0$ is a constant and $\omega\left(\delta, f^{(r)}\right)$ is the continuity modulus of $f^{(r)}$. Recall that if $f \in C(I)$, then

$$\omega(\delta, f) := \sup_{|x-y| \le \delta, \ x,y \in I} |f(x) - f(y)|. \tag{15.91}$$

Let us prove (15.90). There exists a polynomial $S_n(x)$ such that

$$\|f^{(k)} - S_n^{(k)}\| \le cn^{k-r}\omega\left(\frac{1}{n}, f^{(r)}\right), \qquad k \le r. \tag{15.92}$$

We have

$$S_n^{(k)} = L_n^{(k)}(x, f).$$

Therefore

$$\|f^{(k)} - L_n^{(k)}(x, f)\| \le \|f^{(k)} - S_n^{(k)}\| + \|S_n^{(k)} - L_n^{(k)}(x, f)\|$$

$$\le cn^{k-r}\omega\left(n, f^{(r)}\right) + \lambda_n^{(k)}cn^{-r}\omega\left(\frac{1}{n}, f^{(r)}\right)$$

$$\le cn^{k-r}(1 + n^k \ln n)\,\omega\left(\frac{1}{n}, f^{(r)}\right). \tag{15.93}$$

The results of this section are of practical interest. For example, if $f = e^x$, $x \in [-1, 1]$, then

$$\|e^x - L_7(x, e^x)\| \le 10^{-7}, \quad \|e^x - L_6(x, e^x)\| \le 10^{-5}, \tag{15.94}$$

and

$$\begin{aligned} |e^x - L_7'(x, e^x)| &\le 10^{-5}, & |x| \le 0.98; \\ |e^x - L_7'(x, e^x)| &\le 10^{-7}, & 0.98 \le |x| \le 1. \end{aligned} \tag{15.95}$$

15.4 OTHER METHODS OF STABLE DIFFERENTIATION

Let A_n be a sequence of operators which converge strongly on a set S of functions to the identity operator. For example, A_n can be Fejer, S. Bernstein, Vallee-Poussin, or other methods of approximation of f, such as a mollification:

$$\mathcal{M}_\epsilon f = \frac{1}{\epsilon}\int_I g\left(\frac{x-y}{\epsilon}\right)f(y)\,dy, \quad I = [-1, 1]. \tag{15.96}$$

where

$$0 \le g(x) \in C_0^\infty(I), \quad \int_{-1}^1 g(x)\,dx = 1. \tag{15.97}$$

The function g is called the mollification kernel. Specifically, one may choose

$$g(x) = \begin{cases} ce^{-\frac{1}{1-x^2}}, & |x| < 1, \\ 0, & |x| \ge 1, \end{cases}$$

where $c = const > 0$ is chosen so that

$$\int_{-1}^1 g(x)\,dx = 1.$$

One has

$$\lim_{\epsilon \to 0} \|\mathcal{M}_\epsilon f - f\| = 0, \quad \|\mathcal{M}_\epsilon f\|_{L^p(\mathbb{R})} \le \|f\|_{L^p(I)}, \quad p \ge 1, \qquad (15.98)$$

$$\left(\frac{d^m}{dx^m} \mathcal{M}_\epsilon f \right)(x) = \mathcal{M}_\epsilon \left(\frac{d^m}{dx^m} f \right)(x), \quad x \in (-1, 1),$$

$$\lim_{\epsilon \to 0} \left\| \frac{d^m}{dx^m} \mathcal{M}_\epsilon f - \frac{d^m}{dx^m} f \right\|_{L^p(\tilde{I})} = 0, \qquad (15.99)$$

where $\tilde{I} \subset I$ is an interval inside $(-1, 1)$.

If $f \in C(I)$, then, taking $\epsilon = \frac{1}{n}$ and $-1 < x < 1$, one gets

$$\mathcal{M}_\epsilon f = n \int_{x-1}^{x+1} g(nt) f(x - t) \, dt$$

$$= \int_{n(x-1)}^{n(x+1)} g(u) f\left(x - \frac{u}{n} \right) du \to f(x) \quad \text{as} \quad n \to \infty. \qquad (15.100)$$

The convergence in (15.100) is uniform on any compact subset of I. To avoid the discussion of the convergence of $\mathcal{M}_\epsilon f$ at the end points of the interval $[-1, 1]$, let us assume that $f \in C(\mathbb{R})$ is periodic with period 2.

Let us estimate f' given noisy data f_δ, $\quad \|f_\delta - f\| \le \delta$, and assuming that $f \in C^2(\mathbb{R})$. First, note that if $f \in C^2(\mathbb{R})$ and $f(x + 2) = f(x)$, then (see (15.100)):

$$\|(\mathcal{M}_\epsilon f)' - f'\| = \|\mathcal{M}_\epsilon(f') - f'\|$$

$$\le \left\| \int_{-1}^{1} g(u) \left| f'\left(x - \frac{u}{n} \right) - f'(x) \right| du \right\| \le \frac{M_2 c_1}{n}, \qquad (15.101)$$

where

$$\|f''\| \le M_2, \quad c_1 = \int_{-1}^{1} |u| g(u) \, du, \quad \epsilon = \frac{1}{n}.$$

Using the inequality (15.101) and assuming $f \in C^2(\mathbb{R})$, $\quad \epsilon = \frac{1}{n}$, we obtain

$$\|(\mathcal{M}_\epsilon f_\delta)' - f'\| \le \|(\mathcal{M}_\epsilon(f_\delta) - f)'\| + \|(\mathcal{M}_\epsilon f_\delta)' - f'\|$$

$$\le \frac{1}{\epsilon^2} \left\| \int_{-\infty}^{\infty} \left| g'\left(\frac{x - y}{\epsilon} \right) \right| dy \right\| + \frac{M_2 c_1}{n}. \qquad (15.102)$$

Note that

$$\int_{-\infty}^{\infty} \left| g'\left(\frac{x - y}{\epsilon} \right) \right| dy = \epsilon \int_{-1}^{1} |g'(t)| \, dt = c_2 \epsilon = \frac{c_2}{n}. \qquad (15.103)$$

Thus (15.102) yields

$$\|(\mathcal{M}_{\frac{1}{n}} f_\delta)' - f'\| \le c_2 \delta n + \frac{M_2 c_1}{n} := \eta(n, \delta). \qquad (15.104)$$

Minimizing the right-hand side of (15.104) with respect to n for a fixed $\delta > 0$, one gets

$$n(\delta) = \frac{\gamma}{\sqrt{\delta}}, \quad \gamma := \sqrt{\frac{M_2 c_1}{c_2}} \tag{15.105}$$

and

$$\eta(\delta) := \eta(n(\delta), \delta) = \gamma_1 \sqrt{\delta}, \quad \gamma_1 := c_2 \gamma + \frac{M_2 c_1}{\gamma}. \tag{15.106}$$

We have proved the following result.

Theorem 15.4.1 *Assume that*

$$f \in C^2(\mathbb{R}), \quad f(x+2) = f(x), \quad \|f_\delta - f\| \le \delta, \quad \|f\| := ess \ sup_{x \in \mathbb{R}} |f(x)|.$$

Then the operator

$$R_\delta f_\delta := (\mathcal{M}_{\frac{1}{n(\delta)}} f_\delta)'$$

gives a stable approximation of f' provided that $n(\delta) = \frac{\gamma}{\sqrt{\delta}}$ with the error $\gamma_1 \sqrt{\delta}$, where the constants γ and γ_1 are defined in (15.105) and (15.106).

Remark 15.4.2 The method of the proof of Theorem 15.4.1 can be used for other (than mollification) methods of approximation of functions. For example, the Vallee–Poussin approximation method is

$$A_n f = \frac{(2n)!!}{(2n-1)!!} \frac{1}{2\pi} \int_{-\pi}^{\pi} f(y) \cos^{2n} \frac{x-y}{2} \, dy, \quad f(x+2\pi) = f(x), \tag{15.107}$$

yields a uniform approximation with the error

$$\|A_n f - f\| \le 3\omega\left(\frac{1}{\sqrt{n}}, f\right). \tag{15.108}$$

Many results on approximation of functions one finds in [5], [91], [187], [185], and [188].

The method for stable numerical differentiation given noisy data f_δ, based on the approximating the identity sequence of operators A_n, can be summarized as follows: one constructs the formula for stable differentiation $R_\delta f_\delta = (A_{n(\delta)} f_\delta)'$ satisfies relation (15.2) if the sequence of operators A_n has the properties

$$A_n f \to f, \quad (A_n f)' \to f' \quad \text{as} \quad n \to \infty, \tag{15.109}$$

where convergence is in L^∞ (or L^p) norm, and one has estimates of the type

$$\|(A_n f)' - f'\| \le \omega(n), \tag{15.110}$$

where

$$\omega(n) \le \frac{c}{n^b},$$

and $b > 0$ if $f \in C^{\mu}$ with $\mu > 1$. Moreover,

$$\|(A_n f)' - (A_n f_\delta)'\| \leq \|f - f_\delta\| b(n) = \delta b(n), \tag{15.111}$$

where $\lim_{n \to \infty} b(n) = \infty$. The $n(\delta)$ is found by minimizing the quantity

$$\delta b(n) + w(n) = \min, \tag{15.112}$$

or by solving the equation

$$\delta = \frac{w(n)}{b(n)} \tag{15.113}$$

for n. As $\delta \to 0$ one has $n(\delta) \to \infty$.

Consider the operator

$$R_h f = \frac{f(x+h) - f(x-h)}{2h} \tag{15.114}$$

on the space of $C^2(\mathbb{R})$ periodic functions, $f(x+2) = f(x)$, as an integral operator

$$R_h f = \int_{-1}^{1} \frac{\delta(y-h) - \delta(y+h)}{2h} f(x+y) \, dy \tag{15.115}$$

with the distributional kernel. One has

$$\lim_{h \to 0} \frac{\delta(y-h) - \delta(y+h)}{2h} = -\delta'(y) \tag{15.116}$$

in the sense of distributions. Thus

$$\lim_{h \to 0} R_h f = -\int_{-1}^{1} \delta'(y) f(x+y) \, dy = f'(x). \tag{15.117}$$

One may use other kernels to approximate $f'(x)$. For example, consider the operator

$$T_h f := \int_{-1}^{1} w(yh) f(x+yh) \, dy, \tag{15.118}$$

where w is an entire function. Assuming that $f \in C^2(-1, 1)$, we have

$$T_h f = \int_{-1}^{1} \left[w(0) + yhw'(0) + O(y^2 h^2) \right] \left[f(x) + yh f'(x) + O(y^2 h^2) \right] dy. \tag{15.119}$$

Let us require

$$w(0) = 0, \quad w'(0) := c_1, \quad w(y) = c_1 y,$$

then (15.119) yields

$$T_h f = c_1 h^2 \int_{-1}^{1} y^2 \, dy f'(x) + O(h^3) = \frac{2h^2}{3} c_1 f'(x) + O(h^3). \tag{15.120}$$

Choose $c_1 = \frac{3}{2h^2}$, that is, $w(yh) = \frac{3y}{2h}$. Then

$$\lim_{h \to 0} T_h f = f'(x). \tag{15.121}$$

Therefore, formula (15.118) with $w(yh) = \frac{3y}{2h}$ gives an approximation to $f'(x)$. This formula requires an integration, so it is more complicated than formula (15.114). In Section 15.1 we have proved that operator (15.114) with $h = h(\delta)$, defined in (15.8), is optimal in the sense (15.5); that is, among all linear and nonlinear operators acting from $L_\pi^\infty(\mathbb{R})$ into $L_\pi^\infty(\mathbb{R})$, where $L_\pi^\infty(\mathbb{R})$ is the space of 2-periodic functions on $\mathbb{R}$ with the norm $\|f\|_{L_\pi^\infty(\mathbb{R})} = \mathrm{ess\,sup}_{x \in \mathbb{R}} |f(x)|$, the operator $R_h(\delta)$, defined in (15.114), with $h = h(\delta)$, defined in (15.8), gives the best possible approximation of f' in the sense (15.5).

15.5 DSM AND STABLE DIFFERENTIATION

Consider the problem of stable differentiation as the problem of solving Volterra integral equation of the first kind:

$$Au := \int_0^x u\,dt = f(x), \quad 0 \le x \le 1. \tag{15.122}$$

Without loss of generality we assume that f is defined on the interval $[0, 1]$ and $f(0) = 0$. If $f(0) \ne 0$, then the function $f(x) - f(0)$ vanishes at $x = 0$ and has the same derivative as f.

Let us assume that noisy data f_δ are given, $\|f_\delta - f\| \le \delta$, and the norm is $L^2(0, 1) := H$ norm. The Hilbert space H is assumed *real-valued*.

We have

$$\langle Au, u \rangle = \int_0^1 dx \int_0^x u(t)\,dt u(x) = \frac{1}{2} \left(\int_0^x u(t)\,dt \right)^2 \Big|_0^1$$

$$= \frac{1}{2} \left(\int_0^1 u(t)\,dt \right)^2 \ge 0. \tag{15.123}$$

Since A is a linear bounded operator in H, conditions (1.10) hold. We have

$$\|Au\|^2 = \int_0^1 dx \left(\int_0^x u\,dt \right)^2 dx \le \int_0^1 x \int_0^x u^2\,dt dx \le \frac{1}{2}\|u\|^2.$$

Thus $\|A\| \le \frac{1}{\sqrt{2}}$. Consider a DSM scheme for solving equation (15.122):

$$\dot{u}_\delta(t) = -A_{a(t)}^{-1}[Au_\delta(t) + a(t)u_\delta(t) - f_\delta] = -u_\delta(t) + A_{a(t)}^{-1} f_\delta,$$
$$u_\delta(0) = u_0, \tag{15.124}$$

where $a(t)$ satisfies (6.30). Inequality (15.123) and the linearity of A imply

$$\langle Au - Av, u - v \rangle \ge 0, \quad \forall u, v \in H. \tag{15.125}$$

Thus, we may use Theorems 6.2.1 and 6.3.2 for calculating a stable approximation to f' given the noisy data f_δ.

Calculating A_a^{-1} amounts to solving the problem

$$\int_0^x u(y)\,dy + au(x) = g, \quad g \in H. \tag{15.126}$$

This problem is solved analytically:

$$u := u_a(x) = \frac{d}{dx}\frac{1}{a}\int_0^x e^{-\frac{x-y}{a}}g(y)\,dy = \frac{g}{a} - \frac{1}{a^2}\int_0^x e^{-\frac{x-y}{a}}g(y)\,dy. \tag{15.127}$$

If $a = a(t)$, formula (15.127) remains valid. One can use formula (15.127) for constructing a stable approximation of f' given noisy data f_δ. This corresponds to solving the equation

$$Au_{a,\delta} + au_{a,\delta} = f_\delta, \tag{15.128}$$

and choosing $a = a(\delta)$ so that

$$u_\delta := u_{a(\delta),\delta}$$

would converge to f':

$$\lim_{\delta \to 0} \|u_\delta - f'\| = 0. \tag{15.129}$$

To choose $a(\delta)$, we take $g = f_\delta$ in (15.127) and estimate the difference $u_{a,\delta} - f'$:

$$\|u_{a,\delta} - f'\| \le \|u_{a,\delta} - u_a\| + \|u_a - f'\|, \tag{15.130}$$

where u_a is the right-hand side of (15.127) with f in place of g. Using (15.127), we get

$$\|u_{a,\delta} - u_a\| \le \frac{\delta}{a} + \delta\|\frac{1}{a^2}\int_0^x e^{-\frac{x-y}{a}}(f_\delta - f)\,dy\| \le \delta\left(\frac{1}{a} + \frac{1}{a^{\frac{3}{2}}\sqrt{2}}\right), \tag{15.131}$$

where $\|f_\delta - f\| \le \delta$. Furthermore, integrating by parts, we get

$$\|u_a - f'\| = \|\frac{f}{a} - \frac{1}{a^2}\int_0^x e^{-\frac{x-y}{a}}f(y)\,dy - f'\|$$

$$\le \|\frac{f}{a} - \frac{1}{a^2}\left[ae^{-\frac{x-y}{a}}f\Big|_0^x - a\int_0^x e^{-\frac{x-y}{a}}f'(y)\,dy\right] - f'\|$$

$$\le \|\frac{1}{a}f(0) + \frac{1}{a}\int_0^x e^{-\frac{x-y}{a}}f'(y)\,dy - f'(x)\|$$

$$\le \|\frac{1}{a}\int_0^x e^{-\frac{s}{a}}f'(x-s)\,ds - f'(x)\|. \tag{15.132}$$

Here we have used the assumption

$$f(0) = 0.$$

We have

$$\lim_{a \to 0} \left\| \frac{1}{a} \int_0^x e^{-\frac{s}{a}} f'(x-s) \, ds - f'(x) \right\| = 0. \tag{15.133}$$

If $T(a) : H \to H$ is defined as

$$T(a)h = \frac{1}{a} \int_0^x e^{-\frac{x-y}{a}} h(y) \, dy, \tag{15.134}$$

then

$$\begin{aligned}
\|T(a)h\|^2 &= \int_0^1 dx \left| \frac{1}{a} \int_0^x e^{-\frac{x-y}{a}} h(y) \, dy \right|^2 \\
&\leq \int_0^1 dx \frac{1}{a^2} \int_0^x e^{-\frac{2(x-y)}{a}} \, dy \int_0^x h^2 dy \\
&\leq \frac{1}{a^2} \int_0^1 dx \frac{a}{2} \int_0^1 h^2 dy = \frac{1}{2a} \|h\|^2.
\end{aligned}$$

Thus

$$\|T(a)\| \leq \frac{1}{\sqrt{2a}}. \tag{15.135}$$

We have

$$\begin{aligned}
\|T(a)h - h\| &= \left\| \frac{1}{a} \int_0^x e^{-\frac{x-y}{a}} [h(y) - h(x)] \, dy - e^{-\frac{x}{a}} h(x) \right\| \\
&\leq \left\| \frac{1}{a} \int_0^x e^{-\frac{s}{a}} [h(x-s) - h(x)] \, ds \right\| \\
&\quad + \|e^{-\frac{x}{a}} h(x)\|. \tag{15.136}
\end{aligned}$$

Assume that

$$\sup_{0 \leq x \leq 1} |h^{(j)}(x)| \leq M_j, \quad 0 \leq j \leq 2, \quad M_j = const. \tag{15.137}$$

Then (15.136) implies

$$\|T(a)h - h\| \leq M_1 a + M_0 \left(\frac{a}{2} \right)^{\frac{1}{2}}, \tag{15.138}$$

and (15.132) implies

$$\|u_a - f'\| \leq M_2 a + M_1 \left(\frac{a}{2} \right)^{\frac{1}{2}}. \tag{15.139}$$

If $h(x)$ in (15.136) is an arbitrary element of $H = L^2(0,1)$ then by the Lebesgue's dominated convergence theorem we have

$$\lim_{a \to 0} \|e^{-\frac{x}{a}} h(x)\| = 0, \tag{15.140}$$

and, setting $\frac{s}{a} = z$, we obtain

$$\lim_{a\to 0} \|\frac{1}{a}\int_0^x e^{-\frac{s}{a}}[h(x-s) - h(x)]\,ds\|^2$$

$$= \lim_{a\to 0} \|\int_0^{\frac{x}{a}} e^{-z}[h(x-az) - h(x)]\,dz\|^2$$

$$\leq \lim_{a\to 0}\int_0^1 dx \int_0^{\frac{1}{a}} e^{-z}\,dz \int_0^{\frac{1}{a}} e^{-z}|h(x-az) - h(x)|^2 dz$$

$$= 0. \tag{15.141}$$

Therefore

$$\lim_{a\to 0}\|T(a)h - h\| = 0, \quad \forall h \in H. \tag{15.142}$$

This conclusion is a consequence of (15.138) as well.

It follows from (15.130), (15.131), and (15.139) that

$$\|u_{a,\delta} - f'\| \leq \frac{\delta}{a^{\frac{3}{2}}\sqrt{2}}\left(1 + \sqrt{2}a^{\frac{1}{2}}\right) + \frac{M_1 a^{\frac{1}{2}}}{\sqrt{2}}\left(1 + \frac{M_2 a^{\frac{1}{2}}}{M_1}\sqrt{2}\right). \tag{15.143}$$

Minimizing the right-hand side of (15.143) with respect to a, we get

$$a_\delta = O\left(\delta^{\frac{1}{2}}\right). \tag{15.144}$$

Denoting $u_\delta := u_{a_\delta,\delta}$ and using (15.143), we obtain

$$\|u_\delta - f'\| = O\left(\delta^{\frac{1}{2}}\right), \quad \delta \to 0. \tag{15.145}$$

Numerical results for stable differentiation, based on the equation (15.128), are given in [1].

Let us formulate the result we have proved.

Theorem 15.5.1 *Assume (15.137), and let*

$$u_\delta = T(a(\delta))f_\delta,$$

where $\|f_\delta - f\| \leq \delta$, T_a *is defined in (15.134) and* $a(\delta) = O(\delta^{\frac{1}{2}})$. *Then (15.145) holds.*

Let us return to DSM (15.124) and give the stopping rule, that is, the choice of t_δ such that

$$\lim_{\delta\to 0}\|u_\delta(t_\delta) - f'\| = 0. \tag{15.146}$$

The solution to (15.124) is

$$u_\delta(t) = u_0 e^{-t} + \int_0^t e^{-(t-s)}A_{a(s)}^{-1}f_\delta\,ds := u(t, f_\delta). \tag{15.147}$$

Thus

$$\|u_\delta(t) - f'\| \leq \|u(t, f_\delta) - u(t, f)\| + \|u(t, f) - f'\|$$

$$\leq \frac{\delta}{a(t)} + \|u(t, f) - f'\|. \tag{15.148}$$

We have

$$\|u(t, f) - f'\| \leq \|u_0\|e^{-t} + \|\int_0^t e^{-(t-s)} A_{a(s)}^{-1} f \, ds - f'\|. \tag{15.149}$$

From (15.139) it follows that

$$\|A_{a(s)}^{-1} f - f'\| = O(a^{\frac{1}{2}}), \quad a(s) \to 0. \tag{15.150}$$

Since

$$\lim_{t \to \infty} \int_0^t e^{-(t-s)} h(s) \, ds = h(\infty),$$

provided $h(\infty)$ exists, we conclude from (15.149) and (15.150) that

$$\|u(t, f) - f'\| = O(a^{\frac{1}{2}}(t)), \quad t \to \infty. \tag{15.151}$$

From (15.148) and (15.151) we obtain

$$\|u_\delta(t) - f'(x)\| \leq \frac{\delta}{a(t)} + O(a^{\frac{1}{2}}(t)). \tag{15.152}$$

Thus, if t_δ minimizes (15.152) then $\lim_{\delta \to 0} t_\delta = \infty$ and (15.146) holds. The minimizer a_δ of the right-hand side of (15.152) with respect to a is $a_\delta = O(\delta^{\frac{2}{3}})$. Therefore t_δ is found from the relation $a_\delta = a(t)$, and

$$\|u_\delta(t_\delta) - f'(x)\| \leq O(\delta^{\frac{1}{3}}), \quad \text{as} \quad \delta \to 0. \tag{15.153}$$

Consider now another DSM:

$$\dot{u}_\delta = -(Au_\delta + a(t)u_\delta - f_\delta), \quad u(0) = 0. \tag{15.154}$$

Our arguments are valid for an arbitrary initial data $u(0) = u_0$, and we took $u_0 = 0$ just for simplicity of writing.

Denote by

$$u_\delta(t) = u(t, f_\delta)$$

the solution to (15.154), and denote by $V(t)$ the solution to the equation

$$AV + a(t)V - f_\delta = 0. \tag{15.155}$$

Let

$$u_\delta - V(t) := w. \tag{15.156}$$

Then
$$\dot{w} = -\dot{V} - [Aw + a(t)w], \quad w(0) = 0. \tag{15.157}$$

We have proved above (see (15.145)) that
$$\lim_{\delta \to 0} \|V(t_\delta) - f'\| = 0, \tag{15.158}$$

where t_δ is defined by the equation $a_\delta = a(t)$ and $\lim_{\delta \to 0} t_\delta = \infty$. Thus, (15.146) holds if we prove that
$$\lim_{\delta \to 0} w(t_\delta) = 0. \tag{15.159}$$

Multiply (15.157) by w, denote
$$g(t) := \|w(t)\|,$$

and use the inequality $\langle Aw, w \rangle \geq 0$ to get
$$\dot{g} \leq -a(t)g + \|\dot{V}\|. \tag{15.160}$$

Let us estimate $\|\dot{V}\|$. Differentiate (15.155) with respect to t and get
$$A\dot{V} + a(t)\dot{V} = -\dot{a}V. \tag{15.161}$$

Multiply (15.161) by $\dot{V}$, use inequality $\langle A\dot{V}, \dot{V}) \rangle \geq 0$, and get
$$\|\dot{V}\| \leq \frac{|\dot{a}|}{a(t)}\|V\|. \tag{15.162}$$

Similarly, multiply (15.155) by V and get
$$\|V\| \leq \frac{\|f_\delta\|}{a(t)}.$$

This and (15.125) imply
$$\|\dot{V}\| \leq \frac{|\dot{a}|}{a^2}\|f_\delta\|.$$

Assume
$$\lim_{t \to \infty} \frac{|\dot{a}(t)|}{a^3(t)} = 0, \quad \int_0^\infty a(t)\, dt = \infty. \tag{15.163}$$

From (15.160) we obtain
$$g(t) = g_0 e^{-\int_0^t a(s)ds} + e^{-\int_0^t a(s)ds} \int_0^t e^{\int_0^s a(p)dp}\|\dot{V}(s)\|ds := J_1 + J_2.$$

From the second condition (15.163) it follows that
$$\lim_{t \to \infty} \|g_0\| e^{-\int_0^t a(s)ds} = 0.$$

We have

$$J_2(t) \le \frac{\int_0^t e^{\int_0^s a(p)dp} \frac{|\dot{a}(s)|}{a^2(s)} \, ds \| f_\delta \|}{e^{\int_0^t a(s)ds}}. \tag{15.164}$$

Applying L'Hospital's rule to (15.164) and using the first condition (15.163), we conclude that

$$\lim_{t \to \infty} J_2(t) = 0.$$

Thus,

$$\lim_{t \to \infty} g(t) = 0. \tag{15.165}$$

Let us summarize the result.

Theorem 15.5.2 *Assume that* $f \in C^2$, $0 < a(t) \searrow 0$, *(15.163) holds, and* $\|f_\delta - f\|_{L^2(0,1)} \le \delta$. *Then problem (15.154) has a unique global solution* $u_\delta(t)$, *and (15.146) holds with* t_δ *chosen from the equation* $a_\delta = a(t)$, *where* a_δ *is given in (15.144).*

Remark 15.5.3 One can construct an iterative method for stable numerical differentiation using the general results from Sections 2.4 and 4.4.

15.6 STABLE CALCULATING SINGULAR INTEGRALS

Let us denote the hypersingular integral by the symbol

$$\fint_0^b f(x) \, dx.$$

We assume that

$$f \in C^k([0,b]), \quad k > \mu - 1 > 0, \quad b > 0,$$

and define

$$\fint_0^b \frac{f(x) \, dx}{x^\mu} := \int_0^b \frac{f(x) - f_k(x)}{x^\mu} \, dx + \sum_{j=0}^k \frac{f^{(j)}(0) b^{-\mu+j+1}}{j!(-\mu+j+1)}, \tag{15.166}$$

where

$$f_k(x) := \sum_{j=0}^k \frac{f^{(j)}(0) x^j}{j!}. \tag{15.167}$$

If $k \ge \mu - 1$ and $f \in C^{k+1}([0,b])$, then the term in (15.166), corresponding to $\mu = j + 1$, is replaced by $\frac{f^{(j)}(0) \ln b}{j!}$. If the integral is given on the interval $[a,d]$, it can be reduced to the integral over $[0,b]$ by a change of variables. The definition (15.166) does not depend on the choice of k in (15.166) in the region $k > \mu - 1$. Although there are many papers on calculating hypersingular integrals, [113] was the first one, as far as the author knows, in which the ill-possedness of this problem was discussed and a stable approximation of hypersingular integrals was proposed. We present here the results of this paper. Let us first explain why the usual quadrature formulas lead to possibly very large errors in computing hypersingular integrals.

In other words, we explain why computing hypersingular integrals is an ill-posed problem.

Consider a quadrature formula

$$Q_n f = \sum_{j=1}^{n} w_{n,j} f(x_{n,j}), \tag{15.168}$$

where $w_{n,j}$ are the weights and $x_{n,j}$ are the nodes. Let us assume that

$$\lim_{n \to \infty} Q_n f = \int_0^b f \, dx, \quad \forall f \in C([0,b]). \tag{15.169}$$

If one applies formula (15.168) to calculating the right-hand side of (15.166), and if the function f is given with some error, so that f_δ is given in place of f,

$$\|f_\delta(x) - f(x)\| \le \delta, \quad \|\cdot\| = \|\cdot\|_{L^\infty(0,b)}, \tag{15.170}$$

then, even if we assume that $f^{(j)}(0)$, $0 \le j \le k$, are known exactly, we have

$$\lim_{n \to \infty} \sup_{\{f_\delta : \|f_\delta - f\| \le \delta\}} \left| Q_n \left[\frac{f_\delta(x) - f_k(x)}{x^\mu} \right] - Q_n \left[\frac{f(x) - f_k(x)}{x^\mu} \right] \right| = \infty. \tag{15.171}$$

Thus, the problem of calculating hypersingular integrals is ill-posed. Let us check (15.171). Take, for example,

$$f_\delta(x) = \begin{cases} f(x) + \delta, & \text{if } 0 < a \le x \le b, \\ f(x) + \delta \left(\frac{x}{a}\right)^\mu, & \text{if } 0 < x \le a. \end{cases} \tag{15.172}$$

Then the function $f_\delta(x) - f(x)$ is continuous on $[0, b]$, and

$$\lim_{n \to \infty} \sum_{j=1}^{n} w_{n,j} \frac{[f_\delta(x_{n,j}) - f(x_{n,j})]}{x_{n_j}^\mu} = \delta \int_0^a \left(\frac{x}{a}\right)^\mu \frac{dx}{x^\mu}$$

$$+ \delta \int_a^b dx > \delta \frac{1}{a^{\mu-1}}. \tag{15.173}$$

The right-hand side of (15.173) tends to infinity as $a \to 0$, so that (15.171) is verified.

The problem is to construct a quadrature formula for stable computation of integral (15.166) given noisy data f_δ. We will not assume that $f^{(j)}(0)$ are known, but estimate them stably from noisy data f_δ. This can be done by the method developed in Section 5.1, provided that upper bounds on the derivatives of f are known. To estimate stably $f^{(j)}(0)$, $0 \le j \le k$, we use the bounds

$$\|f^{(j)}\| := \sup_{0 \le x \le b} |f^{(j)}(x)| := M_j, \quad 0 \le j \le k + 2. \tag{15.174}$$

One may look for an estimate of $f^{(j)}(0)$ of the form

$$f^{(j)}(0) = \sum_{m=0}^{j} c_{m,j} f(mh) := L_{j,h} f, \tag{15.175}$$

where $c_{m,j}$ are found from the condition

$$f^{(j)}(0) - \sum_{m=0}^{j} c_{m,j} f(mh) = O(h^{j+1}). \tag{15.176}$$

This leads to a linear algebraic system for finding $c_{m,j}$:

$$\sum_{m=0}^{j} \frac{m^p h^p}{p!} = \delta_{jp}, \quad 0 \le p \le j. \tag{15.177}$$

This system is uniquely solvable because its matrix has nonzero Vandermonde determinant.

If f_δ is given in place of f,

$$\|f_\delta - f\| \le \delta,$$

then one uses formula (15.175) with f_δ in place of f and finds $h = h(\delta)$ such that

$$\lim_{\delta \to 0} [L_{j,h(\delta)} f_\delta - f^{(j)}(0)] = 0. \tag{15.178}$$

This $h(\delta)$ is found from the estimate:

$$[L_{j,h} f_\delta - f^{(j)}(0)| \le |L_{j,h}(f_\delta - f)| + |L_{j,h} f - f^{(j)}(0)|$$
$$\le \delta \varphi_j(h) + O(h^{j+1}), \tag{15.179}$$

where

$$\varphi_j(h) = \sum_{m=0}^{j} |c_{m,j}|.$$

The coefficients

$$c_{m,j} = c_{m,j}(h)$$

and

$$\varphi_j(h) \to \infty \quad \text{as} \quad h \to 0.$$

One finds $h(\delta)$ by minimizing the right-hand side of (15.179) with respect to h for a given $\delta > 0$. This gives also an estimate of the error of the approximation (15.175).

Assume now that stable approximations $f_\delta^{(j)}(0)$ to $f^{(j)}(0)$, $0 \le j \le k$, have been calculated. Denote

$$f_{k\delta}(x) := \sum_{j=0}^{k} \frac{f_\delta^{(j)}(0)}{j!} x^j \qquad (15.180)$$

and let

$$Q_\delta f_\delta := Q_{n,a,b}\left(\frac{f_\delta(x) - f_{k\delta}(x)}{x^\mu}\right) + \sum_{j=0}^{k} \frac{f_\delta^{(j)}(0) b^{-\mu+j+1}}{j!(-\mu+j+1)}, \qquad (15.181)$$

Here $Q_{n,a,b}$, $n = 1, 2, ...$, are quadrature formulas with positive weights which converge, as $n \to \infty$, to $\int_a^b f(x)\,dx$ for any $f \in C([a,b])$.

For example, quadrature formulas of Gaussian type have these properties; they are exact for polynomials of degree $\le 2n - 1$ and converge for any $f \in C([a,b])$ (see, e.g., [24]).

Let us formulate our result.

Theorem 15.6.1 *Assume $k > \mu - 1$, $f \in C^{k+2}([0,b])$. Then there is an $a = a(\delta, k)$ such that*

$$J := \left| \int_0^b f\,dx - Q_\delta f_\delta \right| = O\left(\delta^{\frac{k+2-\mu}{k+1}}\right), \qquad \delta \to 0, \qquad (15.182)$$

for all $n \ge n(\delta, k, \mu, f)$.

Proof: Denote

$$J_1 = \int_0^a \frac{f(x) - f_k(x)}{x^\mu}\,dx, \qquad J_1 = O(a^{k+2-\mu}), \qquad (15.183)$$

$$J_2 = \int_a^b \frac{f(x) - f_k(x)}{x^\mu}\,dx - Q_{n,a,b}\left(\frac{f(x) - f_k(x)}{x^\mu}\right) \to 0 \quad \text{as} \quad n \to \infty, \qquad (15.184)$$

$$J_3 = Q_{n,a,b}\left(\frac{f(x) - f_\delta(x)}{x^\mu}\right),$$
$$|J_3| \le O(\delta a^{-\mu+1}) \quad \text{as} \quad n \to \infty, \qquad (15.185)$$

$$J_4 = Q_{n,a,b}\left(\frac{f_{k\delta}(x) - f_k(x)}{x^\mu}\right) + \sum_{j=0}^{k} \frac{f^{(j)}(0) - f_\delta^{(j)}(0)}{j!} \frac{b^{-\mu+j+1}}{(-\mu+j+1)}, \qquad (15.186)$$

$$J_4 = O\left(\delta^{\frac{k+2-\mu}{k+1}}\right) \qquad \text{as} \qquad n \to \infty. \qquad (15.187)$$

From these estimates the relation (15.182) follows.

Theorem 15.6.1 is proved. ∎

PART II

CHAPTER 16

SOLVING LINEAR OPERATOR EQUATIONS BY A NEWTON-TYPE DSM

16.1 AN ITERATIVE SCHEME FOR SOLVING LINEAR OPERATOR EQUATIONS

In this section we will formulate and justify an iterative scheme with a Discrepancy Principle (DP) for solving linear operator equations. Throughout this chapter we denote by $N = N(A) := \{u : Au = 0\}$ the null space of linear operator A, and we will denote by y the unique minimal-norm solution to the equation $Au = f$, so $y \in N^\perp$.

Assume that

$$0 < a(t) \searrow 0, \qquad \lim_{t \to \infty} \frac{|\dot{a}(t)|}{a(t)} = 0. \tag{16.1}$$

Denote $a_n = a(nh)$, $h = const > 0$, $h \in (0, 1)$.

Consider the following iterative scheme:

$$v_{n+1} = (1 - h)v_n + hT_{a_n}^{-1}A^*f_\delta, \qquad v_0 := u_0. \tag{16.2}$$

Remark 16.1.1 From (16.1) one derives

$$\lim_{n \to \infty} (1 - h)^{-n} a_n = \infty, \qquad \lim_{n \to \infty} \frac{a_{n-1} - a_n}{a_n} = 0, \qquad \lim_{n \to \infty} \frac{a_n}{a_{n-1}} = 1. \tag{16.3}$$

Dynamical Systems Method and Applications: Theoretical Developments and Numerical Examples, First Edition. A. G. Ramm and N. S. Hoang.
Copyright © 2012 John Wiley & Sons, Inc.

This and (4.36) imply

$$\lim_{t \to \infty} (1 - h)^{-n} a_n \left\| Q_{a_n}^{-1} f_\delta \right\| = \infty. \tag{16.4}$$

Let us formulate a DP for the iterative scheme (16.2).

Theorem 16.1.2 *Let $a(t)$ satisfy (16.1), v_n be defined by (16.2), and*

$$\| Au_0 - f_\delta \| > c\delta. \tag{16.5}$$

Then there exists a unique $n_\delta > 0$ such that

$$\| Av_{n_\delta} - f_\delta \| \le c\delta, \quad \| Av_n - f_\delta \| > c\delta, \qquad 0 \le n \le n_\delta. \tag{16.6}$$

This n_δ satisfies the relation $\lim_{\delta \to 0} n_\delta = \infty$, and

$$\lim_{\delta \to 0} \| v_{n_\delta} - y \| = 0. \tag{16.7}$$

Proof: The uniqueness of n_δ follows from its definition (16.6).

Let us prove the existence of t_δ satisfying (16.6). From (16.2) one gets

$$v_{n+1} = (1 - h)^{n+1} u_0 + \sum_{i=0}^{n} (1 - h)^{n-i} h T_{a_i}^{-1} A^* f_\delta. \tag{16.8}$$

Thus,

$$Av_{n+1} - f_\delta = (1 - h)^{n+1} Au_0 + \sum_{i=0}^{n} (1 - h)^i h A T_{a_i}^{-1} A^* f_\delta - f_\delta$$

$$= (1 - h)^{n+1} (Au_0 - f_\delta) + \sum_{i=0}^{n} (1 - h)^{n-i} h (QQ_{a_i}^{-1} - I) f_\delta$$

$$= (1 - h)^{n+1} (Au_0 - f_\delta) - \sum_{i=0}^{n} (1 - h)^{n-i} h a_i Q_{a_i}^{-1} f_\delta. \tag{16.9}$$

Let us prove that

$$\lim_{n \to \infty} \frac{\left\| \sum_{i=0}^{n} (1 - h)^{n-i} h a_i Q_{a_i}^{-1} f_\delta \right\|}{a_n \| Q_{a_n}^{-1} f_\delta \|} = 1. \tag{16.10}$$

By the spectral theorem one gets

$$\left\| \sum_{i=0}^{n} (1 - h)^{n-i} h a_i Q_{a_i}^{-1} f_\delta \right\|^2 = \int_0^{\|Q\|} \left(\sum_{i=0}^{n} \frac{(1 - h)^{n-i} h a_i}{a_i + \lambda} \right)^2 d\langle F_\lambda f_\delta, f_\delta \rangle, \tag{16.11}$$

$$\| a_n Q_{a_n}^{-1} f_\delta \|^2 = \int_0^{\|Q\|} \frac{a_n^2}{(a_n + \lambda)^2} d\langle F_\lambda f_\delta, f_\delta \rangle. \tag{16.12}$$

Since $0 < a_n \searrow 0$ one has

$$\frac{a_n}{a_i} \leq \frac{a_n + \lambda}{a_i + \lambda} \leq \frac{a_n + \|Q\|}{a_i + \|Q\|}, \qquad 0 \leq i \leq n, 0 \leq \lambda \leq \|Q\|. \qquad (16.13)$$

Let

$$g_n(\lambda) = \frac{\sum_{i=0}^{n} \frac{(1-h)^{n-i} h a_i}{a_i + \lambda}}{\frac{a_n}{a_n + \lambda}}, \qquad i \geq 0, \lambda \geq 0. \qquad (16.14)$$

From (16.13) one gets

$$1 - (1-h)^n = g_n(0) \leq g_n(\lambda) \leq g_\lambda(\|Q\|) = \frac{\sum_{i=0}^{n} \frac{(1-h)^{-i} h a_i}{a_i + \|Q\|}}{\frac{(1-h)^{-n} a_n}{a_n + \|Q\|}}, \qquad (16.15)$$

for $0 \leq \lambda \leq \|Q\|$. By a discrete version of L'Hospital's rule, one gets

$$\lim_{n\to\infty} g_n(\|Q\|) = \lim_{n\to\infty} \frac{\frac{(1-h)^{-n} h a_n}{a_n + \|Q\|}}{\frac{(1-h)^{-n} a_n}{a_n + \|Q\|} - \frac{(1-h)^{-n+1} a_{n-1}}{a_{n-1} + \|Q\|}}$$

$$= \lim_{n\to\infty} \frac{h a_n}{a_n - (1-h)a_{n-1}} = \lim_{n\to\infty} \frac{1}{\frac{a_n - a_{n-1}}{h a_n} + \frac{a_{n-1}}{a_n}}$$

$$= 1. \qquad (16.16)$$

It follows from (16.15) and (16.16) that for an arbitrary small $\epsilon > 0$ there exists n_ϵ such that

$$1 - \epsilon < g_n(\lambda) < 1 + \epsilon, \qquad \forall n \geq n_\epsilon, \quad 0 \leq \lambda \leq \|Q\|. \qquad (16.17)$$

From (16.11) (16.12), (16.14), (16.15) and (16.17) one gets

$$(1-\epsilon)^2 \leq \frac{\left\| \sum_{i=0}^{n} \frac{(1-h)^{n-i} h a_i}{a_i + \lambda} \right\|^2}{a_n^2 \|Q_{a_n}^{-1}\|^2} \leq (1+\epsilon)^2, \qquad \forall n \geq n_\epsilon. \qquad (16.18)$$

This implies that relation (16.10) holds.
From (16.4) one gets

$$\lim_{n\to\infty} \frac{(1-h)^n \|A u_0 - f_\delta\|}{a_n \|Q_{a_n}^{-1} f_\delta\|} = \lim_{n\to\infty} \frac{\|A u_0 - f_\delta\|}{(1-h)^{-n} a_n \|Q_{a_n}^{-1} f_\delta\|} = 0. \qquad (16.19)$$

This, (16.9) and (16.10) imply

$$\lim_{n\to\infty} \frac{\|A v_n - f_\delta\|}{a_n \|Q_{a_n}^{-1} f_\delta\|} = \lim_{t\to\infty} \frac{\left\| \sum_{i=0}^{n} (1-h)^{n-i} a_i Q_{a_i}^{-1} f_\delta ds \right\|}{a_n \|Q_{a_n}^{-1} f_\delta\|} = 1. \qquad (16.20)$$

It follows from (16.20) that $\lim_{n\to\infty} \|Av_n - f_\delta\| \leq \delta$ since $\lim_{n\to\infty} a_n \|Q_{a_n}^{-1} f_\delta\| \leq \delta$ as we have proved earlier (see (4.40)). This implies the existence of n_δ.

Let us prove that
$$\lim_{\delta\to 0} n_\delta = \infty. \tag{16.21}$$

It follows from (16.5) that $n_\delta \geq 1$. This, (16.15) and (16.11)–(16.12) imply

$$\|a_{n_\delta} Q_{a_{n_\delta}}^{-1} f_\delta\| \leq \frac{1}{1 - (1 - h)^{n_\delta}} \| \sum_{i=0}^{n_\delta} (1 - h)^{n_\delta - i} h a_i Q_{a_i}^{-1} f_\delta\| \leq \frac{1}{h} c\delta. \tag{16.22}$$

From (16.22) and the arguments similar to the ones, given in the proof of Theorem 2.2.5, one gets $\lim_{\delta\to 0} a_{n_\delta} = 0$. This implies (16.21).

It follows from (16.20) that the first equation in (16.6) can be written as

$$a_n \|Q_{a_n}^{-1} f_\delta\| = c_1(n)\delta, \tag{16.23}$$

where
$$c_1(n) = c[1 + o(1)], \quad n \to \infty. \tag{16.24}$$

This and (16.21) imply

$$\lim_{\delta\to 0} \|w_{\delta,n_\delta} - y\| = 0, \qquad w_{\delta,n} := T_{a_n}^{-1} A^* f_\delta. \tag{16.25}$$

Thus, there exists $C = const > 0$ independent of δ such that

$$\|w_{\delta,n_\delta}\| \leq C \quad \text{as} \quad \delta \to 0. \tag{16.26}$$

Let us prove that
$$\lim_{n\to\infty} \frac{\|v_n - w_{\delta,n}\|}{\|w_{\delta,n}\|} = 0. \tag{16.27}$$

One has

$$\|A^* f_\delta\| \leq \|T_{a_n}\| \|T_{a_n}^{-1} A^* f_\delta\| \leq (\|T\| + a_n)\|T_{a_n}^{-1} A^* f_\delta\|.$$

Thus,
$$\|w_{\delta,n}\| = \|T_{a_n}^{-1} A^* f_\delta\| \geq \frac{\|A^* f_\delta\|}{\|T\| + a_0}.$$

This implies

$$0 \leq \lim_{n\to\infty} \frac{(1 - h)^n \|u_0\|}{\|w_{\delta,n}\|} \leq \lim_{n\to\infty} \frac{(1 - h)^n \|u_0\|(\|T\| + a_n)}{\|A^* f_\delta\|} = 0. \tag{16.28}$$

To prove (16.27), it suffices to show that

$$J := \lim_{n\to\infty} \frac{\| \sum_{i=0}^{n-1} (1 - h)^{n-1-i} h T_{a_i}^{-1} A^* f_\delta - (1 - (1 - h)^n) T_{a_n}^{-1} A^* f_\delta \|^2}{\|T_{a_n}^{-1} A^* f_\delta\|^2} = 0. \tag{16.29}$$

Using the commutation formula $T_{a_i}^{-1} A^* = A^* Q_{a_i}^{-1}$ and the spectral theorem, one obtains

$$J = \lim_{t \to \infty} \frac{\int_0^{\|Q\|} \left(\sum_{i=0}^{n-1} (1-h)^{n-1-i} h \frac{\sqrt{\lambda}}{a_i + \lambda} - \frac{\sqrt{\lambda}}{a_n + \lambda} \right)^2 d\langle F_\lambda f_\delta, f_\delta \rangle}{\int_0^{\|Q\|} \frac{\lambda}{(a_n + \lambda)^2} d\langle F_\lambda f_\delta, f_\delta \rangle}. \tag{16.30}$$

Using (16.13) and the arguments used in the derivation of the relations (16.13)–(16.17), one concludes that for any arbitrary small $\epsilon > 0$, there exists n_ϵ such that

$$-\epsilon \le \frac{\sum_{i=0}^{n-1} (1-h)^{n-1-i} h \frac{\sqrt{\lambda}}{a_i + \lambda} - \frac{\sqrt{\lambda}}{a_n + \lambda}}{\frac{\sqrt{\lambda}}{a_n + \lambda}} \le \epsilon, \qquad 0 \le \lambda \le \|Q\|, \tag{16.31}$$

for all $n \ge n_\epsilon$. This and (16.30) imply that (16.29) holds. Therefore, (16.27) holds. From our arguments it follows that (16.27) holds uniformly with respect to δ. This implies

$$\|v_n - w_{\delta,n}\| = \|w_{\delta,n}\| o(1) \quad \text{as} \quad n \to \infty, \quad \forall \delta > 0. \tag{16.32}$$

This, relation (16.26), and (16.21) imply

$$0 \le \lim_{\delta \to 0} \|v_{n_\delta} - w_{\delta,n_\delta}\| = 0. \tag{16.33}$$

This, (16.25), and the triangle inequality

$$\|v_{n_\delta} - y\| \le \|v_{n_\delta} - w_{\delta,n_\delta}\| + \|w_{\delta,n_\delta} - y\|$$

imply (16.7). Theorem 16.1.2 is proved.   ∎

16.2   DSM WITH FAST DECAYING REGULARIZING FUNCTION

Assume that

$$0 < a(t) \searrow 0, \qquad \lim_{t \to \infty} \frac{|\dot{a}(t)|}{a(t)} = \alpha \in (0,1). \tag{16.34}$$

Note that the function $a(t) = a(0)e^{-\alpha t}$ satisfies relations (16.34). This $a(t)$ decays exponentially fast.

Consider the following DSM:

$$\dot{v}(t) = -v(t) + T_{a(t)}^{-1} A^* f_\delta, \qquad v(0) := u_0. \tag{16.35}$$

Remark 16.2.1 From (16.34) one derives

$$\lim_{t \to \infty} e^t a(t) = \infty. \tag{16.36}$$

This and (4.36) imply

$$\lim_{t \to \infty} e^t a(t) \left\| Q_{a(t)}^{-1} f_\delta \right\| = \infty. \tag{16.37}$$

Let $u(t)$ solve the following problem:

$$\dot{u}(t) = -u(t) + T_{a(t)}^{-1} A^* f, \qquad u(0) := u_0. \tag{16.38}$$

Let us prove a lemma.

Lemma 16.2.2 *Let $0 < a(t) \searrow 0$ and $u(t)$ be defined by (16.38). Then*

$$\lim_{t \to \infty} \|u(t) - y\| = 0. \tag{16.39}$$

Proof: From (16.38) one gets

$$u(t) = e^{-t} u_0 + \int_0^t e^{s-t} T_{a(t)}^{-1} A^* f \, ds. \tag{16.40}$$

Thus

$$u(t) - y = e^{-t}(u_0 - y) + \int_0^t e^{s-t}(T_{a(t)}^{-1} A^* A - I) y \, ds,$$
$$= e^{-t}(u_0 - y) + \int_0^{\|T\|} \int_0^t e^{s-t} \frac{a(s)}{\lambda + a(s)} \, ds \, dE_\lambda y. \tag{16.41}$$

Therefore,

$$\lim_{t \to \infty} \|u(t) - y\|^2 = \lim_{t \to \infty} \int_0^{\|T\|} \int_0^t e^{s-t} \frac{a^2(s)}{(\lambda + a(s))^2} \, ds \, d\langle E_\lambda y, y \rangle,$$
$$= \|P_N y\|^2 = 0. \tag{16.42}$$

Here we have used the relation

$$\lim_{t \to \infty} \frac{\int_0^t e^s \frac{a^2(s)}{(a(s)+\lambda)^2} \, ds}{e^t} = 0, \tag{16.43}$$

which follows from L'Hospital's rule and the assumption that $a(t) \searrow 0$. Lemma 16.2.2 is proved. ∎

Let us formulate another version of the discrepancy principle.

Theorem 16.2.3 *Let $a(t)$ satisfy (16.34), $v(t)$ be the solution to (16.35), and*

$$u_0 := T_{a_0}^{-1} A^* f_\delta, \qquad \|A u_0 - f_\delta\| > c\delta, \qquad c > \frac{1}{1 - \alpha}. \tag{16.44}$$

Then there exists a unique $t_\delta > 0$ such that

$$\|A v(t_\delta) - f_\delta\| = c\delta, \qquad \|A v(t) - f_\delta\| > c\delta, \qquad 0 \le t \le t_\delta. \tag{16.45}$$

This t_δ satisfies the relation $\lim_{\delta \to 0} t_\delta = \infty$, and

$$\lim_{\delta \to 0} \|v(t_\delta) - y\| = 0. \tag{16.46}$$

Proof: The uniqueness of t_δ follows from its definition (16.45).

Let us prove the existence of t_δ satisfying (16.45). From (16.35) one gets

$$v(t) = e^{-t}u_0 + \int_0^t e^{s-t} T_{a(s)}^{-1} A^* f_\delta. \tag{16.47}$$

Thus,

$$Av(t) - f_\delta = e^{-t} A u_0 + e^{-t} \int_0^t e^s A T_{a(s)}^{-1} A^* f_\delta \, ds - f_\delta$$

$$= e^{-t}(Au_0 - f_\delta) + e^{-t} \int_0^t e^s (QQ_{a(s)}^{-1} - I) f_\delta \, ds$$

$$= e^{-t}(Au_0 - f_\delta) - \int_0^t e^{-(t-s)} a(s) Q_{a(s)}^{-1} f_\delta \, ds. \tag{16.48}$$

From L'Hospital's rule and (4.40) it follows that

$$\lim_{t \to \infty} \frac{\int_0^t e^s a(s) \|Q_{a(s)}^{-1} f_\delta\| \, ds}{e^t} = \lim_{t \to \infty} \frac{e^t a(t) \|Q_{a(t)}^{-1} f_\delta\|}{e^t} \leq \delta. \tag{16.49}$$

From (16.48) and (16.49) one obtains

$$\lim_{\delta \to 0} \|Av(t) - f_\delta\| = \delta. \tag{16.50}$$

This implies the existence of t_δ.

By the arguments similar to the ones given in the proof of Theorem 4.1.11 (see (4.57)) one gets

$$\lim_{\delta \to 0} t_\delta = \infty. \tag{16.51}$$

By the spectral theorem, one gets

$$\left\| \int_0^t e^{s-t} a(s) Q_{a(s)}^{-1} f_\delta \, ds \right\|^2 = \int_0^{\|Q\|} \left(\int_0^t \frac{e^{s-t} a(s)}{a(s) + \lambda} \, ds \right)^2 d\langle F_\lambda f_\delta, f_\delta \rangle, \tag{16.52}$$

$$\|a(t) Q_{a(t)}^{-1} f_\delta\|^2 = \int_0^{\|Q\|} \frac{a^2(t)}{(a(t) + \lambda)^2} d\langle F_\lambda f_\delta, f_\delta \rangle. \tag{16.53}$$

Since $0 < a(t) \searrow 0$, one has

$$\frac{a(t)}{a(s)} \leq \frac{a(t) + \lambda}{a(s) + \lambda} \leq \frac{a(t) + \|Q\|}{a(s) + \|Q\|}, \qquad 0 \leq s \leq t, \tag{16.54}$$

where $0 \leq \lambda \leq \|Q\|$. Let

$$g_\lambda(t) = \frac{\int_0^t \frac{e^s a(s)}{a(s)+\lambda} ds}{\frac{e^t a(t)}{a(t)+\lambda}}, \qquad t \geq 0, \lambda \geq 0. \tag{16.55}$$

From (16.54) one gets

$$\frac{e^t - 1}{e^t} = g_0(t) \leq g_\lambda(t) \leq g_{\|Q\|}(t) = \frac{\int_0^t \frac{e^s a(s)}{a(s)+\|Q\|} ds}{\frac{e^t a(t)}{a(t)+\|Q\|}}, \tag{16.56}$$

where $0 \leq \lambda \leq \|Q\|$. By L'Hospital's rule, one gets

$$\lim_{t \to \infty} g_{\|Q\|}(t) = \lim_{t \to \infty} \frac{\frac{e^t a(t)}{a(t)+\|Q\|}}{\frac{e^t a(t)}{a(t)+\|Q\|} + \frac{e^t \dot{a}(t)\|Q\|}{(a(t)+\|Q\|)^2}}$$

$$= \lim_{t \to \infty} \frac{1}{1 + \frac{\dot{a}(t)\|Q\|}{a(t)(a(t)+\|Q\|)}} = \frac{1}{1-\alpha}. \tag{16.57}$$

Let $\epsilon > 0$ be sufficiently small so that

$$\frac{1}{1-\alpha} + \epsilon \leq c, \tag{16.58}$$

where c is the constant from (16.44).

From (16.52)–(16.53), (16.55), (16.56), and (16.57) there exists $t_\epsilon > 0$ such that

$$(1 - \frac{\epsilon}{2})^2 \leq \frac{\left\| e^{-t} \int_0^t e^s a(s) Q_{a(s)}^{-1} f_\delta ds \right\|^2}{a^2(t) \|Q_{a(t)}^{-1} f_\delta\|^2} \leq (\frac{1}{1-\alpha} + \frac{\epsilon}{2})^2, \qquad \forall t \geq t_\epsilon. \tag{16.59}$$

This, the relation

$$\lim_{t \to \infty} \frac{\|Au_0 - f_\delta\|}{e^t a(t) \|Q_{a(t)}^{-1} f_\delta\|} = 0, \tag{16.60}$$

the triangle inequality, and formula (16.48) imply that there exists t'_ϵ such that the following condition holds:

$$1 - \epsilon \leq \frac{\|Au(t) - f_\delta\|}{a(t) \|Q_{a(t)}^{-1} f_\delta\|} \leq \frac{1}{1-\alpha} + \epsilon, \qquad \forall t \geq t'_\epsilon.$$

This, (16.45) and (16.51) imply that the following inequalities hold for all sufficiently small $\delta > 0$:

$$\delta \leq (\frac{1}{1-\alpha} + \epsilon)^{-1} c\delta \leq \|Aw_\delta(t_\delta) - f_\delta\| \leq \frac{c}{1-\epsilon}\delta,$$

where
$$w_\delta(t) = T_{a(t)}^{-1} A^* f_\delta. \tag{16.61}$$

This and Theorem 2.2.6 imply
$$\lim_{\delta \to 0} \|w_\delta(t_\delta) - y\| = 0. \tag{16.62}$$

We have proved in Chapter 4 that
$$\lim_{t \to \infty} \|w(t) - y\| = 0, \qquad w(t) := T_{a(t)}^{-1} A^* f.$$

Therefore
$$\lim_{\delta \to 0} \|w_\delta(t_\delta) - w(t_\delta)\| = 0. \tag{16.63}$$

Using the spectral theorem, one gets
$$\|w_\delta(t) - w(t)\|^2 = \int_0^{\|Q\|} \frac{\lambda}{(a(t) + \lambda)^2} d\langle F_s(f_\delta - f), f_\delta - f \rangle. \tag{16.64}$$

Since $0 < a(t) \searrow 0$, it follows from (16.64) that $\|w_\delta(t) - w(t)\|^2$ is an increasing function of t. This and (16.63) imply

$$\lim_{\delta \to 0} \int_0^{t_\delta} e^{s - t_\delta} \|w_\delta(s) - w(s)\| \, ds$$

$$\leq \lim_{\delta \to 0} \|w_\delta(t_\delta) - w(t_\delta)\| \int_0^{t_\delta} e^{s - t_\delta} \, ds = 0. \tag{16.65}$$

From (16.40) and (16.47), one gets

$$\|v(t_\delta) - u(t_\delta)\| \leq \int_0^{t_\delta} e^{s - t_\delta} \|w_\delta(s) - w(s)\| \, ds. \tag{16.66}$$

This and (16.65) imply
$$\lim_{\delta \to 0} \|v(t_\delta) - u(t_\delta)\| = 0. \tag{16.67}$$

From (16.51) and Lemma 16.2.2, one has
$$\lim_{\delta \to 0} \|u(t_\delta) - y\| = 0. \tag{16.68}$$

From the triangle inequality, (16.67) and (16.68) one gets (16.46).
Theorem 16.2.3 is proved.   ∎

From the proof of Theorem 16.2.3, one has the following corollary:

Corollary 16.2.4 *Let $a(t)$ satisfy (16.34), let $v(t)$ be the solution to (16.35), and let*

$$u_0 := T_{a_0}^{-1} A^* f_\delta, \qquad \|A u_0 - f_\delta\| > c\delta^\gamma > \delta, \qquad \gamma \in (0, 1]. \tag{16.69}$$

Then there exists a unique $t_\delta > 0$ such that

$$\|Av(t_\delta) - f_\delta\| = c\delta^\gamma, \quad \|Av(t) - f_\delta\| > c\delta^\gamma, \qquad 0 \le t \le t_\delta. \qquad (16.70)$$

If $\gamma \in (0,1)$ and t_δ satisfies (16.70), then

$$\lim_{\delta \to 0} \|v(t_\delta) - y\| = 0. \qquad (16.71)$$

Denote $a_n = a(nh)$, $h = const > 0$.

Remark 16.2.5 It follows from (16.34) that

$$\lim_{n \to \infty} \frac{a_n}{a_{n-1}} = e^{-h\alpha}. \qquad (16.72)$$

Let us verify (16.72). Let $\epsilon > 0$ be arbitrarily small. From (16.34) one gets

$$(-\alpha - \epsilon)a(t) \le \dot{a}(t) \le (-\alpha + \epsilon)a(t), \qquad (16.73)$$

for all sufficiently large t, say, $t \ge \tau$. This implies

$$e^{-\epsilon h} \le \frac{e^{\alpha h}a(t+h)}{a(t)} \le e^{\epsilon h}, \qquad t \ge \tau. \qquad (16.74)$$

Relation (16.72) follows from (16.74).

Let u_n be defined by

$$u_{n+1} = (1 - h)u_n + hT_{a_n}^{-1}A^* f_\delta, \qquad u_0 \in N^\perp, \qquad n \ge 0. \qquad (16.75)$$

Then one gets

$$u_{n+1} = (1 - h)^{n+1}u_0 + \sum_{i=0}^{n}(1 - h)^{n-i}hT_{a_i}^{-1}A^* f, \qquad n \ge 0. \qquad (16.76)$$

By the arguments similar to the given in the proof of Lemma 16.2.2, one gets the following result

Lemma 16.2.6 *Let $0 < a_n \searrow 0$ and u_n be defined by (16.75). Then*

$$\lim_{n \to \infty} \|u_n - y\| = 0. \qquad (16.77)$$

Consider the following iterative scheme:

$$v_{n+1} = (1 - h)v_n + hT_{a_n}^{-1}A^* f_\delta, \qquad v_0 = u_0. \qquad (16.78)$$

Theorem 16.2.7 *Let $a(t)$ satisfy (16.34), let v_n be defined by (16.78), and let*

$$u_0 := T_{a_0}^{-1}A^* f_\delta, \qquad \|Au_0 - f_\delta\| > c\delta, \qquad c \ge \frac{1}{1 - \alpha}. \qquad (16.79)$$

Then there exists a unique $n_\delta > 0$ such that

$$\|Av_{n_\delta} - f_\delta\| \le c\delta, \quad \|Av_n - f_\delta\| > c\delta, \qquad 0 \le n \le n_\delta. \tag{16.80}$$

This n_δ satisfies the relation $\lim_{\delta \to 0} n_\delta = \infty$, and

$$\lim_{\delta \to 0} \|v_{n_\delta} - y\| = 0. \tag{16.81}$$

Proof: The uniqueness of n_δ follows from its definition (16.80).

Let us prove the existence of n_δ satisfying (16.80). From (16.78) one gets

$$v_{n+1} = (1 - h)^{n+1} u_0 + \sum_{i=0}^{n} (1 - h)^{n-i} h T_{a_i}^{-1} A^* f_\delta. \tag{16.82}$$

Thus,

$$Av_n - f_\delta = (1 - h)^n Au_0 + \sum_{i=0}^{n-1} (1 - h)^{n-i-1} h A T_{a_i}^{-1} A^* f_\delta ds - f_\delta$$

$$= (1 - h)^n (Au_0 - f_\delta) + \sum_{i=0}^{n-1} (1 - h)^{n-i-1} h (QQ_{a_i}^{-1} - I) f_\delta$$

$$= (1 - h)^n (Au_0 - f_\delta) - \sum_{i=0}^{n-1} (1 - h)^{n-i-1} h a_i Q_{a_i}^{-1} f_\delta. \tag{16.83}$$

From a discrete analogue of L'Hospital's rule, one gets

$$\lim_{n \to \infty} \frac{\sum_{i=0}^{n-1} (1 - h)^{-i-1} h a_i \|Q_{a_i}^{-1} f_\delta\|}{(1 - h)^{-n}} \tag{16.84}$$

$$\le \lim_{n \to \infty} \frac{(1 - h)^{-n} h a_n \|Q_{a_n}^{-1} f_\delta\|}{(1 - h)^{-n} - (1 - h)^{-n+1}} \le \delta.$$

This, (16.83), and the triangle inequality yield

$$\lim_{n \to \infty} \|Av_n - f_\delta\| \le \lim_{n \to \infty} \sum_{i=0}^{n-1} (1 - h)^{n-i-1} h a_i \|Q_{a_i}^{-1} f_\delta\| \le \delta.$$

This implies the existence of n_δ.

From the proof of Theorem 16.1.2 (see (16.21)) one gets

$$\lim_{\delta \to 0} n_\delta = \infty. \tag{16.85}$$

Let

$$g_n(\lambda) = \frac{\sum_{i=0}^{n} \frac{(1-h)^{n-i} h a_i}{a_i + \lambda}}{\frac{a_n}{a_n + \lambda}}, \qquad n \ge 0, \lambda \ge 0. \tag{16.86}$$

From (16.13) one gets

$$1 - (1-h)^{n+1} = g_n(0) \leq g_n(\lambda) \leq g_n(\|Q\|) = \frac{\sum_{i=0}^{n} \frac{(1-h)^{-i} h a_i}{a_i + \|Q\|}}{\frac{(1-h)^{-n} a_n}{a_n + \|Q\|}}, \qquad (16.87)$$

where $0 \leq \lambda \leq \|Q\|$. By L'Hospital's rule one obtains

$$\lim_{t \to \infty} g_n(\|Q\|) = \lim_{n \to \infty} \frac{\frac{(1-h)^{-n} h a_n}{a_n + \|Q\|}}{\frac{(1-h)^{-n} a_n}{a_n + \|Q\|} - \frac{(1-h)^{-n+1} a_{n-1}}{a_{n-1} + \|Q\|}}$$

$$= \lim_{t \to \infty} \frac{1}{1 - (1-h)\frac{a_{n-1}}{a_n} \frac{a_n + \|Q\|}{a_{n-1} + \|Q\|}} = \frac{1}{1 - (1-h)e^{h\alpha}}. \qquad (16.88)$$

It follows from (16.87) and (16.88) that for any arbitrarily small $\epsilon > 0$ there exists $n_\epsilon > 0$ such that the following inequality holds:

$$1 - \frac{\epsilon}{2} < g_n(\lambda) < \beta + \frac{\epsilon}{2}, \qquad \beta := \frac{1}{1 - (1-h)e^{\alpha h}}, \qquad \forall n \geq n_\epsilon, \qquad (16.89)$$

where $0 \leq \lambda \leq \|Q\|$. From (16.11) (16.12), (16.86), (16.87), and (16.89) one gets

$$\left(1 - \frac{\epsilon}{2}\right)^2 \leq \frac{\left\| \sum_{i=0}^{n} (1-h)^{n-i} a_i Q_{a_i}^{-1} f_\delta \right\|^2}{a_n^2 \|Q_{a_n}^{-1}\|^2} \leq \left(\beta + \frac{\epsilon}{2}\right)^2, \qquad \forall t \geq t_\epsilon. \; (16.90)$$

This, (16.60), the triangle inequality, and (16.48) imply that there exists n'_ϵ such that

$$1 - \epsilon \leq \frac{\|Au_n - f_\delta\|}{a_n \|Q_{a_n}^{-1} f_\delta\|} \leq \beta + \epsilon, \qquad \forall n \geq n'_\epsilon. \qquad (16.91)$$

This, (16.80), and (16.85) imply that the following inequalities hold for all sufficiently small $\delta > 0$:

$$\delta \leq (\frac{1}{1-\alpha} + \epsilon)^{-1} c\delta \leq \|Aw_{\delta, n_\delta} - f_\delta\| \leq \frac{c}{1-\epsilon}\delta.$$

This and Theorem 2.2.6 imply

$$\lim_{\delta \to 0} \|w_{\delta, n_\delta} - y\| = 0, \qquad w_{\delta, n} := T_{a_n}^{-1} A^* f_\delta. \qquad (16.92)$$

Since

$$\lim_{n \to \infty} \|w_{n_\delta} - y\| = 0, \qquad w_n := T_{a_n}^{-1} A^* f, \qquad (16.93)$$

one gets from (16.82) the relation

$$\lim_{\delta \to 0} \|w_{\delta, n_\delta} - w_{n_\delta}\| = 0. \qquad (16.94)$$

By the spectral theorem one has

$$\|w_{\delta,n} - w_n\|^2 = \int_0^{\|Q\|} \frac{\lambda}{(a_n + \lambda)^2} d\langle F_s(f_\delta - f), f_\delta - f \rangle. \tag{16.95}$$

Since $0 < (a_n)_{n=1}^\infty \searrow 0$, it follows from (16.95) that $\|w_{\delta,n} - w_n\|^2$ is an increasing function of n. This and (16.94) imply

$$\lim_{\delta \to 0} \sum_{i=0}^{n_\delta} (1 - h)^{n_\delta - i} h \|w_{\delta,i} - w_i\|$$

$$\leq \lim_{\delta \to 0} \|w_{\delta,n_\delta} - w_{n_\delta}\| \sum_{i=0}^{n_\delta} (1 - h)^{n_\delta - i} h = 0. \tag{16.96}$$

From (16.76), (16.82), and (16.96) one obtains

$$\lim_{\delta \to 0} \|v_{\delta,n_\delta} - u_{n_\delta}\| = 0. \tag{16.97}$$

Relation (16.81) follows from the triangle inequality, (16.97), and Lemma 16.2.6.

Theorem 16.2.7 is proved. ∎

From the proof of Theorem 16.2.7 one has the following result

Corollary 16.2.8 *Let $a(t)$ satisfy (16.34), let v_n be defined by (16.78), and let*

$$u_0 := T_{a_0}^{-1} A^* f_\delta, \qquad \|Au_0 - f_\delta\| > c\delta^\gamma > \delta, \qquad \gamma \in (0, 1]. \tag{16.98}$$

Then there exists a unique $n_\delta > 0$ such that

$$\|Av_{n_\delta} - f_\delta\| \leq c\delta^\gamma, \quad \|Av_n - f_\delta\| > c\delta^\gamma, \qquad 0 \leq n \leq n_\delta. \tag{16.99}$$

If $\gamma \in (0, 1)$ and n_δ satisfies (16.99), then

$$\lim_{\delta \to 0} \|v_{n_\delta} - y\| = 0. \tag{16.100}$$

CHAPTER 17

DSM OF GRADIENT TYPE FOR SOLVING LINEAR OPERATOR EQUATIONS

The DSM version we study in this chapter consists of solving the Cauchy problem

$$\dot{u}(t) = -A^*(Au(t) - f), \quad u(0) = u_0, \quad u_0 \perp N, \quad \dot{u} := \frac{du}{dt}, \qquad (17.1)$$

where A^* is the adjoint to operator A, and proving the existence of the limit $\lim_{t \to \infty} u(t) = u(\infty)$ and the relation $u(\infty) = y$, that is,

$$\lim_{t \to \infty} \|u(t) - y\| = 0. \qquad (17.2)$$

If the noisy data f_δ are given, then we solve the problem

$$\dot{u}_\delta(t) = -A^*(Au_\delta(t) - f_\delta), \quad u_\delta(0) = u_0, \qquad (17.3)$$

and prove that, for a suitable stopping time t_δ, and $u_\delta := u_\delta(t_\delta)$, one has

$$\lim_{\delta \to 0} \|u_\delta - y\| = 0. \qquad (17.4)$$

In Section 17.1 these results are formulated precisely and recipes for choosing t_δ are proposed.

Dynamical Systems Method and Applications: Theoretical Developments and Numerical Examples, First Edition. A. G. Ramm and N. S. Hoang.
Copyright © 2012 John Wiley & Sons, Inc.

The novel results in this chapter include the proof of the discrepancy principle (Theorem 17.1.3), an efficient method for computing $u_\delta(t_\delta)$ (Section 17.2), and an a priori stopping rule (Theorem 17.1.2).

This chapter is based on paper [54].

17.1 FORMULATIONS AND RESULTS

Suppose $A : H \to H$ is a linear bounded operator in a Hilbert space H. Assume that equation

$$Au = f \tag{17.5}$$

has a solution not necessarily unique. Denote by y the unique minimal-norm solution, that is, $y \perp \mathcal{N} := \mathcal{N}(A)$. Consider the following DSM:

$$\begin{aligned} \dot{u} &= -A^*(Au - f), \\ u(0) &= u_0, \end{aligned} \tag{17.6}$$

where $u_0 \perp \mathcal{N}$ is arbitrary. Denote $T := A^*A$, $Q := AA^*$. The unique solution to (17.6) is

$$u(t) = e^{-tT}u_0 + e^{-tT}\int_0^t e^{sT}\, ds\, A^* f.$$

Let us show that any ill-posed linear equation (17.5) with exact data can be solved by the DSM.

17.1.1 Exact data

Theorem 17.1.1 *Suppose $u_0 \perp \mathcal{N}$. Then problem (17.6) has a unique solution defined on $[0, \infty)$, and $u(\infty) = y$, where $u(\infty) = \lim_{t\to\infty} u(t)$.*

Proof: Denote $w := u(t) - y$, $w_0 = w(0)$. Note that $w_0 \perp \mathcal{N}$. One has

$$\dot{w} = -Tw, \quad T = A^*A. \tag{17.7}$$

The unique solution to (17.7) is $w = e^{-tT}w_0$. Thus,

$$\|w\|^2 = \int_0^{\|T\|} e^{-2t\lambda} d\langle E_\lambda w_0, w_0\rangle.$$

where $\langle u, v\rangle$ is the inner product in H, and E_λ is the resolution of the identity of the selfadjoint operator T. Thus,

$$\|w(\infty)\|^2 = \lim_{t\to\infty}\int_0^{\|T\|} e^{-2t\lambda} d\langle E_\lambda w_0, w_0\rangle = \|P_\mathcal{N} w_0\|^2 = 0,$$

where $P_\mathcal{N} = E_0 - E_{-0}$ is the orthogonal projector onto $\mathcal{N}$. Theorem 17.1.1 is proved. ■

17.1.2 Noisy data f_δ

Let us solve equation (17.5) stably, assuming that f is not known but that f_δ, the noisy data, are known, where $\|f_\delta - f\| \leq \delta$. Consider the following DSM:

$$\dot{u}_\delta = -A^*(Au_\delta - f_\delta), \quad u_\delta(0) = u_0.$$

Denote

$$w_\delta := u_\delta - y, \quad T := A^*A, \quad w_\delta(0) = w_0 := u_0 - y \in \mathcal{N}^\perp.$$

Let us prove the following result:

Theorem 17.1.2 *If* $\lim_{\delta \to 0} t_\delta = \infty$, $\lim_{\delta \to 0} t_\delta \delta = 0$, *and* $w_0 \perp \mathcal{N}$, *then*

$$\lim_{\delta \to 0} \|w_\delta(t_\delta)\| = 0.$$

Proof: One has

$$\dot{w}_\delta = -Tw_\delta + \eta_\delta, \quad \eta_\delta = A^*(f_\delta - f), \quad \|\eta_\delta\| \leq \|A\|\delta. \tag{17.8}$$

The unique solution of equation (17.8) is

$$w_\delta(t) = e^{-tT}w_\delta(0) + \int_0^t e^{-(t-s)T}\eta_\delta\, ds.$$

Let us show that $\lim_{t \to \infty} \|w_\delta(t)\| = 0$. One has

$$\lim_{t \to \infty} \|w_\delta(t)\| \leq \lim_{t \to \infty} \|e^{-tT}w_\delta(0)\| + \lim_{t \to \infty} \left\| \int_0^t e^{-(t-s)T}\eta_\delta\, ds \right\|. \tag{17.9}$$

One uses the spectral theorem and gets:

$$\begin{aligned}
\int_0^t e^{-(t-s)T} ds\eta_\delta &= \int_0^t \int_0^{\|T\|} dE_\lambda \eta_\delta e^{-(t-s)\lambda} ds \\
&= \int_0^{\|T\|} e^{-t\lambda} \frac{e^{t\lambda} - 1}{\lambda} dE_\lambda \eta_\delta \\
&= \int_0^{\|T\|} \frac{1 - e^{-t\lambda}}{\lambda} dE_\lambda \eta_\delta.
\end{aligned} \tag{17.10}$$

Note that

$$0 \leq \frac{1 - e^{-t\lambda}}{\lambda} \leq t, \quad \forall \lambda > 0, t \geq 0, \tag{17.11}$$

since $1 - x \leq e^{-x}$ for $x \geq 0$. From (17.10) and (17.11), one obtains

$$\begin{aligned}
\left\| \int_0^t e^{-(t-s)T} ds\eta_\delta \right\|^2 &= \int_0^{\|T\|} \left| \frac{1 - e^{-t\lambda}}{\lambda} \right|^2 d\langle E_\lambda \eta_\delta, \eta_\delta \rangle \\
&\leq t^2 \int_0^{\|T\|} d\langle E_\lambda \eta_\delta, \eta_\delta \rangle \\
&= t^2 \|\eta_\delta\|^2.
\end{aligned} \tag{17.12}$$

Since $\|\eta_\delta\| \leq \|A\|\delta$, from (17.9) and (17.12), one gets

$$\lim_{\delta \to 0} \|w_\delta(t_\delta)\| \leq \lim_{\delta \to 0} \left(\|e^{-t_\delta T} w_\delta(0)\| + t_\delta \delta \|A\| \right) = 0.$$

Here we have used the relation

$$\lim_{\delta \to 0} \|e^{-t_\delta T} w_\delta(0)\| = \|P_\mathcal{N} w_0\| = 0,$$

and the last equality holds because $w_0 \in \mathcal{N}^\perp$. Theorem 17.1.2 is proved.   ∎

From Theorem 17.1.2, it follows that the relations $t_\delta = \frac{C}{\delta^\gamma}$, $\gamma = const$, $\gamma \in (0, 1)$, and $C > 0$ is a constant can be used as an a priori stopping rule, that is, for such t_δ one has

$$\lim_{\delta \to 0} \|u_\delta(t_\delta) - y\| = 0. \tag{17.13}$$

17.1.3 Discrepancy principle

Let us consider equation (17.5) with noisy data f_δ, and consider a DSM of the form

$$\dot{u}_\delta = -A^* A u_\delta + A^* f_\delta, \quad u_\delta(0) = u_0 \tag{17.14}$$

for solving this equation. Equation (17.14) has been used in Section 17.1.2. Recall that y denotes the minimal-norm solution of equation (17.5).

Theorem 17.1.3 *Assume that $\|A u_0 - f_\delta\| > C\delta$. The solution t_δ to the equation*

$$h(t) := \|A u_\delta(t) - f_\delta\| = C\delta, \quad 1 < C = const \tag{17.15}$$

does exist and is unique, and we have

$$\lim_{\delta \to 0} \|u_\delta(t_\delta) - y\| = 0. \tag{17.16}$$

Proof: Denote

$$v_\delta(t) := A u_\delta(t) - f_\delta, \quad T := A^* A, \quad Q = A A^*$$

and

$$w_\delta(t) := u_\delta(t) - y, \quad w_0 := u_0 - y.$$

One has

$$\begin{aligned}
\frac{d}{dt} \|v_\delta(t)\|^2 &= 2 \operatorname{Re} \langle A \dot{u}_\delta(t), A u_\delta(t) - f_\delta \rangle \\
&= 2 \operatorname{Re} \langle A[-A^*(A u_\delta(t) - f_\delta)], A u_\delta(t) - f_\delta \rangle \\
&= -2 \|A^* v_\delta(t)\|^2 \leq 0.
\end{aligned} \tag{17.17}$$

Thus, $\|v_\delta(t)\|$ is a nonincreasing function. Let us prove that equation (17.15) has a solution for $C > 1$. Recall the known commutation formulas:

$$e^{-sT} A^* = A^* e^{-sQ}, \ Ae^{-sT} = e^{-tQ} A.$$

Using these formulas and the representation

$$u_\delta(t) = e^{-tT} u_0 + \int_0^t e^{-(t-s)T} A^* f_\delta \, ds,$$

one gets

$$
\begin{aligned}
v_\delta(t) &= Au_\delta(t) - f_\delta \\
&= Ae^{-tT} u_0 + A \int_0^t e^{-(t-s)T} A^* f_\delta \, ds - f_\delta \\
&= e^{-tQ} Au_0 + e^{-tQ} \int_0^t e^{sQ} \, ds Q f_\delta - f_\delta \\
&= e^{-tQ} A(u_0 - y) + e^{-tQ} f + e^{-tQ} (e^{tQ} - I) f_\delta - f_\delta \\
&= e^{-tQ} Aw_0 + e^{-tQ} f - e^{-tQ} f_\delta.
\end{aligned}
\tag{17.18}
$$

Note that

$$\lim_{t\to\infty} e^{-tQ} Aw_0 = \lim_{t\to\infty} Ae^{-tT} w_0 = AP_N w_0 = 0.$$

Here the continuity of A, and the following relations

$$\lim_{t\to\infty} e^{-tT} w_0 = \lim_{t\to\infty} \int_0^{\|T\|} e^{-st} dE_s w_0 = (E_0 - E_{-0}) w_0 = P_N w_0,$$

were used. Therefore,

$$\lim_{t\to\infty} \|v_\delta(t)\| = \lim_{t\to\infty} \|e^{-tQ}(f - f_\delta)\| \le \|f - f_\delta\| \le \delta,
\tag{17.19}$$

because $\|e^{-tQ}\| \le 1$. The function $h(t)$ is continuous on $[0, \infty)$, $h(0) = \|Au_0 - f_\delta\| > C\delta$, $h(\infty) \le \delta$. Thus, equation (17.15) must have a solution t_δ.

Let us prove the uniqueness of t_δ. Without loss of generality, we can assume that there exists $t_1 > t_\delta$ such that $\|Au_\delta(t_1) - f_\delta\| = C\delta$. Since $\|v_\delta(t)\|$ is nonincreasing and $\|v_\delta(t_\delta)\| = \|v_\delta(t_1)\|$, one has

$$\|v_\delta(t)\| = \|v_\delta(t_\delta)\|, \quad \forall t \in [t_\delta, t_1].$$

Thus,

$$\frac{d}{dt} \|v_\delta(t)\|^2 = 0, \quad \forall t \in (t_\delta, t_1).
\tag{17.20}$$

Using (17.17) and (17.20), one obtains

$$A^* v_\delta(t) = A^*(Au_\delta(t) - f_\delta) = 0, \quad \forall t \in [t_\delta, t_1].$$

This and (17.14) imply

$$\dot{u}_\delta(t) = 0, \quad \forall t \in (t_\delta, t_1). \tag{17.21}$$

One has

$$\begin{aligned}
\dot{u}_\delta(t) &= -Tu_\delta(t) + A^* f_\delta \\
&= -T\left(e^{-tT} u_0 + \int_0^t e^{-(t-s)T} A^* f_\delta \, ds\right) + A^* f_\delta \\
&= -Te^{-tT} u_0 - (I - e^{-tT})A^* f_\delta + A^* f_\delta \\
&= -e^{-tT}(Tu_0 - A^* f_\delta).
\end{aligned} \tag{17.22}$$

From (17.22) and (17.21), one gets

$$Tu_0 - A^* f = e^{tT} e^{-tT}(Tu_0 - A^* f) = 0.$$

Note that the operator e^{tT} is an isomorphism for any fixed t since T is self-adjoint and bounded. Since $Tu_0 - A^* f = 0$, by (17.22) one has $\dot{u}_\delta(t) = 0$, $u_\delta(t) = u_\delta(0)$, $\forall t \geq 0$. Consequently,

$$C\delta < \|Au_\delta(0) - f_\delta\| = \|Au_\delta(t_\delta) - f_\delta\| = C\delta.$$

This is a contradiction which proves the uniqueness of t_δ.

Let us prove (17.16). First, we have the following estimate:

$$\begin{aligned}
\|Au(t_\delta) - f\| &\leq \|Au(t_\delta) - Au_\delta(t_\delta)\| + \|Au_\delta(t_\delta) - f_\delta\| + \|f_\delta - f\| \\
&\leq \left\|e^{-t_\delta Q} \int_0^{t_\delta} e^{sQ} Q \, ds\right\| \|f_\delta - f\| + C\delta + \delta.
\end{aligned} \tag{17.23}$$

Let us use the inequality

$$\left\|e^{-t_\delta Q} \int_0^{t_\delta} e^{sQ} Q \, ds\right\| = \|I - e^{-t_\delta Q}\| \leq 2$$

and conclude from (17.23) that

$$\lim_{\delta \to 0} \|Au(t_\delta) - f\| = 0. \tag{17.24}$$

Secondly, we claim that

$$\lim_{\delta \to 0} t_\delta = \infty. \tag{17.25}$$

Assume the contrary. Then there exist $t_0 > 0$ and a sequence $(t_{\delta_n})_{n=1}^\infty$, $t_{\delta_n} < t_0$, such that

$$\lim_{n \to \infty} \|Au(t_{\delta_n}) - f\| = 0. \tag{17.26}$$

Analogously to (17.17), one proves that

$$\frac{d\|v\|^2}{dt} \leq 0,$$

where $v(t) := Au(t) - f$. Thus, $\|v(t)\|$ is nonincreasing. This and (17.26) imply the relation $\|v(t_0)\| = \|Au(t_0) - f\| = 0$. Thus,

$$0 = v(t_0) = e^{-t_0 Q} A(u_0 - y).$$

This implies $A(u_0 - y) = e^{t_0 Q} e^{-t_0 Q} A(u_0 - y) = 0$, so $u_0 - y \in \mathcal{N}$. Since $u_0 - y \in \mathcal{N}^\perp$, it follows that $u_0 = y$. This is a contradiction because

$$C\delta \leq \|Au_0 - f_\delta\| = \|f - f_\delta\| \leq \delta, \quad 1 < C.$$

Thus, $\lim_{\delta \to 0} t_\delta = \infty$.

Let us continue the proof of (17.16). Let

$$w_\delta(t) := u_\delta(t) - y.$$

We claim that $\|w_\delta(t)\|$ is nonincreasing on $[0, t_\delta]$. One has

$$\begin{aligned}
\frac{d}{dt}\|w_\delta(t)\|^2 &= 2\operatorname{Re}\langle \dot{u}_\delta(t), u_\delta(t) - y \rangle \\
&= 2\operatorname{Re}\langle -A^*(Au_\delta(t) - f_\delta), u_\delta(t) - y \rangle \\
&= -2\operatorname{Re}\langle Au_\delta(t) - f_\delta, Au_\delta(t) - f_\delta + f_\delta - Ay \rangle \\
&\leq -2\|Au_\delta(t) - f_\delta\|\Big(\|Au_\delta(t) - f_\delta\| - \|f_\delta - f\| \Big) \\
&\leq 0.
\end{aligned}$$

Here we have used the inequalities

$$\|Au_\delta(t) - f_\delta\| \geq C\delta > \|f_\delta - Ay\| = \delta, \quad \forall t \in [0, t_\delta].$$

Let $\epsilon > 0$ be arbitrary small. Since $\lim_{t \to \infty} u(t) = y$, there exists $t_0 > 0$, independent of δ, such that

$$\|u(t_0) - y\| \leq \frac{\epsilon}{2}. \tag{17.27}$$

Since $\lim_{\delta \to 0} t_\delta = \infty$ (see (17.25)), there exists $\delta_0 > 0$ such that $t_\delta > t_0$, $\forall \delta \in (0, \delta_0)$. Since $\|w_\delta(t)\|$ is nonincreasing on $[0, t_\delta]$, one has

$$\|w_\delta(t_\delta)\| \leq \|w_\delta(t_0)\| \leq \|u_\delta(t_0) - u(t_0)\| + \|u(t_0) - y\|, \quad \forall \delta \in (0, \delta_0). \tag{17.28}$$

Note that

$$\begin{aligned}
\|u_\delta(t_0) - u(t_0)\| &= \|e^{-t_0 T} \int_0^{t_0} e^{sT} ds A^*(f_\delta - f)\| \\
&\leq \|e^{-t_0 T} \int_0^{t_0} e^{sT} ds A^*\|\delta.
\end{aligned} \tag{17.29}$$

Since $e^{-t_0 T} \int_0^{t_0} e^{sT} ds A^*$ is a bounded operator for any fixed t_0, one concludes from (17.29) that $\lim_{\delta \to 0} \|u_\delta(t_0) - u(t_0)\| = 0$. Hence, there exists $\delta_1 \in (0, \delta_0)$ such that

$$\|u_\delta(t_0) - u(t_0)\| \leq \frac{\epsilon}{2}, \quad \forall \delta \in (0, \delta_1). \tag{17.30}$$

From (17.27)–(17.30), one obtains

$$\|u_\delta(t_\delta) - y\| = \|w_\delta(t_\delta)\| \leq \frac{\epsilon}{2} + \frac{\epsilon}{2} = \epsilon, \quad \forall \delta \in (0, \delta_1).$$

This means that $\lim_{\delta \to 0} u_\delta(t_\delta) = y$. Theorem 17.1.3 is proved. ∎

17.2 IMPLEMENTATION OF THE DISCREPANCY PRINCIPLE

17.2.1 Systems with known spectral decomposition

One way to solve the Cauchy problem (17.14) is to use explicit Euler or Runge–Kutta methods with a constant or adaptive stepsize h. However, stepsize h for solving (17.14) by explicit numerical methods is often smaller than 1 and the stopping time $t_\delta = nh$ may be large. Therefore, the computation time, characterized by the number of iterations n, for this approach may be large. This fact is also reported in [45], where one of the most efficient numerical methods for solving ordinary differential equations (ODEs), the DOPRI45 (see [41]), is used for solving a Cauchy problem in a DSM. Indeed, the use of explicit Euler method leads to a Landweber iteration which is known for slow convergence. Thus, it may be computationally expensive to compute $u_\delta(t_\delta)$ by numerical methods for ODEs.

However, when A in (17.14) is a matrix and a decomposition $A = USV^*$, where U and V are unitary matrices and S is a diagonal matrix, is known, it is possible to compute $u_\delta(t_\delta)$ at a speed comparable to other methods such as the variational regularization (VR) as it will be shown below.

We have

$$u_\delta(t) = e^{-tT} u_0 + e^{-tT} \int_0^t e^{sT} ds A^* f_\delta, \quad T := A^* A. \tag{17.31}$$

Suppose that a decomposition

$$A = USV^*, \tag{17.32}$$

where U and V are unitary matrices and S is a diagonal matrix, is known. These matrices possibly contain complex entries. Thus, $T = A^* A = V \bar{S} S V^*$ and $e^T = e^{V \bar{S} S V^*}$. Using the formula $e^{V \bar{S} S V^*} = V e^{\bar{S} S} V^*$, which is valid if V is unitary and $\bar{S} S$ is diagonal, equation (17.31) can be rewritten as

$$u_\delta(t) = V e^{-t \bar{S} S} V^* u_0 + V \int_0^t e^{(s-t) \bar{S} S} ds \bar{S} U^* f_\delta. \tag{17.33}$$

Here, the overbar stands for complex conjugation. Choose $u_0 = 0$. Then

$$u_\delta(t) = V \int_0^t e^{(s-t)\bar{S}S} ds \bar{S} h_\delta, \quad h_\delta := U^* f_\delta. \tag{17.34}$$

Let us assume that

$$\delta < \|f\|. \tag{17.35}$$

This is a natural assumption. Let us check that

$$A^* f_\delta \neq 0. \tag{17.36}$$

Indeed, if $A^* f_\delta = 0$, then one gets

$$\langle f_\delta, f \rangle = \langle f_\delta, Ay \rangle = \langle A^* f_\delta, y \rangle = 0. \tag{17.37}$$

This implies

$$\delta^2 \geq \|f - f_\delta\|^2 = \|f\|^2 + \|f_\delta\|^2 > \delta^2. \tag{17.38}$$

This contradiction implies (17.36).

We choose the stopping time t_δ by the following discrepancy principle:

$$\|Au_\delta(t_\delta) - f_\delta\| = \left\| \int_0^{t_\delta} e^{(s-t_\delta)\bar{S}S} ds \bar{S} S h_\delta - h_\delta \right\| = \|e^{-t_\delta \bar{S}S} h_\delta\| = C\delta,$$

where $1 < C$.

Let us find t_δ from the equation

$$\phi(t) := \psi(t) - C\delta = 0, \qquad \psi(t) := \|e^{-t\bar{S}S} h_\delta\|. \tag{17.39}$$

The existence and uniqueness of the solution t_δ to equation (17.39) follow from Theorem 17.1.3.

We claim that *equation (17.39) can be solved by using Newton's iteration (17.47) for any initial value t_0 such that $\phi(t_0) > 0$.*

Let us prove this claim. It is sufficient to prove that $\phi(t)$ is a monotone strictly convex function. This is proved below.

Without loss of generality, we can assume that h_δ (see (17.39)) is a vector with real components. The proof remained essentially the same for h_δ with complex components.

First, we claim that

$$\sqrt{\bar{S}S} h_\delta \neq 0 \qquad \text{and} \qquad \|\sqrt{\bar{S}S} e^{-t\bar{S}S} h_\delta\| \neq 0, \tag{17.40}$$

so $\psi(t) > 0$.

Indeed, since $e^{-t\bar{S}S}$ is an isomorphism and $e^{-t\bar{S}S}$ commutes with $\sqrt{\bar{S}S}$, one concludes that $\|\sqrt{\bar{S}S} e^{-t\bar{S}S} h_\delta\| = 0$ iff $\sqrt{\bar{S}S} h_\delta = 0$. If $\sqrt{\bar{S}S} h_\delta = 0$, then $\bar{S} h_\delta = 0$ and, therefore,

$$0 = \bar{S} h_\delta = \bar{S} U^* f_\delta = V^* V \bar{S} U^* f_\delta = V^* A^* f_\delta. \tag{17.41}$$

Since V is a unitary matrix, it follows from (17.41) that $A^* f_\delta = 0$. This contradicts relation (17.36).

Let us now prove that ϕ monotonically decays and is strictly convex. Then our claim will be proved.

One has

$$\frac{d}{dt}\langle e^{-t\bar{S}S}h_\delta, e^{-t\bar{S}S}h_\delta\rangle = -2\langle e^{-t\bar{S}S}h_\delta, \bar{S}Se^{-t\bar{S}S}h_\delta\rangle.$$

Thus,

$$\dot{\psi}(t) = \frac{d}{dt}\|e^{-t\bar{S}S}h_\delta\| = \frac{\frac{d}{dt}\|e^{-t\bar{S}S}h_\delta\|^2}{2\|e^{-t\bar{S}S}h_\delta\|} = -\frac{\langle e^{-t\bar{S}S}h_\delta, \bar{S}Se^{-t\bar{S}S}h_\delta\rangle}{\|e^{-t\bar{S}S}h_\delta\|}. \quad (17.42)$$

Equation (17.42), relation (17.40), and

$$\langle e^{-t\bar{S}S}h_\delta, \bar{S}Se^{-t\bar{S}S}h_\delta\rangle = \|\sqrt{\bar{S}S}e^{-t\bar{S}S}h_\delta\|^2$$

imply

$$\dot{\psi}(t) < 0. \quad (17.43)$$

From equation (17.42) and the definition of ψ in (17.39), one gets

$$\psi(t)\dot{\psi}(t) = -\langle e^{-t\bar{S}S}h_\delta, \bar{S}Se^{-t\bar{S}S}h_\delta\rangle \quad (17.44)$$

Differentiating equation (17.44) with respect to t, one obtains

$$\psi(t)\ddot{\psi}(t) + \dot{\psi}^2(t) = \langle \bar{S}Se^{-t\bar{S}S}h_\delta, \bar{S}Se^{-t\bar{S}S}h_\delta\rangle + \langle e^{-t\bar{S}S}h_\delta, \bar{S}S\bar{S}Se^{-t\bar{S}S}h_\delta\rangle$$
$$= 2\|\bar{S}Se^{-t\bar{S}S}h_\delta\|^2.$$

This equation and equation (17.42) imply

$$\psi(t)\ddot{\psi}(t) = 2\|\bar{S}Se^{-t\bar{S}S}h_\delta\|^2 - \frac{\langle e^{-t\bar{S}S}h_\delta, \bar{S}Se^{-t\bar{S}S}h_\delta\rangle^2}{\|e^{-t\bar{S}S}h_\delta\|^2}$$
$$\geq \|\bar{S}Se^{-t\bar{S}S}h_\delta\|^2 > 0. \quad (17.45)$$

Here the inequality

$$\langle e^{-t\bar{S}S}h_\delta, \bar{S}Se^{-t\bar{S}S}h_\delta\rangle \leq \|e^{-t\bar{S}S}h_\delta\|\|\bar{S}Se^{-t\bar{S}S}h_\delta\|$$

was used. Since $\psi > 0$, inequality (17.45) implies

$$\ddot{\psi}(t) > 0. \quad (17.46)$$

It follows from inequalities (17.43) and (17.46) that $\phi(t)$ is a strictly convex and decreasing function on $(0, \infty)$. Therefore, t_δ can be found by Newton's iterations:

$$t_{n+1} = t_n - \frac{\phi(t_n)}{\dot{\phi}(t_n)}$$
$$= t_n + \frac{\|e^{-t_n\bar{S}S}h_\delta\| - C\delta}{\langle \bar{S}Se^{-t_n\bar{S}S}h_\delta, e^{-t_n\bar{S}S}h_\delta\rangle}\|e^{-t_n\bar{S}S}h_\delta\|, \quad n = 0, 1, ..., \quad (17.47)$$

for any initial guess t_0 of t_δ such that $\phi(t_0) > 0$. Once t_δ is found, the solution $u_\delta(t_\delta)$ is computed by (17.34).

Remark 17.2.1 In the decomposition $A = VSU^*$ we do not assume that U, V, and S are matrices with real entries. The singular value decomposition (SVD) is a particular case of this decomposition.

It is computationally expensive to get the SVD of a matrix in general. However, there are many problems in which the decomposition (17.32) can be computed fast using the fast Fourier transform (FFT). Examples include image restoration problems with circulant block matrices (see [86]) and deconvolution problems.

17.2.2   On the choice of t_0

Let us discuss a strategy for choosing the initial value t_0 in Newton's iterations for finding t_δ. We choose t_0 satisfying condition

$$0 < \phi(t_0) = \|e^{-t_0 \bar{S}S} h_\delta\| - \delta \leq \delta \tag{17.48}$$

by the following strategy:

1. Choose $t_0 := 10 \frac{\|h_\delta\|}{\delta}$ as an initial guess for t_0.

2. Compute $\phi(t_0)$. If t_0 satisfies (17.48) we are done. Otherwise, we go to step 3.

3. If $\phi(t_0) < 0$ and the inequality $\phi(t_0) > \delta$ has not occurred in iteration, we replace t_0 by $\frac{t_0}{10}$ and go back to step 2. If $\phi(t_0) < 0$ and the inequality $\phi(t_0) > \delta$ has occurred in iteration, we replace t_0 by $\frac{t_0}{3}$ and go back to step 2. If $\phi(t_0) > \delta$, we go to step 4.

4. If $\phi(t_0) > \delta$ and the inequality $\phi(t_0) < 0$ has not occurred in iterations, we replace t_0 by $3t_0$ and go back to step 2. If the inequality $\phi(t_0) < 0$ has occurred in some iteration before, we stop the iteration and use t_0 as an initial guess in Newton's iterations for finding t_δ.

CHAPTER 18

DSM FOR SOLVING LINEAR EQUATIONS WITH FINITE-RANK OPERATORS

In this chapter we study DSM for solving stably the equation

$$Au = f, \tag{18.1}$$

where A is a linear bounded operator in a Hilbert space H. The DSM version we study in this chapter consists of solving the Cauchy problem

$$\dot{u}(t) = -P(Au(t) - f), \quad u(0) = u_0, \quad u_0 \perp \mathcal{N}, \quad \dot{u} := \frac{du}{dt} \tag{18.2}$$

and proving the existence of the limit $\lim_{t \to \infty} u(t) = u(\infty)$ and the relation $u(\infty) = y$, that is,

$$\lim_{t \to \infty} \|u(t) - y\| = 0. \tag{18.3}$$

Here P is a bounded operator such that $T := PA \geq 0$ is self-adjoint, $\mathcal{N}(T) = \mathcal{N}(A)$.

For any linear (not necessarily bounded) operator A there exists a bounded operator P such that $T = PA \geq 0$. For example, if $A = U|A|$ is the polar decomposition of A, then $|A| := (A^*A)^{\frac{1}{2}}$ is a self-adjoint operator, $T := |A| \geq 0$, U is a partial isometry, $\|U\| = 1$; and if $P := U^*$, then $\|P\| = 1$ and

Dynamical Systems Method and Applications: Theoretical Developments and Numerical Examples, First Edition. A. G. Ramm and N. S. Hoang.
Copyright © 2012 John Wiley & Sons, Inc.

$PA = T$. Another choice of P is $P = (A^*A + aI)^{-1}A^*$, $a = const > 0$. For this choice $Q := AP \geq 0$.

If the noisy data f_δ are given, $\|f_\delta - f\| \leq \delta$, then we solve the problem

$$\dot{u}_\delta(t) = -P(Au_\delta(t) - f_\delta), \quad u_\delta(0) = u_0 \tag{18.4}$$

and prove that, for a suitable stopping time t_δ and for $u_\delta := u_\delta(t_\delta)$, one has

$$\lim_{\delta \to 0} \|u_\delta - y\| = 0. \tag{18.5}$$

An a priori and an a posteriori methods for choosing t_δ are given.

In Section 18.1 these results are formulated and recipes for choosing t_δ are proposed.

This chapter is based on paper [49].

18.1   FORMULATION AND RESULTS

Suppose $A : H \to H$ is a linear bounded operator in a real Hilbert space H. Assume that equation (18.1) has a solution not necessarily unique. Denote by y the unique minimal-norm solution, that is, $y \perp \mathcal{N} := \mathcal{N}(A)$. Consider the DSM (18.2) where $u_0 \perp \mathcal{N}$ is arbitrary. Denote

$$T := PA, \quad Q := AP. \tag{18.6}$$

The unique solution to (18.2) is

$$u(t) = e^{-tT}u_0 + e^{-tT} \int_0^t e^{sT} ds P f. \tag{18.7}$$

Let us first show that any ill-posed linear equation (18.1) with exact data can be solved by the DSM. We assume below that $P = (A^*A + aI)^{-1}A^*$, where $a = const > 0$. With this choice of P one has $\mathcal{N}(T) = \mathcal{N}(A)$, $\|T\| \leq 1$.

18.1.1   Exact data

Theorem 18.1.1 *Suppose $u_0 \perp \mathcal{N}$ and $T^* = T \geq 0$. Then problem (18.2) has a unique solution defined on $[0, \infty)$; and $u(\infty) = y$, where $u(\infty) = \lim_{t \to \infty} u(t)$.*

Proof: Denote $w := u(t) - y$, $w_0 := w(0) = u_0 - y$. Note that $w_0 \perp \mathcal{N}$. One has

$$\dot{w} = -Tw, \quad T := PA, \quad w(0) = u_0 - y. \tag{18.8}$$

The unique solution to (18.8) is $w = e^{-tT}w_0$. Thus,

$$\|w\|^2 = \int_0^{\|T\|} e^{-2t\lambda} d\langle E_\lambda w_0, w_0 \rangle.$$

where $\langle u, v \rangle$ is the inner product in H, and E_λ is the resolution of the identity of T. Thus,

$$\|w(\infty)\|^2 = \lim_{t \to \infty} \int_0^{\|T\|} e^{-2t\lambda} d\langle E_\lambda w_0, w_0 \rangle = \|P_\mathcal{N} w_0\|^2 = 0,$$

where $P_\mathcal{N} = E_0 - E_{-0}$ is the orthogonal projector onto $\mathcal{N}$. Theorem 18.1.1 is proved. ∎

18.1.2   Noisy data f_δ

Let us solve equation (18.1) stably, assuming that f is not known but that f_δ, the noisy data, are known, where $\|f_\delta - f\| \leq \delta$. Consider the following DSM:

$$\dot{u}_\delta = -P(Au_\delta - f_\delta), \quad u_\delta(0) = u_0. \tag{18.9}$$

Denote

$$w_\delta := u_\delta - y, \quad T := PA, \quad w_\delta(0) = w_0 := u_0 - y \in \mathcal{N}^\perp.$$

Let us prove the following result:

Theorem 18.1.2 *If $T = T^* \geq 0$, $\lim_{\delta \to 0} t_\delta = \infty$, $\lim_{\delta \to 0} t_\delta \delta = 0$, and $w_0 \in \mathcal{N}^\perp$, then*

$$\lim_{\delta \to 0} \|w_\delta(t_\delta)\| = 0.$$

Proof: One has

$$\dot{w}_\delta = -Tw_\delta + \zeta_\delta, \quad \zeta_\delta = P(f_\delta - f), \quad \|\zeta_\delta\| \leq \|P\|\delta. \tag{18.10}$$

The unique solution of equation (18.10) is

$$w_\delta(t) = e^{-tT} w_\delta(0) + \int_0^t e^{-(t-s)T} \zeta_\delta \, ds.$$

Let us show that $\lim_{\delta \to 0} \|w_\delta(t_\delta)\| = 0$. One has

$$\lim_{t \to \infty} \|w_\delta(t)\| \leq \lim_{t \to \infty} \|e^{-tT} w_\delta(0)\| + \lim_{t \to \infty} \left\| \int_0^t e^{-(t-s)T} \zeta_\delta \, ds \right\|. \tag{18.11}$$

Let E_λ be the resolution of identity corresponding to T. One uses the spectral theorem and gets

$$\int_0^t e^{-(t-s)T} ds \zeta_\delta = \int_0^t \int_0^{\|T\|} dE_\lambda \zeta_\delta e^{-(t-s)\lambda} \, ds$$

$$= \int_0^{\|T\|} e^{-t\lambda} \frac{e^{t\lambda} - 1}{\lambda} dE_\lambda \zeta_\delta$$

$$= \int_0^{\|T\|} \frac{1 - e^{-t\lambda}}{\lambda} dE_\lambda \zeta_\delta. \tag{18.12}$$

Note that

$$0 \leq \frac{1 - e^{-t\lambda}}{\lambda} \leq t, \quad \forall \lambda > 0, \quad t \geq 0, \tag{18.13}$$

since $1 - x \leq e^{-x}$ for $x \geq 0$. From (18.12) and (18.13), one obtains

$$\left\| \int_0^t e^{-(t-s)T} ds \zeta_\delta \right\|^2 = \int_0^{\|T\|} \left| \frac{1 - e^{-t\lambda}}{\lambda} \right|^2 d\langle E_\lambda \zeta_\delta, \zeta_\delta \rangle$$

$$\leq t^2 \int_0^{\|T\|} d\langle E_\lambda \zeta_\delta, \zeta_\delta \rangle$$

$$= t^2 \|\zeta_\delta\|^2. \tag{18.14}$$

This estimate follows also from the inequality: $\|e^{-(t-s)T}\| \leq 1$, which holds for $T^* = T \geq 0$ and $t \geq s$. Indeed, one has $\| \int_0^t e^{-(t-s)T} ds \| \leq t$, and estimate (18.14) follows.

Since $\|\zeta_\delta\| \leq \|P\|\delta$, from (18.11) and (18.14), one gets

$$\lim_{\delta \to 0} \|w_\delta(t_\delta)\| \leq \lim_{\delta \to 0} \left(\|e^{-t_\delta T} w_\delta(0)\| + t_\delta \delta \|P\| \right) = 0.$$

Here we have used the relation

$$\lim_{\delta \to 0} \|e^{-t_\delta T} w_\delta(0)\| = \|P_\mathcal{N} w_0\| = 0,$$

and the last equality holds because $w_0 \in \mathcal{N}^\perp$. Theorem 18.1.2 is proved. ∎

From Theorem 18.1.2, it follows that the relation

$$t_\delta = \frac{C}{\delta^\gamma}, \quad \gamma = const, \quad \gamma \in (0, 1),$$

where $C > 0$ is a constant, can be used as an a priori stopping rule; that is, for such t_δ one has

$$\lim_{\delta \to 0} \|u_\delta(t_\delta) - y\| = 0. \tag{18.15}$$

18.1.3 Discrepancy principle

In this section we assume that A is a linear finite-rank operator. Thus, it is a linear bounded operator. Let us consider equation (18.1) with noisy data f_δ, and also consider a DSM of the form

$$\dot{u}_\delta = -PAu_\delta + Pf_\delta, \quad u_\delta(0) = u_0, \tag{18.16}$$

for solving this equation. Equation (18.16) has been used in Section 18.1.2. Recall that y denotes the minimal-norm solution of equation (18.1) and that $\mathcal{N}(T) = \mathcal{N}(A)$ with our choice of P.

Theorem 18.1.3 *Let* $T := PA$, $Q := AP$. *Assume that* $\|Au_0 - f_\delta\| > C\delta$, $Q = Q^* \geq 0$, $T^* = T \geq 0$, *and* T *is a finite-rank operator. Then the solution* t_δ *to the equation*

$$h(t) := \|Au_\delta(t) - f_\delta\| = C\delta, \quad C = const > 1, \tag{18.17}$$

does exist and is unique, $\lim_{\delta \to 0} t_\delta = \infty$, *and*

$$\lim_{\delta \to 0} \|u_\delta(t_\delta) - y\| = 0, \tag{18.18}$$

where y *is the unique minimal-norm solution to* (18.1).

Proof: Denote

$$v_\delta(t) := Au_\delta(t) - f_\delta, \quad w(t) := u(t) - y, \quad w_0 := u_0 - y.$$

One has

$$\begin{aligned}
\frac{d}{dt}\|v_\delta(t)\|^2 &= 2\operatorname{Re}\langle A\dot{u}_\delta(t), Au_\delta(t) - f_\delta\rangle \\
&= 2\operatorname{Re}\langle A[-P(Au_\delta(t) - f_\delta)], Au_\delta(t) - f_\delta\rangle \\
&= -2\langle AP(Au_\delta - f_\delta), Au_\delta - f_\delta\rangle \leq 0, \tag{18.19}
\end{aligned}$$

where the last inequality holds because $AP = Q \geq 0$. Thus, $\|v_\delta(t)\|$ is a nonincreasing function.

Let us prove that equation (18.17) has a solution for $C > 1$. One has the following commutation formulas:

$$e^{-sT}P = Pe^{-sQ}, \quad Ae^{-sT} = e^{-sQ}A.$$

Using these formulas and the representation

$$u_\delta(t) = e^{-tT}u_0 + \int_0^t e^{-(t-s)T}Pf_\delta\, ds,$$

one gets

$$\begin{aligned}
v_\delta(t) &= Au_\delta(t) - f_\delta \\
&= Ae^{-tT}u_0 + A\int_0^t e^{-(t-s)T}Pf_\delta\, ds - f_\delta \\
&= e^{-tQ}Au_0 + e^{-tQ}\int_0^t e^{sQ}\, ds\, Qf_\delta - f_\delta \\
&= e^{-tQ}A(u_0 - y) + e^{-tQ}f + e^{-tQ}(e^{tQ} - I)f_\delta - f_\delta \\
&= e^{-tQ}Aw_0 - e^{-tQ}f_\delta + e^{-tQ}f = e^{-tQ}Au_0 - e^{-tQ}f_\delta. \tag{18.20}
\end{aligned}$$

Note that

$$\lim_{t \to \infty} e^{-tQ}Aw_0 = \lim_{t \to \infty} Ae^{-tT}w_0 = AP_\mathcal{N}w_0 = 0.$$

Here the continuity of A and the relation

$$\lim_{t\to\infty} e^{-tT} w_0 = \lim_{t\to\infty} \int_0^{\|T\|} e^{-st} dE_s w_0 = (E_0 - E_{-0}) w_0 = P_{\mathcal{N}} w_0$$

were used. Therefore,

$$\lim_{t\to\infty} \|v_\delta(t)\| = \lim_{t\to\infty} \|e^{-tQ}(f - f_\delta)\| \le \|f - f_\delta\| \le \delta, \tag{18.21}$$

where $\|e^{-tQ}\| \le 1$ because $Q \ge 0$. The function $h(t)$ is continuous on $[0, \infty)$, $h(0) = \|Au_0 - f_\delta\| > C\delta$, $h(\infty) \le \delta$. Thus, equation (18.17) must have a solution t_δ.

Let us prove the uniqueness of t_δ. If t_δ is nonunique, then without loss of generality we can assume that there exists $t_1 > t_\delta$ such that $\|Au_\delta(t_1) - f_\delta\| = C\delta$. Since $\|v_\delta(t)\|$ is nonincreasing and $\|v_\delta(t_\delta)\| = \|v_\delta(t_1)\|$, one has

$$\|v_\delta(t)\| = \|v_\delta(t_\delta)\|, \quad \forall t \in [t_\delta, t_1].$$

Thus,

$$\frac{d}{dt} \|v_\delta(t)\|^2 = 0, \quad \forall t \in (t_\delta, t_1). \tag{18.22}$$

Using (18.19) and (18.22), one obtains

$$\|\sqrt{AP}(Au_\delta(t) - f_\delta)\|^2 = \langle AP(Au_\delta(t) - f_\delta), Au_\delta(t) - f_\delta \rangle = 0, \quad \forall t \in [t_\delta, t_1],$$

where $\sqrt{AP} = Q^{\frac{1}{2}} \ge 0$ is well-defined since $Q = Q^* \ge 0$. This implies $Q^{\frac{1}{2}}(Au_\delta - f_\delta) = 0$. Thus

$$Q(Au_\delta(t) - f_\delta) = 0, \quad \forall t \in [t_\delta, t_1]. \tag{18.23}$$

From (18.20) one gets

$$v_\delta(t) = Au_\delta(t) - f_\delta = e^{-tQ} Au_0 - e^{-tQ} f_\delta. \tag{18.24}$$

Since $Qe^{-tQ} = e^{-tQ}Q$ and e^{-tQ} is an isomorphism, equalities (18.23) and (18.24) imply

$$Q(Au_0 - f_\delta) = 0.$$

This and (18.24) imply

$$AP(Au_\delta(t) - f_\delta) = e^{-tQ}(QAu_0 - Qf_\delta) = 0, \quad t \ge 0.$$

This and (18.19) imply

$$\frac{d}{dt} \|v_\delta\|^2 = 0, \quad t \ge 0. \tag{18.25}$$

Consequently,

$$C\delta < \|Au_\delta(0) - f_\delta\| = \|v_\delta(0)\| = \|v_\delta(t_\delta)\| = \|Au_\delta(t_\delta) - f_\delta\| = C\delta.$$

This is a contradiction which proves the uniqueness of t_δ.

Let us prove (18.18). First, we have the following estimate:

$$\|Au(t_\delta) - f\| \le \|Au(t_\delta) - Au_\delta(t_\delta)\| + \|Au_\delta(t_\delta) - f_\delta\| + \|f_\delta - f\|$$

$$\le \left\| e^{-t_\delta Q} \int_0^{t_\delta} e^{sQ} Q ds \right\| \|f_\delta - f\| + C\delta + \delta, \qquad (18.26)$$

where $u(t)$ solves (18.2) and $u_\delta(t)$ solves (18.9). One uses the inequality

$$\left\| e^{-t_\delta Q} \int_0^{t_\delta} e^{sQ} Q ds \right\| = \|I - e^{-t_\delta Q}\| \le 2,$$

and concludes from (18.26) that

$$\lim_{\delta \to 0} \|Au(t_\delta) - f\| = 0. \qquad (18.27)$$

Secondly, we claim that

$$\lim_{\delta \to 0} t_\delta = \infty.$$

Assume the contrary. Then there exist $t_0 > 0$ and a sequence $(t_{\delta_n})_{n=1}^\infty$, $t_{\delta_n} < t_0$, $\lim_{n \to \infty} \delta_n = 0$, such that

$$\lim_{n \to \infty} \|Au(t_{\delta_n}) - f\| = 0. \qquad (18.28)$$

Analogously to (18.19), one proves that

$$\frac{d}{dt} \|v\|^2 \le 0,$$

where $v(t) := Au(t) - f$. Thus, $\|v(t)\|$ is nonincreasing. This and (18.28) imply the relation $\|v(t_0)\| = \|Au(t_0) - f\| = 0$. Thus,

$$0 = v(t_0) = e^{-t_0 Q} A(u_0 - y).$$

This implies $A(u_0 - y) = e^{t_0 Q} e^{-t_0 Q} A(u_0 - y) = 0$, so $u_0 - y \in \mathcal{N}$. Since $u_0 - y \in \mathcal{N}^\perp$, it follows that $u_0 = y$. This is a contradiction because

$$C\delta \le \|Au_0 - f_\delta\| = \|f - f_\delta\| \le \delta, \quad C > 1.$$

Thus,

$$\lim_{\delta \to 0} t_\delta = \infty. \qquad (18.29)$$

Let us continue the proof of (18.18). From (18.20) and the relation $\|Au_\delta(t_\delta) - f_\delta\| = C\delta$, one has

$$C\delta t_\delta = \|t_\delta e^{-t_\delta Q} Aw_0 - t_\delta e^{-t_\delta Q}(f_\delta - f)\|$$

$$\le \|t_\delta e^{-t_\delta Q} Aw_0\| + \|t_\delta e^{-t_\delta Q}(f_\delta - f)\|$$

$$\le \|t_\delta e^{-t_\delta Q} Aw_0\| + t_\delta \delta. \qquad (18.30)$$

We claim that

$$\lim_{\delta \to 0} t_\delta e^{-t_\delta Q} A w_0 = \lim_{\delta \to 0} t_\delta A e^{-t_\delta T} w_0 = 0. \tag{18.31}$$

Note that (18.31) holds if $T \geq 0$ *has finite rank, and* $w_0 \in \mathcal{N}^\perp$. It also holds if $P = A^*$, $T = A^* A \geq 0$ is compact, and the Fourier coefficients $w_{0j} := \langle w_0, \phi_j \rangle$, where $T\phi_j = \lambda_j \phi_j$, decay sufficiently fast. In this case

$$\|A e^{-tT} w_0\|^2 \leq \|T^{\frac{1}{2}} e^{-tT} w_0\|^2 = \sum_{j=1}^{\infty} \lambda_j e^{-2\lambda_j t} |w_{0j}|^2 := S = o\left(\frac{1}{t^2}\right), \quad t \to \infty,$$

provided that $\sum_{j=1}^{\infty} |w_{0j}| \lambda_j^{-2} < \infty$. Indeed,

$$S = \sum_{\lambda_j \leq \frac{1}{t^{\frac{2}{3}}}} + \sum_{\lambda_j > \frac{1}{t^{\frac{2}{3}}}} := S_1 + S_2.$$

One has

$$S_1 \leq \frac{1}{t^2} \sum_{\lambda_j \leq t^{-\frac{2}{3}}} \frac{|w_{0j}|^2}{\lambda_j^2} = o\left(\frac{1}{t^2}\right), \quad S_2 \leq ce^{-2t^{\frac{1}{3}}} = o\left(\frac{1}{t^2}\right), \quad t \to \infty,$$

where $c > 0$ is a constant.

From (18.31) and (18.30), one gets

$$0 \leq \lim_{\delta \to 0} (C - 1)\delta t_\delta \leq \lim_{\delta \to 0} \|t_\delta e^{-t_\delta Q} A w_0\| = 0.$$

Thus,

$$\lim_{\delta \to 0} \delta t_\delta = 0 \tag{18.32}$$

Now, the desired conclusion (18.18) follows from (18.29), (18.32), and Theorem 18.1.2. Theorem 18.1.3 is proved. ∎

18.1.4   An iterative scheme

Let us solve equation (18.1) stably, assuming that f is not known but that f_δ, the noisy data, are known, where $\|f_\delta - f\| \leq \delta$. Consider the following discrete version of the DSM:

$$u_{n+1,\delta} = u_{n,\delta} - hP(Au_{n,\delta} - f_\delta), \quad u_{\delta,0} = u_0. \tag{18.33}$$

Let us denote $u_n := u_{n,\delta}$ when $\delta \neq 0$, and set

$$w_n := u_n - y, \quad T := PA, \quad w_0 := u_0 - y \in \mathcal{N}^\perp.$$

Let $n = n_\delta$ be the stopping rule for iterations (18.33). Let us prove the following result:

Theorem 18.1.4 *Assume that* $T = T^* \geq 0$, $h\|T\| < 2$, $\lim_{\delta \to 0} n_\delta h = \infty$, $\lim_{\delta \to 0} n_\delta h\delta = 0$, *and* $w_0 \in \mathcal{N}^\perp$. *Then*

$$\lim_{\delta \to 0} \|w_{n_\delta}\| = \lim_{\delta \to 0} \|u_{n_\delta} - y\| = 0. \tag{18.34}$$

Proof: One has

$$w_{n+1} = w_n - hTw_n + h\zeta_\delta, \quad \zeta_\delta = P(f_\delta - f), \quad \|\zeta_\delta\| \leq \|P\|\delta, \quad w_0 = u_0 - y. \tag{18.35}$$

The unique solution of equation (18.35) is

$$w_{n+1} = (I - hT)^{n+1}w_0 + h\sum_{i=0}^{n}(I - hT)^i\zeta_\delta.$$

Let us show that $\lim_{\delta \to 0} \|w_{n_\delta}\| = 0$. One has

$$\|w_n\| \leq \|(I - hT)^n w_0\| + \left\| h\sum_{i=0}^{n-1}(I - hT)^i\zeta_\delta \right\|. \tag{18.36}$$

Let E_λ be the resolution of the identity corresponding to T. One uses the spectral theorem and gets

$$h\sum_{i=0}^{n-1}(I - hT)^i = h\sum_{i=0}^{n-1}\int_0^{\|T\|}(1 - h\lambda)^i \, dE_\lambda$$

$$= h\int_0^{\|T\|}\frac{1 - (1 - \lambda h)^n}{1 - (1 - h\lambda)} \, dE_\lambda$$

$$= \int_0^{\|T\|}\frac{1 - (1 - \lambda h)^n}{\lambda} \, dE_\lambda. \tag{18.37}$$

Note that

$$0 \leq \frac{1 - (1 - h\lambda)^n}{\lambda} \leq hn, \quad \forall \lambda > 0, \quad t \geq 0, \tag{18.38}$$

since $1 - (1 - \alpha)^n \leq \alpha n$ for all $\alpha \in [0, 2]$. From (18.37) and (18.38), one obtains

$$\left\| h\sum_{i=0}^{n-1}(I - hT)^i\zeta_\delta \right\|^2 = \int_0^{\|T\|}\left|\frac{1 - (1 - \lambda h)^n}{\lambda}\right|^2 d\langle E_\lambda\zeta_\delta, \zeta_\delta\rangle$$

$$\leq (hn)^2\int_0^{\|T\|} d\langle E_\lambda\zeta_\delta, \zeta_\delta\rangle$$

$$= (nh)^2\|\zeta_\delta\|^2. \tag{18.39}$$

Alternatively, this estimate follows from the inequality $\|(I - hT)^i\| \leq 1$, provided that $0 \leq hT < 2$. Indeed, in this case one has $\|\sum_{i=0}^{n-1}(I - hT)^i\| \leq n$, and this implies estimate (18.39).

Since $\|\zeta_\delta\| \leq \|P\|\delta$, from (18.36) and (18.39), one gets

$$\lim_{\delta \to 0} \|w_{n_\delta}\| \leq \lim_{\delta \to 0} \left(\|(I - hT)^{n_\delta} w_\delta(0)\| + hn_\delta \delta \|P\| \right) = 0.$$

Here we have used the relation

$$\lim_{\delta \to 0} \|(I - hT)^{n_\delta} w_\delta(0)\| = \|P_{\mathcal{N}} w_0\| = 0,$$

and the last equality holds because $w_0 \in \mathcal{N}^\perp$. Theorem 18.1.4 is proved. ∎

From Theorem 18.1.4, it follows that the relation

$$n_\delta = \frac{C}{h\delta^\gamma}, \quad \gamma = const, \quad \gamma \in (0,1)$$

where $C > 0$ is a constant, can be used as an a priori stopping rule; that is, for such n_δ one has

$$\lim_{\delta \to 0} \|u_{n_\delta} - y\| = 0. \tag{18.40}$$

18.1.5 An iterative scheme with a stopping rule based on a discrepancy principle

In this section we assume that A is a linear finite-rank operator. Thus, it is a linear bounded operator. Let us consider equation (18.1) with noisy data f_δ, and also consider an iterative scheme of the form

$$u_{n+1} = u_n - hP(Au_n - f_\delta), \quad u_0 = u_0, \tag{18.41}$$

for solving this equation. Equation (18.41) has been used in Section 18.1.4. Recall that y denotes the minimal-norm solution of equation (18.1).

Note that $\mathcal{N} := \mathcal{N}(T) = \mathcal{N}(A)$.

Theorem 18.1.5 *Let* $T := PA$, $Q := AP$. *Assume that* $\|Au_0 - f_\delta\| > C\delta$, $Q = Q^* \geq 0$, $T^* = T \geq 0$, $h\|T\| < 2$, $h\|Q\| < 2$, *and* T *is a finite-rank operator. Then there exists a unique* n_δ *such that*

$$\|Au_{n_\delta} - f_\delta\| \leq C\delta < \|Au_{n_\delta-1} - f_\delta\|, \quad C = const > 1. \tag{18.42}$$

For this n_δ *one has*

$$\lim_{\delta \to 0} \|u_{n_\delta} - y\| = 0. \tag{18.43}$$

Proof: Denote

$$v_n := Au_n - f_\delta, \quad w_n := u_n - y, \quad w_0 := u_0 - y.$$

From (18.41) one gets

$$v_{n+1} = Au_{n+1} - f_\delta = Au_n - f_\delta - hAP(Au_n - f_\delta) = v_n - hQv_n.$$

This implies

$$\begin{aligned}
\|v_{n+1}\|^2 - \|v_n\|^2 &= \mathrm{Re}\langle v_{n+1} - v_n, v_{n+1} + v_n\rangle \\
&= \mathrm{Re}\langle -hQv_n, v_n - hQv_n + v_n\rangle \\
&= -\langle v_n, hQ(2 - hQ)v_n\rangle \leq 0, \qquad (18.44)
\end{aligned}$$

where the last inequality holds because $AP = Q \geq 0$ and $\|hQ\| < 2$. Thus, $(\|v_n\|)_{n=1}^\infty$ is a nonincreasing sequence.

Let us prove that equation (18.42) has a solution for $C > 1$. One has the following commutation formulas:

$$(I - hT)^n P = P(I - hQ)^n, \quad A(I - hT)^n = (I - hQ)^n A.$$

Using these formulas, the representation

$$u_n = (I - hT)^n u_0 + h \sum_{i=0}^{n-1} (I - hT)^i P f_\delta,$$

and the identity $(I - B)\sum_{i=0}^{n-1} B^i = I - B^n$, with $B = I - hQ$ and $I - B = hQ$, one gets

$$\begin{aligned}
v_n &= Au_n - f_\delta \\
&= A(I - hT)^n u_0 + Ah \sum_{i=0}^{n-1} (I - hT)^i P f_\delta - f_\delta \\
&= (I - hQ)^n Au_0 + \sum_{i=0}^{n-1} (I - hQ)^i hQ f_\delta - f_\delta \\
&= (I - hQ)^n Au_0 - (I - (I - hQ)^n) f_\delta - f_\delta \\
&= (I - hQ)^n (Au_0 - f) + (I - hQ)^n (f - f_\delta) \\
&= (I - hQ)^n Aw_0 + (I - hQ)^n (f - f_\delta). \qquad (18.45)
\end{aligned}$$

If $V = V^* \geq 0$ is an operator with $\|V\| \leq 2$, then

$$\|I - V\| = \sup_{0 \leq s \leq 2} |1 - s| \leq 1.$$

Note that

$$\lim_{n \to \infty} (I - hQ)^n Aw_0 = \lim_{n \to \infty} A(I - hT)^n w_0 = AP_\mathcal{N} w_0 = 0,$$

where $P_\mathcal{N}$ is the orthoprojection onto the null space $\mathcal{N}$ of the operator T, and the continuity of A and the relation

$$\lim_{n \to \infty} (I - hT)^n w_0 = \lim_{n \to \infty} \int_0^{\|T\|} (1 - sh)^n dE_s w_0 = (E_0 - E_{-0}) w_0 = P_\mathcal{N} w_0,$$

for $0 \le sh < 2$, were used. Therefore,

$$\lim_{n \to \infty} \|v_n\| = \lim_{n \to \infty} \|(I - hQ)^n (f - f_\delta)\| \le \|f - f_\delta\| \le \delta, \qquad (18.46)$$

where $\|I - hQ\| \le 1$ because $Q \ge 0$ and $\|hQ\| < 2$. The sequence $\{\|v_n\|\}_{n=1}^\infty$ is nonincreasing with $\|v_0\| > C\delta$ and $\lim_{n \to \infty} \|v_n\| \le \delta$. Thus, there exists $n_\delta > 0$ such that (18.42) holds.

Let us prove (18.43). Let $u_{n,0}$ be the sequence defined by the relations

$$u_{n+1,0} = u_{n,0} - hP(Au_{n,0} - f), \quad u_{0,0} = u_0.$$

First, we have the following estimate:

$$\|Au_{n_\delta,0} - f\| \le \|Au_{n_\delta} - Au_{n_\delta,0}\| + \|Au_{n_\delta} - f_\delta\| + \|f_\delta - f\|$$

$$\le \left\| \sum_{i=0}^{n_\delta - 1} (I - hQ)^i hQ \right\| \|f_\delta - f\| + C\delta + \delta. \qquad (18.47)$$

Since $0 \le hQ < 2$, one has $\|I - hQ\| \le 1$. This implies the inequality

$$\left\| \sum_{i=0}^{n_\delta - 1} (I - hQ)^i hQ \right\| = \|I - (I - hQ)^{n_\delta}\| \le 2,$$

and one concludes from (18.47) that

$$\lim_{\delta \to 0} \|Au_{n_\delta,0} - f\| = 0. \qquad (18.48)$$

Secondly, we claim that

$$\lim_{\delta \to 0} n_\delta = \infty.$$

Assume the contrary. Then there exist $n_0 > 0$ and a sequence $(n_{\delta_n})_{n=1}^\infty$, $n_{\delta_n} < n_0$, such that

$$\lim_{n \to \infty} \|Au_{n_\delta,0} - f\| = 0. \qquad (18.49)$$

Analogously to (18.44), one proves that

$$\|v_{n,0}\| \le \|v_{n-1,0}\|,$$

where $v_{n,0} = Au_{n,0} - f$. Thus, the sequence $\|v_{n,0}\|$ is nonincreasing. This and (18.49) imply the relation $\|v_{n_0,0}\| = \|Au_{n_0,0} - f\| = 0$. Thus,

$$0 = v_{n_0,0} = (I - hQ)^{n_0} A(u_0 - y).$$

This implies $A(u_0 - y) = (I - hQ)^{-n_0}(I - hQ)^{n_0} A(u_0 - y) = 0$, so $u_0 - y \in \mathcal{N}$. Since, by the assumption, $u_0 - y \in \mathcal{N}^\perp$, it follows that $u_0 = y$. This is a contradiction because

$$C\delta \le \|Au_0 - f_\delta\| = \|f - f_\delta\| \le \delta, \quad C > 1.$$

Thus,

$$\lim_{\delta \to 0} n_\delta = \infty. \tag{18.50}$$

Let us continue the proof of (18.43). From (18.45) and $\|Au_{n_\delta} - f_\delta\| = C\delta$, one has

$$
\begin{aligned}
C\delta n_\delta h &= \|n_\delta h(I - hQ)^{n_\delta} Aw_0 - n_\delta h(I - hQ)^{n_\delta}(f_\delta - f)\| \\
&\leq \|n_\delta h(I - hQ)^{n_\delta} Aw_0\| + \|n_\delta h(I - hQ)^{n_\delta}(f_\delta - f)\| \\
&\leq \|n_\delta h(I - hQ)^{n_\delta} Aw_0\| + n_\delta h\delta. \tag{18.51}
\end{aligned}
$$

We claim that if $w_0 \in \mathcal{N}^\perp$, $0 \leq hT < 2$, and T is a finite-rank operator, then

$$\lim_{\delta \to 0} n_\delta h(I - hQ)^{n_\delta} Aw_0 = \lim_{\delta \to 0} n_\delta hA(I - hT)^{n_\delta} w_0 = 0. \tag{18.52}$$

From (18.51) and (18.52) one gets

$$0 \leq \lim_{\delta \to 0}(C - 1)\delta h n_\delta \leq \lim_{\delta \to 0} \|n_\delta h(I - hQ)^{n_\delta} Aw_0\| = 0.$$

Thus,

$$\lim_{\delta \to 0} \delta n_\delta h = 0 \tag{18.53}$$

Now (18.43) follows from (18.50), (18.53), and Theorem 18.1.4. Theorem 18.1.5 is proved. ∎

18.1.6 Computing $u_\delta(t_\delta)$

In Chapter 17 the DSM (18.9) was investigated with $P = A^*$ and the singular value decomposition (SVD) of A was assumed known. In general, it is computationally expensive to get the SVD of large-scale matrices. In this chapter, we have derived an iterative scheme for solving ill-conditioned linear algebraic systems $Au = f_\delta$ without using SVD of A.

Choose $P = (A^*A + aI)^{-1}A^*$, where a is a fixed positive constant. This choice of P satisfies all the conditions in Theorem 18.1.3. In particular, $Q = AP = A(A^*A + aI)^{-1}A^* = AA^*(AA^* + aI)^{-1} \geq 0$ is a self-adjoint operator, and $T = PA = (A^*A + aI)^{-1}A^*A \geq 0$ is a self-adjoint operator. Since

$$\|T\| = \left\| \int_0^{\|A^*A\|} \frac{\lambda}{\lambda + a} dE_\lambda \right\| = \sup_{0 \leq \lambda \leq \|A^*A\|} \frac{\lambda}{\lambda + a} < 1,$$

where E_λ is the resolution of the identity of A^*A, and the condition $h\|T\| < 2$ in Theorem 18.1.5 is satisfied for all $0 < h \leq 1$. Set $h = 1$ and $P = (A^*A + a)^{-1}A^*$ in (18.41). Then one gets the following iterative scheme:

$$u_{n+1} = u_n - (A^*A + aI)^{-1}(A^*Au_n - A^*f_\delta), \quad u_0 = 0. \tag{18.54}$$

For simplicity we have chosen $u_0 = 0$. However, one may choose $u_0 = v_0$ if v_0 is known to be a better approximation to y than 0 and $v_0 \in \mathcal{N}^\perp$. In iterations

(18.54) we use a stopping rule of discrepancy type. Indeed, we stop iterations if u_n satisfies the following condition:

$$\|Au_n - f_\delta\| \leq 1.01\delta. \tag{18.55}$$

The choice of a affects both the accuracy and the computation time of the method. If a is too large, one needs more iterations to approach the desired accuracy, so the computation time will be large. If a is too small, then the results become less accurate because for too small a the inversion of the operator $A^*A + aI$ is an ill-posed problem since the operator A^*A is not boundedly invertible. Using the idea of the choice of the initial guess of the regularization parameter in [45], we choose a to satisfy the following condition:

$$\delta \leq \phi(a) := \|A(A^*A + a)^{-1}A^*f_\delta - f_\delta\| \leq 2\delta. \tag{18.56}$$

This can be done by using the following strategy:

1. Choose $a := \frac{\delta\|A\|^2}{3\|f_\delta\|}$ as an initial guess for a.

2. Compute $\phi(a)$. If a satisfies (18.56), then we are done. Otherwise, we go to step 3.

3. If $c = \frac{\phi(a)}{\delta} > 3$, we replace a by $\frac{a}{2(c-1)}$ and go back to step 2. If $2 < c \leq 3$, then we replace a by $\frac{a}{2(c-1)}$ and go back to step 2. Otherwise, we go to step 4.

4. If $c = \frac{\phi(a)}{\delta} < 1$, we replace a by $3a$. If the inequality $c < 1$ has occurred in an earlier iteration, we stop the iterations and use $3a$ as our choice for a in iterations (18.54). Otherwise we go back to step 2.

CHAPTER 19

A DISCREPANCY PRINCIPLE FOR EQUATIONS WITH MONOTONE CONTINUOUS OPERATORS

Consider the equation

$$F(u) = f, \tag{19.1}$$

where F is a monotone operator in a Hilbert space H. Recall that F is called monotone if

$$\langle F(u) - F(v), u - v \rangle \geq 0, \quad \forall u, v \in H. \tag{19.2}$$

Assume that F is continuous.

Equations with monotone operators are important in many applications and were studied extensively (see, e.g., [88], [179], and [190] and references therein). There are many technical and physical problems leading to equations with such operators in the cases when dissipation of energy occurs. For example, in [103] and in [93, pp. 156-189] a wide class of nonlinear dissipative systems is studied, and the basic equations of such systems can be reduced to equation (19.1) with monotone operators. Many examples of equations with monotone operators can be found in [79] and in references mentioned above. In [94] and [95] it is proved that any solvable linear operator equation with a closed densely defined operator in a Hilbert space H can be reduced to

Dynamical Systems Method and Applications: Theoretical Developments and Numerical Examples, First Edition. A. G. Ramm and N. S. Hoang.
Copyright © 2012 John Wiley & Sons, Inc.

an equation with a monotone operator and solved by a convergent iterative process.

In this chapter a discrepancy principle for solving equation (19.3) with noisy data (see Section 19.1) is proved under natural assumptions. No smallness assumptions on the nonlinearity, no global restrictions on its growth, or other special properties of the nonlinearity, except the monotonicity and continuity, are imposed. No source-type assumptions are used. Our result is widely applicable. It is well known that without extra assumptions, usually source-type assumption concerning the right-hand side, or some equivalent assumption concerning the smoothness of the solution, one cannot get a rate of convergence even for linear ill-posed equations. On the other hand, such assumptions are usually not algorithmically verifiable and often they do not hold. By this reason we do not make such assumptions and do not give estimates of the rate of convergence.

In [183] a stationary equation $F(u) = f$ with a nonlinear monotone operator F was studied. The assumptions A1–A3 on p. 197 in [183] are more restrictive than ours, and the Rule R2 on p. 199, formula (4.1) in [183], for the choice of the regularization parameter is more difficult to use computationally: One has to solve nonlinear equation (4.1) in [183] for the regularization parameter. Moreover, to use this equation one has to invert an ill-conditioned linear operator $A + aI$ for small values of a. Assumption A1 in [183] is not verifiable, because the solution $x^\dagger$ is not known. Assumption A3 in [183] requires F to be constant in a ball $B_r(x^\dagger)$ if $F'(x^\dagger) = 0$. Our discrepancy principle does not require these assumptions, and, in contrast to equation (4.1) in [183], it does not require inversion of ill-conditioned linear operators.

Main results in this chapter include Theorem 19.2.1 in Section 19.2 and Theorem 19.3.2 in Section 19.3. In Theorem 19.2.1 a new discrepancy principle is proposed and justified, assuming only the monotonicity and continuity of F. Implementing the discrepancy principle in Theorem 19.2.1 requires solving equation (19.3) and then solving nonlinear equation (19.16) for the regularization parameter $a(\delta)$. Theorem 19.3.2 allows one to solve equations (19.3) and (19.16) approximately. Thus, when δ is not too small, one can save a large amount of computations in solving equations (19.3) and (19.16) by applying Theorem 19.3.2 and using our new stopping rule. These results allow one to solve numerically stably equation (19.1) if F is locally Lipschitz and monotone. Based on Theorem 19.3.2, an algorithm for stable solution of equation (19.1) is formulated for locally Lipschitz monotone operators.

This chapter is based on paper [51].

19.1   AUXILIARY RESULTS

Let us consider the equation

$$F(V_{\delta,a}) + aV_{\delta,a} - f_\delta = 0, \qquad a > 0, \tag{19.3}$$

where $a = const$. It is known (see Lemma 6.1.7) that equation (19.3) with monotone continuous operator F has a unique solution for any $f_\delta \in H$.

Below, the word decreasing means strictly decreasing and increasing means strictly increasing.

Let us first prove the following results:

Lemma 19.1.1 *Assume* $\|F(0) - f_\delta\| > 0$. *Let* $a > 0$, *and* F *be monotone. Denote*

$$\psi(a) := \|V_{\delta,a}\|, \qquad \phi(a) := a\psi(a) = \|F(V_{\delta,a}) - f_\delta\|,$$

where $V_{\delta,a}$ *solves* (19.3). *Then* $\psi(a)$ *is decreasing, and* $\phi(a)$ *is increasing.*

Proof: Since $\|F(0) - f_\delta\| > 0$, one has $\psi(a) \neq 0$, $\forall a \geq 0$. Indeed, if $\psi(a)\big|_{a=\tau} = 0$, then $V_{\delta,a} = 0$, and equation (19.3) implies $\|F(0) - f_\delta\| = 0$, which is a contradiction. Note that $\phi(a) = a\|V_{\delta,a}\|$. From (19.2) and (19.3), one gets

$$
\begin{aligned}
0 &\leq \langle F(V_{\delta,a}) - F(V_{\delta,b}), V_{\delta,a} - V_{\delta,b} \rangle \\
&= \langle -aV_{\delta,a} + bV_{\delta,b}, V_{\delta,a} - V_{\delta,b} \rangle \\
&= a\langle V_{\delta,a}, V_{\delta,b} \rangle + b\langle V_{\delta,b}, V_{\delta,a} \rangle - a\|V_{\delta,a}\|^2 - b\|V_{\delta,b}\|^2 \\
&= \operatorname{Re}\left[a\langle V_{\delta,a}, V_{\delta,b} \rangle + b\langle V_{\delta,b}, V_{\delta,a} \rangle \right] - a\|V_{\delta,a}\|^2 - b\|V_{\delta,b}\|^2. \quad (19.4)
\end{aligned}
$$

Thus,

$$
\begin{aligned}
0 &\leq a\operatorname{Re}\langle V_{\delta,a}, V_{\delta,b} \rangle + b\operatorname{Re}\langle V_{\delta,b}, V_{\delta,a} \rangle - a\|V_{\delta,a}\|^2 - b\|V_{\delta,b}\|^2 \\
&\leq (a+b)\|V_{\delta,a}\|\|V_{\delta,b}\| - a\|V_{\delta,a}\|^2 - b\|V_{\delta,b}\|^2 \\
&= (a\|V_{\delta,a}\| - b\|V_{\delta,b}\|)(\|V_{\delta,b}\| - \|V_{\delta,a}\|) \\
&= (\phi(a) - \phi(b))(\psi(b) - \psi(a)). \quad (19.5)
\end{aligned}
$$

If $\psi(b) > \psi(a)$, then (19.5) implies $\phi(a) \geq \phi(b)$, so

$$a\psi(a) \geq b\psi(b) > b\psi(a).$$

Therefore, if $\psi(b) > \psi(a)$, then $b < a$.

Similarly, if $\psi(b) < \psi(a)$, then $\phi(a) \leq \phi(b)$. This implies $b > a$.

Suppose $\psi(a) = \psi(b)$, that is, $\|V_{\delta,a}\| = \|V_{\delta,b}\|$. From (19.4) one has

$$\|V_{\delta,a}\|^2 \leq \frac{a\langle V_{\delta,a}, V_{\delta,b} \rangle + b\langle V_{\delta,b}, V_{\delta,a} \rangle}{a+b} \leq \|V_{\delta,a}\|\|V_{\delta,b}\| = \|V_{\delta,a}\|^2.$$

This implies $V_{\delta,a} = V_{\delta,b}$, and then equation (19.3) implies $a = b$.

Therefore ϕ is increasing and ψ is decreasing. ∎

Lemma 19.1.2 *If* F *is monotone and continuous, then* $\|V_{\delta,a}\| = O(\frac{1}{a})$ *as* $a \to \infty$, *and*

$$\lim_{a \to \infty} \|F(V_{\delta,a}) - f_\delta\| = \|F(0) - f_\delta\|. \quad (19.6)$$

Proof: Rewrite (19.3) as

$$F(V_{\delta,a}) - F(0) + aV_{\delta,a} = f_\delta - F(0).$$

Multiply this equation by $V_{\delta,a}$, use the monotonicity of F, and get

$$a\|V_{\delta,a}\|^2 \leq \langle aV_{\delta,a} + F(V_{\delta,a}) - F(0), V_{\delta,a}\rangle$$
$$= \langle f_\delta - F(0), V_{\delta,a}\rangle \leq \|f_\delta - F(0)\|\|V_{\delta,a}\|.$$

Therefore, $\|V_{\delta,a}\| = O(\frac{1}{a})$. This and the continuity of F imply (19.6).   ∎

Remark 19.1.3 If $\|F(0) - f_\delta\| > C\delta^\gamma, 0 < \gamma \leq 1$, then relation (19.6) implies

$$\|F(V_{\delta,a}) - f_\delta\| \geq C\delta^\gamma, \qquad 0 < \gamma \leq 1, \tag{19.7}$$

for sufficiently large $a > 0$.

Lemma 19.1.4 *Let $C > 0$ and $\gamma \in (0,1]$ be constants such that $C\delta^\gamma > \delta$. Suppose that $\|F(0) - f_\delta\| > C\delta^\gamma$. Then, there exists a unique $a(\delta) > 0$ such that $\|F(V_{\delta,a(\delta)}) - f_\delta\| = C\delta^\gamma$.*

Proof: We have $F(y) = f$, and

$$0 = \langle F(V_{\delta,a}) + aV_{\delta,a} - f_\delta, F(V_{\delta,a}) - f_\delta\rangle$$
$$= \|F(V_{\delta,a}) - f_\delta\|^2 + a\langle V_{\delta,a} - y, F(V_{\delta,a}) - f_\delta\rangle + a\langle y, F(V_{\delta,a}) - f_\delta\rangle$$
$$= \|F(V_{\delta,a}) - f_\delta\|^2 + a\langle V_{\delta,a} - y, F(V_{\delta,a}) - F(y)\rangle + a\langle V_{\delta,a} - y, f - f_\delta\rangle$$
$$\quad + a\langle y, F(V_{\delta,a}) - f_\delta\rangle$$
$$\geq \|F(V_{\delta,a}) - f_\delta\|^2 + a\langle V_{\delta,a} - y, f - f_\delta\rangle + a\langle y, F(V_{\delta,a}) - f_\delta\rangle.$$

Here the monotonicity of F was used. Therefore

$$\|F(V_{\delta,a}) - f_\delta\|^2 \leq -a\langle V_{\delta,a} - y, f - f_\delta\rangle - a\langle y, F(V_{\delta,a}) - f_\delta\rangle$$
$$\leq a\|V_{\delta,a} - y\|\|f - f_\delta\| + a\|y\|\|F(V_{\delta,a}) - f_\delta\|$$
$$\leq a\delta\|V_{\delta,a} - y\| + a\|y\|\|F(V_{\delta,a}) - f_\delta\|. \tag{19.8}$$

From (19.3), relation $F(y) = f$, and the monotonicity of F, one gets

$$0 = \langle F(V_{\delta,a}) - F(y) + aV_{\delta,a} + f - f_\delta, V_{\delta,a} - y\rangle$$
$$= \langle F(V_{\delta,a}) - F(y), V_{\delta,a} - y\rangle + a\|V_{\delta,a} - y\|^2 + a\langle y, V_{\delta,a} - y\rangle$$
$$\quad + \langle f - f_\delta, V_{\delta,a} - y\rangle$$
$$\geq a\|V_{\delta,a} - y\|^2 + a\langle y, V_{\delta,a} - y\rangle + \langle f - f_\delta, V_{\delta,a} - y\rangle,$$

where the monotonicity of F was used again. Therefore,

$$a\|V_{\delta,a} - y\|^2 \leq a\|y\|\|V_{\delta,a} - y\| + \delta\|V_{\delta,a} - y\|.$$

This implies

$$a\|V_{\delta,a} - y\| \leq a\|y\| + \delta. \tag{19.9}$$

From (19.8), (19.9), and an elementary inequality $ab \leq \epsilon a^2 + \frac{b^2}{4\epsilon}$, $\forall \epsilon > 0$, one gets

$$\|F(V_{\delta,a}) - f_\delta\|^2 \leq \delta^2 + a\|y\|\delta + a\|y\|\|F(V_{\delta,a}) - f_\delta\|$$
$$\leq \delta^2 + a\|y\|\delta + \epsilon\|F(V_{\delta,a}) - f_\delta\|^2 + \frac{1}{4\epsilon}a^2\|y\|^2, \tag{19.10}$$

where $\epsilon > 0$ is arbitrary small, fixed, and independent of a and can be chosen arbitrary small. Let $a \searrow 0$. Then (19.10) implies

$$\lim_{a \to 0}(1 - \epsilon)\|F(V_{\delta,a}) - f_\delta\|^2 \leq \delta^2, \qquad \forall \epsilon > 0. \tag{19.11}$$

Thus,

$$\lim_{a \to 0}\|F(V_{\delta,a}) - f_\delta\| \leq \delta < C\delta^\gamma, \qquad 0 < \gamma \leq 1. \tag{19.12}$$

This, the continuity of F, the continuity of $V_{\delta,a}$ with respect to $a \in [0,\infty)$, and inequality (19.7) imply that equation $\|F(V_{\delta,a}) - f_\delta\| = C\delta^\gamma$ must have a solution $a(\delta) > 0$.

The uniqueness of $a(\delta) > 0$ follows from Lemma 19.1.1.

Lemma 19.1.4 is proved.   ■

Remark 19.1.5 Let $V_a := V_{\delta,a}|_{\delta=0}$, so $F(V_a) + aV - f = 0$. From the proof of Lemma 6.3.1 one gets (see (6.53))

$$\|V_{\delta,a} - V_a\| \leq \frac{\delta}{a}. \tag{19.13}$$

It follows from (6.19) and the relation $V_a = u_a$ that

$$\|V_a\| \leq \|y\|, \qquad \forall a > 0. \tag{19.14}$$

From (19.13) and (19.14), one gets the following estimate:

$$\|V_{\delta,a}\| \leq \|V_a\| + \frac{\delta}{a} \leq \|y\| + \frac{\delta}{a}. \tag{19.15}$$

19.2   A DISCREPANCY PRINCIPLE

Our standing assumptions are the monotonicity and continuity of F and the solvability of equation (19.1). They are not repeated below. We assume without loss of generality that $\delta \in (0,1)$.

Theorem 19.2.1 *Let $\gamma \in (0,1]$ and $C > 0$ be some constants such that $C\delta^\gamma > \delta$. Assume that $\|F(0) - f_\delta\| > C\delta^\gamma$. Let y be the minimal-norm solution to equation (19.1). Then there exists a unique $a(\delta) > 0$ such that*

$$\|F(V_{\delta,a(\delta)}) - f_\delta\| = C\delta^\gamma, \tag{19.16}$$

where $V_{\delta,a(\delta)}$ solves (19.3) with $a = a(\delta)$.
If $0 < \gamma < 1$, then

$$\lim_{\delta \to 0} \|V_{\delta,a(\delta)} - y\| = 0. \qquad (19.17)$$

Proof: The existence and uniqueness of $a(\delta)$ follow from Lemma 19.1.4. Let us show that

$$\lim_{\delta \to 0} a(\delta) = 0. \qquad (19.18)$$

The triangle inequality, inequality (19.13) and equality (19.16) imply

$$a(\delta)\|V_{a(\delta)}\| \leq a(\delta)\big(\|V_{\delta,a(\delta)} - V_{a(\delta)}\| + \|V_{\delta,a(\delta)}\|\big)$$
$$\leq \delta + a(\delta)\|V_{\delta,a(\delta)}\| = \delta + C\delta^{\gamma}. \qquad (19.19)$$

From inequality (19.19), one gets

$$\lim_{\delta \to 0} a(\delta)\|V_{a(\delta)}\| = 0. \qquad (19.20)$$

It follows from Lemma 19.1.1 with $f_\delta = f$, i.e., $\delta = 0$, that the function $\phi_0(a) := a\|V_a\|$ is nonnegative and strictly increasing on $(0, \infty)$. This and relation (19.20) imply

$$\lim_{\delta \to 0} a(\delta) = 0. \qquad (19.21)$$

From (19.16) and (19.15), one gets

$$C\delta^{\gamma} = a\|V_{\delta,a}\| \leq a(\delta)\|y\| + \delta. \qquad (19.22)$$

Thus, one gets

$$C\delta^{\gamma} - \delta \leq a(\delta)\|y\|. \qquad (19.23)$$

If $\gamma < 1$, then $C - \delta^{1-\gamma} > 0$ for sufficiently small δ. This implies

$$0 \leq \lim_{\delta \to 0} \frac{\delta}{a(\delta)} \leq \lim_{\delta \to 0} \frac{\delta^{1-\gamma}\|y\|}{C - \delta^{1-\gamma}} = 0. \qquad (19.24)$$

By the triangle inequality and inequality (19.13), one has

$$\|V_{\delta,a(\delta)} - y\| \leq \|V_{a(\delta)} - y\| + \|V_{a(\delta)} - V_{\delta,a(\delta)}\| \leq \|V_{a(\delta)} - y\| + \frac{\delta}{a(\delta)}. \ (19.25)$$

Relation (19.17) follows from (19.24), (19.25), and Lemma 6.1.9. ∎

Instead of using (19.3), one may use the following equation:

$$F(V_{\delta,a}) + a(V_{\delta,a} - \bar{u}) - f_\delta = 0, \qquad a > 0, \qquad (19.26)$$

where $\bar{u}$ is an element of H. Denote $F_1(u) := F(u + \bar{u})$. Then F_1 is monotone and continuous. Equation (19.3) can be written as

$$F_1(U_{\delta,a}) + aU_{\delta,a} - f_\delta = 0, \qquad U_{\delta,a} := V_{\delta,a} - \bar{u}, \quad a > 0. \qquad (19.27)$$

By applying Theorem 19.2.1 with $F = F_1$ one gets the following result:

Corollary 19.2.2 *Let $\gamma \in (0, 1]$ and $C > 0$ be some constants such that $C\delta^\gamma > \delta$. Let $\bar{u} \in H$ and z be the solution to (19.1) with minimal distance to $\bar{u}$. Assume that $\|F(\bar{u}) - f_\delta\| > C\delta^\gamma$. Then there exists a unique $a(\delta) > 0$ such that*

$$\|F(\tilde{V}_{\delta,a(\delta)}) - f_\delta\| = C\delta^\gamma, \tag{19.28}$$

where $\tilde{V}_{\delta,a(\delta)}$ solves the following equation:

$$F(\tilde{V}_{\delta,a}) + a(\delta)(\tilde{V}_{\delta,a} - \bar{u}) - f_\delta = 0.$$

If $\gamma \in (0, 1)$, then this $a(\delta)$ satisfies

$$\lim_{\delta \to 0} \|\tilde{V}_{\delta,a(\delta)} - z\| = 0. \tag{19.29}$$

Remark 19.2.3 It is an open problem to choose γ and C optimal in some sense.

Remark 19.2.4 Theorem 19.2.1 and Theorem 19.3.2 do not hold, in general, for $\gamma = 1$. Indeed, let $Fu = \langle u, p \rangle p$, $\|p\| = 1$, $p \perp \mathcal{N}(F) := \{u \in H : Fu = 0\}$, $f = p$, $f_\delta = p + q\delta$, where $\langle p, q \rangle = 0$, $\|q\| = 1$, $Fq = 0$, $\|q\delta\| = \delta$. One has $Fy = p$, where $y = p$ is the minimal-norm solution to the equation $Fu = p$. Equation $Fu + au = p + q\delta$ has the unique solution $V_{\delta,a} = q\delta/a + p/(1 + a)$. Equation (19.16) is $C\delta = \|q\delta + (ap)/(1 + a)\|$. This equation yields $a = a(\delta) = c\delta/(1 - c\delta)$, where $c := (C^2 - 1)^{1/2}$, and we assume $c\delta < 1$. Thus, $\lim_{\delta \to 0} V_{\delta,a(\delta)} = p + c^{-1}q := v$, and $Fv = p$. Therefore $v = \lim_{\delta \to 0} V_{\delta,a(\delta)}$ is not p; that is, is not the minimal-norm solution to the equation $Fu = p$. One can find similar arguments in [137, p. 29].

19.3 APPLICATIONS

In this section we discuss methods for solving equations (19.3) and (19.1) using the new discrepancy principle, that is, Theorem 19.2.1. Implementing this principle [i.e., solving equation (19.16)] requires solving equation (19.3). *If F is linear*, then equation (19.3) has the form

$$(F + aI)u = f_\delta. \tag{19.30}$$

Since $F \geq 0$ the operator $F + aI$ is boundedly invertible, $\|(F + aI)^{-1}\| \leq \frac{1}{a}$, and equation (19.30) is well-posed if $a > 0$ is not too small. There are many methods for solving efficiently well-posed linear equations with positive-definite operators. For this reason we *mainly discuss some methods for stable solution of equation (19.1) with nonlinear operators*. In this section a method is developed for a stable solution of equation (19.1) with locally Lipschitz *monotone operator F*, so we assume that

$$\|F(u) - F(v)\| \leq L\|u - v\|, \quad u, v \in B(u_0, R), \quad L = L(R). \tag{19.31}$$

Here $u_0 \in H$ is an arbitrary fixed element and

$$B(u_0, R) := \{u : \|u - u_0\| \le R\}.$$

Consider the operator

$$G(u) := u - \lambda[F(u) + au - f_\delta], \quad \lambda > 0.$$

We claim that G is a contraction mapping in H, provided that λ is sufficiently small. Let $F_1 := F + aI$. Then (19.31) implies $\|F_1(u) - F_1(v)\| \le (a+L)\|u-v\|$. Using the monotonicity of F, one gets

$$\begin{aligned} \|G(u) - G(v)\|^2 &= \|(u - v) - \lambda(F_1(u) - F_1(v))\|^2 \\ &= \|u - v\|^2 - 2\lambda \operatorname{Re}\langle u - v, F_1(u) - F_1(v)\rangle \\ &\quad + \lambda^2 \|F_1(u) - F_1(v)\|^2 \\ &\le \|u - v\|^2[1 - 2\lambda a + \lambda^2(a + L)^2]. \end{aligned} \tag{19.32}$$

This implies that G is a contraction mapping if

$$0 < \lambda < \frac{2a}{(a + L)^2}.$$

For these λ the solution $V_{\delta,a}$ of equation (19.3) can be found by fixed-point iteration as follows:

$$u_{n+1} = u_n - \lambda[F(u_n) + au_n - f_\delta], \quad u_0 := u_0. \tag{19.33}$$

After finding $V_{\delta,a}$, one finds $a(\delta)$ from the discrepancy principle (19.16)—that is, by solving the nonlinear equation:

$$\phi(a(\delta)) := \|F(V_{\delta,a(\delta)}) - f_\delta\| = C\delta^\gamma. \tag{19.34}$$

There are many methods for solving this equation. For example, one can use the bisection method or the golden section method. If $a(\delta)$ is found, one solves equation (19.3) with $a = a(\delta)$ for $V_{\delta,a(\delta)}$ and takes its solution as an approximate solution to (19.1).

Although the sequence u_n, defined by (19.33), converges to the solution of equation (19.3) at the rate of a geometrical series with a denominator $q \in (0, 1)$, it is very time-consuming to try to solve equation (19.3) with high accuracy if q is close to 1. Theorem 19.3.2 (see below) allows one to stop iterations (19.33) at the smallest value of n which satisfies the following condition:

$$\|F(u_n) + au_n - f_\delta\| \le \theta\delta, \quad \theta > 0, \tag{19.35}$$

where θ is a fixed constant. This saves the time of computation. While Theorem 19.3.2 is justified for any $\theta > 0$, it is recommended to choose θ to be "small" so that u_n is "close" to the solution to the equation

$$F(u) + au - f_\delta = 0.$$

Lemma 19.3.1 *Let F be a monotone operator in a Hilbert space H. Then the following inequalities hold $\forall u, v \in H$ and $\forall a > 0$:*

$$a\|u - v\| \leq \|F(u) - F(v) + au - av\|, \tag{19.36}$$
$$\|F(u) - F(v)\| \leq \|F(u) - F(v) + a(u - v)\|. \tag{19.37}$$

Proof: From the monotonicity of F one gets

$$a\|u - v\|^2 \leq \langle u - v, F(u) - F(v) + au - av \rangle$$
$$\leq \|u - v\| \|F(u) - F(v) + au - av\|, \qquad \forall a > 0. \tag{19.38}$$

This implies inequality (19.36).

Similarly, from the monotonicity of F one gets

$$\|F(u) - F(v)\|^2 \leq \langle F(u) - F(v), F(u) - F(v) + a(u - v) \rangle$$
$$\leq \|F(u) - F(v)\| \|F(u) - F(v) + a(u - v)\|, \quad \forall u, v \in H.$$

This implies inequality (19.37). ∎

Let us formulate a relaxed discrepancy principle.

Theorem 19.3.2 *Let δ, F, f_δ, and y be as in Theorem 19.2.1 and $0 < \gamma < 1$. Assume that $v_\delta \in H$ and $\alpha(\delta) > 0$ satisfy the following conditions:*

$$\|F(v_\delta) + \alpha(\delta)v_\delta - f_\delta\| \leq \theta\delta, \qquad \theta > 0, \tag{19.39}$$

and

$$C_1\delta^\gamma \leq \|F(v_\delta) - f_\delta\| \leq C_2\delta^\gamma, \qquad 0 < C_1 < C_2. \tag{19.40}$$

Then one has

$$\lim_{\delta \to 0} \|v_\delta - y\| = 0. \tag{19.41}$$

Proof: Using inequality (19.36) with $v = v_\delta$ and $u = V_{\delta,\alpha(\delta)}$, equation (19.3) with $a = \alpha(\delta)$, and inequality (19.39), one gets

$$\alpha(\delta)\|v_\delta - V_{\delta,\alpha(\delta)}\| \leq \|F(v_\delta) - F(V_{\delta,\alpha(\delta)}) + \alpha(\delta)v_\delta - \alpha(\delta)V_{\delta,\alpha(\delta)}\|$$
$$= \|F(v_\delta) + \alpha(\delta)v_\delta - f_\delta\| \leq \theta\delta. \tag{19.42}$$

Therefore,

$$\|v_\delta - V_{\delta,\alpha(\delta)}\| \leq \frac{\theta\delta}{\alpha(\delta)}. \tag{19.43}$$

Using (19.15) and (19.43), one gets

$$\alpha(\delta)\|v_\delta\| \leq \alpha(\delta)\|V_{\delta,\alpha(\delta)}\| + \alpha(\delta)\|v_\delta - V_{\delta,\alpha(\delta)}\| \leq \theta\delta + \alpha(\delta)\|y\| + \delta. \tag{19.44}$$

From the triangle inequality and inequalities (19.39) and (19.40), one obtains

$$\alpha(\delta)\|v_\delta\| \geq \|F(v_\delta) - f_\delta\| - \|F(v_\delta) + \alpha(\delta)v_\delta - f_\delta\| \geq C_1\delta^\gamma - \theta\delta. \tag{19.45}$$

Inequalities (19.44) and (19.45) imply

$$C_1 \delta^\gamma - \theta\delta \leq \theta\delta + \alpha(\delta)\|y\| + \delta. \tag{19.46}$$

This inequality and the fact that $C_1 - \delta^{1-\gamma} - 2\theta\delta^{1-\gamma} > 0$ for sufficiently small δ and $0 < \gamma < 1$ imply

$$\frac{\delta}{\alpha(\delta)} \leq \frac{\delta^{1-\gamma}\|y\|}{C_1 - \delta^{1-\gamma} - 2\theta\delta^{1-\gamma}}, \qquad 0 < \delta \ll 1. \tag{19.47}$$

Thus, one obtains

$$\lim_{\delta \to 0} \frac{\delta}{\alpha(\delta)} = 0. \tag{19.48}$$

From the triangle inequality and inequalities (19.39), (19.40), and (19.43), one gets

$$\begin{aligned}
\alpha(\delta)\|V_{\delta,\alpha(\delta)}\| &\leq \|F(v_\delta) - f_\delta\| + \alpha(\delta)\|v_\delta - V_{\delta,\alpha(\delta)}\| \\
&\quad + \|F(v_\delta) + \alpha(\delta)v_\delta - f_\delta\| \\
&\leq C_2\delta^\gamma + \theta\delta + \theta\delta.
\end{aligned} \tag{19.49}$$

This inequality implies

$$\lim_{\delta \to 0} \alpha(\delta)\|V_{\delta,\alpha(\delta)}\| = 0. \tag{19.50}$$

The triangle inequality and inequality (19.13) imply

$$\alpha\|V_\alpha\| \leq \alpha\big(\|V_{\delta,\alpha} - V_\alpha\| + \|V_{\delta,\alpha}\|\big) \leq \delta + \alpha\|V_{\delta,\alpha}\|. \tag{19.51}$$

From formulas (19.51) and (19.50), one gets

$$\lim_{\delta \to 0} \alpha(\delta)\|V_{\alpha(\delta)}\| = 0. \tag{19.52}$$

It follows from Lemma 19.1.1 with $f_\delta = f$ (i.e., $\delta = 0$) that the function $\phi_0(a) := a\|V_a\|$ is nonnegative and strictly increasing on $(0, \infty)$. This and relation (19.52) imply

$$\lim_{\delta \to 0} \alpha(\delta) = 0. \tag{19.53}$$

From the triangle inequality and inequalities (19.43) and (19.13), one obtains

$$\begin{aligned}
\|v_\delta - y\| &\leq \|v_\delta - V_{\delta,\alpha(\delta)}\| + \|V_{\delta,\alpha(\delta)} - V_{\alpha(\delta)}\| + \|V_{\alpha(\delta)} - y\| \\
&\leq \frac{\theta\delta}{\alpha(\delta)} + \frac{\delta}{\alpha(\delta)} + \|V_{\alpha(\delta)} - y\|,
\end{aligned} \tag{19.54}$$

where $V_{\alpha(\delta)}$ solves equation (19.3) with $a = \alpha(\delta)$ and $f_\delta = f$.

The conclusion (19.41) follows from inequalities (19.48), (19.53), and (19.54) and Lemma 6.1.9. Theorem 19.3.2 is proved.   ∎

Remark 19.3.3 Inequalities (19.39) and (19.40) are used as stopping rules for finding approximations:

$$\alpha(\delta) \approx a(\delta), \quad \text{and} \quad v(\delta) \approx V_{\delta, a(\delta)}.$$

Fix $\delta > 0$ and $\theta > 0$. Let C be as in Theorem 19.2.1. Choose C_1 and C_2 such that

$$C_1 \delta^\gamma + \theta \delta < C \delta^\gamma < C_2 \delta^\gamma - \theta \delta. \tag{19.55}$$

Suppose α_i and v_i, $i = 1, 2$, satisfy condition (19.39) and

$$\|F(v_1) - f_\delta\| < C_1 \delta^\gamma, \qquad C_2 \delta^\gamma < \|F(v_2) - f_\delta\|. \tag{19.56}$$

Let us show that

$$\alpha_{low} := \alpha_1 < a(\delta) < \alpha_2 := \alpha_{up}, \tag{19.57}$$

where $a(\delta)$ satisfies conditions of Theorem 19.2.1. Using inequality (19.37) for v_i and V_{δ, α_i}, $i = 1, 2$, along with inequality (19.39), one gets

$$\|F(v_i) - F(V_{\delta, \alpha_i})\| \le \|F(v_i) - F(V_{\delta, \alpha_i}) + \alpha_i v_i - \alpha_i V_{\delta, \alpha_i}\|$$
$$\le \|F(v_i) + \alpha_i v_i - f_\delta\| \le \theta \delta. \tag{19.58}$$

From inequalities (19.56) and (19.58) and the triangle inequality, one derives

$$\|F(V_{\delta, \alpha_1}) - f_\delta\| < C_1 \delta^\gamma + \theta \delta \quad \text{and} \quad C_2 \delta^\gamma - \theta \delta < \|F(V_{\delta, \alpha_2}) - f_\delta\|. \tag{19.59}$$

Recall that $\|F(V_{\delta, a(\delta)}) - f_\delta\| = C \delta^\gamma$. Inequality (19.57) is obtained from inequalities (19.55) and (19.59) and the fact that the function

$$\phi(\alpha) = \|F(V_{\delta, \alpha}) - f_\delta\|$$

is strictly increasing (see Lemma 19.1.1).

Let $f_\delta, F, C, \theta, \gamma$, and δ be as in Theorem 19.2.1 and 19.3.2, and let C_1 and C_2 satisfy inequality (19.55). Let us formulate an algorithm (see Algorithm 1 below) for finding $\alpha(\delta) \approx a(\delta)$ and $v(\delta) \approx V_{\delta, a(\delta)}$, using the bisection method and assuming that F is a locally Lipschitz monotone operator and α_{low} and α_{up} are known. By Theorem 19.3.2, $v(\delta)$ can be considered as a stable solution to equation (19.1).

Algorithm 1: *Finding $\alpha(\delta) \approx a(\delta)$ and $v_\delta \approx V_{\delta, a(\delta)}$, given α_{low} and α_{up}.*

1. Let $a := \frac{\alpha_{up} + \alpha_{low}}{2}$ and u_0 be an initial guess for $V_{\delta, a}$. Compute u_n by formula (19.33) and stop at n_{stop}, where n_{stop} is the smallest $n > 0$ for which condition (19.39) is satisfied. Then go to step 2.

2. If $C_2 \delta^\gamma < \|F(u_{n_{stop}}) - f_\delta\|$, then set $\alpha_{up} := a$ and go to step 4. Otherwise, go to step 3.

3. If $C_1 \delta^\gamma \le \|F(u_{n_{stop}}) - f_\delta\|$, then stop the process and take $v(\delta) := u_{n_{stop}}$ as a solution to (19.1). If $\|F(u_{n_{stop}}) - f_\delta\| < C_1 \delta^\gamma$, then set $\alpha_{low} := a$ and go to step 4.

4. Check if $\|a - \alpha_{low}\|$ is less than a desirable small value $\epsilon > 0$. If it is, then take $v(\delta) := u_{n_{stop}}$ as a solution to (19.1). If is is not, then go back to step 1.

Let us formulate algorithms for finding α_{up} and α_{low}.

Algorithm 2: *Finding α_{up}.*

1. Let $a = \alpha$ be an initial guess for $\alpha(\delta)$ and let u_0 be an initial guess for v_δ. Compute u_n by formula (19.33) with a and stop at n_{stop}, the smallest $n > 0$ for which condition (19.39) is satisfied. Then go to step 2.

2. If condition (19.40) holds for $v_\delta := u_{n_{stop}}$, then stop the process and take $u_{n_{stop}}$ as a solution to (19.1). Otherwise, go to step 3.

3. If $C_2 \delta^\gamma < \|F(u_{n_{stop}}) - f_\delta\|$, then set $\alpha_{up} := a$. Otherwise, set $\alpha := 2a$ and go back step 1.

Algorithm 3: *Finding α_{low}.*

1. Let $a = \alpha$ be an initial guess for $\alpha(\delta)$ and let u_0 be an initial guess for v_δ. Compute u_n by formula (19.33) with a and stop at n_{stop}, the smallest $n > 0$ for which condition (19.39) is satisfied. Then go to step 2.

2. If condition (19.40) holds for $v_\delta := u_{n_{stop}}$, then stop the process and take $u_{n_{stop}}$ as a solution to (19.1). Otherwise, go to step 3.

3. If $\|F(u_{n_{stop}}) - f_\delta\| < C_1 \delta^\gamma$, then set $\alpha_{low} := a$. Otherwise, set $\alpha := \frac{a}{2}$ and go back step 1.

In practice, these algorithms are often implemented at the same time to avoid repetition calculations.

Remark 19.3.4 The sequence $(\|u_n - V_{\delta, a(\delta)}\|)_{n=0}^\infty$, where u_n is computed by formula (19.33) and $V_{\delta, a(\delta)}$ is the solution to (19.3) with $a = a(\delta)$, is decreasing. Thus, the sequence u_n will stay inside a ball $B(0, R)$, assuming that $R > 0$ is chosen sufficiently large, so that $y, u_0 \in B(0, R)$.

Remark 19.3.5 Theorem 19.3.2 and the above algorithms are useful not only for solving nonlinear equations with monotone operators but also for solving linear equations with monotone operators. If one uses iterative methods to solve equation (19.30), then, by using Theorem 19.3.2, one can stop iterations whenever inequality (19.39) holds. By using stopping rule (19.39), one saves time of computations compared to solving (19.30) exactly. If F is a positive matrix, then one can solve (19.30) by conjugate gradient, or Jacobi, or Gauss-Seidel, or successive over-relaxation methods, with stopping rule (19.39).

Remark 19.3.6 If F is twice Fréchet differentiable, there are more options for solving equations (19.3) and (19.34): They can be solved by gradient-type methods, Newton-type methods, or a combination of these methods.

CHAPTER 20

DSM OF NEWTON-TYPE FOR SOLVING OPERATOR EQUATIONS WITH MINIMAL SMOOTHNESS ASSUMPTIONS

In this chapter we study the DSM for solving operator equation

$$F(u) = f, \tag{20.1}$$

where $F : H \to H$ is a Fréchet differentiable operator and H is a Hilbert space. We assume that equation (20.1) has a solution, possibly nonunique.

The DSM for solving equation (20.1) consists of finding a nonlinear operator $\Phi(t, u)$ such that the Cauchy problem

$$\dot{u} = \Phi(t, u), \qquad u(0) = u_0, \tag{20.2}$$

has a unique global solution $u = u(t; u_0)$, there exists $u(\infty) = \lim_{t \to \infty} u(t; u_0)$, and $F(u(\infty)) = f$:

$$\exists! u(t), \quad \forall t \geq 0; \quad \exists u(\infty); \quad F(u(\infty)) = f. \tag{20.3}$$

The problem is to find a Φ such that conclusions (20.3) hold. Various choices of Φ for which these properties hold have been proposed, where the DSM is justified for wide classes of operator equations—in particular, for wide classes of nonlinear ill-posed equations (i.e., equations $F(u) = f$ for which the

Dynamical Systems Method and Applications: Theoretical Developments and Numerical Examples, First Edition. A. G. Ramm and N. S. Hoang.
Copyright © 2012 John Wiley & Sons, Inc.

linear operator $F'(u)$ is not boundedly invertible). By $F'(u)$ we denote the Fréchet derivative of the nonlinear map F at the element u.

Several versions of the DSM has been studied for solving (20.1). In [56] various versions of the DSM with stopping rules of Discrepancy Principle-type are proposed and justified.

This chapter consists of five sections. In Section 20.1 the Newton-type DSM is discussed from various points of view. A version of an abstract inverse function theorem is used in a proof of the existence of the global solution to the Cauchy problem used in the Newton-type DSM. A novel feature of the proof is the lack of the assumption that $F'(u)$ is Lipschitz-continuous. A novel version of the convergence theorem for the classical Newton method is proved. In this theorem (Theorem 20.1.10) there is no assumption about smoothness of $F'(u)$, only continuity of $F'(u)$ is assumed. If $F'(u)$ is Hölder continuous, then a faster rate of convergence is established. In Section 20.2 a justification of the Newton-type DSM is given for the maps F which are global homeomorphisms. In Section 20.3, convergence of the DSM is proved for monotone operators, the case of noisy data is discussed, and a discrepancy principle is justified for a stable solution by the DSM for ill-posed problems with monotone operators. The operator $F'(u)$ is assumed continuous, and no Lipschitz continuity is assumed in this section and in the chapter. In Section 20.4 a version of the Newton-type DSM is studied under the assumption that $F'(u)$ is a smoothing injective operator, so that the operator $[F'(u)]^{-1}$ acts as a differential operator of finite order, so that the "loss of derivatives" phenomenon occurs. The cases when only continuity of $F'(u)$ is assumed and when $F'(u)$ is Hölder-continuous are considered and convergence of the DSM is proved.

This chapter is based on papers [53], [57], [58], [163], [173], [174], and [175].

20.1 DSM OF NEWTON-TYPE

Let us assume that equation (20.1) has a solution y and that the Fréchet derivative $F'(y)$ exists and is boundedly invertible:

$$\|[F'(y)]^{-1}\| \leq m, \qquad m = const > 0. \tag{20.4}$$

This assumption is relaxed in Remark 20.1.6 (see also Theorem 20.1.7), where it is assumed that the operator $[F'(y)]^{-1}$ is unbounded and causes loss of smoothness: It acts as a differential operator.

Let us also assume that $F'(u)$ exists in the ball $B(y, R) := \{u : \|u-y\| \leq R\}$ and depends continuously on u, and $\omega(r), r \geq 0$, is its modulus of continuity in the ball $B(y, R)$:

$$\omega(r) := \sup_{u,v \in B(y,R), \|u-v\| \leq r} \|F'(u) - F'(v)\|. \tag{20.5}$$

The function $\omega(r) \geq 0$ is assumed to be continuous on the interval $[0, 2R]$, strictly increasing, and $\omega(0) = 0$.

A widely used method for solving equation (20.1) is the Newton method:

$$u_{n+1} = u_n - [F'(u_n)]^{-1}[F(u_n) - f], \qquad u_0 = z, \tag{20.6}$$

where z is an initial approximation. Sufficient condition for the convergence of the iterative scheme (20.6) to the solution y of equation (20.1) are proposed in [22], [69], [73], and [87] and references therein. These conditions in most cases require a Lipschitz condition for $F'(u)$, a sufficient closeness of the initial approximation u_0 to the solution y, and other conditions (see, for example, [22, p. 157]).

Let us consider, instead of (20.1), the following equation:

$$F(u) = h, \tag{20.7}$$

where $h \in H$ is "sufficiently close" to f. The meaning of "sufficiently close" is made precise in Assumption A3 in Section 20.1.1. Consider the following continuous analog of the Newton method:

$$\dot{u}(t) = -[F'(u(t))]^{-1}(F(u(t)) - h), \qquad u(0) = u_0; \qquad \dot{u}(t) = \frac{du(t)}{dt}. \tag{20.8}$$

The question of general interest is:

Under what assumptions on F, h, and u_0 can one establish conclusions (20.3), that is, the global existence and uniqueness of the solution to problem (20.8), the existence of $u(\infty)$, and the relation $F(u(\infty)) = h$?

The usual condition, sufficient for the local existence and uniqueness of the solution to the Cauchy problem (20.8) is the local Lipschitz condition on the right-hand side of (20.8). Such condition can be satisfied, in general, only if $F'(u)$ satisfies a Lipschitz condition.

In [58] a *novel approach* was developed to a study of equation (20.8). The approach does not require a Lipschitz condition for $F'(u)$, and it leads to a justification of the conclusions (20.3) for the solution to problem (20.8) under natural assumptions on h and u_0.

Apparently for the first time a proof of convergence of the continuous analog (20.8) of the Newton method (20.6) is given without any smoothness assumptions on $F'(u)$, only the local continuity of $F'(u)$ is assumed; see (20.5).

This approach uses the special structure of equation (20.8), which corresponds to the Newton-type methods. The Newton-type methods are widely used in theoretical, numerical, and applied research, and by this reason our results are of general interest.

Our results demonstrate the universality of the Newton-type methods in the following sense: We prove that any operator equation (20.7) can be solved by the DSM Newton method (20.8), provided that conditions (20.4)–(20.5) hold, the initial approximation u_0 is sufficiently close to y, where y is the solution of equation (20.1), and the right-hand side h in (20.7) is sufficiently close to f.

A generalization of the classical results on the Newton method is given in Theorem 20.1.10, where the usual assumption about Lipschitz condition for $F'(u)$ is replaced by the continuity of $F'(u)$ or by a Hölder condition for $F'(u)$. Under these conditions the rate of convergence of the Newton method depends on the properties of the modulus of continuity of $F'(u)$ with respect to u.

Precise formulations of the results are given in Theorems 20.1.1, 20.1.3, 20.1.5, 20.1.7, 20.1.8, and 20.1.10.

The basic tool in this section is a new version of the inverse function theorem. The novelty of this version is in a specification of the region in which the inverse function exists. This is done in terms of the modulus of continuity of the operator $F'(u)$ in the ball $B(y, R)$.

In Section 20.1.1 we formulate and prove this version of the inverse function theorem. The result is stated as Theorem 20.1.1.

In Section 20.1.2 we justify the DSM for equation (20.8). The result is stated in Theorem 20.1.3. Morover, we generalize the result to the case when assumption (20.4) is not valid, and the operator $[F'(u)]^{-1}$ is unbounded, acting similar to a differential operator and causing the "loss of derivatives." The result is stated in Theorem 20.1.7.

In Section 20.1.3 we prove convergence of the usual Newton method (20.6). The result is stated in Theorem 20.1.8.

20.1.1   Inverse function theorem

Consider equation (20.7).

Let us make the following assumptions.

Assumptions A:

1. *Equation* (20.1) *and estimates* (20.4) *and* (20.5) *hold in* $B(y, R)$,

2. $m\omega(R) = q, \qquad q \in (0, 1),$

3. $h \in B(f, \rho), \qquad \rho = \frac{(1-q)R}{m}, \qquad q \in (0, 1).$

Assumption $A2$ defines R uniquely because $\omega(r)$ is assumed to be strictly increasing. We assume that equation $m\omega(R) = q$ has a solution. This assumption is always satisfied if $q \in (0, 1)$ is sufficiently small. The constant m is defined in (20.4).

Our first result, Theorem 20.1.1, says that under *Assumptions A*, equation (20.7) is uniquely solvable for any h in a sufficiently small neighborhood of f.

Theorem 20.1.1 *If Assumptions A hold, then equation* (20.7) *has a unique solution u for any $h \in B(f, \rho)$, and*

$$\|[F'(u)]^{-1}\| \le \frac{m}{1-q}, \qquad \forall u \in B(y, R). \tag{20.9}$$

Proof: Let us denote

$$Q := [F'(y)]^{-1}, \qquad \|Q\| \leq m.$$

Then equation (20.7) is equivalent to

$$u = T(u), \qquad T(u) := u - Q(F(u) - h). \tag{20.10}$$

Let us check that T maps the ball $B(y, R)$ into itself:

$$TB(y, R) \subset B(y, R); \tag{20.11}$$

and check that T is a contraction mapping in this ball:

$$\|T(u) - T(v)\| \leq q\|u - v\|, \qquad \forall u, v \in B(y, R), \tag{20.12}$$

where $q \in (0, 1)$ is defined in *Assumptions A*.

If (20.10) and (20.11) are verified, then the contraction mapping principle guarantees existence and uniqueness of the solution to equation (20.10) in $B(y, R)$, where R is defined in *Assumptions A2*.

Let us check the inclusion (20.11). One has

$$J_1 := \|u - y - Q(F(u) - h)\| = \|u - y - Q[F(u) - F(y) + f - h]\| \tag{20.13}$$

and

$$\begin{aligned}
F(u) - F(y) &= \int_0^1 F'(y + s(u - y)) \, ds (u - y) \\
&= \int_0^1 [F'(y + s(u - y)) - F'(y)] \, ds (u - y) \\
&\quad + F'(y)(u - y).
\end{aligned} \tag{20.14}$$

Note that

$$\|Q(f - h)\| \leq m\rho$$

and

$$\sup_{s \in [0,1]} \|F'(y + s(u - y)) - F'(y)\| \leq \omega(R).$$

Therefore, for any $u \in B(y, R)$ one gets from (20.4), (20.12), and (20.13) the following estimate:

$$J_1 \leq m\rho + m\omega(R)R \leq (1 - q)R + qR = R, \tag{20.15}$$

where the inequalities

$$\|f - h\| \leq \rho, \qquad \|u - y\| \leq R, \tag{20.16}$$

and Assumptions A3 and A2 in *Assumptions A* were used.

Let us establish inequality (20.12):

$$J_2 := \|T(u) - T(v)\| = \|u - v - Q(F(u) - F(v))\|. \tag{20.17}$$

One has

$$F(u) - F(v) = F'(y)(u - v) + \int_0^1 [F'(v + s(u - v)) - F'(y)] \, ds(u - v). \tag{20.18}$$

Note that

$$\|v + s(u - v) - y\| = \|(1 - s)(v - y) + s(u - y)\| \leq (1 - s)R + sR = R.$$

Thus, from (20.17) and (20.18) one gets

$$J_2 \leq m\omega(R)\|u - v\| \leq q\|u - v\|, \qquad \forall u, v \in B(y, R). \tag{20.19}$$

Therefore, both conditions (20.11) and (20.12) are verified. Consequently, the existence of the unique solution to (20.7) in $B(y, R)$ is proved.

Let us prove estimate (20.9). One has

$$\begin{aligned}[F'(u)]^{-1} &= [F'(y) + F'(u) - F'(y)]^{-1} \\ &= [I + (F'(y))^{-1}(F'(u) - F'(y))]^{-1}[F'(y)]^{-1} \end{aligned} \tag{20.20}$$

and

$$\|(F'(y))^{-1}(F'(u) - F'(y))\| \leq m\omega(R) \leq q, \qquad u \in B(y, R). \tag{20.21}$$

It is well known that if a linear operator A satisfies the estimate $\|A\| \leq q$, where $q \in (0, 1)$, then the inverse operator $(I + A)^{-1}$ does exist, and $\|(I + A)^{-1}\| \leq \frac{1}{1-q}$. Thus, the operator $[I + (F'(y))^{-1}(F'(u) - F'(y))]^{-1}$ exists and its norm can be estimated as follows:

$$\|[I + (F'(y))^{-1}(F'(u) - F'(y))]^{-1}\| \leq \frac{1}{1 - q}. \tag{20.22}$$

Consequently, (20.20) and (20.22) imply (20.9).

Theorem 20.1.1 is proved.   ∎

Remark 20.1.2 If $h = h(t) \in C^1([0, T])$, then the solution $u = u(t)$ of equation (20.7) is in $C^1([0, T])$, provided that Assumptions A hold.

20.1.2   Convergence of the DSM

Let us discuss the convergence of the DSM (20.8).

Consider the following equation:

$$F(u) = h + v(t), \tag{20.23}$$

where

$$u = u(t), \quad v(t) = e^{-t}v_0, \quad v_0 := F(u_0) - h, \quad r = \|v_0\|. \tag{20.24}$$

At $t = 0$ equation (20.23) has a unique solution u_0.

Let us make the following assumptions:

Assumptions B:

1. *Assumptions A hold,*

2. $h \in B(f, \delta), \qquad \delta + r \leq \rho := \frac{(1-q)R}{m}.$

Theorem 20.1.3 *If Assumptions B hold, then conclusions* (20.3), *with* f *replaced by* h, *hold for the solution of problem* (20.8).

Proof: The proof is divided into 3 parts.

Part 1. Proof of the global existence and uniqueness of the solution to problem (20.8).

One has

$$\|h + v(t) - f\| \leq \|h - f\| + \|v_0 e^{-t}\| \leq \delta + r \leq \rho, \qquad \forall t \geq 0.$$

Thus, it follows from Theorem 20.1.1 that equation (20.23) has a unique solution

$$u = u(t) \in B(y, R)$$

defined on the interval $t \in [0, \infty)$, and $u(t) \in C^1([0, \infty))$.

Differentiation of (20.23) with respect to t yields

$$F'(u)\dot{u} = \dot{v} = -v = -(F(u(t)) - h). \tag{20.25}$$

Since $u(t) \in B(y, R)$, the operator $F'(u(t))$ is boundedly invertible, so equation (20.25) is equivalent to (20.8). The initial condition $u(0) = u_0$ is satisfied, as was mentioned below (20.24). Therefore, the existence of the unique global solution to (20.8) is proved.

Part 2. Proof of the existence of $u(\infty)$.

From (20.23), (20.24), (20.9), and (20.8) it follows that

$$\|\dot{u}\| \leq \frac{mr}{1 - q}e^{-t}, \qquad q \in (0, 1). \tag{20.26}$$

This and the Cauchy criterion for the existence of the limit $u(\infty)$ imply that $u(\infty)$ exists.

Integrating (20.26), one gets

$$\|u(t) - u_0\| \leq \frac{mr}{1 - q} \tag{20.27}$$

and

$$\|u(\infty) - u(t)\| \leq \frac{mr}{1 - q}e^{-t}. \tag{20.28}$$

Part 3. Proof of the relation $F(u(\infty)) = h$.
Let us now prove that

$$F(u(\infty)) = h. \tag{20.29}$$

Relation (20.29) follows from (20.23) and (20.24) as $t \to \infty$, because $v(\infty) = 0$, $u(t) \in B(y, R)$, and F is continuous in $B(y, R)$.

Theorem 20.1.3 is proved. ∎

Remark 20.1.4 Let us explain why there is no assumption on the location of u_0 in Theorem 20.1.3. The reason is simple: In the proof of Theorem 20.1.3 it was established that $u(t) \in B(y, R)$ for all $t \geq 0$. Therefore, it follows that the assumptions of Theorem 20.1.3 imply the location of u_0, namely $u_0 = u(0) \in B(y, R)$.

From the proof of Theorem 20.1.3 the following result follows.

Theorem 20.1.5 *Assume that F is a global homeomorphism, that $m(u) \geq \|[F'(u)]^{-1}\|$, where $m(u) > 0$ is a constant which depends on u, and that $F'(u)$ is continuous with respect to u. Then problem (20.8) has a unique global solution for any h and any u_0, there exists $u(\infty)$, and $F(u(\infty)) = h$.*

Proof: If F is a global homeomorphism, then equation (20.23) is uniquely solvable for any $v(t)$. Differentiation of this equation with respect to t yields equation (20.25), and this equation is equivalent to (20.8) because of the bounded invertibility of $F'(u)$ at any u. The existence of $u(\infty)$ and the equality $F(u(\infty)) = h$ follow from the relation $\lim_{t \to \infty}(h + v(t)) = h$ and from the assumption that F is a global homeomorphism. Theorem 20.1.5 is proved. ∎

A practically important example of equations (20.7) with a global homeomorphism F is the equation $F(u) := G(u) + bu = h$, where $b = const > 0$ and G is a monotone Fréchet differentiable operator. One has $F'(u) = G'(u) + bI$ and $\|[G'(u) + bI]^{-1}\| \leq \frac{1}{b}$, because the monotonicity of G implies $G'(u) \geq 0$. Recall that if a linear operator $A \geq 0$, and $b = const > 0$, then $\|(A+bI)^{-1}\| \leq \frac{1}{b}$.

It is known that such F are global homeomorphisms (see, e.g., [22]). For such F, equation (20.7) can be solved for any h by the DSM Newton-type method (20.8) with any initial approximation u_0. In this sense, convergence of the DSM Newton method (20.8) is *global* for the F, satisfying the above assumptions.

Remark 20.1.6 Our arguments can be generalized to the case when $F'(u)$ *is unbounded*. For example, let H_a be a Hilbert scale of spaces, $H_a \subset H_b$ if $a > b \geq 0$. A typical example is the case of Sobolev spaces H_a, $H_0 := H = L^2(D)$. Let us assume that the following assumptions hold:
Assumptions C:

1. $F : H_a \rightarrow H_{a+b}$, $b > 0$; the Fréchet derivative $A(u) := F'(u) : H_a \rightarrow H_{a+b}$ exists and is continuous with respect to u in the following sense:

$$\|A(u) - A(v)\|_{a+b} \leq \omega(\|u - v\|_a), \qquad \forall u, v \in B(y, R),$$

where y solves the equation $F(y) = f$, the function $\omega(r)$ is continuous and strictly increasing for $r \in (0, R_1)$, $\omega(r) \geq 0$, $\omega(0) = 0$, and $R_1 > 0$ is a sufficiently large constant, so that equation

$$m\omega(R) = q, \qquad q \in (0, 1),$$

has a unique solution $R < R_1$, and $m > 0$ is a constant in the following inequality:

$$\|[A(y)]^{-1}g\|_a \leq m\|g\|_{a+b}, \qquad \forall g \in H_{a+b},$$

2. $h \in B(f, \rho)$, where $\rho = \frac{(1-q)R}{m}$.

Theorem 20.1.7 If Assumptions C hold, then for any $h \in B(f, \rho)$ there exists a unique $u \in B(y, R)$ such that $F(u) = h$. The operator F^{-1} is continuous on $B(f, \rho)$.

Proof: The proof is similar to the proof of Theorem 20.1.1. ∎

This theorem deals with the case when the operator $A^{-1}(u)$ causes loss of $b > 0$ derivatives; it acts similarly to a differentiation operator of order $b > 0$.

One can prove that conclusions (20.3) hold for (20.8) if *Assumptions C* hold and $h \in B(f, \delta)$, where $\delta + r \leq \rho$, $r = \|F(u_0) - h\|_{a+b}$, and ρ is the same as in *Assumptions C*. The proof is similar to the proof of Theorem 20.1.3.

20.1.3   The Newton method

The goal in this section is to give a novel result on the convergence of the classical Newton method. We drop the usual assumption that $F'(u)$ satisfies a Lipschitz condition, and we assume that $F'(u)$ is continuous. We also consider the case when $F'(u)$ satisfies a Hölder condition. The rate of the convergence is estimated. The result is formulated in Theorem 20.1.10.

We start with a proof of the convergence of the Newton method (20.6) to the solution y of equation (20.1) without any additional assumptions on the smoothness of $F'(u)$. Only the continuity of $F'(u)$ with respect to $u \in B(y, R)$ is assumed. The notations are the same as in *Assumptions A* in Section 20.1.1.

Theorem 20.1.8 *Assume that $F(y) = f$, conditions (20.1)–(20.5) and Assumptions A hold, and*

$$m\omega(R) = q \in (0, \frac{1}{2}), \qquad q_1\|z - y\| \leq R, \qquad q_1 := \frac{q}{1-q}. \tag{20.30}$$

Then process (20.6) *converges to* y.

Proof: Let $a_n := \|u_n - y\|$. From our assumptions one can derive that the sequence $(a_n)_{n=0}^n$ is decreasing at the rate not slower than that of a geometric sequence with ratio $r \in (0, 1)$. This implies the conclusion of the theorem. ∎

Remark 20.1.9 In general, the global convergence of the Newton method (20.6) does not hold under the assumptions of Theorem 20.1.5. That is, there exists F satisfying assumptions of Theorem 20.1.5 so that the Newton method (20.6) does not converge for some f and z.

The following result is an extension of the standard result (see, e.g., Theorem 15.6 in [22, p. 157]) on the convergence of the Newton method. In [22] it is assumed that $F'(u)$ satisfies a Lipschitz condition. In Theorem 20.1.10 it is assumed only that $F'(u)$ is continuous, which is a much weaker assumption. Of course, the rate of convergence depends on the behavior of the modulus of continuity of $F'(u)$ in a neighborhood of zero. One can prove the quadratic rate of convergence, characteristic for the Newton method if $F'(u)$ satisfies the Lipschitz condition. In Theorem 20.1.10 it is proved that if $F'(u)$ is continuous, then convergence is at the rate of a geometric series; and if $F'(u)$ satisfies a Hölder condition, then the rate of convergence is superlinear and depends on the Hölder exponent p.

Theorem 20.1.10 *Let F be a Fréchet differentiable operator in a Banach space X, and let $F'(u)$ be continuous with respect to u. Assume that*

$$\|F'(u) - F'(v)\| \le \omega(\|u - v\|), \qquad \forall u, v \in B(u_0, r), \tag{20.31}$$

$$K := \sup_{0 < \xi \le \alpha} \frac{\int_0^1 \omega(s\xi)\, ds}{\omega(\xi)} < 1, \tag{20.32}$$

$$\frac{\alpha}{1 - K} < r, \qquad \beta\omega(r) < 1, \tag{20.33}$$

$$\beta\omega(\alpha) < \min\left[\frac{1}{2}, 1 - \sup_{0 < \xi \le \alpha} \frac{\omega(K\xi)}{\omega(\xi)}, 1 - K^p\right], \tag{20.34}$$

where

$$\alpha := \|F'(u_0)^{-1}(F(u_0) - f)\|, \tag{20.35}$$

$$\beta := \|F'(u_0)^{-1}\|, \tag{20.36}$$

$\omega(s) \ge 0$ *is strictly increasing on the segment* $[0, r]$, *and* $\omega(0) = 0$. *Then equation* (20.1) *has a unique solution z in $B(u_0, r)$. Let u_n be defined by* (20.6). *Then*

$$\lim_{n \to \infty} u_n = z \tag{20.37}$$

and

$$\|u_n - z\| \le \frac{K^n}{1 - K}, \qquad n \ge 0. \tag{20.38}$$

If in addition

$$\frac{\omega(t)}{\omega(t')} \leq \left(\frac{t}{t'}\right)^p, \qquad 0 < t \leq t', \qquad 0 < p \leq 1, \tag{20.39}$$

then

$$\|u_n - z\| \leq \frac{\alpha q^{\frac{(p+1)^n - 1}{p}}}{1 - \kappa}, \qquad n > 0, \tag{20.40}$$

where

$$\kappa := (1 - \beta\omega(\alpha))^{\frac{1}{p}} < 1, \qquad q := \left(\frac{K^p}{1 - \beta\omega(\alpha)}\right)^{\frac{1}{p}} \frac{\beta\omega(\alpha)}{1 - \beta\omega(\alpha)} < 1. \tag{20.41}$$

Proof: The proof consists of 3 parts.

Part 1. Proof of the uniqueness of z.

Assume that z and $\bar{z}$ are two solution to (20.1) in $B(u_0, r)$. Then one gets, using (20.31), the following inequalities:

$$\|z - \bar{z}\| \leq \beta \|F(z) - F(\bar{z}) - F'(u_0)(z - \bar{z})\|$$

$$\leq \beta \|z - \bar{z}\| \int_0^1 \|F'(\bar{z} + t(z - \bar{z})) - F'(u_0)\| \, dt$$

$$\leq \|z - \bar{z}\| \beta\omega(r). \tag{20.42}$$

This and the second inequality in (20.33) imply that $z = \bar{z}$.

Part 2. Proof of the relations (20.37) and (20.38).

Let

$$\alpha_n := \|u_{n+1} - u_n\|, \quad \beta_n := \|F'(u_n)^{-1}\|, \quad \gamma_n := \beta_n \omega(\alpha_n), \quad n \geq 1, \tag{20.43}$$

and

$$\alpha_0 := \alpha, \qquad \beta_0 := \beta, \qquad \gamma_0 := \beta\omega(\alpha). \tag{20.44}$$

From (20.6) and (20.31) one gets

$$\alpha_n \leq \beta_n \|F(u_n) - f - [F(u_{n-1}) - f + F'(u_{n-1})(u_n - u_{n-1})]\|$$

$$\leq \beta_n \alpha_{n-1} \int_0^1 \|F'(u_{n-1} + t(u_n - u_{n-1})) - F'(u_{n-1})\| \, dt$$

$$\leq \beta_n \alpha_{n-1} \omega(\alpha_{n-1}) K(\alpha_{n-1}), \qquad n \geq 1, \tag{20.45}$$

where

$$K(\xi) := \frac{\int_0^1 \omega(t\xi) \, dt}{\omega(\xi)}, \qquad \xi > 0. \tag{20.46}$$

From (20.6) one gets

$$F'(u_n) = F'(u_{n-1})\left[I + F'(u_{n-1})^{-1}(F'(u_n) - F'(u_{n-1}))\right], \qquad n \geq 1. \tag{20.47}$$

This and (20.43) imply

$$\beta_n \le \beta_{n-1}(1 - \beta_{n-1}\omega(\alpha_{n-1}))^{-1} = \beta_{n-1}(1 - \gamma_{n-1})^{-1}, \qquad \forall n \ge 1. \quad (20.48)$$

It follows from (20.43) and (20.48) that

$$\gamma_n = \beta_n \omega(\alpha_n) \le \beta_{n-1}(1 - \gamma_{n-1})^{-1}\omega(\alpha_n)$$

$$\le \frac{\gamma_{n-1}}{1 - \gamma_{n-1}} \frac{\omega(\alpha_n)}{\omega(\alpha_{n-1})}, \qquad n \ge 1. \quad (20.49)$$

Inequalities (20.45) and (20.48) imply

$$\alpha_n \le \alpha_{n-1} \frac{\gamma_{n-1}}{1 - \gamma_{n-1}} K(\alpha_{n-1}), \qquad \forall n \ge 1. \quad (20.50)$$

From (20.34)–(20.32) one gets

$$\gamma_0 = \beta\omega(\alpha) < \frac{1}{2}, \quad \frac{\gamma_0}{1 - \gamma_0} < 1, \quad K(\xi) \le K < 1, \quad \forall \xi \in (0, \alpha]. \quad (20.51)$$

It follows from (20.34) that

$$\frac{K^p}{1 - \beta\omega(\alpha)} < 1, \quad \frac{\omega(K\xi)}{\omega(\xi)} < 1 - \gamma_0, \quad \forall \xi \in (0, \alpha]. \quad (20.52)$$

From the first inequality in (20.52) and the second inequality in (20.51), one obtains

$$q := \left(\frac{K^p}{1 - \beta\omega(\alpha)}\right)^{\frac{1}{p}} \frac{\beta\omega(\alpha)}{1 - \beta\omega(\alpha)} < 1. \quad (20.53)$$

Thus, inequalities (20.41) hold.

Let us prove by induction that

$$\alpha_n \le \alpha_{n-1}K < \alpha_{n-1}, \qquad \gamma_n < \gamma_{n-1} \le \gamma_0, \quad (20.54)$$

for all $n \ge 1$.

The first inequality in (20.54) for $n = 1$ follows from (20.50) and (20.51). The second inequality in (20.54) for $n = 1$ follows from (20.49), the second inequality in (20.52) and the first inequality in (20.54) for $n = 1$. Thus, (20.54) holds for $n = 1$. Assume that (20.54) holds for $n \ge 1$. From (20.50), (20.51), and the induction hypothesis, one gets

$$\alpha_{n+1} \le \alpha_n \frac{\gamma_n}{1 - \gamma_n} K(\alpha_n) < \alpha_n \frac{\gamma_0}{1 - \gamma_0} K \le \alpha_n K. \quad (20.55)$$

From (20.49), the induction hypothesis, and (20.55), one obtains

$$\gamma_{n+1} \le \frac{\gamma_n}{1 - \gamma_n} \frac{\omega(\alpha_{n+1})}{\omega(\alpha_n)} \le \gamma_n \frac{1}{1 - \gamma_0} \frac{\omega(K\alpha_n)}{\omega(\alpha_n)} < \gamma_n. \quad (20.56)$$

Here, we have used the second inequality in (20.52). From (20.55)–(20.56) one concludes that (20.54) holds for all $n \geq 1$.

It follows from the first inequality in (20.54) that

$$a_n \leq K^n a_0 = K^n a, \qquad n \geq 0. \tag{20.57}$$

This implies

$$\|u_{n+m} - u_n\| \leq \sum_{i=n}^{n+m-1} \|u_{i+1} - u_i\| \leq \sum_{i=n}^{n+m-1} aK^i \leq a\frac{K^n}{1-K}, \tag{20.58}$$

for all $m, n \geq 0$. This and the Cauchy criterion for convergence imply (20.37). Letting $m \to \infty$ in (20.58), one gets (20.38). It follows from inequality (20.38) with $n = 0$ that

$$\|z - u_0\| \leq a\frac{1}{1-K} < r, \tag{20.59}$$

where (20.33) was used. Therefore, $z \in B(u_0, r)$. Thus, (20.37) and (20.38) are proved.

Part 3. Proof of the estimate (20.40).

From (20.39) and (20.49)–(20.50), one obtains

$$\gamma_n \leq \frac{\gamma_{n-1}}{1-\gamma_{n-1}} \frac{\omega(a_n)}{\omega(a_{n-1})} \leq \frac{\gamma_{n-1}}{1-\gamma_{n-1}} \left(\frac{\gamma_{n-1}K}{1-\gamma_{n-1}}\right)^p, \qquad \forall n \geq 1. \tag{20.60}$$

This and (20.51)–(20.54) imply

$$\frac{\gamma_n K}{1-\gamma_n} \leq \frac{\gamma_n K}{1-\gamma_0} \leq \frac{1}{1-\gamma_0} \left(\frac{\gamma_{n-1}K}{1-\gamma_{n-1}}\right)^{p+1}, \qquad \forall n \geq 1. \tag{20.61}$$

From (20.61) one gets

$$\frac{\gamma_n K}{1-\gamma_n} \leq \left(\frac{1}{1-\gamma_0}\right)^{\frac{(p+1)^n - 1}{p}} \left(\frac{\gamma_0 K}{1-\gamma_0}\right)^{(p+1)^n} = \kappa q^{(p+1)^n}, \tag{20.62}$$

where

$$\kappa := (1-\gamma_0)^{\frac{1}{p}}, \qquad q := \left(\frac{1}{1-\gamma_0}\right)^{\frac{1}{p}} \frac{\gamma_0 K}{1-\gamma_0}. \tag{20.63}$$

From (20.50) and (20.62) one obtains, by induction, the following inequality:

$$a_n \leq \kappa^n q^{\frac{(p+1)^n - 1}{p}} a, \qquad n \geq 1. \tag{20.64}$$

Therefore

$$\|u_n - z\| \leq \sum_{i=n}^{\infty} a_i \leq q^{\frac{(p+1)^n - 1}{p}} a \sum_{i=0}^{\infty} \kappa^i \leq \frac{a q^{\frac{(p+1)^n - 1}{p}}}{1-\kappa}, \qquad n > 0. \tag{20.65}$$

Thus, (20.40) holds and Theorem 20.1.10 is proved.   ∎

Remark 20.1.11 If $F'(u)$ satisfies the Hölder condition

$$\|F'(u) - F'(v)\| \leq c\|u - v\|^p, \qquad 0 < p \leq 1,$$

then the continuity modulus of $F'(u)$ is $\omega(r) = cr^p$, and condition (20.39) in Theorem 20.1.10 is satisfied. If $p = 1$, that is, $F'(u)$ satisfies a Lipschitz condition, then one obtains the quadratic rate of convergence of Newton method.

20.2   A JUSTIFICATION OF THE DSM FOR GLOBAL HOMEOMORPHISMS

We assume in this section that F is a global homeomorphism.

For instance, F may be a hemicontinuous monotone operator operator such that a coercivity condition is satisfied:

$$\lim_{\|u\| \to \infty} \frac{\langle F(u), u \rangle}{\|u\|} = \infty, \tag{20.66}$$

where $\langle ., . \rangle$ denotes the inner product in H. We assume that $F \in C^1_{loc}$, that is, the Fréchet derivative $F'(u)$ exists for every u and depends continuously on u. Furthermore, we assume that

$$\|[F'(u)]^{-1}\| \leq m(u), \tag{20.67}$$

where $m(u)$ is a constant depending on u. This assumption implies that F is a local homeomorphism. If $m(u) < m$, where $m > 0$ is a constant independent of u, then it was proved in [146] that F is a global homeomorphism.

In Remark 20.2.3 at the end of this section, the following condition is mentioned:

$$\|F(u)\| < c \Rightarrow \|u\| < c_1, \qquad c, c_1 = const > 0, \tag{20.68}$$

which means that the preimages of bounded sets under the map F are bounded sets. This condition does not hold for the operator $F(u) := e^u$, $u \in \mathbb{R}$, $H = \mathbb{R}$, and that is why this monotone operator F is not surjective: Equation $e^u = 0$ does not have a solution in H.

By $c > 0$ we denote various constants.

Our main result, Theorem 20.2.1, says that if $F \in C^1_{loc}$ is a global homeomorphism and condition (20.67) holds, then the DSM Newton-type method (20.69) converges globally, that is, it converges for any initial approximation $u_0 \in H$ and any right-hand side $f \in H$. One of the novel features of our result is the absence of any smoothness assumptions on $F'(u)$: Only the continuity of $F'(u)$ with respect to u is assumed. In the earlier work it was often assumed that $F'(u)$ is Lipschitz continuous or, at least, Hölder-continuous.

Results in this section are taken from [164].

Let us formulate the result:

Theorem 20.2.1 *If $F \in C^1_{loc}$ is a global homeomorphism and condition (20.67) holds, then the problem*

$$\dot{u} = -[F'(u)]^{-1}(F(u) - f), \quad u(0) = u_0; \qquad \dot{u} = \frac{du}{dt}, \tag{20.69}$$

is solvable for any f and u_0 in H, the solution $u(t)$ exists for all $t \geq 0$, there exists the limit $u(\infty) = \lim_{t\to\infty} u(t)$, and $F(u(\infty)) = f$.

Proof: Denote

$$v := F(u(t)) - f. \tag{20.70}$$

If $u(t)$ solves (20.69), then

$$\dot{v} = F'(u(t))\dot{u} = -v.$$

Thus, problem (20.69) is reduced to the following problem:

$$\dot{v} = -v, \quad v(0) = F(u_0) - f. \tag{20.71}$$

Problem (20.71) obviously has a unique global solution:

$$v(t) = (F(u_0) - f)e^{-t}, \qquad \lim_{t\to\infty} v(t) := v(\infty) = 0. \tag{20.72}$$

Therefore, problem (20.69) has a unique global solution.

Indeed, consider an interval $[0, T]$, where $T > 0$ is arbitrarily large. The equation

$$F(u(t)) - f = v(t) \qquad 0 \leq t \leq T, \tag{20.73}$$

is uniquely solvable for $u(t)$ for any $v(t)$ because F is a global homeomorphism. The assumption (20.67), the continuity of $F'(u)$ with respect to u, and the inverse function theorem imply that the solution $u(t)$ to equation (20.73) is continuously differentiable with respect to t, because v and F are. Differentiating (20.73) and using (20.71) and (20.70), one gets

$$F'(u(t))\dot{u} = \dot{v} = -v = -(F(u(t)) - f). \tag{20.74}$$

Using assumption (20.67), one concludes from (20.74) that $u = u(t)$ solves (20.69) in the interval $t \in [0, T]$. Since $T > 0$ is arbitrary, $u = u(t)$ is a global solution to (20.69).

Since $\lim_{t\to\infty} v(t) := v(\infty)$ exists, and F is a global homeomorphism, one concludes that $\lim_{t\to\infty} u(t) := u(\infty)$ exists.

Since $v(\infty) = 0$, it follows that $F(u(\infty)) = f$.

Theorem 20.2.1 is proved. ∎

Remark 20.2.2 Theorem 20.2.1 implies that any equation (20.1) with F being a global homeomorphism and $F \in C^1_{loc}$, such that (20.67) holds, can be solved by the DSM method (20.69).

Remark 20.2.3 The equation $e^u = 0$, $u \in \mathbb{R}$, $H = \mathbb{R}$, does not have a solution, although $F(u) = e^u$ is monotone, that is, $F'(u) \geq 0$, $F'(u) = e^u > 0$ is boundedly invertible for every $u \in \mathbb{R}$ and $\|[e^u]^{-1}\| = e^{-u} \leq m_u < \infty$ for every $u \in \mathbb{R}$. The assumption (20.68) is not satisfied in this example, and this is the reason for the unsolvability of the equation $e^x = 0$. Note that $e^x \leq c$ as $x \to -\infty$, so assumption (20.68) does not hold.

20.3   DSM OF NEWTON-TYPE FOR SOLVING NONLINEAR EQUATIONS WITH MONOTONE OPERATORS

In this section we study a version of the DSM for solving the equation (20.1) where F is a nonlinear Fréchet differentiable monotone operator in a Hilbert space H, and equation (20.1) is assumed solvable. Monotonicity means that inequality (6.2) holds. It is proved in chapter 6 that the set $\mathcal{N} := \{u : F(u) = f\}$ is closed and convex if F is monotone and continuous. A closed and convex set in a Hilbert space has a unique minimal-norm element. This element in $\mathcal{N}$ we denote y, $F(y) = f$. We assumed in earlier chapters that $F'(u)$ is locally Lipschitz. This assumption is considerably weakened in this work: We assume now only the continuity of $F'(u)$. Since F is monotone, one has $F'(u) \geq 0$, so $\|[F'(u) + a(t)I]^{-1}\| \leq \frac{1}{a(t)}$ if $a(t) > 0$. The local and global existence and uniqueness of the solution to the Cauchy problem (20.107) (see below) was proved under these weak assumptions (see [53] and [58]).

The theory of monotone operators is presented, for example, in [88] and [190]. In [87] methods for solving well-posed nonlinear equations in a finite-dimensional space are discussed.

Methods for solving equation (20.1) with monotone operators are quite important in many applications. As we have discussed earlier, solving any solvable linear operator equation $Au = f$ with a closed densely defined linear operator A can be reduced to solving equation (20.1) with a monotone operator. Equations (20.1) with monotone operators arise often when the physical system is dissipative. In the earlier chapters in this book it was assumed that F is locally twice Fréchet differentiable, and a nonlinear differential inequality (see Theorem 5.1.1) was used in a study of the behavior of the solution to the DSM (20.107). The smoothness assumptions on F are weakened in this section, the method of our proofs is new, and, as a result, the proofs are shorter and simpler than the earlier ones. The assumptions on the "regularizing function" $a(t)$ are also weakened.

In this section we propose and justify a stopping rule for solving ill-posed equation (20.1) based on a discrepancy principle (DP) for the DSM (20.107). The main result of this section is Theorem 20.3.11 in which a DP is formulated, the existence of the stopping time t_δ is proved, and the convergence of the DSM (20.107) with the proposed DP is justified under some natural assumptions for a wide class of nonlinear equations with monotone operators.

The novelties in this section include a justification of the convergence of the DSM under less restrictive assumptions on F and an establishment of the discrepancy principle for problem (20.107). Moreover, the rate of decay of the function $a(t)$ as $t \to \infty$ can be arbitrary in the power scale, while $a(t)$ was often assumed to satisfy the condition $\int_0^\infty a(t)\, dt = \infty$ in the literature, which implies the rate of decay in the power scale not faster than $O(\frac{1}{t})$ as $t \to \infty$.

A few remarks about the history of the method (20.107) may be useful for the reader. Probably the first paper in which a continuous analog of the Newton's method was proposed for solving well-posed operator equation (20.1) was paper [34]. Method (20.107) has been studied in the literature earlier by several authors, usually under the assumption that $F'(u)$ satisfies a Lipschitz condition. Iterative versions of the method (20.107) were also studied, for example, in [67], and in some of the cited papers by the authors [46], [48], [50], and [61], also under some smoothness assumptions on $F'(u)$. A discrepancy principle for linear ill-posed problems was proposed by V. A. Morozov (see [83]).

A justification of the convergence of the method (20.107) is proved in this section under the minimal assumption of the continuity of $F'(u)$. The method of the proof is novel and can be used in a study of other problems. The justification of the discrepancy principle for stable solution of (20.1) with noisy data by the method (20.107) is also given under the minimal assumption of the continuity of $F'(u)$.

Results in this section have been published in [53], [57], and [173].

20.3.1   Existence of solution and a justification of the DSM for exact data

One of the versions of the DSM for solving nonlinear operator equation (20.1) with monotone continuously Fréchet differentiable operator F in a Hilbert space is based on a regularized DSM of Newton-type method, which consists of solving the following Cauchy problem:

$$\dot{u} = -\big(F'(u) + a(t)I\big)^{-1}\big(F(u) + a(t)u - f\big), \quad u(0) = u_0. \qquad (20.75)$$

Here $F : H \to H$ is a monotone continuously Fréchet differentiable operator in a Hilbert space H, $u_0, f \in H$ are arbitrary, and $a(t) > 0$ is a continuously differentiable function, defined for all $t \geq 0$ and monotonically decaying to zero as $t \to \infty$. This function is a regularizing function: If $F'(u)$ is not a boundedly invertible operator and f is monotone, then $F'(u) \geq 0$ and the operator $F'(u) + a(t)I$ is boundedly invertible if $a(t) > 0$.

Throughout this section we denote by I the identity operator, by y the minimal-norm solution to (20.1), and by $c > 0$ various estimation constants.

If F is monotone and continuous, then the minimal-norm solution to (20.1) exists and is unique (see, e.g., Chapter 6). Monotonicity of F is understood as in (6.2). The DSM is a basis for developing efficient numerical methods for solving operator equations, both linear and nonlinear, especially when the problems are ill-posed, when $F'(u)$ is not a boundedly invertible operator.

If one has a general evolution problem with a nonlinear operator in a Hilbert (or Banach) space

$$\dot{u} = B(u), \qquad u(0) = u_0, \tag{20.76}$$

then the local existence of the solution to this problem is usually established by assuming that $B(u)$ satisfies a Lipschitz condition, and the global existence is usually established by proving a uniform bound on the solution:

$$\sup_{t \geq 0} \|u(t)\| < c, \tag{20.77}$$

where $c > 0$ is a constant.

In (20.75) the operator

$$B(u) = -\big(F'(u) + a(t)I\big)^{-1}\big(F(u) + a(t)u - f\big)$$

is Lipschitz if one assumes that

$$\sup_{\{u:\|u-u_0\|\leq R\}} \|F^{(j)}(u)\| \leq M_j(R), \quad 0 \leq j \leq 2.$$

This assumption was used in many cases in earlier chapters, and a bound (20.77) was established under suitable assumptions earlier chapters.

There are many results (see, e.g., [22] and [83] and references therein) concerning the properties and global existence of the solution to (20.76) if $-B(u)$ is a maximal monotone operator. However, even when F is a monotone operator, the operator $-B$ in the right-hand side of (20.75) is not monotone. *Therefore these known results are not applicable. Even the proof of local existence is an open problem if one makes only the following assumption:*

Assumption D:

F is monotone and $F'(u)$ is continuous with respect to u.

The main result of this section is a proof that under Assumption D problem (20.75) has a unique local solution $u(t)$ and that under assumptions (20.80) on $a(t)$ (see below) this local solution exists for all $t \geq 0$, so it is a global solution. These results are formulated in Theorems 20.3.1 and 20.3.2.

Moreover, if the equation $F(y) = f$ has a solution and y is its (unique) minimal-norm solution, and if $\lim_{t \to \infty} a(t) = 0$ and $\lim_{t \to \infty} \frac{\dot{a}(t)}{a(t)} = 0$, then there exists $u(\infty)$, and $u(\infty) = y$. This justifies the DSM for solving the equation $F(u) = 0$ with a monotone continuously Fréchet differentiable operator F, for the first time under the weak *Assumption D*. The result is formulated in Theorem 20.3.4.

20.3.1.1 Local existence   Let us prove the following theorem.

Theorem 20.3.1 *If Assumption D holds, then problem* (20.75) *has a unique local solution.*

Proof: Let us prove the local existence of the solution to (20.75).
Let

$$\psi(t) = F(u) + a(t)u - f := \Psi(u, t) := \Psi(u). \tag{20.78}$$

If $a(t) > 0$ and F is monotone and hemicontinuous, then it is known (see, e.g., [22], p. 100) that the operator $F(u) + a(t)u$ is surjective. If $F'(u)$ is continuous, then, clearly, F is hemicontinuous. If F is monotone and $a(t) > 0$ then, clearly, the operator $F(u) + a(t)u$ is injective. Thus, Assumption D implies that the operator $F(u) + a(t)u$ is injective and surjective, it is continuously Fréchet differentiable, as well as its inverse, so the map $u \mapsto F(u) + a(t)u$ is a diffeomorphism. Therefore equation (20.78) is uniquely solvable for u for any ψ at any $t \geq 0$, and the map $\psi = \Psi(u)$ is a diffeomorphism. The inverse map $u = U(\psi)$, is continuously differentiable by the inverse function theorem since the operator $\Psi'_u = F'(u) + a(t)I$ is boundedly invertible if $a(t) > 0$, $\|(\Psi'_u)^{-1}\| \leq \frac{1}{a(t)}$. Recall that $F'(u) \geq 0$, because F is monotone. If $a(t) \in C^1([0, \infty))$ then the solution $u = u(t)$ of equation (20.78) is continuously differentiable with respect to t (see Section A.6), and if $u = u(t)$ is continuously differentiable with respect to t, then so is $\psi(t) = \Psi(u(t))$. The differentiability of $u(t) = U(\psi(t))$ also follows from a consequence of the classical inverse function theorem (see, e.g., [25, p. 147]). Therefore, equation (20.75) can be written in an equivalent form as

$$\dot\psi(t) = \dot a(t)u(t) - \psi(t) := Q(t, \psi), \qquad \psi(0) := \psi(u_0), \tag{20.79}$$

where $u(t) = U(\psi(t))$ is continuously differentiable with respect to t, and $\psi(t)$ is continuously differentiable with respect to t. The map $Q(t, \psi)$ is Lipschitz with respect to ψ, and the local existence of the solution to problem (20.79) follows from the standard result (see Section A.2). Since the map $U(\psi)$ is continuously differentiable and $\dot\psi$ is a continuous function of t, the function $\dot u$ is a continuous function of t, and problem (20.79) is equivalent to problem (20.75). Thus, problem (20.75) has a unique local solution. Theorem 20.3.1 is proved. ∎

20.3.1.2 Global existence Assuming that $0 < a(t) \in C^1(0, \infty)$ satisfying the following condition

$$\limsup_{t \to \infty} \frac{|\dot a(t)|}{a(t)} < q < 1. \tag{20.80}$$

We have the following result:

Theorem 20.3.2 *If Assumption D and* (20.80) *hold, then problem* (20.75) *has a unique global solution.*

Proof: Since $G(t, \psi)$ is Lipschitz with respect to ψ and continuously differentiable with respect to t, the solution to (20.79) exists globally, i.e., for all

$t \geq 0$, if

$$\sup_{t \geq 0} \|\psi(t)\| \leq c < \infty. \tag{20.81}$$

If the solution ψ to problem (20.79) exists globally, then the solution $u(t)$ to the equivalent problem (20.75) exists globally because the map $\psi \mapsto u$ is a diffeomorphism for $t \in [0, T]$, where $T > 0$ is an arbitrary large number.

Denote $h(t) := \|\psi(t)\|$. The function $\psi(t)$ is continuously differentiable with respect to $t > 0$. The function $h(t)$ is continuously differentiable with respect to t at any point at which $h(t) > 0$. If $h(t) = 0$ on an open interval, then $\dot{h}(t) = 0$ on this interval. If $h(s) = 0$, then we understand by $\dot{h}(s)$ the one-sided derivative from the right,

$$\dot{h}(s) = \lim_{\tau \to +0} \frac{h(s + \tau)}{\tau}.$$

This limit does exist if $\psi(t)$ is continuously differentiable. Indeed,

$$\lim_{\tau \to +0} \frac{h(s + \tau)}{\tau} = \lim_{\tau \to +0} \frac{\|\dot{\psi}(s)\tau + o(\tau)\|}{\tau} = \|\dot{\psi}(s)\|.$$

The left-sided derivative of $h(t)$ also exists and is equal to $-\|\dot{\psi}(s)\|$.

Multiply both sides of (20.79) by $\psi(t)$ and get

$$h\dot{h} = -h^2 + \mathrm{Re}\langle \dot{a}(t)u(t), \psi \rangle. \tag{20.82}$$

Let $w(t)$ solve the equation:

$$F(w(t)) + a(t)w(t) - f = 0, \qquad t \geq 0. \tag{20.83}$$

It is known (see Section 6.1) that if F is monotone and continuous and equation (20.1) is solvable and $\lim_{t \to \infty} a(t) = 0$, then there exists $w(\infty)$, and $w(\infty) = y$, where y is the unique minimal-norm solution to (20.1). So

$$\sup_{t \geq 0} \|w(t)\| < c,$$

where $c > 0$ denotes various constants. Equation (20.82) implies

$$\dot{h} \leq -h + |\dot{a}|\|u(t) - w(t)\| + |\dot{a}(t)|\|w(t)\|. \tag{20.84}$$

Let us prove the following estimate:

$$\|u(t) - w(t)\| \leq \frac{h(t)}{a(t)}, \qquad \forall t \geq 0. \tag{20.85}$$

Using (6.2), one gets

$$\langle F(u) - F(w) + a(u - w), u - w \rangle \geq a\|u - w\|^2. \tag{20.86}$$

Thus,

$$\|u(t) - w(t)\| \leq \frac{\|F(u(t)) - F(w(t)) + a(t)(u(t) - w(t))\|}{a(t)} = \frac{h(t)}{a(t)}. \quad (20.87)$$

From (20.85) and (20.84) one obtains

$$\dot{h} \leq -h\left(1 - \frac{|\dot{a}(t)|}{a(t)}\right) + |\dot{a}|\|w(t)\|. \quad (20.88)$$

From (20.80) there exists $T > 0$ such that

$$\frac{|\dot{a}(t)|}{a(t)} \leq q, \qquad \forall t \geq T. \quad (20.89)$$

Therefore,

$$h(t) \leq h(T)e^{-(t-T)(1-q)} + ce^{-t(1-q)} \int_T^t e^{s(1-q)}|\dot{a}(s)|\,ds, \quad \forall t \geq T. \quad (20.90)$$

From (20.90) and (20.89) one gets

$$h(t) \leq h(T)e^{-(t-T)(1-q)} + qce^{-t(1-q)} \int_T^t e^{s(1-q)}a(s)\,ds, \quad \forall t \geq T. \quad (20.91)$$

Since we have assumed that $a(t) > 0$ is a $C^1([0,\infty))$ function, such that $a(t) \to 0$ as $t \to \infty$, we have $\sup_{t\geq 0} a(t) < c$. Thus, from inequality (20.91) one gets

$$h(t) \leq h(T)e^{-(t-T)(1-q)} + c(1 - e^{-(t-T)(1-q)}), \qquad \forall t \geq T. \quad (20.92)$$

One the other hand, one gets from (20.88) and the Gronwall inequality the following estimate:

$$h(t) \leq h(0)e^{-\phi(t)} + e^{-\phi(t)}c \int_0^t e^{\phi(s)}|a(s)|\,ds, \qquad 0 \leq t \leq T, \quad (20.93)$$

where

$$\phi(t) := \int_0^t \left(1 - \frac{|\dot{a}(s)|}{a(s)}\right)\,ds.$$

Estimate (20.81) follows from (20.92) and (20.93).
    Theorem 20.3.2 is proved.   ∎

20.3.1.3   Justification of the DSM for exact data   By the justification of the DSM for solving equation (20.1), we mean the statements (20.3).
    In Theorem 20.3.2 the first of these statements is proved. Let us assume

$$\lim_{t\to\infty} a(t) = 0, \qquad \lim_{t\to\infty} \frac{\dot{a}(t)}{a(t)} = 0 \quad (20.94)$$

and prove the remaining two statements from (20.3).

Remark 20.3.3 Actually our argument allows for the following generalization of the results: Assumption (20.80) can be weakened to $\frac{|\dot{a}(t)|}{a(t)} \leq q$, $\forall t \geq 0$, $q \in (0,1)$ and the second assumption (20.94) can be weakened to

$$\limsup_{t\to\infty} \frac{|\dot{a}(t)|}{a(t)} \leq q', \qquad q + q' < 1.$$

Theorem 20.3.4 *If Assumption D and (20.94) hold, and equation (20.1) has a solution, then (20.3) hold, and $u(\infty) = y$, where y is the unique minimal-norm solution to (20.1).*

Proof: Let $w(t)$ be defined by (20.83). It is proved in Section 6.1 that

$$\lim_{t\to\infty} w(t) = y, \tag{20.95}$$

so $\limsup_{t\to\infty} \|w(t)\| < c$. Inequality (20.85) implies

$$a(t)\|u(t)\| \leq a(t)\|w(t)\| + h(t) \leq ca(t) + h(t). \tag{20.96}$$

Inequalities (20.96) and (20.82) imply

$$\dot{h} \leq -h + \frac{|\dot{a}(t)|}{a(t)}[ca(t) + h(t)]. \tag{20.97}$$

Assumptions (20.80) and (20.94) imply that

$$\lim_{t\to\infty} |\dot{a}(t)| = 0. \tag{20.98}$$

From the second assumption (20.94) it follows that

$$\frac{|\dot{a}(t)|}{a(t)} < \delta, \qquad \forall t > t_\delta, \tag{20.99}$$

where $\delta > 0$ is an arbitrary small fixed number. From (20.97)–(20.99) it follows that

$$\lim_{t\to\infty} h(t) = 0. \tag{20.100}$$

Indeed, (20.97) implies

$$\dot{h} \leq -(1-\delta)h + c|\dot{a}(t)|, \qquad t > t_\delta. \tag{20.101}$$

Thus

$$h(t) \leq h(t_\delta)e^{-(1-\delta)(t-t_\delta)} + ce^{-(1-\delta)t}\int_{t_\delta}^{t} e^{(1-\delta)s}|\dot{a}(s)|\,ds, \quad t \geq t_\delta. \tag{20.102}$$

Clearly, $\lim_{t\to\infty} h(t_\delta)e^{-(1-\delta)t} = 0$. The L'Hospital rule yields

$$0 \le \lim_{t\to\infty} \frac{\int_{t_\delta}^{t} e^{(1-\delta)s}|\dot{a}(s)|ds}{e^{(1-\delta)t}} = \lim_{t\to\infty}(1-\delta)^{-1}|\dot{a}(t)| = 0. \tag{20.103}$$

Thus, (20.100) is proved.

Let us prove that (20.100) implies the existence of the limit $u(\infty) := \lim_{t\to\infty} u(t)$, the relation

$$F(u(\infty)) = f, \tag{20.104}$$

and the relation $u(\infty) = y$, where y is the minimal-norm solution of the equation $F(u) = f$.

In Section 6.1, it is proved that the limit $w(\infty)$, as $a = a(t) \to 0$ (i.e., $t \to \infty$), of the solution w_a to the equation

$$F(w_a) + aw_a - f = 0, \qquad a > 0, \tag{20.105}$$

with a hemicontinuous monotone operator F, exists, and $w(\infty) = y$, provided that equation (20.1) is solvable.

Thus, the existence of $u(\infty)$ follows from (20.85) if one proves that

$$\lim_{t\to\infty} \frac{h(t)}{a(t)} = 0, \tag{20.106}$$

To verify (20.106), we claim that assumption (20.80) implies that

$$\lim_{t\to\infty} \frac{e^{-(1-\delta)t}}{a(t)} = 0.$$

Indeed, the inequality $\dot{a}(t) \ge -qa(t)$ implies $a(t) \ge ce^{-qt}$, where $c > 0$ is a constant. Thus, the claim follows if $\delta < 1 - q$.

Let us now prove (20.106). Divide both sides of (20.102) by $a(t)$ and let $t \to \infty$. The first term on the right tends to zero, and the second term by L'Hospital's rule tends also to zero because of the second assumption (20.94). Thus, (20.106) is established.

Since the limit $w(\infty) = y$ exists, it follows from (20.85) and (20.106) that $u(\infty)$ exists and $u(\infty) = y$.

Theorem 20.3.4 is proved. ∎

20.3.2 Solving equations with monotone operators when the data are noisy

Assume that f is not known but f_δ, the noisy data, are known, and $\|f_\delta - f\| \le \delta$. If $F'(u)$ is not boundedly invertible, then solving equation (20.1) for u, given noisy data f_δ, is often (but not always) an ill-posed problem. When F is a linear bounded operator, many methods for stable solution of (20.1) were proposed (see [49], [64], [83], and [160], and references therein). However, when F is nonlinear, then the theory is less complete.

The DSM for solving equation (20.1) was studied extensively in [48]–[50] and [148], where also numerical examples, illustrating efficiency of the algorithms, based on the DSM methods, were given. In Chapter 6 the following version of the DSM for solving equation (20.1) was studied:

$$\dot{u}_\delta = -\big(F'(u_\delta) + a(t)I\big)^{-1}\big(F(u_\delta) + a(t)u_\delta - f_\delta\big), \quad u_\delta(0) = u_0. \quad (20.107)$$

Here F is a monotone operator, and $a(t) > 0$ is a continuous function, defined for all $t \geq 0$, strictly monotonically decaying, $\lim_{t\to\infty} a(t) = 0$. These assumptions on $a(t)$ hold throughout the section and are not repeated. Additional assumptions on $a(t)$ will appear in Theorem 20.3.11. Convergence of the above DSM was proved in Chapter 6 for any initial value u_0 with an a priori choice of stopping time t_δ, provided that $a(t)$ is suitably chosen. In this section an a posteriori choice of t_δ is formulated and justified.

20.3.2.1 Auxiliary results   Let us consider the following equation:

$$F(V_{\delta,a}) + aV_{\delta,a} - f_\delta = 0, \qquad a > 0, \qquad (20.108)$$

where $a = const$. It is known (see Chapter 6) that equation (20.108) with monotone continuous operator F has a unique solution for any $f_\delta \in H$.

Let $a = a(t)$ and $0 < a(t) \searrow 0$, and assume $a \in C^1[0,\infty)$. Then the solution $V_\delta(t) := V_{\delta,a(t)}$ of (20.108) is a function of t.

Remark 20.3.5 Assume $0 < a(t) \searrow 0$. Then it follows from (19.12) that

$$\lim_{t\to\infty} \|F(V_\delta(t)) - f_\delta\| \leq \delta. \qquad (20.109)$$

Lemma 20.3.6 *Let* $0 < a(t)$ *satisfy the following relations:*

$$0 < a(t) \searrow 0, \qquad \frac{\dot{a}(t)}{a(t)} \searrow 0. \qquad (20.110)$$

Let

$$\phi(t) := \int_0^t \left(1 - \frac{|\dot{a}(s)|}{a(s)}\right) ds. \qquad (20.111)$$

Then

$$\lim_{t\to\infty} e^{\frac{t}{2}} a(t) = \infty, \qquad (20.112)$$

$$\lim_{t\to\infty} \frac{\int_0^t e^{\frac{s}{2}} \frac{|\dot{a}(s)|}{a(s)} ds}{e^{\frac{t}{2}}} = 0, \qquad (20.113)$$

$$\lim_{t\to\infty} \frac{\int_0^t e^{\frac{s}{2}} |\dot{a}(s)| ds}{e^{\frac{t}{2}} a(t)} = 0. \qquad (20.114)$$

and

$$\lim_{t\to\infty} \phi(t) = \infty, \tag{20.115}$$

$$\lim_{t\to\infty} e^{-\phi(t)} \int_0^t e^{\phi(s)} \frac{|\dot{a}(s)|}{a(s)} \, ds = 0, \tag{20.116}$$

$$\lim_{t\to\infty} e^{\phi(t)} a(t) = \infty, \tag{20.117}$$

$$\lim_{t\to\infty} \frac{\int_0^t e^{\phi(s)} |\dot{a}(s)| \, ds}{e^{\phi(t)} a(t)} = 0. \tag{20.118}$$

Proof: Let us prove (20.112)–(20.114). Relations (20.115)–(20.118) are obtained similarly.

First, let us prove (20.112). We claim that, for sufficiently large $t > 0$, the following inequality holds:

$$\frac{t}{2} > \ln \frac{1}{a^2(t)}. \tag{20.119}$$

By L'Hospital's rule and (20.110), one obtains

$$\lim_{t\to\infty} \frac{t}{2\ln\frac{1}{a^2(t)}} = \lim_{t\to\infty} \frac{a(t)}{4|\dot{a}(t)|} = \infty. \tag{20.120}$$

This implies that (20.119) holds for $t > 0$ sufficiently large. From (20.119) one concludes

$$\lim_{t\to\infty} e^{\frac{t}{2}} a(t) \geq \lim_{t\to\infty} e^{\ln \frac{1}{a^2(t)}} a(t) = \lim_{t\to\infty} \frac{1}{a(t)} = \infty. \tag{20.121}$$

Thus, relation (20.112) is proved.

Let us prove (20.113). If $I := \int_0^\infty e^{\frac{s}{2}} \frac{|\dot{a}(s)|}{a(s)} \, ds < \infty$, then (20.113) is obvious. If $I = \infty$, then (20.113) follows from L'Hospital's rule.

Let us prove (20.114). The denominator of (20.114) tends to ∞ as $t \to \infty$ by (20.112). Thus, if the numerator of (20.114) is bounded then (20.114) holds. Otherwise, L'Hospital's rule yields

$$\lim_{t\to\infty} \frac{\int_0^t e^{\frac{s}{2}} |\dot{a}(s)| \, ds}{e^{\frac{t}{2}} a(t)} = \lim_{t\to\infty} \frac{e^{\frac{t}{2}} |\dot{a}(t)|}{\frac{1}{2} e^{\frac{t}{2}} a(t) - e^{\frac{t}{2}} |\dot{a}(t)|} = 0. \tag{20.122}$$

Lemma 20.3.6 is proved.   ∎

Let us assume that

$$\frac{1}{6} \geq \frac{|\dot{a}(t)|}{a(t)}, \qquad t \geq 0. \tag{20.123}$$

Lemma 20.3.7 *Let $a(t)$ satisfy (20.123). Then*

$$e^{-\frac{t}{2}} \int_0^t e^{\frac{s}{2}} |\dot{a}(s)| \|V_\delta(s)\| \, ds \leq \frac{1}{2} a(t) \|V_\delta(t)\|, \qquad t \geq 0. \tag{20.124}$$

Proof: Let us check that

$$e^{\frac{t}{2}}|\dot{a}(t)| \leq \frac{d}{dt}\left(\frac{1}{2}a(t)e^{\frac{t}{2}}\right), \qquad t > 0. \tag{20.125}$$

One has

$$\frac{d}{dt}\left(\frac{1}{2}a(t)e^{\frac{t}{2}}\right) = \frac{a(t)e^{\frac{t}{2}}}{4} + \frac{\dot{a}(t)e^{\frac{t}{2}}}{2} = \frac{a(t)e^{\frac{t}{2}}}{4} - \frac{|\dot{a}(t)|e^{\frac{t}{2}}}{2}. \tag{20.126}$$

Thus, inequality (20.125) is equivalent to

$$\frac{3}{2}|\dot{a}(t)| \leq \frac{1}{4}a(t), \qquad \forall t > 0. \tag{20.127}$$

Inequality (20.127) holds since $a(t)$ satisfies (20.123) by our assumptions. Integrating both sides of (20.125) from 0 to t, one gets

$$\int_0^t e^{\frac{s}{2}}|\dot{a}(s)|\,ds \leq \frac{1}{2}a(t)e^{\frac{t}{2}} - \frac{1}{2}a(0)e^0 < \frac{1}{2}a(t)e^{\frac{t}{2}}, \qquad t \geq 0. \tag{20.128}$$

Multiplying (20.128) by $e^{-\frac{t}{2}}\|V_\delta(t)\|$ and using the fact that $\|V_\delta(t)\|$ is increasing (see Lemma 19.1.1), one gets (20.124). Note that $V_\delta(t) = V_{\delta,a(t)}$. Lemma 20.3.7 is proved. ∎

20.3.2.2 Main results   Denote

$$A := F'(u_\delta(t)), \qquad A_a := A + aI,$$

where I is the identity operator, and $u_\delta(t)$ solves the following Cauchy problem:

$$\dot{u}_\delta = -A_{a(t)}^{-1}[F(u_\delta) + a(t)u_\delta - f_\delta], \qquad u_\delta(0) = u_0, \tag{20.129}$$

where $u_0 \in H$.

Theorem 20.3.8 *Let F be a Fréchet differentiable monotone operator. Assume that F' is continuous. Let $0 < a(t)$ satisfy conditions (20.110). Then problem (20.129) has a unique global solution.*

Proof: The uniqueness and local existence of u_δ follows from the arguments similar to the ones in the proof of Theorem 20.3.1. Let us prove that $u_\delta(t)$ is defined globally.

From (20.129), one obtains

$$\frac{d}{dt}\left(F(u_\delta) + au_\delta - f_\delta\right) = A_a\dot{u}_\delta + \dot{a}u_\delta = -\left(F(u_\delta) + au_\delta - f_\delta\right) + \dot{a}u_\delta.$$

This and (20.108) imply

$$\frac{d}{dt}\left[F(u_\delta) - F(V_\delta) + a(u_\delta - V_\delta)\right]$$
$$= -\left[F(u_\delta) - F(V_\delta) + a(u_\delta - V_\delta)\right] + \dot{a}u_\delta. \tag{20.130}$$

Denote

$$v := v(t) := F(u_\delta(t)) - F(V_\delta(t)) + a(t)(u_\delta(t) - V_\delta(t)),$$

and

$$h := h(t) := \|v(t)\|.$$

Multiply (20.130) by v and get

$$\begin{aligned} h\dot{h} &= -h^2 + \mathrm{Re}\langle v, \dot{a}(u_\delta - V_\delta)\rangle + \dot{a}\,\mathrm{Re}\langle v, V_\delta\rangle \\ &\le -h^2 + h|\dot{a}|\|u_\delta - V_\delta\| + |\dot{a}|h\|V_\delta\|. \end{aligned} \tag{20.131}$$

This implies

$$\dot{h} \le -h + |\dot{a}|\|u_\delta - V_\delta\| + |\dot{a}|\|V_\delta\|. \tag{20.132}$$

Since $\langle F(u_\delta) - F(V_\delta), u_\delta - V_\delta\rangle \ge 0$, one obtains from two equations

$$\langle v, u_\delta - V_\delta\rangle = \langle F(u_\delta) - F(V_\delta) + a(t)(u_\delta - V_\delta), u_\delta - V_\delta\rangle,$$

and

$$\langle v, F(u_\delta) - F(V_\delta)\rangle = \|F(u_\delta) - F(V_\delta)\|^2 + a(t)\langle u_\delta - V_\delta, F(u_\delta) - F(V_\delta)\rangle,$$

the following two inequalities:

$$a\|u_\delta - V_\delta\|^2 \le \langle v, u_\delta - V_\delta\rangle \le \|u_\delta - V_\delta\|h, \tag{20.133}$$

and

$$\|F(u_\delta) - F(V_\delta)\|^2 \le \langle v, F(u_\delta) - F(V_\delta)\rangle \le h\|F(u_\delta) - F(V_\delta)\|. \tag{20.134}$$

Inequalities (20.133) and (20.134) imply

$$a\|u_\delta - V_\delta\| \le h, \quad \|F(u_\delta) - F(V_\delta)\| \le h. \tag{20.135}$$

Inequalities (20.132) and (20.135) imply

$$\dot{h} \le -h\left[1 - \frac{|\dot{a}|}{a}\right] + |\dot{a}|\|V_\delta\|. \tag{20.136}$$

From (20.136) and the Gronwall inequality, one obtains

$$h(t) \le h(0)e^{-\phi(t)} + e^{-\phi(t)}\int_0^t e^{\phi(s)}|\dot{a}(s)|\|V_\delta(s)\|\,ds, \tag{20.137}$$

where $\phi(t)$ is defined in (20.111).

From the inequality (20.137) it follows that $h(t)$ is bounded for every $t \ge 0$. Therefore, $v(t)$ is defined globally, that is, for every $t \ge 0$. Consequently, $u_\delta(t)$ is defined globally (see Lemma 6.1.7).

Theorem 20.3.8 is proved. ∎

Assume that equation (20.1) has a solution, possibly nonunique, and y is the minimal-norm solution to this equation. Let f be unknown but f_δ be given, $\|f_\delta - f\| \le \delta$.

Theorem 20.3.9 *Let $a(t)$ satisfy (20.110). Let $C > 0$ and $\zeta \in (0, 1]$ be constants such that $C\delta^\zeta > \delta$. Assume that $F : H \to H$ is a Fréchet differentiable monotone operator, and u_0 is an element of H, satisfying the following inequality:*

$$\|F(u_0) - f_\delta\| > C\delta^\zeta. \tag{20.138}$$

Then there exists a unique t_δ, such that

$$\|F(u_\delta(t_\delta)) - f_\delta\| = C\delta^\zeta, \quad \|F(u_\delta(t)) - f_\delta\| > C\delta^\zeta, \quad \forall t \in [0, t_\delta). \tag{20.139}$$

If $\zeta \in (0, 1)$ and

$$\lim_{\delta \to 0} t_\delta = \infty, \tag{20.140}$$

then

$$\lim_{\delta \to 0} \|u_\delta(t_\delta) - y\| = 0. \tag{20.141}$$

Remark 20.3.10 Inequality (20.138) is a natural assumption because if this inequality does not hold and $\|u_0\|$ is not "too large," then u_0 can be considered an approximate solution to (20.1).

In Theorem 20.3.9 the existence of t_δ satisfying (20.139) is guaranteed for any $\zeta \in (0, 1]$. However, we prove relation (20.141) for $\zeta \in (0, 1)$. If $\zeta = 1$ it is possible to prove that $u_\delta(t_\delta)$ converges to a solution to (20.1), but it is not known whether this solution is the minimal-norm solution of (20.1) if (20.1) has more than one solution.

Further results on the choices of ζ require extra assumptions on F and y. Since the minimal-norm solution y satisfies the relation $\|F(y) - f_\delta\| = \|f - f_\delta\| \le \delta$, it is natural to choose $C > 0$ and $\zeta \in (0, 1)$ so that $C\delta^\zeta$ is close to δ.

Proof: The uniqueness of t_δ follows from (20.139). Indeed, if t_δ and $\tau_\delta > t_\delta$ both satisfy (20.139), then the second inequality in (20.139) does not hold on the interval $[0, \tau_\delta)$.

From (20.135) and (20.137) one obtains

$$\|F(u_\delta) - F(V_\delta)\| \le h(0)e^{-\phi(t)} + e^{-\phi(t)} \int_0^t e^{\phi(s)} |\dot{a}(s)| \|V_\delta(s)\| \, ds, \tag{20.142}$$

This and the triangle inequality imply

$$\|F(u_\delta(t)) - f_\delta\| \le \|F(V_\delta(t)) - f_\delta\|$$
$$+ h(0)e^{-\phi(t)} + e^{-\phi(t)} \int_0^t e^{\phi(s)} |\dot{a}| \|V_\delta\| \, ds. \tag{20.143}$$

Since $a(s)\|V_\delta(s)\| = \|F(V_\delta(s)) - f_\delta\|$ is decreasing (see Lemma 19.1.1), one obtains

$$\lim_{t\to\infty} e^{-\phi(t)} \int_0^t e^{\phi(s)} |\dot{a}| \|V_\delta\| \, ds$$

$$\leq \lim_{t\to\infty} e^{-\phi(t)} \int_0^t e^{\phi(s)} \frac{|\dot{a}(s)|}{a(s)} a(0) \|V_\delta(0)\| \, ds. \tag{20.144}$$

It follows from (20.109), (20.143)–(20.144) and (20.115)–(20.116) that

$$\lim_{t\to\infty} \|F(u_\delta(t)) - f_\delta\| \leq \lim_{t\to\infty} \|F(V_\delta(t)) - f_\delta\|$$

$$+ \lim_{t\to\infty} e^{-\phi(t)} \int_0^t e^{\phi(s)} |\dot{a}| \|V_\delta\| \, ds \leq \delta. \tag{20.145}$$

The assumption $\|F(u_0) - f_\delta\| > C\delta^\zeta > \delta$ and inequality (20.145) imply the existence of a $t_\delta > 0$ such that (20.139) holds because $\|F(u_\delta(t)) - f_\delta\|$ is a continuous function of t.

From (20.117), (20.140), and the fact that the function $\|V_\delta(t)\|$ is increasing, one gets the following inequality for all sufficiently small $\epsilon > 0$:

$$h(0)e^{-\phi(t_\delta)} \leq a(t_\delta) \|V_\delta(t_\delta)\|. \tag{20.146}$$

Similarly, from (20.118) and (20.140) one obtains

$$e^{-\phi(t_\delta)} \int_0^{t_\delta} e^{\phi(s)} |\dot{a}(s)| \|V_\delta(s)\| \, ds \leq a(t_\delta) \|V_\delta(t_\delta)\|, \tag{20.147}$$

for all sufficiently small $\delta > 0$.

From (20.139), (20.143), (20.146)–(20.147), and (19.15), one gets

$$C\delta^\zeta = \|F(u_\delta(t_\delta)) - f_\delta\|$$

$$\leq a(t_\delta) \|V_\delta(t_\delta)\| (1 + 1 + 1) \leq 3\left(a(t_\delta)\|y\| + \delta\right). \tag{20.148}$$

This and the relation $\lim_{\delta\to 0} \frac{\delta}{\delta^\zeta} = 0$, for a fixed $\zeta \in (0,1)$, imply

$$\lim_{\delta\to 0} \frac{\delta^\zeta}{a(t_\delta)} \leq \frac{3\|y\|}{C}. \tag{20.149}$$

It follows from inequality (20.137) and the first inequality in (20.135) that

$$a(t)\|u_\delta(t) - V_\delta(t)\| \leq h(0)e^{-\phi(t)} + e^{-\phi(t)} \int_0^t e^{\phi(s)} |\dot{a}(s)| \|V_\delta(s)\| \, ds. \tag{20.150}$$

From (20.149) and the first inequality in (19.15), one gets, for sufficiently small δ, the following inequality:

$$\|V_\delta(t)\| \leq \|y\| + \frac{\delta}{a(t)} < \|y\| + \frac{C\delta^\zeta}{a(t)} < 4\|y\|, \qquad 0 \leq t \leq t_\delta. \tag{20.151}$$

Therefore,

$$\lim_{\delta \to 0} \frac{\int_0^{t_\delta} e^{\phi(s)} |\dot{a}(s)| \|V_\delta(s)\| \, ds}{e^{\phi(t)} a(t_\delta)} \leq 4\|y\| \lim_{\delta \to 0} \frac{\int_0^{t_\delta} e^{\phi(s)} |\dot{a}(s)| \, ds}{e^{\phi(t)} a(t_\delta)}. \qquad (20.152)$$

It follows from (20.152) and (20.118) that

$$\lim_{\delta \to 0} \frac{\int_0^{t_\delta} e^{\phi(s)} |\dot{a}(s)| \|V_\delta(s)\| \, ds}{e^{\phi(t)} a(t_\delta)} = 0. \qquad (20.153)$$

From (20.153), (20.150), and (20.140), one gets

$$0 \leq \lim_{\delta \to 0} \|u_\delta(t_\delta) - V_\delta(t_\delta)\| = \lim_{\delta \to 0} \frac{h(t_\delta)}{a(t_\delta)} = 0. \qquad (20.154)$$

It is now easy to finish the proof of Theorem 20.3.9.
From the triangle inequality and inequality (19.13), one obtains

$$\|u_\delta(t_\delta) - y\| \leq \|u_\delta(t_\delta) - V_\delta(t_\delta)\| + \|V(t_\delta) - V_\delta(t_\delta)\| + \|V(t_\delta) - y\|$$

$$\leq \|u_\delta(t_\delta) - V_\delta(t_\delta)\| + \frac{\delta}{a(t_\delta)} + \|V(t_\delta) - y\|, \qquad (20.155)$$

where $V(t_\delta) = V_{0,a(t_\delta)}$ (see equation (20.108)). From (20.140), (20.154), inequality (20.155), and Lemma 6.1.9, one obtains (20.141).
Theorem 20.3.9 is proved. ∎

The following result gives sufficient conditions for (20.140) to hold:

Theorem 20.3.11 *Let $a(t)$ satisfy (20.110) and (20.123). Assume that u_0 satisfies either inequality*

$$\|F(u_0) + a(0)u_0 - f_\delta\| \leq \frac{1}{4} a(0) \|V_\delta(0)\|, \qquad (20.156)$$

or inequality

$$\|F(u_0) + a(0)u_0 - f_\delta\| \leq \theta \delta^\zeta, \qquad 0 < \theta < C, \qquad (20.157)$$

where $V_\delta(t) := V_{\delta,a(t)}$ solves (20.108) with $a = a(t)$. Then

$$\lim_{\delta \to 0} t_\delta = \infty. \qquad (20.158)$$

Remark 20.3.12 One can choose u_0 satisfying inequality (20.156). Indeed, if u_0 approximates $V_\delta(0)$, the solution to equation (20.108), with a small error, then the first inequality in (20.156) is satisfied. Inequality (20.156) is a sufficient condition for the following inequality:

$$e^{-\frac{t}{2}} \|F(u_0) + a(0)u_0 - f_\delta\| \leq \frac{1}{4} a(t) \|V_\delta(t)\|, \quad t \geq 0, \qquad (20.159)$$

to hold. In our proof inequality (20.159) is used at $t = t_\delta$. The stopping time t_δ is often sufficiently large for the quantity $e^{\frac{t_\delta}{2}} a(t_\delta)$ to be large. This follows from the fact that $\lim_{t\to\infty} e^{\frac{t}{2}} a(t) = \infty$ (see (20.112)). In this case, inequality (20.159) with $t = t_\delta$ is satisfied for a wide range of u_0.

Proof (Proof of Theorem 20.3.11):

From (20.136) and the assumption $1 - \frac{|\dot a|}{a} \geq \frac{1}{2}$, one gets

$$\dot h \leq -\frac{1}{2}h + |\dot a|\|V_\delta\|. \tag{20.160}$$

Inequality (20.160) implies

$$h(t) \leq h(0)e^{-\frac{t}{2}} + e^{-\frac{t}{2}} \int_0^t e^{\frac{s}{2}}|\dot a(s)|\|V_\delta(s)\|ds. \tag{20.161}$$

From (20.161) and (20.135), one gets

$$\|F(u_\delta(t)) - F(V_\delta(t))\| \leq h(0)e^{-\frac{t}{2}} + e^{-\frac{t}{2}} \int_0^t e^{\frac{s}{2}}|\dot a|\|V_\delta\|ds. \tag{20.162}$$

From the triangle inequality and (20.162), one gets

$$\|F(u_\delta(t)) - f_\delta\| \geq \|F(V_\delta(t)) - f_\delta\| - \|F(V_\delta(t)) - F(u_\delta(t))\|$$
$$\geq a(t)\|V_\delta(t)\| - h(0)e^{-\frac{t}{2}} - e^{-\frac{t}{2}} \int_0^t e^{\frac{s}{2}}|\dot a|\|V_\delta\|ds. \tag{20.163}$$

Recall that $a(t)$ satisfies (20.123) by our assumptions. From (20.123) and Lemma 20.3.7, one obtains

$$\frac{1}{2}a(t)\|V_\delta(t)\| \geq e^{-\frac{t}{2}} \int_0^t e^{\frac{s}{2}}|\dot a|\|V_\delta(s)\|ds. \tag{20.164}$$

From (20.156) we have

$$h(0)e^{-\frac{t}{2}} \leq \frac{1}{4}a(0)\|V_\delta(0)\|e^{-\frac{t}{2}}, \qquad t \geq 0. \tag{20.165}$$

It follows from (20.123) that

$$e^{-\frac{t}{2}}a(0) \leq a(t). \tag{20.166}$$

Specifically, inequality (20.166) is obviously true for $t = 0$, and

$$\left(a(t)e^{\frac{t}{2}}\right)'_t = a(t)e^{\frac{t}{2}}\left[\frac{1}{2} - \frac{|\dot a(t)|}{a(t)}\right] > 0,$$

by (20.123). Therefore, one gets from (20.166) and (20.165) the following inequality:

$$e^{-\frac{t}{2}}h(0) \leq \frac{1}{4}a(t)\|V_\delta(0)\| \leq \frac{1}{4}a(t)\|V_\delta(t)\|, \qquad t \geq 0, \tag{20.167}$$

Here, the inequality $\|V_\delta(t')\| \leq \|V_\delta(t)\|$ for $t' < t$ (see Lemma 19.1.1) was used. From (20.139) and (20.163)–(20.167), one gets

$$C\delta^\varsigma = \|F(u_\delta(t_\delta)) - f_\delta\| \geq \frac{1}{4} a(t_\delta) \|V_\delta(t_\delta)\|. \tag{20.168}$$

From (19.13) and the triangle inequality, one derives

$$a(t)\|V(t)\| \leq a(t)\|V(t) - V_\delta(t)\| + a(t)\|V_\delta(t)\| \leq \delta + a(t)\|V_\delta(t)\|, \qquad \forall t \geq 0. \tag{20.169}$$

It follows from (20.168) and (20.169) that

$$0 \leq \lim_{\delta \to 0} a(t_\delta)\|V(t_\delta)\| \leq \lim_{\delta \to 0} \left(\delta + 4C\delta^\varsigma\right) = 0. \tag{20.170}$$

Since $\|V(t)\|$ increases (see Lemma 19.1.1), the above formula implies

$$\lim_{\delta \to 0} a(t_\delta) = 0.$$

Since $0 < a(t) \searrow 0$, it follows that $\lim_{\delta \to 0} t_\delta = \infty$, that is, (20.158) holds. Theorem 20.3.11 is proved. ∎

20.4 IMPLICIT FUNCTION THEOREM AND THE DSM

The aim of this section is to demonstrate the power of the DSM as a tool for proving theoretical results. The DSM was systematically developed and applied to solving nonlinear operator equations where the emphasis was on convergence and stability of the DSM-based algorithms for solving operator equations, especially nonlinear and ill-posed equations. In this section the DSM is used as a tool for proving a "hard" implicit function theorem.

Results in this section are published in [163].

Let us first recall the usual implicit function theorem. Let U solve the equation $F(U) = f$.

Proposition 20.4.1 *If $F(U) = f$, F is a C^1-map in a Hilbert space H, and $F'(U)$ is a boundedly invertible operator (i.e., $\|[F'(U)]^{-1}\| \leq m$), then the equation*

$$F(u) = h \tag{20.171}$$

is uniquely solvable for every h sufficiently close to f.

For convenience of the reader we include a proof of this known result.

Proof: First, one can reduce the problem to the case $u = 0$ and $h = 0$. This is done as follows. Let $u = U + z$, $h - f = p$, $F(U + z) - F(U) := \phi(z)$. Then $\phi(0) = 0$, $\phi'(0) = F'(U)$, and equation (20.171) is equivalent to the equation

$$\phi(z) = p, \tag{20.172}$$

with the assumptions

$$\phi(0) = 0, \quad \lim_{z \to 0} \|\phi'(z) - \phi'(0)\| = 0, \quad \|[\phi'(0)]^{-1}\| \le m. \qquad (20.173)$$

We want to prove that equation (20.172) under the assumptions (20.173) has a unique solution $z = z(p)$, such that $z(0) = 0$, and $\lim_{p \to 0} z(p) = 0$. To prove this, consider the equation

$$z = z - [\phi'(0)]^{-1}(\phi(z) - p) := B(z), \qquad (20.174)$$

and check that the operator B is a contraction in a ball $\mathcal{B}_\epsilon := \{z : \|z\| \le \epsilon\}$ if $\epsilon > 0$ is sufficiently small, and B maps $\mathcal{B}_\epsilon$ into itself. If this is proved, then the desired result follows from the contraction mapping principle.

One has

$$\|B(z)\| = \|z - [\phi'(0)]^{-1}(\phi'(0)z + \eta - p)\| \le m\|\eta\| + m\|p\|, \qquad (20.175)$$

where $\|\eta\| = o(\|z\|)$. If ϵ is so small that $m\|\eta\| < \frac{\epsilon}{2}$ and p is so small that $m\|p\| < \frac{\epsilon}{2}$, then $\|B(z)\| < \epsilon$, so $B : \mathcal{B}_\epsilon \to \mathcal{B}_\epsilon$.

Let us check that B is a contraction mapping in $\mathcal{B}_\epsilon$. One has

$$\|Bz - By\| = \|z - y - [\phi'(0)]^{-1}(\phi(z) - \phi(y))\|$$

$$= \|z - y - [\phi'(0)]^{-1} \int_0^1 \phi'(y + t(z - y)) \, dt(z - y)\|$$

$$\le m \int_0^1 \|\phi'(y + t(z - y)) - \phi'(0)\| \, dt \|z - y\|. \qquad (20.176)$$

If $y, z \in \mathcal{B}_\epsilon$, then

$$\sup_{0 \le t \le 1} \|\phi'(y + t(z - y)) - \phi'(0)\| = o(1), \qquad \epsilon \to 0.$$

Therefore, if ϵ is so small that $mo(1) < 1$, then B is a contraction mapping in $\mathcal{B}_\epsilon$, and equation (20.172) has a unique solution $z = z(p)$ in $\mathcal{B}_\epsilon$, such that $z(0) = 0$. The proof is complete. ∎

The crucial assumptions, on which this proof is based, are assumptions (20.173).

Suppose now that $\phi'(0)$ *is not boundedly invertible, so that the last assumption in* (20.173) *is not valid.* Then a theorem which still guarantees the existence of a solution to equation (20.172) for some set of p is called a "hard" implicit function theorem. One may find examples of such theorems in [7], [13], and [196].

Our goal in this section is to establish a new theorem of this type using a new method of proof, based on the Dynamical Systems Method (DSM). In [146] we have demonstrated a theoretical application of the DSM by establishing some surjectivity results for nonlinear operators.

The result, presented in this section, is a new illustration of the applicability of the DSM as a tool for proving theoretical results.

To formulate the result, let us introduce the notion of a scale of Hilbert spaces H_a (see [70]). Let $H_a \subset H_b$ and $\|u\|_b \leq \|u\|_a$ if $a \geq b$. Example of spaces H_a is the scale of Sobolev spaces $H_a = W^{a,2}(D)$, where $D \subset \mathbb{R}^n$ is a bounded domain with a sufficiently smooth boundary.

Consider equation (20.171). Assume that

$$F(U) = f; \qquad F : H_a \to H_{a+\delta}, \qquad u \in B(U, R) := B_a(U, R), \qquad (20.177)$$

where $B_a(U, R) := \{u : \|u - U\|_a \leq R\}$, $\delta = const > 0$, and the operator $F : H_a \to H_{a+\delta}$ is continuous. Furthermore, assume that $A := A(u) := F'(u)$ exists and is an isomorphism of H_a onto $H_{a+\delta}$:

$$c_0\|v\|_a \leq \|A(u)v\|_{a+\delta} \leq c_0'\|v\|_a, \qquad u, v \in B(U, R). \qquad (20.178)$$

Assume also that

$$\|A^{-1}(v)A(w)\|_a \leq c, \qquad v, w \in B(U, R), \qquad (20.179)$$

and

$$\|A^{-1}(u)[A(u) - A(v)]\|_a \leq c\|u - v\|_a, \qquad u, v \in B(U, R). \qquad (20.180)$$

By $c > 0$ we denote various constants. Note that (20.178) implies

$$\|A^{-1}(u)\psi\|_a \leq c_0^{-1}\|\psi\|_{a+\delta}, \qquad \psi = A(u)[F(v) - h], \qquad v \in B(U, R).$$

Assumption (20.178) implies that $A(u)$ is a smoothing operator similar to a smoothing integral operator, and its inverse is similar to the differentiation operator of order $\delta > 0$. Therefore, the operator $A^{-1}(u) = [F'(u)]^{-1}$ causes the "loss of the derivatives." In general, this may lead to a breakdown of the Newton process (method) (20.6) in a finitely many steps. Our assumptions (20.177)–(20.180) guarantee that this will not happen.

Assume that
$$u_0 \in B_a(U, \rho), \qquad h \in B_{a+\delta}(f, \rho), \qquad (20.181)$$

where $\rho > 0$ is a sufficiently small number:

$$\rho \leq \rho_0 := \frac{R}{1 + c_0^{-1}(1 + c_0')},$$

and c_0 and c_0' are the constants from (20.178). Then $F(u_0) \in B_{a+\delta}(f, c_0'\rho)$, because

$$\|F(u_0) - F(U)\| \leq c_0'\|u_0 - U\| \leq c_0'\rho.$$

Consider the problem

$$\dot{u} = -[F'(u)]^{-1}(F(u) - h), \qquad u(0) = u_0. \qquad (20.182)$$

Our basic result is:

Theorem 20.4.2 *If the assumptions* (20.177)–(20.181) *hold, and* $0 < \rho \le \rho_0 := \frac{R}{1+c_0^{-1}(1+c_0')}$, *where* c_0, c_0' *are the constants from* (20.178), *then problem* (20.182) *has a unique global solution* $u(t)$, *there exists* $V := u(\infty)$,

$$\lim_{t\to\infty} \|u(t) - V\|_a = 0, \tag{20.183}$$

and

$$F(V) = h. \tag{20.184}$$

Theorem 20.4.2 says that if $F(U) = f$ and $\rho \le \rho_0$, then for any $h \in B_{a+\delta}(f, \rho)$ equation (20.171) is solvable and a solution to (20.171) is $u(\infty)$, where $u(t)$ solves problem (20.182).

Let us prove Theorem 20.4.2.

Proof: Let us outline the ideas of the proof. The local existence and uniqueness of the solution to (20.182) will be established if one verifies that the operator $A^{-1}(u)[F(u) - h]$ is locally Lipschitz in H_a. The global existence of this solution $u(t)$ will be established if one proves the uniform boundedness of $u(t)$:

$$\sup_{t \ge 0} \|u(t)\|_a \le c. \tag{20.185}$$

Let us first prove (in paragraph a below) estimate (20.185), the existence of $u(\infty)$, and the relation (20.184), *assuming the local existence* of the solution to (20.182).

In paragraph b below, the local existence of the solution to (20.182) is proved.

(a) *Proof of* (20.184), (20.185), *and the existence of* $u(\infty)$.
If $u(t)$ exists locally, then the function

$$g(t) := \|\phi\|_{a+\delta} := \|F(u(t)) - h\|_{a+\delta} \tag{20.186}$$

satisfies the relation

$$g\dot{g} = \langle F'(u(t))\dot{u}, \phi \rangle_{a+\delta} = -g^2, \tag{20.187}$$

where equation (20.182) was used. Since $g \ge 0$, it follows from (20.187) that

$$g(t) \le g(0)e^{-t}, \qquad g(0) = \|F(u_0) - h\|_{a+\delta}. \tag{20.188}$$

From (20.182), (20.187), and (20.178), one gets

$$\|\dot{u}\|_a \le \frac{1}{c_0}\|\phi\|_{a+\delta} = \frac{g(0)}{c_0}e^{-t} := re^{-t}, \qquad r := \frac{\|F(u_0) - h\|_{a+\delta}}{c_0}. \tag{20.189}$$

Therefore,

$$\lim_{t\to\infty} \|\dot{u}(t)\|_a = 0, \tag{20.190}$$

and

$$\int_0^\infty \|\dot{u}(t)\|_a \, dt < \infty. \tag{20.191}$$

This inequality implies

$$\|u(\tau) - u(s)\| \le \int_s^\tau \|\dot{u}(t)\|_a \, dt < \epsilon, \qquad \tau > s > s(\epsilon),$$

where $\epsilon > 0$ is an arbitrary small fixed number, and $s(\epsilon)$ is a sufficiently large number. Thus, the limit $V := \lim_{t\to\infty} u(t) := u(\infty)$ exists by the Cauchy criterion, and (20.183) holds.

Assumptions (20.177) and (20.178) and relations (20.182), (20.183), and (20.190) imply (20.184).

Integrating inequality (20.189) yields

$$\|u(t) - u_0\|_a \le r, \tag{20.192}$$

and

$$\|u(t) - u(\infty)\|_a \le re^{-t}. \tag{20.193}$$

Inequality (20.192) implies (20.185).

(b) *Let us now prove the local existence of the solution to* (20.182).

We prove that the operator in (20.182) $A^{-1}(u)[F(u)-h]$ is locally Lipschitz in H_a. This implies the local existence of the solution to (20.182).

One has

$$\|A^{-1}(u)(F(u) - h) - A^{-1}(v)(F(v) - h)\|_a$$
$$\le \|[A^{-1}(u) - A^{-1}(v)](F(u) - h)\|_a$$
$$+ \|A^{-1}(v)(F(u) - F(v))\|_a := I_1 + I_2. \tag{20.194}$$

Write

$$F(u) - F(v) = \int_0^1 A(v + t(u - v))(u - v) \, dt, \tag{20.195}$$

and use assumption (20.179) with $w = v + t(u - v)$ to conclude that

$$I_2 \le c\|u - v\|_a. \tag{20.196}$$

Write

$$A^{-1}(u) - A^{-1}(v) = A^{-1}(u)[A(v) - A(u)]A^{-1}(v), \tag{20.197}$$

and use the estimate

$$\|A^{-1}(v)[F(u) - h]\|_a \le c, \tag{20.198}$$

which is a consequence of assumptions (20.177) and (20.178). Then use assumption (20.180) to conclude that

$$I_1 \le c\|u - v\|_a. \tag{20.199}$$

From (20.194), (20.196), and (20.199) it follows that the operator $A^{-1}(u)[F(u) - h]$ is locally Lipschitz.

Note that

$$\|u(t) - U\|_a \leq \|u(t) - u_0\|_a + \|u_0 - U\|_a \leq r + \rho \qquad (20.200)$$

and

$$\|F(u(t)) - h\|_{a+\delta} \leq \|F(u_0) - h\|_{a+\delta}$$
$$\leq \|F(u_0) - f\|_{a+\delta} + \|f - h\|_{a+\delta} \leq (1 + c_0')\rho, \qquad (20.201)$$

so, from (20.189) one gets

$$r \leq \frac{(1 + c_0')\rho}{c_0}. \qquad (20.202)$$

Choose

$$R \geq r + \rho. \qquad (20.203)$$

Then the trajectory $u(t)$ stays in the ball $B(U, R)$ for all $t \geq 0$, and, therefore, assumptions (20.177)–(20.180) hold in this ball for all $t \geq 0$.

Condition (20.203) and inequality (20.202) imply

$$\rho \leq \rho_0 = \frac{R}{1 + c_0^{-1}(1 + c_0')}. \qquad (20.204)$$

This is the "smallness" condition on ρ.

Theorem 20.4.2 is proved. ∎

20.4.1 Example

Let

$$F(u) = \int_0^x u^2(s)\, ds, \qquad x \in [0, 1].$$

Then

$$A(u)q = 2 \int_0^x u(s)q(s)\, ds.$$

Let $f = x$ and $U = 1$. Then $F(U) = x$. Choose $a = 1$ and $\delta = 1$. Denote by $H_a = H_a(0, 1)$ the usual Sobolev space. Assume that

$$h \in B_2(x, \rho) := \{h \; : \; \|h - x\|_2 \leq \rho\},$$

and $\rho > 0$ is sufficiently small. One can verify that

$$A^{-1}(u)\psi = \frac{\psi'(x)}{2u(x)}$$

for any $\psi \in H_1$.

Let us check conditions (20.177)–(20.181) for this example.

Condition (20.177) holds, because if $u_n \to u$ in H_1, then

$$\int_0^x u_n^2(s)\,ds \to \int_0^x u^2(s)\,ds$$

in H_2. To verify this, it is sufficient to check that

$$\frac{d^2}{dx^2}\int_0^x u_n^2(s)\,ds \to 2uu',$$

where $\to$ means the convergence in $H := H_0 := L^2(0,1)$. In turn, this is verified if one checks that $u_n' u_n \to u'u$ in $L^2(0,1)$, provided that $u_n' \to u'$ in $L^2(0,1)$.

One has

$$I_n := \|u_n' u_n - u'u\|_0 \le \|(u_n' - u')u_n\|_0 + \|u'(u_n - u)\|_0.$$

Since $\|u_n'\|_0 \le c$, one concludes that $\|u_n\|_{L^\infty(0,1)} \le c_1$ and $\lim_{n\to\infty}\|u_n - u\|_{L^\infty} = 0$. Thus,

$$\lim_{n\to\infty} I_n = 0.$$

Condition (20.178) holds because $\|u\|_{L^\infty(0,1)} \le c\|u\|_1$, and

$$\left\|\int_0^x u(s)q(s)\,ds\right\|_2 \le c\|u'q + uq'\|_0 \le c(\|q\|_{L^\infty(0,1)}\|u\|_1 + \|u\|_{L^\infty(0,1)}\|q\|_1),$$

so

$$\left\|\int_0^x u(s)q(s)\,ds\right\|_2 \le c_0'\|u\|_1\|q\|_1,$$

and

$$\left\|\int_0^x uq\,ds\right\|_2 \ge \|uq\|_1 \ge c_0\|q\|_1,$$

provided that $u \in B_1(1,\rho)$ and $\rho > 0$ is sufficiently small.

Condition (20.179) holds because

$$\|A^{-1}(v)A(w)q\|_1 = \left\|\frac{1}{v(x)}w(x)q\right\|_1 \le c\|q\|_1,$$

provided that $u, w \in B_1(1,\rho)$ and $\rho > 0$ is sufficiently small.

Condition (20.180) holds because

$$\left\|A^{-1}(u)\int_0^x (u-v)q\,ds\right\|_1 = \left\|\frac{u-v}{2u}q\right\|_1 \le c\|u-v\|_1\|q\|_1,$$

provided that $u, v \in B_1(1,\rho)$ and $\rho > 0$ is sufficiently small.

By Theorem 20.4.2 the equation

$$F(u) := \int_0^x u^2(s)\,ds = h,$$

where $\|h - x\|_2 \leq \rho$ and $\rho > 0$ is sufficiently small, has a solution V,

$$F(V) = h.$$

This solution can be obtained as $u(\infty)$, where $u(t)$ solves problem (20.182) and conditions (20.181) and (20.204) hold.

CHAPTER 21

DSM OF GRADIENT TYPE

In this chapter we study the DSM of gradient type for solving the equation

$$F(u) = f, \tag{21.1}$$

where F is a nonlinear, Fréchet differentiable, monotone operator in a Hilbert space H. We assume equation (21.1) is solvable, possibly nonuniquely. We assume in addition that

$$\|F'(u)\| \leq M_R, \qquad \forall u \in B(0, R). \tag{21.2}$$

If $F'(u)$ is not boundedly invertible, then solving equation (21.1) for u given noisy data f_δ is often (but not always) an ill-posed problem. For solving nonlinear operator equations, various methods are studied and discussed in the recent book [67] and references therein. However, the theory for nonlinear operators is less complete compared to the one for linear operators.

Results in literature can be divided into two groups:

1. Convergence results that guarantee that a certain regularized solution u_δ converges (for $\delta \to 0$) to the minimal-norm solution y of the nonlinear operator equation $F(u) = f$ and

Dynamical Systems Method and Applications: Theoretical Developments and Numerical Examples, First Edition. A. G. Ramm and N. S. Hoang.
Copyright © 2012 John Wiley & Sons, Inc.

2. Convergence rate results which, under certain additional assumptions on F and on the smoothness of y, guarantee certain error bounds for the error $\|u_\delta - y\|$ in dependence on the noise level δ.

Much work has been devoted to the study of the convergence rate. This often requires some source-type assumptions. It is well known that without extra assumptions, usually source-type assumptions about the right-hand side or some assumption concerning the smoothness of the solution, one cannot get a specific rate of convergence even for linear ill-posed problems. On the other hand, such assumptions are often difficult to verify and often they do not hold. By this reason we do not make such assumptions.

In [67] the Landweber and modified Landweber methods are studied under the assumption that

$$\|F(\tilde{x}) - F(x) - F'(x)(\tilde{x} - x)\| \le \eta \|F(x) - F(\tilde{x})\|, \qquad \eta < \frac{1}{2}, \qquad (21.3)$$

for all x, $\tilde{x}$ in some ball $B(x_0, R) \subset H$. While condition (21.3) holds for some operators F which are not monotone, it does not hold for some monotone operators F. For example, (21.3) does not hold for $F(x) = x^3$, $x_0 = 0$ and any $R > 0$. In addition, if η is close to $\frac{1}{2}$, then the constant $\tau = 2\frac{1+\eta}{1-2\eta}$ used in the discrepancy in [67] becomes very large. If this is the case, then the stopping rule proposed there is not applicable.

An a posteriori choice of stopping rule for solving equations with monotone operators was proposed in [183]. In [183] the regularization parameter α is chosen from the following equation

$$\|\alpha(F'(u_\alpha^\delta) + \alpha I)^{-1}[F(u_\alpha^\delta) - f_\delta]\| = C\delta, \qquad C = const > 1, \qquad (21.4)$$

where u_α^δ is the solution to the regularized equation (19.3) with $a = \alpha$, for example, $u_\alpha^\delta := V_{\delta,\alpha}$. Optimal rate of convergence was obtained with this choice of stopping rule in [183]. However, numerical methods for finding α from equation (21.4) is not studied there. Moreover, the convergence to the minimal norm solution was obtained under some source-type assumptions (cf. [183]).

In [46] the following version of the DSM for solving equation (21.1):

$$\dot{u}_\delta = -\left(F'(u_\delta)^* + a(t)I\right)\left(F(u_\delta) + a(t)u_\delta - f_\delta\right), \qquad u_\delta(0) = u_0, \qquad (21.5)$$

was studied under the assumption that F is a monotone operator. There a stopping rule of discrepancy principle type was proposed and justified mathematically. If F is monotone, then $F'(\cdot) := A \ge 0$. If a bounded linear operator A is defined on all of the complex Hilbert space H and $A \ge 0$ (i.e., $\langle Au, u \rangle \ge 0$, $\forall u \in H$), then $A = A^*$, so A is self-adjoint. In a real Hilbert space H a bounded linear operator defined on all of H and satisfying the inequality $\langle Au, u \rangle \ge 0$, $\forall u \in H$ is not necessary selfadjoint. Example: $H = \mathbb{R}^2$, $A = \begin{pmatrix} 2 & 1 \\ 0 & 2 \end{pmatrix}$, $\langle Au, u \rangle = 2u_1^2 + u_1u_2 + u_2^2 \ge 0$, but $A^* = \begin{pmatrix} 2 & 0 \\ 1 & 2 \end{pmatrix} \ne A$.

The advantage of method (21.5), a modified version of the gradient method, over the Gauss–Newton method and the DSM of Newton type is the following: No inversion of matrices is needed in (21.5). Although the convergence rate of the DSM (21.5) may be slower than that of the DSM of Newton type, the DSM (21.5) might be faster than the DSM of Newton type for large-scale systems due to its lower computation cost at each iteration.

In this chapter we investigate a stopping rule based on a discrepancy principle (DP) for the DSM (21.5) under a weaker condition on the differentiability of F. The main results of this chapter are Theorem 21.2.2–21.3.10. In Theorem 21.2.2 and Theorem 21.2.4 a DP is formulated, the existence of a stopping time t_δ is proved, and the convergence of the DSM with the proposed DP is justified under some weak and natural assumptions.

Based on the DSM (21.5) the following iterative scheme with a stopping rule of discrepancy type is also studied in this chapter:

$$u_{n+1} = u_n - \gamma_n (F'(u_n)^* + a_n I)(F(u_n) + a_n u_n - f_\delta), \qquad u_0 = u_0. \quad (21.6)$$

Convergence of u_{n_δ}, chosen by our stopping rule, to the minimal-norm solution to (21.1) is proved in Theorem 21.3.7 and Theorem 21.3.10 for suitably chosen u_0, γ_n and a_n.

The novelty of the results, presented in this chapter compared to [46] is: Convergence results are obtained under a weaker assumption on the differentiability of F for a larger class of regularizing functions. Specifically, in this chapter the iterative scheme (21.6) is studied under the assumption that F is Fréchet differentiable. In [46] it was assumed F is twice Fréchet differentiable. The DSM (21.5) in this chapter is also justified under a weaker assumption on F and for a larger class of regularizing function $a(t)$ than in [46]. The regularizing function $a(t)$ in this chapter can decay faster than the one in [46]. This results in less computation time in practice. Moreover, sufficient conditions for the convergence of iterative scheme (21.6) with the proposed DP to the minimal-norm solution are formulated in this chapter (see Theorem 21.3.10).

This chapter is based on paper [42].

21.1 AUXILIARY RESULTS

Let $0 < a = a(t) \in C^1[0, \infty)$ be a strictly decaying function. This assumption holds throughout the chapter and often are not repeated.

From Lemma 19.1.1 one gets the following corollary.

Corollary 21.1.1 *Assume* $\|F(0) - f_\delta\| > 0$. *Let* $0 < a(t) \searrow 0$, *and* F *be monotone. Denote*

$$\sigma(t) := \|V_\delta(t)\|, \qquad \rho(t) := a(t)\sigma(t) = \|F(V_\delta(t)) - f_\delta\|,$$

where $V_\delta(t)$ *solves* (19.3) *with* $a = a(t)$. *Then* $\rho(t)$ *is decreasing, and* $\sigma(t)$ *is increasing.*

It follows from (19.12) that

$$\lim_{t \to \infty} \|F(V_\delta(t)) - f_\delta\| \leq \delta. \tag{21.7}$$

From Remark 19.1.5 one gets

$$\|V_\delta(t) - V(t)\| \leq \frac{\delta}{a(t)}, \qquad \|V_\delta\| \leq \|V\| + \frac{\delta}{a} \leq \|y\| + \frac{\delta}{a}, \tag{21.8}$$

Lemma 21.1.2 *Let $F : H \to H$ be a monotone operator, let $0 < a(t)$ be a decreasing function, and let $V_\delta(t)$ solve equation (19.3) with $a = a(t)$. Then the following inequalities hold:*

$$\|V_\delta(t_1) - V_\delta(t_2)\| \leq \frac{a(t_1) - a(t_2)}{a(t_1)} \|V_\delta(t_2)\|, \qquad 0 < t_1 \leq t_2, \tag{21.9}$$

$$\alpha\|\tilde{u} - \tilde{v}\| \leq \|F(\tilde{u}) - F(\tilde{v}) + \alpha(\tilde{u} - \tilde{v})\|, \qquad \forall \tilde{u}, \tilde{v} \in H, \quad \forall \alpha \geq 0, \tag{21.10}$$

$$\|F(\tilde{u}) - F(\tilde{v})\| \leq \|F(\tilde{u}) - F(\tilde{v}) + \alpha(\tilde{u} - \tilde{v})\|, \qquad \forall \tilde{u}, \tilde{v} \in H, \tag{21.11}$$

where $\alpha = const > 0$.

Proof: From equation (19.3) and the monotonicity of F, one gets

$$\begin{aligned}
0 &\leq \langle F(V_\delta(t_1)) - F(V_\delta(t_2)), V_\delta(t_1) - V_\delta(t_2) \rangle \\
&= \langle -a(t_1)V_\delta(t_1) + a(t_2)V_\delta(t_2), V_\delta(t_1) - V_\delta(t_2) \rangle \\
&= -a(t_1)\|V_\delta(t_1) - V_\delta(t_2)\|^2 + (a(t_2) - a(t_1))\langle V_\delta(t_2), V_\delta(t_1) - V_\delta(t_2) \rangle \\
&\leq -a(t_1)\|V_\delta(t_1) - V_\delta(t_2)\|^2 + (a(t_1) - a(t_2))\|V_\delta(t_2)\|\|V_\delta(t_1) - V_\delta(t_2)\|.
\end{aligned}$$

This implies inequality (21.9).

From the monotonicity of F, one gets

$$\begin{aligned}
\alpha\|\tilde{u} - \tilde{v}\|^2 &\leq \langle \alpha(\tilde{u} - \tilde{v}), F(\tilde{u}) - F(\tilde{v}) + \tilde{u} - \tilde{v} \rangle \\
&\leq \alpha\|\tilde{u} - \tilde{v}\|\|F(\tilde{u}) - F(\tilde{v}) + \tilde{u} - \tilde{v}\|, \quad \forall \tilde{u}, \tilde{v} \in H, \alpha > 0. \tag{21.12}
\end{aligned}$$

Inequality (21.10) follows from (21.12).

Similarly, one has

$$\begin{aligned}
\|F(\tilde{u}) - F(\tilde{v})\|^2 &\leq \langle F(\tilde{u}) - F(\tilde{v}), F(\tilde{u}) - F(\tilde{v}) + \alpha(\tilde{u} - \tilde{v}) \rangle \\
&\leq \|F(\tilde{u}) - F(\tilde{v})\|\|F(\tilde{u}) - F(\tilde{v}) + \alpha(\tilde{u} - \tilde{v})\|, \tag{21.13}
\end{aligned}$$

for all $\tilde{u}, \tilde{v} \in H$ and $\alpha > 0$. This implies inequality (21.11).

Lemma (21.1.2) is proved. ∎

Let $a(t)$ satisfy the following conditions:

$$0 < a(t) \searrow 0, \qquad 0 < \frac{|\dot{a}(s)|}{a^3(s)} \searrow 0. \tag{21.14}$$

Lemma 21.1.3 *Let* $\psi(t) = r \int_0^t a^2(s)\,ds$ *where* $0 < a(t)$ *satisfy conditions* (21.14) *and* $r = const > 0$. *Then the following relations hold:*

$$\lim_{t \to \infty} \psi(t) = \infty, \tag{21.15}$$

$$\lim_{t \to \infty} e^{\psi(t)} a(t) = \infty, \tag{21.16}$$

$$\lim_{t \to \infty} \frac{\int_0^t e^{\psi(s)} \frac{|\dot{a}(s)|}{a(s)}\,ds}{e^{\psi(t)}} = 0, \tag{21.17}$$

$$\lim_{t \to \infty} \frac{\int_0^t e^{\psi(s)} |\dot{a}(s)|\,ds}{e^{\psi(t)} a(t)} = 0. \tag{21.18}$$

Proof: Let us prove (21.15).

It follows from (21.14) that there exists $t_1 > 0$ such that

$$ra^2(t) \geq -\frac{\dot{a}(t)}{a(t)}, \qquad \forall t \geq t_1. \tag{21.19}$$

This implies

$$\psi(t) \geq r \int_{t_1}^t a^2(s)\,ds \geq \int_{t_1}^t \frac{-\dot{a}(s)}{a(s)}\,ds = -\ln a(s) \Big|_{t_1}^t = \ln a(t_1) - \ln a(t),$$
$$\tag{21.20}$$

for all $t \geq t_1$. Since $\lim_{t \to \infty} a(t) = 0$, relation (21.15) follows from (21.20).

Let us prove (21.16).

We claim that, for sufficiently large $t > 0$, the following inquality holds:

$$\psi(t) = r \int_0^t a^2(s)\,ds > \ln \frac{1}{a^2(t)}. \tag{21.21}$$

Indeed, by L'Hospital's rule and (21.14), one gets

$$\lim_{t \to \infty} \frac{\psi(t)}{\ln \frac{1}{a^2(t)}} = \lim_{t \to \infty} \frac{ra^2(t)}{a^2(t) \frac{-2\dot{a}(t)}{a^3(t)}} = \lim_{t \to \infty} \frac{ra^3(t)}{2|\dot{a}(t)|} = \infty. \tag{21.22}$$

This implies that (21.21) holds for all $t \geq \tilde{T}$, provided that $\tilde{T} > 0$ is sufficiently large. From inequality (21.21) one gets

$$\lim_{t \to \infty} a(t) e^{\psi(t)} \geq \lim_{t \to \infty} a(t) e^{\ln \frac{1}{a^2(t)}} = \lim_{t \to \infty} \frac{1}{a(t)} = \infty. \tag{21.23}$$

Thus, equality (21.16) is proved.

Let us prove relation (21.17).

If $I := \lim_{t \to \infty} \int_0^t e^{\psi(s)} \frac{|\dot{a}(s)|}{a(s)}\,ds < \infty$, then (21.17) holds since $\psi(t)$ goes to ∞ as $t \to \infty$ (see (21.15)). If $I = \infty$, then (21.17) follows from L'Hospital's rule and (21.14).

Let us prove (21.18).

Since $a(t)e^{\psi(t)} \to \infty$ as $t \to \infty$, by (21.16), relation (21.18) holds if the numerator of (21.18) is bounded. Otherwise, L'Hospital's rule and (21.14) yield

$$\lim_{t\to\infty} \frac{\int_0^t e^{\psi(s)}|\dot{a}(s)|\,ds}{e^{\psi(t)}a(t)} = \lim_{t\to\infty} \frac{e^{\psi(t)}|\dot{a}(t)|}{re^{\psi(t)}a^3(t) + e^{\psi(t)}\dot{a}(t)} = 0. \qquad (21.24)$$

Lemma 21.1.3 is proved.   ∎

Assume that $a(t)$ satisfies the following conditions

$$0 < a(t) \searrow 0, \qquad \frac{1}{6} \geq q \geq \frac{|\dot{a}(t)|}{a^3(t)} \searrow 0. \qquad (21.25)$$

Lemma 21.1.4 *Let $a(t)$ satisfy* (21.25) *and* $\varphi(t) := (1-q)\int_0^t a^2(s)\,ds$. *Then the following inequality holds:*

$$e^{-\varphi(t)} \int_0^t e^{\varphi(s)}|\dot{a}(s)|\|V_\delta(s)\|\,ds \leq \frac{q}{1-2q}a(t)\|V_\delta(t)\|, \qquad t \geq 0. \qquad (21.26)$$

Proof: Let us show that

$$e^{\varphi(t)}|\dot{a}(t)| \leq \frac{q}{1-2q}\frac{d}{dt}\left(a(t)e^{\varphi(t)}\right), \qquad \forall t > 0. \qquad (21.27)$$

One has

$$\frac{d}{dt}\left(a(t)e^{\varphi(t)}\right) = \varphi'(t)e^{\varphi(t)}a(t) + e^{\varphi(t)}\dot{a}(t)$$

$$= (1-q)a^2(t)e^{\varphi(t)}a(t) - e^{\varphi(t)}|\dot{a}(t)|. \qquad (21.28)$$

Therefore, inequality (21.27) is equivalent to

$$\left(\frac{q}{1-2q}+1\right)|\dot{a}(t)| \leq \frac{(1-q)q}{1-2q}a^3(t), \qquad \forall t > 0. \qquad (21.29)$$

Inequality (21.29) holds because $a(t)$ satisfies (21.25) by our assumptions. Thus, (21.27) holds.

Integrating both sides of (21.27), one gets

$$\int_0^t e^{\varphi(s)}|\dot{a}(s)|\,ds \leq \frac{q}{1-2q}\left(e^{\varphi(t)}a(t) - e^0 a(0)\right)$$

$$\leq \frac{q}{1-2q}e^{\varphi(t)}a(t), \qquad t \geq 0. \qquad (21.30)$$

Multiplying (21.30) by $e^{-\varphi(t)}\|V_\delta(t)\|$, using the fact that $\|V_\delta(t)\|$ is increasing (see Lemma 19.1.1), one obtains (21.26). Lemma 21.1.4 is proved.   ∎

21.2 DSM GRADIENT METHOD

Throughout this section, let us assume that $F'(u)$ is self-adjoint and satisfies a Lipschitz condition. Note that if H is a complex Hilbert space, then it follows from the monotonicity that F' is self-adjoint. This is not true if H is a real Hilbert space.

Denote

$$A := F'(u_\delta(t)), \quad A_a := A + aI, \quad a = a(t),$$

where I is the identity operator, and $u_\delta(t)$ solves the following Cauchy problem:

$$\dot{u}_\delta = -A^*_{a(t)}[F(u_\delta) + a(t)u_\delta - f_\delta], \quad u_\delta(0) = u_0, \tag{21.31}$$

where $u_0 \in H$. Assume that $a(t)$ satisfies the following conditions (cf. (21.14)):

$$0 < a(t) \searrow 0, \qquad 0 < \frac{|\dot{a}(t)|}{a^3(t)} \searrow 0, \qquad t \geq 0. \tag{21.32}$$

The global existence of a unique solution to (21.31) is guaranteed by the following lemma:

Lemma 21.2.1 *Let F be a Fréchet differentiable and monotone operator. Assume that $F'(u)$ is self-adjoint and satisfies a Lipschitz condition. Let $0 < a(t)$ satisfy conditions (21.32) and $u_\delta(t)$ solve equation (21.31). Then $u_\delta(t)$ exists globally and is unique.*

Proof: The uniqueness and local existence of $u_\delta(t)$ follow from the assumption that F' satisfies a Lipschitz condition.

Let us prove the global existence of $u_\delta(t)$. To prove the global existence of $u_\delta(t)$, it suffices to show that $u_\delta(t)$ is bounded on $[0, T]$ for any $T > 0$.

Let us first prove that

$$\|w(t)\| \leq e^{-\phi(t)}\|w(0)\| + e^{-\phi(t)} \int_0^t e^{\phi(s)} \frac{|\dot{a}(s)|}{a(s)} \|V_\delta(s)\| \, ds, \tag{21.33}$$

for all $t \in [0, T)$, where

$$\phi(t) := \int_0^t a^2(s) \, ds, \qquad w(t) := u_\delta(t) - V_\delta(t). \tag{21.34}$$

One has

$$F(u_\delta) - F(V_\delta) = J(u_\delta - V_\delta), \qquad J := \int_0^1 F'(V_\delta + \xi(u_\delta - V_\delta)) \, d\xi. \tag{21.35}$$

Since F' is nonnegative and self-adjoint, the operator J is also self-adjoint and nonnegative. From (21.35), (21.31), (19.3), and the monotonicity of F, one gets

$$
\begin{aligned}
\langle \dot{u}_\delta(t), w(t) \rangle &= -\langle A^*_{a(t)}[F(u_\delta(t)) + a(t)u_\delta(t) - F(V_\delta(t)) - a(t)V_\delta(t)], w(t) \rangle \\
&= -\langle A^*_{a(t)}[J + a(t)I](u_\delta(t) - V_\delta(t)), w(t) \rangle \\
&\leq -a^2(t)\|w(t)\|^2,
\end{aligned}
\tag{21.36}
$$

where I is the identity operator. It follows from the triangle inequality, (21.36), and (6.32) that

$$
\begin{aligned}
\frac{d}{dt}\|w(t)\|^2 &= 2\operatorname{Re}\langle \dot{u}_\delta(t), w(t) \rangle + 2\operatorname{Re}\langle \dot{V}_\delta(t), w(t) \rangle \\
&\leq 2\operatorname{Re}\langle \dot{u}_\delta(t), w(t) \rangle + 2\|\dot{V}_\delta(t)\|\|w(t)\| \\
&\leq -2a^2(t)\|w(t)\|^2 + 2\|w(t)\|\frac{|\dot{a}(t)|}{a(t)}\|V_\delta(t)\|, \qquad \forall t \geq 0. \quad (21.37)
\end{aligned}
$$

This implies

$$
\frac{d}{dt}\|w(t)\| \leq -a^2(t)\|w(t)\| + \frac{|\dot{a}(t)|}{a(t)}\|V_\delta(t)\|, \qquad \forall t \geq 0.
\tag{21.38}
$$

Inequality (21.33) follows from (21.38) and Gronwall's inequality.

Let

$$
K = 1 + \sup_{t \geq 0} e^{-\phi(t)} \int_0^t e^{\phi(s)} \frac{|\dot{a}(s)|}{a(s)} \, ds,
\tag{21.39}
$$

then K is finite as a consequence of (21.17). From (21.33), (21.39), and the fact that the function $\|V_\delta(t)\|$ is increasing, one obtains

$$
\|w(t)\| \leq e^{-\phi(t)}\|w(0)\| + (K-1)\|V_\delta(t)\|, \qquad \forall t \in [0, T].
\tag{21.40}
$$

This and the triangle inequality imply

$$
\|u_\delta(t)\| \leq \|w(t)\| + \|V_\delta(t)\| \leq \|w(0)\| + K\|V_\delta(T)\|, \qquad \forall t \in [0, T]. \quad (21.41)
$$

Thus, $\|u_\delta(t)\|$ is bounded on $[0, T]$ for any $T > 0$. This implies the global existence of $u_\delta(t)$. Lemma 21.2.1 is proved. ∎

Let us formulate one of our main results.

Theorem 21.2.2 *Let $C > 0$ and $\zeta \in (0, 1)$ be constants such that $C\delta^\zeta > \delta$. Assume that $F : H \to H$ is a monotone and Fréchet differentiable operator, F' is self-adjoint and satisfies a Lipschitz condition, and (21.2) holds. Let $a(t)$ satisfy (21.32). Assume that u_0 satisfies the following inequality:*

$$
\|F(u_0) - f_\delta\| > C\delta^\zeta.
\tag{21.42}
$$

Assume that equation $F(u) = f$ has a solution, possibly nonunique, and y is the minimal-norm solution to this equation. Let f be unknown but f_δ be given, $\|f_\delta - f\| \leq \delta$. Then there exists a unique t_δ such that

$$
\|F(u_\delta(t_\delta)) - f_\delta\| = C\delta^\zeta, \qquad \|F(u_\delta(t)) - f_\delta\| > C\delta^\zeta, \quad \forall t \in (0, t_\delta). \quad (21.43)
$$

If $\zeta \in (0,1)$, t_δ satisfies (21.43) *and*

$$\lim_{\delta \to 0} t_\delta = \infty, \tag{21.44}$$

then

$$\lim_{\delta \to 0} \|u_\delta(t_\delta) - y\| = 0. \tag{21.45}$$

Remark 21.2.3 Inequality (21.42) is a natural assumption. If inequality (21.42) does not hold and $\|u_0\|$ is not large, then u_0 can be considered an approximate solution to (21.1).

Theorem 21.2.2 remains valid when the self-adjointness of F' is dropped provided that $a(t)$ satisfies (21.32) and the following condition

$$1 > \frac{|\dot{a}(t)|}{a^3(t)}, \qquad \forall t \geq 0.$$

Theorem 21.2.4 (see below) gives a sufficient condition for relation (21.44) to hold.

Proof (Proof of Theorem (21.2.2)): It is clear that there exists at most one t_δ satisfying (21.43). Indeed, if t_δ and $\tau_\delta > t_\delta$ both satisfy (21.43), then the second inequality of (21.43) does not hold on the interval $(0, \tau_\delta)$. This verifies the uniqueness of t_δ.

Let us verify the existence of t_δ.

Let

$$v(t) := F(u_\delta(t)) + a(t)u_\delta(t) - f_\delta, \qquad h(t) = \|v(t)\|. \tag{21.46}$$

Since $F(V_\delta) + aV_\delta = f_\delta$, one has

$$v = F(u_\delta) + au_\delta - F(V_\delta) + aV_\delta. \tag{21.47}$$

From (21.31) and (21.46) one gets

$$\dot{v} = A_a\dot{u} + \dot{a}u_\delta = -A_aA_a^*v + \dot{a}u_\delta. \tag{21.48}$$

Multiplying (21.48) by v and using monotonicity of F, one obtains

$$h\dot{h} = -\langle A_aA_a^*v, v \rangle + \mathrm{Re}\langle v, \dot{a}u_\delta \rangle \leq -h^2a^2 + h|\dot{a}|\|u_\delta\|. \tag{21.49}$$

Again, we have used the inequality $A_aA_a^* \geq a^2$, which holds for $A \geq 0$, that is, monotone operators F. Inequalities (21.49) and (21.41) imply

$$\dot{h} \leq -ha^2 + |\dot{a}|\|u_\delta\| \leq -ha^2 + |\dot{a}|(\|w(0)\| + K\|V_\delta\|). \tag{21.50}$$

Inequality (21.50) and Gronwall's inequality imply

$$h(t) \leq h(0)e^{-\phi(t)} + e^{-\phi(t)} \int_0^t e^{\phi(s)}|\dot{a}(s)|\left(\|w(0)\| + K\|V_\delta(s)\| \right) ds, \tag{21.51}$$

where

$$\phi(t) := \int_0^t a^2(s)\,ds.$$

From the triangle inequality, inequality (21.11) with $\tilde{u} = u_\delta$ and $\tilde{v} = V_\delta$, and (21.51), one gets

$$
\begin{aligned}
\|F(u_\delta(t)) - f_\delta\| &\leq \|F(V_\delta(t)) - f_\delta\| + \|F(u_\delta(t)) - F(V_\delta(t))\| \\
&\leq \|F(V_\delta(t)) - f_\delta\| + h(t) \\
&\leq \|F(V_\delta(t)) - f_\delta\| + h(0)e^{-\phi(t)} \\
&\quad + e^{-\phi(t)} \int_0^t e^{\phi(s)} |\dot{a}(s)| \Big(\|w(0)\| + K\|V_\delta(s)\| \Big)\,ds. \quad (21.52)
\end{aligned}
$$

Since $a(t)\|V_\delta(t)\|$ is a decreasing function of t (see Lemma 19.1.1), one gets from (21.17) the following inequalities:

$$
\begin{aligned}
0 \leq \lim_{t\to\infty} \frac{\int_0^t e^{\phi(s)} |\dot{a}(s)| \|V_\delta(s)\|\,ds}{e^{\phi(t)}} \\
\leq \lim_{t\to\infty} \frac{\int_0^t e^{\phi(s)} \frac{|\dot{a}(s)|}{a(s)} a(0)\|V_\delta(0)\|\,ds}{e^{\phi(t)}} = 0. \quad (21.53)
\end{aligned}
$$

From (21.52), (21.53), relation (21.15) with $r = 1$ (i.e., $\psi = \phi$), and relation (21.7), one obtains

$$\lim_{t\to\infty} \|F(u_\delta(t)) - f_\delta\| \leq \delta. \quad (21.54)$$

This, the continuity of F, and the assumption $\|F(u_0) - f_\delta\| > C\delta^\zeta$ imply the existence of t_δ satisfying (21.43).

Let us prove (21.45), *assuming that* (21.44) *holds.*

From relation (21.44), relations (21.16) and (21.18) with $r = 1$, and the inequality $\|V_\delta(t)\| \geq \|V_\delta(0)\| > 0$, $t \geq 0$, one gets, for all sufficiently small $\delta > 0$, the following inequalities:

$$h(0)e^{-\phi(t_\delta)} \leq a(t_\delta)\|V_\delta(0)\| \leq a(t_\delta)\|V_\delta(t_\delta)\|, \quad (21.55)$$

$$Ke^{-\phi(t_\delta)} \int_0^{t_\delta} e^{\phi(s)} |\dot{a}(s)| \|V_\delta(s)\|\,ds \leq a(t_\delta)\|V_\delta(0)\| \leq a(t_\delta)\|V_\delta(t_\delta)\|, \quad (21.56)$$

$$\|w(0)\| e^{-\phi(t_\delta)} \int_0^{t_\delta} e^{\phi(s)} |\dot{a}(s)|\,ds \leq a(t_\delta)\|V_\delta(0)\| \leq a(t_\delta)\|V_\delta(t_\delta)\|. \quad (21.57)$$

From (21.55)–(21.57), (19.3), (21.52) with $t = t_\delta$, and the second inequality in (21.8), one obtains

$$\|F(u_\delta(t_\delta)) - f_\delta\| = C\delta^\zeta \leq 4a(t_\delta)\|V_\delta(t_\delta)\| \leq 4\Big(a(t_\delta)\|y\| + \delta \Big). \quad (21.58)$$

This and the relation $\lim_{\delta\to 0} \frac{\delta^\zeta}{\delta} = \infty$ for a fixed $\zeta \in (0,1)$ imply

$$\limsup_{\delta\to 0} \frac{\delta^\zeta}{a(t_\delta)} < \frac{5\|y\|}{C}. \quad (21.59)$$

The second inequality in (21.8), inequality $C\delta^\zeta > \delta$, and relation (21.59) imply, for sufficiently small $\delta > 0$, the following inequality

$$\|V_\delta(t)\| \leq \|y\| + \frac{\delta}{a(t_\delta)} < \|y\| + \frac{C\delta^\zeta}{a(t_\delta)} < 6\|y\|, \qquad 0 \leq t \leq t_\delta. \qquad (21.60)$$

This and (21.18) with $\psi = \phi$ (i.e., $r = 1$), imply

$$0 \leq \lim_{\delta \to 0} \frac{\int_0^{t_\delta} e^{\phi(s)}|\dot{a}(s)|\|V_\delta(s)\|\,ds}{e^{\phi(t_\delta)}a(t_\delta)} \leq 6\|y\| \lim_{\delta \to 0} \frac{\int_0^{t_\delta} e^{\phi(s)}|\dot{a}(s)|\,ds}{e^{\phi(t_\delta)}a(t_\delta)} = 0. \qquad (21.61)$$

From inequality (21.11) with $\tilde{u} = u_\delta$ and $\tilde{v} = V_\delta$ and inequality (21.51) we have

$$\|u_\delta(t) - V_\delta(t)\| \leq h(0)\frac{e^{-\phi(t)}}{a(t)}$$
$$+ \frac{e^{-\phi(t)}}{a(t)}\int_0^t e^{\phi(s)}|\dot{a}(s)|\Big(\|w(0)\| + K\|V_\delta(s)\|\Big)ds.$$

This, relations (21.16) and (21.18) with $r = 1$ (i.e., $\psi = \phi$), (21.61), and (21.44) imply

$$\lim_{\delta \to 0} \|u_\delta(t_\delta) - V_\delta(t_\delta)\| = 0. \qquad (21.62)$$

It follows from (21.59) that

$$\lim_{\delta \to 0} \frac{\delta}{a(t_\delta)} = 0. \qquad (21.63)$$

From the triangle inequality and inequality (21.8), one obtains

$$\|u_\delta(t_\delta) - y\| \leq \|u_\delta(t_\delta) - V_\delta(t_\delta)\| + \|V(t_\delta) - V_\delta(t_\delta)\| + \|V(t_\delta) - y\|$$
$$\leq \|u_\delta(t_\delta) - V_\delta(t_\delta)\| + \frac{\delta}{a(t_\delta)} + \|V(t_\delta) - y\|. \qquad (21.64)$$

From (21.62)–(21.64) and Lemma 6.1.11, one obtains (21.45). Theorem 21.2.2 is proved. ∎

Assume that $a(t)$ satisfies the following conditions:

$$0 < a(t) \searrow 0, \qquad \frac{1}{6} > q > \frac{|\dot{a}(t)|}{a^3(t)} \searrow 0. \qquad (21.65)$$

The following theorem gives a sufficient condition for relation (21.44) to hold.

Theorem 21.2.4 *Let $a(t)$ satisfy (21.65). Let F, f_δ and $u_\delta(t)$ be as in Theorem 21.2.2. Assume that u_0 satisfies either*

$$\|F(u_0) + a_0 u_0 - f_\delta\| \leq pa(0)\|V_\delta(0)\|, \qquad 0 < p < 1 - \frac{q}{1 - 2q}, \qquad (21.66)$$

or

$$\|F(u_0) + a_0u_0 - f_\delta\| \leq \theta\delta^\varsigma, \qquad 0 \leq \theta < C, \tag{21.67}$$

where $C > 0$ and $\varsigma \in (0,1)$ are the constants from Theorem 21.2.2. Let t_δ be defined by (21.43). Then

$$\lim_{\delta \to 0} t_\delta = \infty. \tag{21.68}$$

Remark 21.2.5 Inequalities (21.66) and (21.67) hold if u_0 is sufficiently close to $V_\delta(0)$. Note that inequality (21.66) is a sufficient condition for (21.78), that is,

$$e^{-\varphi(t)}\|F(u_0) + a(0)u_0 - f_\delta\| \leq pa(t)\|V_\delta(t)\|, \qquad t \geq 0, \tag{21.69}$$

to hold. In our proof, inequality (21.69) is used at $t = t_\delta$. The stopping time t_δ is often sufficiently large for the quantity $e^{\varphi(t_\delta)}a(t_\delta)$ to be large. This follows from the relation that $\lim_{t\to\infty} e^{\varphi(t)}a(t) = \infty$ (see also (21.16)). In this case, inequality (21.69) with $t = t_\delta$ is satisfied for a wide range of u_0.

Proof: Let us prove (21.68), assuming that (21.66) holds. When (21.67) holds instead of (21.66), relation (21.68) is proved similarly.

From the first inequality in (21.50) and the triangle inequality, one gets

$$\dot{h} \leq -ha^2 + |\dot{a}|\|u_\delta - V_\delta\| + |\dot{a}|\|V_\delta\|. \tag{21.70}$$

This and inequality (21.11) with $\tilde{u} = u_\delta$ and $\tilde{v} = V_\delta$ imply

$$\dot{h} \leq -h\left(a^2 - \frac{|\dot{a}|}{a}\right) + |\dot{a}|\|V_\delta\|. \tag{21.71}$$

Since $a^2 - \frac{|\dot{a}|}{a} \geq (1-q)a^2$ by (21.65), it follows from inequality (21.71) that

$$\dot{h} \leq -(1-q)a^2h + |\dot{a}|\|V_\delta\|. \tag{21.72}$$

Inequality (21.72) and Gronwall's inequality imply

$$h(t) \leq e^{-\varphi(t)}h(0) + e^{-\varphi(t)}\int_0^t e^{\varphi(s)}|\dot{a}(s)|\|V_\delta(s)\|\, ds, \tag{21.73}$$

where

$$\varphi(t) := (1-q)\int_0^t a^2(s)\, ds, \qquad t \geq 0.$$

From the triangle inequality, inequality (21.73), and equation (19.3), one gets

$$\|F(u_\delta(t)) - f_\delta\| \geq \|F(V_\delta(t)) - f_\delta\| - \|F(V_\delta(t)) - F(u_\delta(t))\|$$
$$\geq a(t)\|V_\delta(t)\| - h(0)e^{-\varphi(t)}$$
$$- e^{-\varphi(t)}\int_0^t e^{\varphi(s)}|\dot{a}|\|V_\delta\|\, ds. \tag{21.74}$$

It follows from our assumptions and Lemma 21.1.4 that $a(t)$ satisfies inequality (21.26), that is,

$$\frac{q}{1-2q}a(t)\|V_\delta(t)\| \geq e^{-\varphi(t)}\int_0^t e^{\varphi(s)}|\dot{a}|\|V_\delta(s)\|\,ds. \qquad (21.75)$$

From the relation $h(0) = \|F(u_0)+a(0)u_0 - f_\delta\|$ and inequality (21.66), one obtains

$$h(0)e^{-\varphi(t)} \leq pa(0)\|V_\delta(0)\|e^{-\varphi(t)}, \qquad \forall t \geq 0. \qquad (21.76)$$

It follows from (21.65) that

$$e^{-\varphi(t)}a(0) \leq a(t). \qquad (21.77)$$

Indeed, inequality $a(0) \leq a(t)e^{\varphi(t)}$ is obviously true for $t = 0$, and

$$\left(a(t)e^{\varphi(t)}\right)'_t = a^3(t)e^{\varphi(t)}\left(1 - q - \frac{|\dot{a}(t)|}{a^3(t)}\right) \geq a^3(t)e^{\varphi(t)}(1 - 2q) > 0,$$

by (21.65). Here, we have used the relation $\dot{a} = -|\dot{a}|$.

Inequalities (21.76) and (21.77) imply

$$e^{-\varphi(t)}h(0) \leq pa(t)\|V_\delta(0)\| \leq pa(t)\|V_\delta(t)\|, \quad t \geq 0, \qquad (21.78)$$

where we have used the inequality $\|V_\delta(t)\| \leq \|V_\delta(t')\|$ for $t \leq t'$, established in Corollary 21.1.1. From equality $\|F(u_\delta(t_\delta)) - f_\delta\| = C\delta^\zeta$, (21.74), (21.75), and (21.78), one gets

$$C\delta^\zeta = \|F(u_\delta(t_\delta)) - f_\delta\| \geq \left(1 - \frac{q}{1-2q} - p\right)a(t_\delta)\|V_\delta(t_\delta)\|. \qquad (21.79)$$

Recall that $V = V_\delta|_{\delta=0}$. From the triangle inequality and the first inequality in (21.8), one obtains

$$a(t)\|V(t)\| \leq a(t)\|V(t) - V_\delta(t)\| + a(t)\|V_\delta(t)\| \leq \delta + a(t)\|V_\delta(t)\|,$$

for all $t \geq 0$. This and (21.79) imply

$$\lim_{\delta \to 0} a(t_\delta)\|V(t_\delta)\| \leq \lim_{\delta \to 0}\left[(1 - \frac{q}{1-2q} - p)^{-1}C\delta^\zeta + \delta\right] = 0. \qquad (21.80)$$

Since $\|V(t)\|$ is increasing (see Corollary 21.1.1), this implies $\lim_{\delta \to 0} a(t_\delta) = 0$. Since $0 < a(t) \searrow 0$, it follows from (21.80) that (21.68) holds.

Theorem 21.2.4 is proved. ∎

21.3 AN ITERATIVE SCHEME

Let $0 < a(t) \in C^1(\mathbb{R}_+)$ satisfy the following conditions:

$$0 < a(t) \searrow 0, \qquad 0 < \nu(t) := \frac{|\dot{a}(t)|}{a^3(t)} \searrow 0, \qquad t \geq 0. \qquad (21.81)$$

Remark 21.3.1 If $a(t) = \frac{d}{(c+t)^b}$, where $b \in (0, \frac{1}{2})$ and c and d are positive constants, then $a(t)$ satisfies (21.81).

Denote $a_n := a(nh)$, $h = const > 0$, $h \in (0, 1)$. Consider the following iterative scheme:

$$u_{n+1} = u_n - \gamma_n A_n^*[F(u_n) + a_n u_n - f_\delta], \quad u_0 = \tilde{u}_0, \tag{21.82}$$

where $\tilde{u}_0 \in H$, $A_n := F'(u_n) + a_n I$ and

$$0 < h \leq \gamma_n \leq \max\left(1, \frac{2}{a_n^2 + (M_R + a_n)^2}\right), \quad n \geq 0, \tag{21.83}$$

where M_R is the constant from (21.2).
    Let

$$V_n := V_\delta(nh), \quad n = 1, 2, ...,$$

where $V_\delta(t)$ solves (19.3) with $a = a(t)$. Then V_n solves the following regularized equation

$$F(V_n) + a_n V_n - f_\delta = 0, \quad n \geq 0.$$

By similar arguments as in Lemma 21.1.3, one gets the following result:

Lemma 21.3.2 Let $a(t)$ satisfy (21.32). Let $\psi_n := r \sum_{i=0}^{n-1} a_{i+1}^2 \gamma_i$, where γ_i are positive constants, $\gamma_i \geq h > 0, \forall i$, and $r = const > 0$. Then one gets

$$\lim_{n \to \infty} \psi_n = \infty, \tag{21.84}$$

$$\lim_{n \to \infty} e^{\psi_n} a_n = \infty, \tag{21.85}$$

$$\lim_{n \to \infty} \frac{\sum_{i=0}^n e^{\psi_{i+1}} \frac{a_i - a_{i+1}}{a_i}}{e^{\psi_{n+1}}} = 0, \tag{21.86}$$

$$\lim_{n \to \infty} \frac{\sum_{i=0}^n e^{\psi_{i+1}}(a_i - a_{i+1})}{e^{\psi_{n+1}} a_{n+1}} = 0. \tag{21.87}$$

Lemma 21.3.3 Let $0 < h = const$, $a(t)$ satisfy (21.32), $a_n := a(nh)$, and

$$\phi_n = h \sum_{i=0}^n a_i^2, \quad \phi(t) := \int_0^t a^2(s)\, ds. \tag{21.88}$$

Then $\forall n \geq 1$ the following inequalities hold:

$$e^{-\phi_{n-1}} \sum_{i=0}^{n-1} e^{\phi_i}(a_i - a_{i+1}) \leq e^{a^2(0)h} e^{-\phi(nh)} \int_0^{nh} e^{\phi(s)} |\dot{a}(s)|\, ds, \tag{21.89}$$

$$e^{-\phi_{n-1}} \sum_{i=0}^{n-1} e^{\phi_i} \frac{a_i - a_{i+1}}{a_i} \leq e^{a^2(0)h} e^{-\phi(nh)} \int_0^{nh} e^{\phi(s)} \frac{|\dot{a}(s)|}{a(s)}\, ds. \tag{21.90}$$

Proof: Let us prove (21.90). Inequality (21.89) *is proved similarly.*

Since $a_n = a(nh)$ and $0 < a(t) \searrow 0$, one gets

$$\phi_n - \phi_i = \sum_{k=i+1}^{n} a_k^2 h \geq \sum_{k=i+1}^{n} \int_{kh}^{(k+1)h} a^2(s)\, ds$$

$$= \int_{(i+1)h}^{(n+1)h} a^2(s)\, ds = \phi((n+1)h) - \phi((i+1)h), \qquad (21.91)$$

for $0 \leq i \leq n$. This and the inequalities

$$\phi((i+1)h) - \phi(s) = \int_{s}^{(i+1)h} a^2(s)\, ds \leq \int_{s}^{(i+1)h} a^2(0)\, ds \leq a^2(0)h,$$

for all $s \in [ih, (i+1)h]$, imply

$$-\phi_{n-1} + \phi_i \leq -\phi(nh) + \phi(s) + a^2(0)h, \qquad \forall s \in [ih, (i+1)h], \qquad (21.92)$$

where $0 \leq i \leq n - 1$.

Since $0 < a_n \searrow 0$ and $|\dot{a}(t)| = -a(t)$, one obtains

$$\frac{a_i - a_{i+1}}{a_i} = \int_{ih}^{(i+1)h} \frac{|\dot{a}(s)|}{a_i}\, ds \leq \int_{ih}^{(i+1)h} \frac{|\dot{a}(s)|}{a(s)}\, ds. \qquad (21.93)$$

It follows from (21.93) and (21.92) that

$$e^{-\phi_{n-1}} \sum_{i=0}^{n-1} e^{\phi_i} \frac{a_i - a_{i+1}}{a_i} \leq \sum_{i=0}^{n-1} \int_{ih}^{(i+1)h} e^{-\phi_{n-1}+\phi_i} \frac{|\dot{a}(s)|}{a(s)}\, ds$$

$$\leq e^{a^2(0)h} e^{-\phi(nh)} \int_{0}^{nh} e^{\phi(s)} \frac{|\dot{a}(s)|}{a(s)}\, ds, \qquad (21.94)$$

for all $n \geq 1$.

Lemma 21.3.3 is proved. ∎

Assume that

$$0 < a(t) \searrow 0, \qquad \frac{1}{12} \geq \frac{|\dot{a}(t)|}{a^3(t)} \searrow 0, \qquad (21.95)$$

and

$$a_0^2 h \leq 2. \qquad (21.96)$$

Remark 21.3.4 It follows from (21.81) and (21.96) that

$$0 < \frac{1}{a_{n+1}^2} - \frac{1}{a_n^2} = -2 \int_{nh}^{nh+h} \frac{\dot{a}(s)}{a^3(s)}\, ds \leq 2h\nu(nh) \leq \frac{h}{6}. \qquad (21.97)$$

Inequalities (21.96) and (21.97) imply

$$1 < \frac{a_n^2}{a_{n+1}^2} \leq 1 + \frac{a_n^2 h}{6} \leq \frac{4}{3}. \qquad (21.98)$$

From (21.83), (21.97), and (21.98), one obtains

$$\gamma_n \geq h \geq 6\frac{a_n^2 - a_{n+1}^2}{a_n^2 a_{n+1}^2} > \frac{6(a_n - a_{n+1})}{a_n^2 a_{n+1}}, \qquad \forall n \geq 0. \qquad (21.99)$$

This implies

$$a_{n+1}a_n\gamma_n - \frac{a_n - a_{n+1}}{a_n} \geq \frac{5a_{n+1}a_n\gamma_n}{6} > \frac{a_{n+1}^2\gamma_n}{2}, \qquad \forall n \geq 0. \qquad (21.100)$$

Remark 21.3.5 One has

$$a_0 e^{-\varphi_n} \leq \frac{3}{2}a_{n+1}, \qquad n = 0, 1, ... \qquad (21.101)$$

Indeed, inequality $a_1 \leq a_{n+1}e^{\varphi_n}$ is obviously true for $n = 0$, and $a_{n+1}e^{\varphi_n}$ is an increasing sequence because

$$a_{n+1}e^{\varphi_n} - a_n e^{\varphi_{n-1}} = e^{\varphi_{n-1}}\left(a_{n+1}e^{\frac{\gamma_{n-1}a_n^2}{2}} - a_n\right)$$

$$\geq e^{\varphi_{n-1}}\left(a_{n+1} + \frac{\gamma_{n-1}a_n^2}{2}a_{n+1} - a_n\right)$$

$$= e^{\varphi_{n-1}}a_n^2 a_{n+1}\left(\frac{\gamma_{n-1}}{2} - \frac{a_n - a_{n+1}}{a_n^2 a_{n+1}}\right) \geq 0, \qquad (21.102)$$

by (21.99). From the inequality $a_1 \leq a_{n+1}e^{\varphi_n}$ and (21.98), one gets

$$e^{-\varphi_n}a_0 \leq a_{n+1}\frac{a_0}{a_1} < \frac{2}{\sqrt{3}}a_{n+1} < \frac{3}{2}a_{n+1}, \qquad n \geq 0. \qquad (21.103)$$

Lemma 21.3.6 *Let* $a(t)$ *satisfy* (21.95). *Let* $a_n := a(nh)$ *and*

$$\varphi_n := \sum_{i=0}^{n-1} \frac{a_{i+1}^2\gamma_i}{2}, \qquad n \geq 1, \quad \gamma_n \geq h > 0.$$

Then the following inequality holds:

$$e^{-\varphi_n}\sum_{i=0}^{n-1}e^{\varphi_{i+1}}(a_i - a_{i+1})\|V_i\| \leq \frac{1}{2}a_n\|V_n\|, \qquad n \geq 1. \qquad (21.104)$$

Proof: Let us prove that

$$e^{\varphi_n}(a_{n-1} - a_n) \leq \frac{1}{2}(a_n e^{\varphi_n} - a_{n-1}e^{\varphi_{n-1}}), \qquad \forall n \geq 1. \qquad (21.105)$$

Inequality (21.105) is equivalent to

$$\frac{a_{n-1}}{3a_n} \leq \frac{e^{\varphi_n}}{2e^{\varphi_n} + e^{\varphi_{n-1}}}, \qquad \forall n \geq 1. \qquad (21.106)$$

From our assumptions we get

$$\frac{e^{\varphi_n}}{2e^{\varphi_n} + e^{\varphi_{n-1}}} = \frac{1}{2 + e^{-\frac{\gamma_{n-1} a_n^2}{2}}} \geq \frac{1}{2 + e^{-h\frac{a_n^2}{2}}}, \qquad \forall n \geq 1. \qquad (21.107)$$

From (21.107) and (21.106) it suffices to prove that

$$2 + e^{-h\frac{a_n^2}{2}} \leq \frac{3a_n}{a_{n-1}}, \qquad n \geq 1.$$

This inequality is equivalent to

$$\frac{a_{n-1} - a_n}{a_{n-1} a_n^2} \leq \frac{h(1 - e^{-\frac{h a_n^2}{2}})}{6h\frac{a_n^2}{2}}, \qquad n \geq 1. \qquad (21.108)$$

Let us prove (21.108). From (21.97) one gets

$$\frac{a_{n-1} - a_n}{a_{n-1} a_n^2} = \frac{(a_{n-1}^2 - a_n^2)}{a_{n-1}^2 a_n^2} \frac{a_{n-1}}{a_{n-1} + a_n} \leq 2h\nu(0) \frac{1}{1 + 0.866}. \qquad (21.109)$$

Here, we have used the inequality $\frac{a_n}{a_{n-1}} \geq \sqrt{\frac{3}{4}} > 0.866$ following from (21.98). Note that the function $f(x) := \frac{1 - e^{-x}}{x}$ is decreasing on $(0, \infty)$. Therefore, one gets

$$\frac{h(1 - e^{-\frac{h a_n^2}{2}})}{6h\frac{a_n^2}{2}} = \frac{hf(\frac{h a_n^2}{2})}{6} \geq \frac{hf(\frac{h a_0^2}{2})}{6} \geq \frac{hf(1)}{6} \geq \frac{h\frac{6}{10}}{6} = \frac{h}{10}. \qquad (21.110)$$

We have used inequalities (21.96) and $f(1) > \frac{6}{10}$ in (21.110). From (21.81) one obtains

$$\frac{h}{10} \geq h\frac{2}{12}\frac{1}{1 + 0.866} \geq 2h\nu(0)\frac{1}{1 + 0.866}. \qquad (21.111)$$

Inequality (21.108) follows from (21.109)–(21.111). Inequality (21.105) is proved.

From (21.105) one gets

$$2\sum_{i=0}^{n-1} e^{\varphi_{i+1}}(a_i - a_{i+1}) \leq \sum_{i=0}^{n-1} \left(e^{\varphi_{i+1}} a_{i+1} - e^{\varphi_i} a_i\right) < e^{\varphi_n} a_n, \qquad n \geq 1. \qquad (21.112)$$

Multiplying (21.112) by $\frac{1}{2}\|V_n\|e^{-\varphi_n}$ and recalling the fact that $\|V_i\|$ is increasing (see Corollary 21.1.1), one gets inequality (21.104). Lemma 21.3.6 is proved. ∎

Theorem 21.3.7 *Let $C > 0$ and $\zeta \in (0, 1]$ be constants such that $C\delta^\zeta > \delta$. Assume that $F : H \to H$ is a monotone and Fréchet differentiable operator,*

F' is self-adjoint, and (21.2) holds. Let u_0 be an element of H satisfying inequality

$$\|F(u_0) - f_\delta\| > C\delta^\zeta. \tag{21.113}$$

Assume that equation $F(u) = f$ has a solution, possibly nonunique, and y is the minimal-norm solution to this equation. Let f be unknown but f_δ be given, $\|f_\delta - f\| \leq \delta$. Let u_n be defined by (21.82). Then there exists a unique n_δ such that

$$\|F(u_{n_\delta}) - f_\delta\| \leq C\delta^\zeta, \quad \|F(u_n) - f_\delta\| > C\delta^\zeta, \qquad 0 \leq n < n_\delta. \tag{21.114}$$

If $\zeta \in (0,1)$, n_δ satisfies (21.114), and

$$\lim_{\delta \to 0} n_\delta = \infty, \tag{21.115}$$

then

$$\lim_{\delta \to 0} \|u_{n_\delta} - y\| = 0. \tag{21.116}$$

Proof: The uniqueness of n_δ, satisfying (21.114), follows from its definition (21.114).

Let us prove the existence of n_δ. First, let us prove first that there exists $R > 0$ such that the sequence $(u_n)_{n=1}^{n_\delta}$ remains inside the ball $B(0, R)$.

We assume without loss of generality that $\delta \in (0, 1)$. It follows from Corollary 21.1.1 and the triangle inequality that

$$a_n\|V_n\| \leq a_0\|V_0\| \leq \|F(u_0) - f_\delta\| \leq \|F(u_0) - f\| + \|f_\delta - f\| \leq \Gamma, \tag{21.117}$$

$\forall n \geq 0, \forall \delta \in (0, 1)$, where

$$\Gamma := \|F(0) - f\| + 1. \tag{21.118}$$

From (21.117) one obtains

$$\|V_n\| \leq \frac{\Gamma}{a_n}, \qquad \forall n \geq 0. \tag{21.119}$$

Let

$$\phi(t) := \int_0^t a^2(s)\, ds, \tag{21.120}$$

$$K = 1 + \sup_{t \geq 0} e^{a^2(0)} e^{-\phi(t)} \int_0^t e^{\phi(s)} \frac{|\dot{a}(s)|}{a(s)}\, ds. \tag{21.121}$$

It follows from (21.17) that K is bounded.

Recall that $V_\delta(t) = V_{\delta,a(t)}$ solves equation (19.3) with $a = a(t)$. It follows from Corollary 21.1.1 that

$$\|F(V_\delta(t)) - f_\delta\| \leq \|F(V_\delta(nh)) - f_\delta\| = \|F(V_n) - f_\delta\|, \quad \forall t \geq nh. \tag{21.122}$$

Relations (19.12), (21.15), and (21.17) imply that there exists $T > 0$ such that the following inequality holds $\forall t \geq T$:

$$\|F(V_\delta(t)) - f_\delta\| + e^{-\phi(t)} h_0$$
$$+ e^{a(0)} e^{-\phi(t)} \int_0^t e^{\phi(s)} |\dot{a}(s)| \left(\frac{\Gamma K}{a(s)} + w_0 \right) ds < C\delta^\varsigma, \qquad (21.123)$$

where
$$h_0 = \|F(u_0) + a_0 u_0 - f_\delta\|, \qquad w_0 = \|u_0 - V_0\|. \qquad (21.124)$$

Define
$$R := \|V_\delta(T)\| K + w_0, \qquad (21.125)$$

and let $h > 0$ be a constant such that

$$0 < h \leq \min\left(1, \frac{2}{a_0^2 + (M_R + a_0)^2}\right).$$

Let N be the largest integer such that $Nh \leq T$. Let us prove by induction that the sequence $(u_n)_{n=1}^N$ stays in side the ball $B(0, R)$. To prove this, it suffices to prove that

$$\|u_n - V_n\| \leq w_0 + \|V_n\|(K - 1), \qquad n = 0, 1, ..., N. \qquad (21.126)$$

Inequality (21.126) holds for $n = 0$, by (21.124). Assume that (21.126) holds for $n = 0, 1, ..., m$, where $0 \leq m < N$. Let us prove that (21.126) also holds for $m + 1$.

One has

$$F(u_n) - F(V_n) = J_n(u_n - V_n), \qquad J_n := \int_0^1 F'(V_n + \xi(u_n - V_n)) \, d\xi. \quad (21.127)$$

Since

$$\|V_n + \xi(u_n - V_n)\| \leq (1 - \xi)\|V_n\| + \xi\|u_n\| \leq R, \qquad 0 \leq n \leq m,$$

one gets

$$\|J_n\| \leq M_R, \qquad 0 \leq n \leq m.$$

It follows from (21.127), the self-adjointness of F' and the monotonicity of F that J_n is self-adjoint and $J_n \geq 0$. From equation (21.31) we get

$$u_{n+1} - V_n = u_n - V_n - \gamma_n A_n^* \left[F(u_n) + a_n u_n - F(V_n) - a_n V_n \right]$$
$$= [I - \gamma_n A_n^*(J_n + a_n I)](u_n - V_n). \qquad (21.128)$$

From the Spectral Theorem and the relations

$$a_n I \leq A_n = A_n^* \leq (M_R + a_n)I, \qquad 0 \leq J_n = J_n^* \leq M_R, \qquad 0 \leq n \leq m,$$

one gets

$$\|I - \gamma_n A_n^*(J_n + a_n I)\| \leq 1 - \gamma a_n^2, \qquad 0 \leq n \leq m. \qquad (21.129)$$

From (21.128) and (21.129) one obtains

$$\|u_{n+1} - V_n\| \leq \|u_n - V_n\|(1 - \gamma_n a_n^2), \qquad 0 \leq n \leq m. \qquad (21.130)$$

From (21.130), the triangle inequality, and inequality (21.9) with $t_1 = nh$ and $t_2 = (n+1)h$, one gets

$$\|u_{n+1} - V_{n+1}\| \leq \|u_n - V_n\|(1 - \gamma_n a_n^2) + \|V_{n+1} - V_n\|$$
$$\leq \|u_n - V_n\|e^{-ha_n^2} + \frac{a_n - a_{n+1}}{a_n}\|V_{n+1}\|. \qquad (21.131)$$

Here we have used the inequality $1 - \gamma_n a_n^2 \leq e^{-ha_n^2}$, where $0 < h \leq \gamma_n$ and $n \geq 0$. From (21.131) and a discrete analog of Gronwall' inequality, one gets

$$\|u_{n+1} - V_{n+1}\| \leq w_0 e^{-\phi_n} + e^{-\phi_n} \sum_{i=1}^{n} e^{\phi_i} \frac{a_i - a_{i+1}}{a_i} \|V_{i+1}\|, \qquad (21.132)$$

where $0 \leq n \leq m$ and

$$\phi_n = \sum_{i=0}^{n} ha_i^2, \qquad n \geq 0.$$

From (21.132), (21.90), and (21.121), one obtains

$$\|u_{m+1} - V_{m+1}\| \leq w_0 + \|V_{m+1}\|e^{-\phi_m} \sum_{i=1}^{m} e^{\phi_i} \frac{a_i - a_{i+1}}{a_i}$$
$$\leq w_0 + \|V_{m+1}\|e^{a^2(0)}e^{-\phi((m+1)h)} \int_0^{(m+1)h} e^{\phi(s)} \frac{|\dot{a}(s)|}{a(s)} ds$$
$$\leq w_0 + \|V_{m+1}\|(K - 1), \qquad (21.133)$$

where $\phi(t)$ is defined by (21.120) (see also (21.88)). It follows from (21.133) that (21.126) holds for $m + 1$. Thus, by induction, (21.126) holds for $n = 0, 1, ..., N$.

Inequalities (21.126) and (21.125), and the inequality $\|V_n\| \leq \|V_\delta(T)\|$, $\forall n \leq N$ (see Corollary 21.1.1), imply that the sequence $(u_n)_{n=1}^{N}$ remains inside the ball $B(0, R)$.

Let us prove that there exists $n_\delta \leq N$ satisfying (21.114). Denote

$$g_n := g_{n,\delta} := F(u_n) + a_n u_n - f_\delta.$$

Equation (21.82) can be rewritten as

$$u_{n+1} - u_n = -\gamma_n A_n^* g_n, \qquad n \geq 0. \qquad (21.134)$$

It follows from (21.134) that

$$g_{n+1} = g_n + a_{n+1}(u_{n+1} - u_n) + F(u_{n+1}) - F(u_n) + (a_{n+1} - a_n)u_n$$
$$= (I - a_{n+1}\gamma_n A_n^*)g_n + F(u_{n+1}) - F(u_n) + (a_{n+1} - a_n)u_n. \quad (21.135)$$

One has

$$F(u_{n+1}) - F(u_n) = \tilde{J}_n(u_{n+1} - u_n), \quad (21.136)$$

where

$$\tilde{J}_n := \int_0^1 F'(u_n + \xi(u_{n+1} - u_n))\, d\xi.$$

By our assumptions, one gets

$$0 \le \tilde{J}_n = \tilde{J}_n^* \le M_R I, \qquad 0 \le n \le N.$$

Denote $h_n = \|g_n\|$. It follows from (21.135), (21.136), and the triangle inequality that

$$h_{n+1} \le \|I - a_{n+1}\gamma_n A_n - \gamma_n \tilde{J}_n A_n\| h_n + (a_n - a_{n+1})\|u_n\|. \quad (21.137)$$

Since $0 \le \tilde{J}_n = \tilde{J}_n^* \le M_R I$, $a_n I \le A_n = A_n^* \le (M_R + a_n)I$, and γ_n satisfies (21.83), one gets by the Spectral Theorem the following inequality:

$$\|I - a_{n+1}\gamma_n A_n^* - \gamma_n \tilde{J}_n A_n^*\| \le 1 - \gamma_n a_n a_{n+1}, \qquad \forall n \ge 0. \quad (21.138)$$

From (21.138), (21.137), and (21.126), one gets

$$h_{n+1} \le (1 - \gamma_n a_{n+1} a_n)h_n + (a_n - a_{n+1})\|u_n\|$$
$$\le (1 - \gamma_n a_{n+1}^2)h_n + (a_n - a_{n+1})(w_0 + K\|V_n\|), \quad (21.139)$$

where $0 \le n \le N$. From (21.139) and a discrete analog of Gronwall's inequality, one obtains

$$h_n \le h_0 e^{-\varphi_n} + e^{-\phi_n} \sum_{i=0}^{n-1} e^{\phi_{i+1}}(a_i - a_{i+1})(w_0 + K\|V_i\|), \quad (21.140)$$

where $0 \le n \le N$ and

$$\phi_n = \sum_{i=0}^{n-1} a_{i+1}^2 \gamma_i, \qquad n \ge 1. \quad (21.141)$$

From the triangle inequality, inequality (21.11) with $\tilde{u} = u_n$ and $\tilde{v} = V_n$, and inequality (21.140), one obtains

$$\|F(u_n) - f_\delta\| \le \|F(V_n) - f_\delta\| + \|F(V_n) - F(u_n)\|$$
$$\le \|F(V_n) - f_\delta\| + h_n$$
$$\le \|F(V_n) - f_\delta\| + h_0 e^{-\phi_n}$$
$$+ e^{-\phi_n} \sum_{i=0}^{n-1} e^{\phi_{i+1}}(a_i - a_{i+1})(w_0 + \|V_i\|), \quad (21.142)$$

for $0 \leq n \leq N$. From (21.142) and (21.119) one obtains

$$\|F(u_n) - f_\delta\| \leq \|F(V_n) - f_\delta\| + h_0 e^{-\phi_n}$$
$$+ e^{-\phi_n} \sum_{i=0}^{n-1} e^{\phi_{i+1}}(a_i - a_{i+1})\left(w_0 + \frac{\Gamma K}{a_i}\right), \qquad (21.143)$$

for $0 \leq n \leq N$. From (21.89), (21.90), (21.123), and (21.143) one gets

$$\|F(u_N) - f_\delta\| \leq \|F(V(Nh)) - f_\delta\| + h_0 e^{-\phi(Nh)}$$
$$+ e^{a^2(0)} e^{-\phi(Nh)} \int_0^{Nh} e^{\phi(s)} |\dot{a}(s)|\left(w_0 + \frac{\Gamma K}{a(s)}\right) ds$$
$$< C\delta^\zeta. \qquad (21.144)$$

This implies the existence of n_δ and one gets $n_\delta \leq N$.

Let us prove (21.116) *assuming that* (21.115) *holds.* From (21.85)–(21.87) with $r = 1$ (i.e. $\psi = \phi$), the inequality $\|V_0\| \leq \|V_n\|, \forall n \geq 0$ (see Corollary 21.1.1), and the relation $\lim_{\delta \to 0} n_\delta = \infty$, one gets, for all sufficiently small $\delta > 0$, the following inequalities:

$$h_0 e^{-\phi_{n_\delta - 1}} \leq a_{n_\delta - 1}\|V_0\| \leq a_{n_\delta - 1}\|V_{n_\delta - 1}\|, \qquad (21.145)$$

$$w_0 e^{-\phi_n} \sum_{i=0}^{n-1} e^{\phi_{i+1}}(a_i - a_{i+1}) \leq a_{n_\delta - 1}\|V_0\| \leq a_{n_\delta - 1}\|V_{n_\delta - 1}\|, \qquad (21.146)$$

$$e^{-\phi_n} \sum_{i=0}^{n-1} K\Gamma e^{\phi_{i+1}} \frac{a_i - a_{i+1}}{a_i}) \leq a_{n_\delta - 1}\|V_0\| \leq a_{n_\delta - 1}\|V_{n_\delta - 1}\|. \qquad (21.147)$$

From (21.114), (21.145)–(21.147), (21.143) with $n = n_\delta - 1$, and the second inequality in (21.8), one obtains

$$C\delta^\zeta \leq 4a_{n_\delta - 1}\|V_{n_\delta - 1}\| \leq 4\left(a_{n_\delta - 1}\|y\| + \delta\right). \qquad (21.148)$$

This and the relation $\lim_{\delta \to 0} \frac{\delta^\zeta}{\delta} = \infty$ for a fixed $\zeta \in (0, 1)$ imply

$$\lim_{\delta \to 0} \frac{\delta^\zeta}{a_{n_\delta}} \leq \frac{4\|y\|}{C}. \qquad (21.149)$$

Inequalities (21.149) and (21.8) imply, for sufficiently small $\delta > 0$, the following inequality:

$$\|V_n\| \leq \|y\| + \frac{\delta}{a_{n_\delta}} \leq \|y\| + \frac{C\delta^\zeta}{a_{n_\delta}} \leq 5\|y\|, \qquad 0 \leq n \leq n_\delta. \qquad (21.150)$$

Using estimate (21.150) and relation (21.87), one obtains

$$0 \le \lim_{\delta \to 0} \frac{\sum_0^{n_\delta - 1} e^{\phi_{i+1}}(a_i - a_{i+1})\|V_i\|}{e^{\phi_{n_\delta}} a_{n_\delta}}$$

$$\le 5\|y\| \lim_{\delta \to 0} \frac{\sum_0^{n_\delta - 1} e^{\phi_{i+1}}(a_i - a_{i+1})}{e^{\phi_{n_\delta}} a_{n_\delta}} = 0. \tag{21.151}$$

From (21.10) with $\tilde{v} = V_{n_\delta}$ and $\tilde{u} = u_{n_\delta}$, (21.140), (21.85), and (21.151), one obtains

$$0 \le \lim_{\delta \to 0} \|u_{n_\delta} - V_{n_\delta}\| \le \lim_{\delta \to 0} \frac{h_{n_\delta}}{a_{n_\delta}} = 0. \tag{21.152}$$

It follows from (21.149) that

$$\lim_{\delta \to 0} \frac{\delta}{a_{n_\delta}} = 0. \tag{21.153}$$

Now let us finish the proof of Theorem 21.3.7.
From the triangle inequality and the first inequality in (21.8), one obtains

$$\|u_{n_\delta} - y\| \le \|u_{n_\delta} - V_{n_\delta}\| + \|V_{n_\delta} - V_{0,n_\delta}\| + \|V_{0,n_\delta} - y\|$$

$$\le \|u_{n_\delta} - V_{n_\delta}\| + \frac{\delta}{a_{n_\delta}} + \|V_{0,n_\delta} - y\|. \tag{21.154}$$

From (21.152)–(21.154), (21.162), and Lemma 6.1.11, one obtains (21.116). Theorem 21.3.7 is proved. ∎

Theorem 21.3.8 *Let F, f, f_δ, and u_δ be as in Theorem 21.3.7 and let y be the minimal-norm solution to the equation $F(u) = f$. Let $0 < (\delta_m)_{m=1}^\infty$ be a sequence such that $\delta_m \to 0$ as $m \to \infty$. If the sequence $(n_{\delta_m})_{m=1}^\infty$ is bounded and $(n_{m_j})_{j=1}^\infty$ is a convergent subsequence, then*

$$\lim_{j \to \infty} u_{n_{m_j}} = u^\star, \tag{21.155}$$

where $u^\star$ is a solution to the equation $F(u) = f$.

Proof: If $n > 0$ is fixed, then u_n is a continuous function of f_δ. Denote

$$u^\star := u_N^\star := \lim_{j \to \infty} u_{n_{\delta_{m_j}}}, \tag{21.156}$$

where

$$\lim_{j \to \infty} n_{m_j} = N. \tag{21.157}$$

From the continuity of F, one gets

$$\|F(u^\star) - f\| = \lim_{j \to \infty} \|F(u_{n_{\delta_{m_j}}}) - f_{\delta_{m_j}}\| \le \lim_{j \to \infty} C\delta_{m_j}^\zeta = 0. \tag{21.158}$$

Thus, u^* is a solution to the equation $F(u) = f$.

Theorem 21.3.8 is proved.   ∎

Assume that

$$0 < a(t) \searrow 0, \qquad \frac{1}{12} \geq \frac{|\dot{a}(t)|}{a^3(t)} \searrow 0. \qquad (21.159)$$

Remark 21.3.9 If $a(t) = \frac{d}{(c+t)^b}$, where $b \in (0, \frac{1}{2})$, $c \geq 1$, and $c^{1-2b}d^2 > 12b$, then $a(t)$ satisfies condition (21.159).

The following Theorem gives a sufficient condition for (21.115) to hold.

Theorem 21.3.10 *Let $a(t)$ satisfy conditions (21.159). Let F and f_δ be as in Theorem 21.3.7. Assume that u_0 satisfies either*

$$\|F(u_0) + a_0 u_0 - f_\delta\| \leq \frac{1}{6} a_0 \|V_0\|, \qquad (21.160)$$

or

$$\|F(u_0) + a_0 u_0 - f_\delta\| \leq \theta C^\zeta, \qquad 0 \leq \theta < C. \qquad (21.161)$$

Then

$$\lim_{\delta \to 0} n_\delta = \infty. \qquad (21.162)$$

Remark 21.3.11 One can easily choose u_0 so that inequalities (21.160) and (21.161) hold. Indeed, inequalities (21.160) and (21.161) hold if u_0 is sufficiently close to V_0. Note that inequality (21.160) is a sufficient condition for the following inequality (cf. (21.174)) to hold:

$$e^{-\varphi_n} \|F(u_0) + a_0 u_0 - f_\delta\| \leq \frac{1}{6} a_n \|V_n\|, \qquad t \geq 0. \qquad (21.163)$$

In our proof, inequality (21.163) (see (21.174)) is used at $n = n_\delta$. The stopping time n_δ is often sufficiently large for the quantity $e^{\varphi_{n_\delta}} a_{n_\delta}$ to be large. This follows from the fact that $\lim_{n \to \infty} e^{\varphi_n} a_n = \infty$ (cf. (21.85)). In this case, inequality (21.163) with $n = n_\delta$ is satisfied for a wide range of u_0.

Proof: Inequalities (21.10) and (21.11) with $\tilde{u} = u_n$ and $\tilde{v} = V_n$ imply

$$a_n \|u_n - V_n\| \leq h_n, \qquad \|F(u_n) - F(V_n)\| \leq h_n. \qquad (21.164)$$

From the first inequality in (21.139), the triangle inequality, and the first inequality in (21.164), one gets

$$h_{n+1} \leq (1 - \gamma_n a_{n+1} a_n) h_n + (a_n - a_{n+1}) \|u_n - V_n\| + (a_n - a_{n+1}) \|V_n\|$$

$$\leq (1 - \gamma_n a_{n+1} a_n) h_n + \frac{a_n - a_{n+1}}{a_n} h_n + (a_n - a_{n+1}) \|V_n\|. \qquad (21.165)$$

It follows from (21.100) and (21.165) that

$$h_{n+1} \le (1 - \frac{\gamma_n a_{n+1}^2}{2})h_n + (a_n - a_{n+1})\|V_n\|$$

$$\le e^{-\frac{a_{n+1}^2 \gamma_n}{2}} h_n + (a_n - a_{n+1})\|V_n\|. \tag{21.166}$$

From (21.166) and a discrete analog of Gronwall's inequality, one gets

$$h_n \le h_0 e^{-\varphi_n} + e^{-\varphi_n} \sum_{i=0}^{n-1} e^{\varphi_{i+1}} (a_i - a_{i+1})\|V_i\|, \qquad n \ge 0, \tag{21.167}$$

where

$$\varphi_n := \sum_{i=0}^{n-1} \frac{a_{i+1}^2 \gamma_i}{2}, \qquad n \ge 1. \tag{21.168}$$

From (21.167) and the second inequality in (21.164), one gets

$$\|F(u_n) - F(V_n)\| \le h_0 e^{-\varphi_n} + e^{-\varphi_n} \sum_{i=0}^{n-1} e^{\varphi_{i+1}} (a_i - a_{i+1})\|V_i\|. \tag{21.169}$$

This and the triangle inequality imply

$$\|F(u_n) - f_\delta\| \le \|F(V_n) - f_\delta\| + h_0 e^{-\varphi_n}$$

$$+ e^{-\varphi_n} \sum_{i=0}^{n-1} e^{\varphi_{i+1}} (a_i - a_{i+1})\|V_i\|. \tag{21.170}$$

From inequality (21.169) and the triangle inequality, one gets

$$\|F(u_n) - f_\delta\| \ge \|F(V_n) - f_\delta\| - \|F(V_n) - F(u_n)\|$$

$$\ge a_n \|V_n\| - h_0 e^{-\varphi_n}$$

$$- e^{-\varphi_n} \sum_{i=0}^{n-1} e^{\varphi_{i+1}} (a_i - a_{i+1})\|V_i\|. \tag{21.171}$$

It follows from our assumptions and Lemma 21.3.6 that a_n satisfies (21.104), that is,

$$\frac{1}{2} a_n \|V_n\| \ge e^{-\varphi_n} \sum_{i=0}^{n-1} e^{\varphi_{i+1}} (a_i - a_{i+1})\|V_i\|. \tag{21.172}$$

From inequality (21.160) one gets

$$h_0 e^{-\varphi_n} \le \frac{1}{6} a_0 \|V_0\| e^{-\varphi_n}, \qquad n \ge 0. \tag{21.173}$$

Inequalities (21.173) and (21.101) imply

$$e^{-\varphi_n} h_0 \le \frac{1}{4} a_{n+1} \|V_0\| \le \frac{1}{4} a_n \|V_n\|, \quad n \ge 0, \tag{21.174}$$

where we have used the inequality $\|V_{n'}\| \le \|V_n\|$ for $n' \le n$ (see Lemma 21.1.1). From (21.171), (21.172), and (21.174), one gets

$$\|F(u_n) - f_\delta\| \ge a_n\|V_n\| - \frac{1}{4}a_n\|V_n\| - \frac{1}{2}a_n\|V_n\| = \frac{1}{4}a_n\|V_n\|, \qquad n \ge 0.$$

This and (21.114) imply

$$C\delta^\zeta \ge \|F(u_{n_\delta}) - f_\delta\| \ge \frac{1}{4}a_{n_\delta}\|V_{n_\delta}\|. \tag{21.175}$$

From the triangle inequality and the first inequality in (21.8) we obtain

$$a_{n_\delta}\|V_{0,n_\delta}\| \le a_{n_\delta}\|V_{n_\delta}\| + a_{n_\delta}\|V_{n_\delta} - V_{0,n_\delta}\| \le a_{n_\delta}\|V_{n_\delta}\| + \delta. \tag{21.176}$$

Here $V_{0,n} := V_n|_{\delta=0}$. This and (21.175) imply

$$0 \le \lim_{\delta \to 0} a_{n_\delta}\|V_{0,n_\delta}\| \le \lim_{\delta \to 0} \left(4C\delta^\zeta + \delta\right) = 0. \tag{21.177}$$

Since the sequence $(\|V_{0,n}\|)_{n=0}^\infty$ is increasing (cf. Corollary 21.1.1) and $\|V_{0,0}\| > 0$, relation (21.177) implies $\lim_{\delta \to 0} a_{n_\delta} = 0$. Since $0 < a_n \searrow 0$, it follows that (21.162) holds. ∎

Instead of using iterative scheme (21.82), one may use the following iterative scheme:

$$u_{n+1} = u_n - \gamma_n(F'(u_n) + a_n I)^*[F(u_n) + a_n(u_n - \bar{u}) - f_\delta], \quad u_0 = \tilde{u}_0, \tag{21.178}$$

where $\bar{u}$ and $\tilde{u}_0 \in H$. Denote $\tilde{F}(u) := F(u + \bar{u})$. If F is a monotone operator, then so is $\tilde{F}$. Using Theorem 21.3.7 with $F := \tilde{F}$, one gets the following corollary:

Corollary 21.3.12 *Let $C > 0$ and $\zeta \in (0, 1]$ be constants satisfying $C\delta^\zeta > \delta$. Let a_n and h be as in Theorem 21.3.7. Assume that $F : H \to H$ is a Fréchet differentiable and monotone operator, (21.2) holds, F' is self-adjoint, and u_0 is an element of H, satisfying inequalities*

$$\|F(u_0) + a_0 u_0 - f_\delta\| \le \frac{1}{6}a_0\|V_0\|, \quad \|F(u_0) - f_\delta\| > C\delta^\zeta. \tag{21.179}$$

Assume that equation $F(u) = f$ has a solution, possibly nonunique, and $z \in B(\bar{u}, R)$ is the solution with minimal distance to $\bar{u}$. Let f be unknown but let f_δ be given, $\|f_\delta - f\| \le \delta$. Let u_n be defined by (21.178). Then there exists a unique n_δ such that

$$\|F(u_{n_\delta}) - f_\delta\| \le C\delta^\zeta, \quad \|F(u_n) - f_\delta\| > C\delta^\zeta, \qquad 0 \le n < n_\delta. \tag{21.180}$$

If $\zeta \in (0, 1)$ and n_δ satisfies (21.180), then

$$\lim_{\delta \to 0} \|u_{n_\delta} - z\| = 0. \tag{21.181}$$

CHAPTER 22

DSM OF SIMPLE ITERATION TYPE

22.1 DSM OF SIMPLE ITERATION TYPE

In this section we study a version of the DSM for solving the equation

$$F(u) = f, \qquad (22.1)$$

where F is a nonlinear monotone operator in a Hilbert space H, and equation (22.1) is assumed solvable, possibly nonuniquely. Let y be the minimal norm solution to (22.1). We assume in addition that F is locally Hölder continuous of order $\alpha > 1/2$, that is,

$$\|F(u) - F(v)\| \le C_R \|u - v\|^\alpha, \qquad \forall u, v \in B(y, R). \qquad (22.2)$$

Assume that $f = F(y)$ is not known but that f_δ, the noisy data, are known, and $\|f_\delta - f\| \le \delta$. If $F'(u)$ is not boundedly invertible, then solving equation (22.1) for u given noisy data f_δ is often (but not always) an ill-posed problem. The most frequently used and studied methods are regularized Newton-type and gradient-type methods. These methods requires the knowledge of the Fréchet derivative of F. Therefore, they are not applicable if F is not Fréchet

Dynamical Systems Method and Applications: Theoretical Developments and Numerical Examples, First Edition. A. G. Ramm and N. S. Hoang.
Copyright © 2012 John Wiley & Sons, Inc.

differentiable. Our goal in this section is to study a method for a stable solution to problem (22.1) when *F is not Fréchet differentiable.*

In this section we consider the following version of the DSM for a stable solution to equation (22.1):

$$\dot{u}_\delta = -\big(F(u_\delta) + a(t)u_\delta - f_\delta\big), \quad u_\delta(0) = u_0, \tag{22.3}$$

where F is a monotone continuous operator and $u_0 \in H$. It is known that a local solution to (22.3) exists under the assumption that F is monotone continuous and $a(t) > 0$ (see Chapter 6). When $\delta = 0$ and $a(t)$ satisfies some conditions, then it is known that the solution to (22.3) exists globally (see Chapter 6).

The advantage of the method in (22.3) compared with the one in (20.75) is the absence of the inverse operator in the algorithm, which makes the algorithm (22.3) less expensive than (20.75). On the other hand, algorithm (20.75) converges faster than (22.3) in many cases. Another advantage of the DSM (22.3) is the applicability when F is locally Hölder continuous of order $\alpha \in (\frac{1}{2}, 1)$ but not Fréchet differentiable as shown in this section.

The convergence of the method (22.3) for any initial value u_0 with an a priori choice of stopping rule was justified in Chapter 12. In [50] the DSM (22.3) with a stopping rule of Discrepancy Principle (DP) type was proposed and justified under the assumption that F is Fréchet differentiable. There, convergence of $u_\delta(t_\delta)$, chosen by a stopping rule of Discrepancy Principle type, is proved for the regularizing function $a(t) = d/(c + t)^b$, where $c \geq 1$, $b \in (0, \frac{1}{2})$, and d is sufficiently large. However, how large one should choose the parameter d is not explained in [50].

In this section we study the DSM (22.3) with the stopping rule proposed in [50] under weaker assumption on F and for a larger class of regularizing function $a(t)$. The novel results in this section include a justification of the DSM (22.3) with our stopping rule for a stable solution to (22.1) under the assumption that F is locally Hölder continuous of order $\alpha \in (\frac{1}{2}, 1)$. This condition is much weaker than the Fréchet differentiability of F which was used in [50]. Moreover, our results are justified for a larger class of regularizing function $a(t)$. The main results of this section are Theorem 22.1.9 and Theorem 22.1.12 in which a DP is formulated, the existence of a stopping time t_δ is proved, and the convergence of the DSM with the proposed DP is justified under some weak and natural assumptions.

This section is based on paper [43].

22.1.1   Auxiliary results

Let $a = a(t)$ be a strictly monotonically decaying continuous positive function on $[0, \infty)$, $0 < a(t) \searrow 0$, and assume $a \in C^1[0, \infty)$. These assumptions hold throughout the chapter and often are not repeated. Then the solution V_δ of (19.3) is a function of t, $V_\delta = V_\delta(t)$.

Remark 22.1.1 From the monotonicity of F and (19.3) one gets

$$
\begin{aligned}
0 &\le \langle F(V_\delta(t)) - F(V_\delta(t')), V_\delta(t) - V_\delta(t') \rangle \\
&= \langle a(t')V_\delta(t') - a(t)V_\delta(t), V_\delta(t) - V_\delta(t') \rangle \\
&= -a(t')\|V_\delta(t) - V_\delta(t')\|^2 + (a(t') - a(t))\langle V_\delta(t), V_\delta(t) - V_\delta(t') \rangle \\
&\le -a(t')\|V_\delta(t) - V_\delta(t')\|^2 + |a(t') - a(t)|\|V_\delta(t)\|\|V_\delta(t) - V_\delta(t')\|, \quad (22.4)
\end{aligned}
$$

for all $t, t' > 0$. This implies

$$
\limsup_{\xi \to 0} \frac{\|V_\delta(t + \xi) - V_\delta(t)\|}{|\xi|} \le \frac{|\dot{a}(t)|}{a(t)}\|V_\delta(t)\|, \qquad t > 0. \qquad (22.5)
$$

Let us formulate and prove a new version of the Gronwall's inequality.

Lemma 22.1.2 *Let $\alpha(t)$ and $\beta(t)$ be continuous nonnegative functions on $[0, \infty)$. Let $0 \le g(t)$ be a continuous function on $[0, \infty)$ satisfying the following condition:*

$$
\limsup_{\xi \to 0} \frac{g^2(t + \xi) - g^2(t)}{\xi} \le -2\alpha(t)g^2(t) + 2\beta(t)g(t), \qquad \forall t \ge 0. \qquad (22.6)
$$

Then

$$
g(t) \le g(0)e^{-\tilde{\varphi}(t)} + e^{-\tilde{\varphi}(t)} \int_0^t e^{\tilde{\varphi}(s)}\beta(s)\, ds, \qquad \tilde{\varphi}(t) := \int_0^t \alpha(s)\, ds. \qquad (22.7)
$$

Proof: Let

$$
g_\epsilon(t) := \left[g^2(t) + \epsilon e^{-2\int_0^t \alpha(\xi)\, d\xi} \right]^{\frac{1}{2}}, \qquad t \ge 0, \qquad \epsilon > 0.
$$

From (22.6) one obtains

$$
\begin{aligned}
\limsup_{\xi \to 0} \frac{g_\epsilon^2(t + \xi) - g_\epsilon^2(t)}{\xi} &= \limsup_{\xi \to 0} \frac{g^2(t + \xi) - g^2(t)}{\xi} + \epsilon\frac{d}{dt}e^{-2\int_0^t \alpha(\xi)\, d\xi} \\
&\le -2\alpha(t)g^2(t) + 2\beta(t)g(t) - 2\epsilon\alpha(t)e^{-2\int_0^t \alpha(\xi)\, d\xi} \\
&\le -2\alpha(t)g_\epsilon^2(t) + 2\beta(t)g_\epsilon(t), \qquad \forall t \ge 0. \qquad (22.8)
\end{aligned}
$$

Since $g_\epsilon(t) > 0$, it follows from (22.8) and the continuity of g_ϵ that

$$
\begin{aligned}
\limsup_{\xi \to 0} \frac{g_\epsilon(t + \xi) - g_\epsilon(t)}{\xi} &= \limsup_{\xi \to 0} \frac{g_\epsilon^2(t + \xi) - g_\epsilon^2(t)}{\xi}\frac{1}{g_\epsilon(t + \xi) + g_\epsilon(t)} \\
&= \frac{1}{2g_\epsilon(t)}\limsup_{\xi \to 0} \frac{g_\epsilon^2(t + \xi) - g_\epsilon^2(t)}{\xi} \\
&\le -\alpha(t)g_\epsilon(t) + \beta(t). \qquad (22.9)
\end{aligned}
$$

From the Taylor expansion of $e^{\int_t^{t+\xi} \alpha(s)\,ds}$, we have

$$e^{\tilde{\varphi}(t+\xi)} = e^{\tilde{\varphi}(t)+\int_t^{t+\xi}\alpha(s)\,ds} = e^{\tilde{\varphi}(t)}\left(1 + \int_t^{t+\xi}\alpha(s)\,ds + O(\xi^2)\right), \qquad \xi \to 0.$$

This, (22.9), the mean value theorem for integration, and the continuity of $g_\epsilon(t)$ imply

$$\limsup_{\xi \to 0} \frac{e^{\tilde{\varphi}(t+\xi)}g_\epsilon(t+\xi) - e^{\tilde{\varphi}(t)}g_\epsilon(t)}{\xi} \le e^{\tilde{\varphi}(t)}\beta(t). \tag{22.10}$$

From (22.10) one obtains

$$e^{\tilde{\varphi}(t)}g_\epsilon(t) - e^{\tilde{\varphi}(0)}g_\epsilon(0) \le \int_0^t e^{\tilde{\varphi}(s)}\beta(s)\,ds, \qquad t \ge 0. \tag{22.11}$$

This implies

$$g(t) < g_\epsilon(t) \le \left(g^2(0) + \epsilon\right)^{\frac{1}{2}} e^{-\tilde{\varphi}(t)} + e^{-\tilde{\varphi}(t)}\int_0^t e^{\tilde{\varphi}(s)}\beta(s)\,ds, \tag{22.12}$$

for all $t \ge 0$. Letting $\epsilon \to 0$ in (22.12), one obtains (22.7).
    Lemma 22.1.2 is proved. ∎

Lemma 22.1.3 *Let $a(t) \in C^1[0,\infty)$ satisfy the following conditions*

$$0 < a(t) \searrow 0, \qquad 0 < \frac{|\dot{a}(t)|}{a^2(t)} \searrow 0. \tag{22.13}$$

Let

$$\phi(t) := \int_0^t a(s)\,ds \tag{22.14}$$

and $V_\delta(t)$ be the solution to (19.3) with $a = a(t)$. Then the following relations hold:

$$\lim_{t\to\infty} \phi(t) = \infty, \tag{22.15}$$

$$\lim_{t\to\infty} e^{r\phi(t)}a(t) = \infty, \qquad r = const > 0, \tag{22.16}$$

$$\lim_{t\to\infty} \frac{\int_0^t e^{\phi(s)}|\dot{a}(s)|\,\|V_\delta(s)\|\,ds}{e^{\phi(t)}} = 0, \tag{22.17}$$

$$M := \lim_{t\to\infty} \frac{\int_0^t e^{\phi(s)}|\dot{a}(s)|\,ds}{e^{\phi(t)}a(t)} = 0. \tag{22.18}$$

Proof: Let us first prove (22.15). It follows from (22.13) that there exists $t_0 \ge 0$ such that

$$a(t) \ge -\frac{\dot{a}(t)}{a(t)}, \qquad \forall t \ge t_0.$$

This implies

$$\phi(t) \geq \int_{t_0}^{t} a(s)\,ds \geq \int_{t_0}^{t} \frac{-\dot{a}(s)}{a(s)}\,ds = -\ln a(s)\Big|_{t_0}^{t} = \ln a(t_0) - \ln a(t). \quad (22.19)$$

Relation (22.15) follows from the relation $\lim_{t\to\infty} a(t) = 0$ and (22.19).

Let us prove (22.16). We claim that, for sufficiently large $t > 0$, the following inequality holds:

$$\phi(t) = \int_0^t a(s)\,ds > \frac{1}{r}\ln\frac{1}{a^2(t)}. \quad (22.20)$$

Indeed, by L'Hospital's rule and (22.13), one gets

$$\lim_{t\to\infty} \frac{\phi(t)}{\ln\frac{1}{a^2(t)}} = \lim_{t\to\infty} \frac{a^2(t)}{2|\dot{a}(t)|} = \infty. \quad (22.21)$$

This implies that (22.20) holds for all $t \geq \tilde{T}$ provided that $\tilde{T} > 0$ is sufficiently large. It follows from inequality (22.20) that

$$\lim_{t\to\infty} a(t)e^{r\phi(t)} \geq \lim_{t\to\infty} a(t)e^{\ln\frac{1}{a^2(t)}} = \lim_{t\to\infty} \frac{1}{a(t)} = \infty. \quad (22.22)$$

Thus, equality (22.16) is proved.

Let us prove (22.17). Since $a(t)\|V_\delta(t)\|$ is a decreasing function of t (cf. Lemma 19.1.1), one gets

$$\lim_{t\to\infty} \frac{\int_0^t e^{\phi(s)}|\dot{a}(s)|\,\|V_\delta(s)\|\,ds}{e^{\phi(t)}} \leq \lim_{t\to\infty} \frac{\int_0^t e^{\phi(s)}\frac{|\dot{a}(s)|}{a(s)}a(0)\|V_\delta(0)\|\,ds}{e^{\phi(t)}}. \quad (22.23)$$

We claim that

$$\lim_{t\to\infty} \frac{\int_0^t e^{\phi(s)}\frac{|\dot{a}(s)|}{a(s)}\,ds}{e^{\phi(t)}} = 0. \quad (22.24)$$

Indeed, if $\int_0^t e^{\phi(s)}\frac{|\dot{a}(s)|}{a(s)}\,ds < \infty$, then (22.24) follows from (22.15). Otherwise, relation (22.24) follows from L'Hospital's rule and the relation $\lim_{t\to\infty} \frac{|\dot{a}(t)|}{a^2(t)} = 0$.

From (22.23) and (22.24) one gets (22.17).

Let us prove (22.18). Since (22.15) holds and $a(t)e^{\phi(t)} \to 0$ as $t \to \infty$, by (22.16) with $r = 1$, relation (22.18) holds if the numerator of (22.18) is bounded. Otherwise, L'Hospital's rule yields

$$M = \lim_{t\to\infty} \frac{e^{\phi(t)}|\dot{a}(t)|}{e^{\phi(t)}a^2(t) + e^{\phi(t)}\ddot{a}(t)} = 0. \quad (22.25)$$

Here we have used the relation $\lim_{t\to\infty} \frac{|\dot{a}(t)|}{a^2(t)} = 0$.

Lemma 22.1.3 is proved. ∎

Remark 22.1.4 From (22.17) and the inequality $\|V_\delta(t)\| \geq \|V_\delta(0)\| > 0$, $\forall t \geq 0$, (see Lemma 19.1.1), one gets the following relation:

$$\lim_{t\to\infty} \frac{\int_0^t e^{\phi(s)}|\dot{a}(s)|ds}{e^{\phi(t)}} = 0. \tag{22.26}$$

Let $\epsilon > 0$ be arbitrary. It follows from (22.18) that there exists $t_\epsilon > 0$ such that the following inequality holds:

$$e^{-\phi(t)} \int_0^t e^{\phi(s)}|\dot{a}(s)|\,ds < \epsilon a(t), \qquad \forall t \geq t_\epsilon. \tag{22.27}$$

Let us assume that $a(t)$ satisfies the following conditions:

$$0 < a(t) \searrow 0, \qquad \frac{1}{2} > q > \frac{|\dot{a}(t)|}{a^2(t)} \searrow 0. \tag{22.28}$$

Lemma 22.1.5 *Let $a(t)$ satisfy (22.28) and $\varphi(t) := (1-q)\int_0^t a(s)\,ds$, $q \in (0, 1/2)$. Then one has*

$$e^{-\varphi(t)} \int_0^t e^{\varphi(s)}|\dot{a}(s)|\|V_\delta(s)\|\,ds \leq \frac{q}{1-2q} a(t)\|V_\delta(t)\|, \qquad t \geq 0. \tag{22.29}$$

Proof:

Let us prove that

$$e^{\varphi(t)}|\dot{a}(t)| \leq \frac{q}{1-2q}\left(a(t)e^{\varphi(t)}\right)', \qquad \forall t \geq 0. \tag{22.30}$$

Inequality (22.30) is equivalent to

$$\left(\frac{1}{q} - 2\right)e^{\varphi(t)}|\dot{a}(t)| \leq \dot{a}(t)e^{\varphi(t)} + (1-q)a^2(t)e^{\varphi(t)}, \qquad t \geq 0. \tag{22.31}$$

Note that $\dot{a} = -|\dot{a}|$. Inequality (22.31) holds because from (22.28) one obtains

$$\left(\frac{1}{q} - 2\right)|\dot{a}(t)| < -|\dot{a}(t)| + (1-q)a^2(t), \qquad t \geq 0. \tag{22.32}$$

Thus, inequality (22.30) holds. Integrate (22.30) from 0 to t and get

$$\int_0^t e^{\varphi(s)}|\dot{a}(s)|\,ds \leq \frac{q}{1-2q}\left(a(t)e^{\varphi(t)} - a(0)e^0\right) < \frac{q}{1-2q}a(t)e^{\varphi(t)}, \tag{22.33}$$

for all $t \geq 0$.

Multiplying (22.33) by $e^{-\varphi(t)}\|V_\delta(t)\|$ and using the fact that $\|V_\delta(t)\|$ is increasing, one gets inequality (22.29).

Lemma 22.1.5 is proved. ∎

22.1.2   Main results

Let $u_\delta(t)$ solve the following Cauchy problem:

$$\dot{u}_\delta = -[F(u_\delta) + a(t)u_\delta - f_\delta], \qquad u_\delta(0) = u_0. \tag{22.34}$$

Assume

$$0 < a(t) \searrow 0, \qquad 0 < \frac{|\dot{a}(t)|}{a^2(t)} \searrow 0, \qquad t \ge 0. \tag{22.35}$$

Remark 22.1.6 Let $a(t) = \frac{d}{(c+t)^b}$, where $b \in (0,1)$, $c > 0$, and $d > 0$. Then this $a(t)$ satisfies (22.35).

Remark 22.1.7 It is known that there exists a unique local solution to problem (22.34) for any initial data u_0 if F is monotone continuous and $0 < a(t)$ is a continuous function. Proofs for this are often based on Peano approximations (see Lemma 12.1.3 and [25, p. 99]). When F is monotone and hemicontinuous, then equation (22.34) is understood in the weak sense. When F is monotone and continuous, it is known that equation (22.34) can be understood in the strong sense as discussed in Lemma 12.1.3.

The following lemma guarantees the global existence of a unique solution to (22.34).

Lemma 22.1.8 *Let F be monotone and continuous. Let $0 < a(t)$ be a continuous function satisfying (22.35). Then the unique solution to (22.34) exists globally.*

Proof: Assume the contrary, that is, that $u_\delta(t)$ exists on interval $[0, T)$ but does not exist on $[T, T + d]$, where $d > 0$ is arbitrary small. Let us prove that the limit

$$\lim_{t \to T} u_\delta(t) = u_\delta(T) \tag{22.36}$$

exists and is finite. This contradicts the definition of T since one can consider $u_\delta(T)$ as an initial data and construct the solution $u_\delta(t)$ on interval $[T, T+d]$, for sufficiently small $d > 0$, using the local existence of $u_\delta(t)$.

Let us first prove that

$$\|u_\delta(t) - V_\delta(t)\| \le e^{-\phi(t)}\|w(0)\| + e^{-\phi(t)} \int_0^t e^{\phi(s)} \frac{|\dot{a}(s)|}{a(s)} \|V_\delta(s)\| \, ds, \tag{22.37}$$

for all $t \in [0, T)$ where

$$\phi(t) := \int_0^t a(s) \, ds, \qquad w(t) := u_\delta(t) - V_\delta(t). \tag{22.38}$$

From (22.34), (19.3), and the monotonicity (see (19.2)) of F one gets

$$\langle \dot{u}_\delta(t), w(t)\rangle = -\langle F(u_\delta(t)) + a(t)u_\delta(t) - F(V_\delta(t)) - a(t)V_\delta(t), w(t)\rangle$$
$$\le -a(t)\|w(t)\|^2. \tag{22.39}$$

It follows from (22.5) and (22.39) that

$$\limsup_{\xi \to 0} \frac{\|w(t+\xi)\|^2 - \|w(t)\|^2}{\xi}$$

$$= \operatorname{Re}\limsup_{\xi \to 0} \frac{\langle w(t+\xi) - w(t), w(t+\xi) + w(t)\rangle}{\xi}$$

$$\le 2\operatorname{Re}\langle \dot{u}_\delta(t), w(t)\rangle + \operatorname{Re}\limsup_{\xi \to 0} \frac{\langle V_\delta(t+\xi) - V_\delta(t), w(t+\xi) + w(t)\rangle}{\xi}$$

$$\le -2a(t)\|w(t)\|^2 + 2\|w(t)\|\frac{|\dot{a}(t)|}{a(t)}\|V_\delta(t)\|. \tag{22.40}$$

This and Lemma 22.1.2 imply (22.37).

Let

$$K = 1 + \sup_{t \ge 0} e^{-\phi(t)} \int_0^t e^{\phi(s)} \frac{|\dot{a}(s)|}{a(s)}\, ds. \tag{22.41}$$

It follows from (22.24) that K is finite. From (22.37), (22.41), and the fact that the function $\|V_\delta(t)\|$ is increasing, one obtains

$$\|u_\delta(t)\| \le e^{-\phi(t)}\|w(0)\| + K\|V_\delta(t)\| \le K_T, \qquad \forall t \in [0,T), \tag{22.42}$$

where

$$K_T := \|w(0)\| + K\|V_\delta(T)\|.$$

Let $z_h(t) := u_\delta(t+h) - u_\delta(t)$. It follows from (22.34) that

$$\dot{z}_h(t) = -\left[F(u_\delta(t+h)) - F(u_\delta(t)) + a(t)z_h(t)\right]$$
$$+ (a(t) - a(t+h))u_\delta(t+h), \qquad 0 < t < t+h < T. \tag{22.43}$$

Multiply (22.43) by $z_h(t)$ and use the monotonicity of F to get

$$\|z_h(t)\|\frac{d}{dt}\|z_h(t)\| \le -a(t)\|z_h(t)\|^2$$
$$+ (a(t) - a(t+h))\operatorname{Re}\langle u_\delta(t+h), z_h(t)\rangle. \tag{22.44}$$

This and (22.42) imply

$$\frac{d}{dt}\|z_h(t)\| \le -a(t)\|z_h(t)\| + (a(t) - a(t+h))\|u_\delta(t+h)\|$$
$$\le -a(t)\|z_h(t)\| + (a(t) - a(t+h))K_T, \tag{22.45}$$

for $0 < t < t + h < T$. From (22.45) and Gronwall's inequality one obtains

$$\|z_h(t)\| \le e^{-\phi(t)}\|z_h(0)\| + e^{-\phi(t)}K_T \int_0^t e^{\phi(s)}(a(s) - a(s+h))\,ds$$

$$\le \|z_h(0)\| + \max_{0 \le s \le T-h}(a(s) - a(s+h))K_T e^{-\phi(t)}\int_0^t e^{\phi(s)}ds, \quad (22.46)$$

for $0 < t < t + h < T$. It follows from (22.46) and the uniform continuity of $a(t)$ on $[0, T]$ that

$$\lim_{h \to 0} \|u_\delta(t+h) - u_\delta(t)\| \le \lim_{h \to 0} \|u_\delta(0+h) - u_\delta(0)\| = 0, \quad (22.47)$$

and this relation holds uniformly with respect to t and $t + h$ such that $t < t + h < T$. Here, the last equality in (22.47) follows from the fact that $u_\delta(t)$ is continuous at 0. Relation (22.47) and the Cauchy criterion for convergence imply the existence of the finite limit in (22.36).

Lemma 22.1.8 is proved. ∎

Theorem 22.1.9 *Let $a(t)$ satisfy (22.35). Assume that $F : H \to H$ is a monotone operator satisfying condition (22.2), and u_0 is an element of H, satisfying inequality*

$$\|F(u_0) - f_\delta\| > C\delta^\zeta > \delta, \quad (22.48)$$

where $C > 0$ and $0 < \zeta \le 1$ are constants. Assume that equation $F(u) = f$ has a solution, f is unknown but f_δ is given, and $\|f_\delta - f\| \le \delta$. Let y be the minimal-norm solution to (22.1). Then the solution $u_\delta(t)$ to problem (22.34) exists globally and there exists a unique t_δ such that

$$\|F(u_\delta(t_\delta)) - f_\delta\| = C\delta^\zeta, \quad \|F(u_\delta(t)) - f_\delta\| > C\delta^\zeta, \quad \forall t \in [0, t_\delta). \quad (22.49)$$

If $\zeta \in (0, 1)$ and

$$\lim_{\delta \to 0} t_\delta = \infty, \quad (22.50)$$

then

$$\lim_{\delta \to 0} \|u_\delta(t_\delta) - y\| = 0. \quad (22.51)$$

Remark 22.1.10 Inequality (22.48) is not a restrictive assumption. Indeed, if it does not hold and $\|u_0\|$ is not too large, then u_0 can be considered an approximate solution to (22.1).

Proof (Proof of Theorem 22.1.9): The uniqueness of t_δ follows from (22.49). Indeed, if t_δ and $\tau_\delta > t_\delta$ both satisfy (22.49), then the second inequality in (22.49) does not hold on the interval $[0, \tau_\delta)$.

Let us verify the existence of t_δ.

Denote

$$v := F(u_\delta) + au_\delta - f_\delta, \quad h = \|v\|. \quad (22.52)$$

We have

$$\limsup_{\xi \to 0} \frac{h^2(t + \xi) - h^2(t)}{\xi} = \operatorname{Re} \limsup_{\xi \to 0} \frac{\langle v(t + \xi) - v(t), v(t + \xi) + v(t) \rangle}{\xi}$$

$$\leq \operatorname{Re} \limsup_{\xi \to 0} \frac{\langle F(u_\delta(t + \xi)) - F(u_\delta(t)), v(t + \xi) + v(t) \rangle}{\xi}$$

$$+ 2 \operatorname{Re}\langle a(t)\dot{u}_\delta(t), v(t) \rangle + 2 \operatorname{Re}\langle \dot{a}(t)u_\delta(t), v(t) \rangle. \tag{22.53}$$

From (22.34) and (22.52) one gets $u_\delta(t + \xi) - u_\delta(t) = -\int_t^{t+\xi} v(s)\,ds$. This and the monotonicity of F imply

$$\left\langle F(u_\delta(t + \xi)) - F(u_\delta(t)), \int_t^{t+\xi} v(s)\,ds \right\rangle \leq 0. \tag{22.54}$$

Since F is Hölder continuous of order α and $u_\delta(t)$ is differentiable, one obtains

$$\big\| F(u_\delta(t + \xi)) - F(u_\delta(t)) \big\| = O(|\xi|^\alpha) \tag{22.55}$$

and

$$\left\| 2 \int_t^{t+\xi} v(s)\,ds - \xi\big[v(t + \xi) + v(t)\big] \right\| = O(|\xi|^{1+\alpha}). \tag{22.56}$$

Relations (22.55), (22.56), and the inequality $\alpha > 1/2$ imply

$$\lim_{\xi \to 0} \frac{\langle F(u_\delta(t + \xi)) - F(u_\delta(t)), v(t + \xi) + v(t) - \frac{2}{\xi}\int_t^{t+\xi} v(s)\,ds \rangle}{\xi} = 0. \tag{22.57}$$

From (22.54) and (22.57) we get

$$\limsup_{\xi \to 0} \frac{\langle F(u_\delta(t + \xi)) - F(u_\delta(t)), v(t + \xi) + v(t) \rangle}{\xi} \leq 0. \tag{22.58}$$

This, the relation $\dot{u}_\delta = -v$ (see (22.34)), and (22.53) imply

$$\limsup_{\xi \to 0} \frac{h^2(t + \xi) - h^2(t)}{\xi} \leq -2a(t)h^2(t) + 2|\dot{a}(t)|\|u_\delta(t)\|h(t). \tag{22.59}$$

This, Lemma 22.1.2, and (22.42) imply

$$h(t) \leq e^{-\phi(t)}h(0) + e^{-\phi(t)} \int_0^t e^{\phi(s)}|\dot{a}(s)|\Big(\|w(0)\| + K\|V_\delta(s)\|\Big)ds. \tag{22.60}$$

It follows from (20.135) that

$$a\|u_\delta - V_\delta\| \leq h, \quad \|F(u_\delta) - F(V_\delta)\| \leq h. \tag{22.61}$$

The triangle inequality, the second inequality in (22.61), and (22.60) imply

$$\|F(u_\delta(t)) - f_\delta\| \leq \|F(V_\delta(t)) - f_\delta\| + \|F(u_\delta) - F(V_\delta)\|$$
$$\leq \|F(V_\delta(t)) - f_\delta\| + h(t)$$
$$\leq \|F(V_\delta(t)) - f_\delta\| + h(0)e^{-\phi(t)}$$
$$+ e^{-\phi(t)} \int_0^t e^{\phi(s)} |\dot{a}(s)| \left(\|w(0)\| + K\|V_\delta(s)\| \right) ds. \quad (22.62)$$

This, (22.17), (22.15), (21.7), and (22.26) imply

$$\lim_{t \to \infty} \|F(u_\delta(t)) - f_\delta\| \leq \lim_{t \to \infty} \|F(V_\delta(t)) - f_\delta\| \leq \delta. \quad (22.63)$$

The existence of t_δ satisfying (22.49) follows from (22.63) and the continuity of the function $\|F(u_\delta(t)) - f_\delta\|$.

Let us prove (22.51), given that (22.50) holds.

From (22.50), (22.16) with $r = 1$, and the inequality $\|V_\delta(t)\| \geq \|V_\delta(0)\| > 0$, $t \geq 0$, one gets, for all sufficiently small $\delta > 0$, the following inequality:

$$h(0)e^{-\phi(t_\delta)} \leq a(t_\delta)\|V_\delta(0)\| \leq a(t_\delta)\|V_\delta(t_\delta)\|. \quad (22.64)$$

From the fact that $\|V_\delta(t)\|$ is a nondecreasing function of t, (22.27), and (22.50), one obtains

$$Ke^{-\phi(t_\delta)} \int_0^{t_\delta} e^{\phi(s)} |\dot{a}(s)| \|V_\delta(s)\| ds \leq K\|V_\delta(t_\delta)\| e^{-\phi(t_\delta)} \int_0^{t_\delta} e^{\phi(s)} |\dot{a}(s)| ds$$
$$\leq a(t_\delta)\|V_\delta(t_\delta)\|, \quad (22.65)$$

for all sufficiently small $\delta > 0$. From (22.27) and (22.50) one gets, for all sufficiently small $\delta > 0$, the following inequality:

$$\|w(0)\| e^{-\phi(t_\delta)} \int_0^{t_\delta} e^{\phi(s)} |\dot{a}(s)| ds \leq a(t_\delta)\|V_\delta(0)\| \leq a(t_\delta)\|V_\delta(t_\delta)\|. \quad (22.66)$$

From (22.64)–(22.66), (21.8), and (22.62) with $t = t_\delta$, one obtains

$$C\delta^\zeta \leq 4a(t_\delta)\|V_\delta(t_\delta)\| \leq 4 \left(a(t_\delta)\|y\| + \delta \right). \quad (22.67)$$

This and the relation $\lim_{\delta \to 0} \frac{\delta^\zeta}{\delta} = \infty$ for a fixed $\zeta \in (0, 1)$ imply

$$\lim_{\delta \to 0} \frac{\delta^\zeta}{a(t_\delta)} \leq \frac{4\|y\|}{C}. \quad (22.68)$$

Relation (22.68) and the inequality (21.8) imply, for sufficiently small $\delta > 0$, the following inequality:

$$\|V_\delta(t)\| \leq \|y\| + \frac{\delta}{a(t_\delta)} < \|y\| + \frac{C\delta^\zeta}{a(t_\delta)} < 5\|y\|, \qquad 0 \leq t \leq t_\delta. \quad (22.69)$$

This implies

$$\lim_{\delta \to 0} \frac{\int_0^{t_\delta} e^{\phi(s)} |\dot{a}(s)| \|V_\delta(s)\| \, ds}{e^{\phi(t_\delta)} a(t_\delta)} \le 5\|y\| \lim_{\delta \to 0} \frac{\int_0^{t_\delta} e^{\phi(s)} |\dot{a}(s)| \, ds}{e^{\phi(t_\delta)} a(t_\delta)}. \tag{22.70}$$

It follows from (22.18) and (22.70) that

$$\lim_{\delta \to 0} \frac{\int_0^{t_\delta} e^{\phi(s)} |\dot{a}(s)| \|V_\delta(s)\| \, ds}{e^{\phi(t_\delta)} a(t_\delta)} = 0. \tag{22.71}$$

It follows from (22.61) and (22.60) that

$$\|u_\delta(t) - V_\delta(t)\| \le h(0) \frac{e^{-\phi(t)}}{a(t)} + \frac{e^{-\phi(t)}}{a(t)} \int_0^t e^{\phi(s)} |\dot{a}(s)| \Big(\|w(0)\| + K\|V_\delta(s)\| \Big) ds.$$

This, (22.18), (22.16) with $r = 1$, and (22.71) imply that

$$\lim_{\delta \to 0} \|u_\delta(t_\delta) - V_\delta(t_\delta)\| = 0. \tag{22.72}$$

From (22.68) one gets

$$\lim_{\delta \to 0} \frac{\delta}{a(t_\delta)} = 0. \tag{22.73}$$

Now let us finish the proof of Theorem 22.1.9.

From the triangle inequality and inequality (21.8), one obtains

$$\|u_\delta(t_\delta) - y\| \le \|u_\delta(t_\delta) - V_\delta(t_\delta)\| + \|V(t_\delta) - V_\delta(t_\delta)\| + \|V(t_\delta) - y\|$$

$$\le \|u_\delta(t_\delta) - V_\delta(t_\delta)\| + \frac{\delta}{a(t_\delta)} + \|V(t_\delta) - y\|. \tag{22.74}$$

From (22.72)–(22.74), (22.50), and Lemma 6.1.9, one obtains (22.51). Theorem 22.1.9 is proved. ∎

Assume that $a(t)$ satisfies the following conditions:

$$0 < a(t) \searrow 0, \qquad \frac{1}{3} > q > \frac{|\dot{a}(t)|}{a^2(t)} \searrow 0. \tag{22.75}$$

Remark 22.1.11 Let $a(t) = \frac{d}{(c+t)^b}$, where $b \in (0,1)$, $c > 0$, and $d > bq^{-1}c^{b-1}$. Then this $a(t)$ satisfies (22.75).

Theorem 22.1.12 *Let $a(t)$ satisfy (22.75). Let F, f, and f_δ be as in Theorem 22.1.9. Assume that $u_0 \in H$ satisfies either*

$$\|F(u_0) + a(0)u_0 - f_\delta\| \le pa(0)\|V_\delta(0)\|, \qquad 0 < p < 1 - \frac{q}{1 - 2q}, \tag{22.76}$$

or

$$\|F(u_0) + a(0)u_0 - f_\delta\| \le \theta\delta^\zeta, \qquad 0 \le \theta < C, \tag{22.77}$$

where $C > 0$ is the constant from Theorem 22.1.9. Let t_δ be defined by (22.49). Then

$$\lim_{\delta \to 0} t_\delta = \infty. \tag{22.78}$$

Remark 22.1.13 One can easily choose u_0 satisfying inequality (22.76). Indeed, (22.76) holds if u_0 is sufficiently close to $V_\delta(0)$. Note that inequality (22.76) is a sufficient condition for (22.89), that is,

$$e^{-\varphi(t)}h(0) \le pa(t)\|V_\delta(t)\|, \qquad t \ge 0, \tag{22.79}$$

to hold. In our proof, inequality (22.79) (or (22.89)) is used at $t = t_\delta$. The stopping time t_δ is often sufficiently large for the quantity $e^{\varphi(t_\delta)}a(t_\delta)$ to be large. In this case, inequality (22.79) with $t = t_\delta$ is satisfied for a wide range of u_0. Note that by (22.16) one gets $\lim_{t\to\infty} e^{\varphi(t)}a(t) = \infty$. Here $\varphi(t) = (1-q)\phi(t)$ (see also (22.38) and (22.83)).

Proof (Proof of Theorem 22.1.12): *Let us prove* (22.78), *assuming that* (22.76) *holds.* The proof goes similarly when (22.77) holds instead of (22.76).

From (22.59) and the triangle inequality, one gets

$$\limsup_{\xi \to 0} \frac{h^2(t+\xi) - h^2(t)}{\xi} \le -2a(t)h^2(t) + 2|\dot{a}(t)|\|V_\delta(t)\|h(t)$$
$$+ 2|\dot{a}(t)|\|u_\delta(t) - V_\delta(t)\|h(t). \tag{22.80}$$

This and the first inequality in (22.61) imply

$$\limsup_{\xi \to 0} \frac{h^2(t+\xi) - h^2(t)}{\xi} \le -2\left(a(t) - \frac{|\dot{a}(t)|}{a(t)}\right)h^2(t)$$
$$+ 2|\dot{a}(t)|\|V_\delta(t)\|h(t). \tag{22.81}$$

Since $a - \frac{|\dot{a}|}{a} \ge (1-q)a$, by (22.75), it follows from (22.81) and Lemma 22.1.2 that

$$h(t) \le h(0)e^{-\varphi(t)} + e^{-\varphi(t)}\int_0^t e^{\varphi(s)}|\dot{a}(s)|\|V_\delta(s)\|\,ds, \tag{22.82}$$

where

$$\varphi(t) := \int_0^t (1-q)a(s)\,ds = (1-q)\phi(t), \qquad t > 0. \tag{22.83}$$

From (22.82) and (22.61), one gets

$$\|F(u_\delta(t)) - F(V_\delta(t))\| \le h(0)e^{-\varphi(t)} + e^{-\varphi(t)}\int_0^t e^{\varphi(s)}|\dot{a}(s)|\|V_\delta(s)\|\,ds. \tag{22.84}$$

It follows from inequality (22.84) and the triangle inequality that

$$\|F(u_\delta(t)) - f_\delta\| \ge \|F(V_\delta(t)) - f_\delta\| - \|F(V_\delta(t)) - F(u_\delta(t))\|$$
$$\ge a(t)\|V_\delta(t)\| - h(0)e^{-\varphi(t)}$$
$$- e^{-\varphi(t)}\int_0^t e^{\varphi(s)}|\dot{a}|\|V_\delta\|\,ds. \tag{22.85}$$

Since $a(t)$ satisfies (22.75), one gets by Lemma 22.1.5 the following inequality:

$$\frac{q}{1-2q} a(t) \|V_\delta(t)\| \geq e^{-\varphi(t)} \int_0^t e^{\varphi(s)} |\dot{a}| \|V_\delta(s)\| \, ds. \qquad (22.86)$$

From the relation $h(t) = \|F(u_\delta(t)) + a(t)u_\delta(t) - f_\delta\|$ (cf. (22.52)) and inequality (22.76), one gets

$$h(0)e^{-\varphi(t)} \leq pa(0)\|V_\delta(0)\| e^{-\varphi(t)}, \qquad t \geq 0. \qquad (22.87)$$

It follows from (22.35) that

$$e^{-\varphi(t)} a(0) \leq a(t). \qquad (22.88)$$

Indeed, inequality $a(0) \leq a(t)e^{\varphi(t)}$ is obviously true for $t = 0$, and

$$\left(a(t)e^{\varphi(t)} \right)_t' = a^2(t)e^{\varphi(t)} \left(1 - q - \frac{|\dot{a}(t)|}{a^2(t)} \right) \geq 0,$$

by (22.35). Here, we have used the relation $\dot{a} = -|\dot{a}|$ and the inequality $1 - q > q$.

Inequalities (22.87) and (22.88) imply

$$e^{-\varphi(t)} h(0) \leq pa(t)\|V_\delta(0)\| \leq pa(t)\|V_\delta(t)\|, \qquad t \geq 0. \qquad (22.89)$$

Here the inequality $\|V_\delta(t)\| \leq \|V_\delta(t')\|$ for $t \leq t'$ (see Lemma 19.1.1) was used. From (22.49), (22.85), (22.86), and (22.89), one gets

$$C\delta^\zeta = \|F(u_\delta(t_\delta)) - f_\delta\| \geq \left(1 - p - \frac{q}{1-2q} \right) a(t_\delta) \|V_\delta(t_\delta)\|.$$

Thus,

$$\lim_{\delta \to 0} a(t_\delta) \|V_\delta(t_\delta)\| = 0. \qquad (22.90)$$

From (21.8) and the triangle inequality we obtain

$$a(t_\delta)\|V(t_\delta)\| \leq a(t_\delta)\|V_\delta(t_\delta)\| + a(t_\delta)\|V(t_\delta) - V_\delta(t_\delta)\|$$
$$\leq a(t_\delta)\|V_\delta(t_\delta)\| + \delta. \qquad (22.91)$$

This and (22.90) imply

$$\lim_{\delta \to 0} a(t_\delta) \|V(t_\delta)\| = 0. \qquad (22.92)$$

Since $\|V(t)\|$ is increasing and $\|V(0)\| > 0$, it follows from relation (22.92) that $\lim_{\delta \to 0} a(t_\delta) = 0$. Since $0 < a(t) \searrow 0$, it follows that (22.78) holds.

Theorem 22.1.12 is proved. ∎

If F is a monotone operator, then $F_1(u) = F(u + \bar{u})$, where $\bar{u} \in H$, is also a monotone operator. Consider the following Cauchy problem:

$$\dot{u} = -(F(u) + a(t)(u - \bar{u}) - f_\delta), \qquad t \geq 0. \qquad (22.93)$$

Applying Theorem 22.1.9 and Theorem 22.1.12 for F_1, one gets the following corollaries:

Corollary 22.1.14 *Let $\bar{u} \in H$ be arbitrary and let y^* be the solution to (22.1) with minimal distance to $\bar{u}$. Let $a(t)$ satisfy (22.35). Assume that $F : H \to H$ is a monotone operator satisfying condition (22.2), and u_0 is an element of H, satisfying the inequality*

$$\|F(u_0) - f_\delta\| > C\delta^\zeta > \delta, \qquad (22.94)$$

where $C > 0$ and $0 < \zeta \leq 1$ are constants.

Then the solution $u_\delta(t)$ to problem (22.93) exists globally, and there exists a unique $t_\delta > 0$ such that

$$\|F(u_\delta(t_\delta)) - f_\delta\| = C\delta^\zeta, \quad \|F(u_\delta(t)) - f_\delta\| > C\delta^\zeta, \qquad \forall t \in [0, t_\delta). \quad (22.95)$$

If $\zeta \in (0,1)$ and

$$\lim_{\delta \to 0} t_\delta = \infty, \qquad (22.96)$$

then

$$\lim_{\delta \to 0} \|u_\delta(t_\delta) - y^*\| = 0. \qquad (22.97)$$

Corollary 22.1.15 *Let $a(t)$ satisfy (22.75). Let F and f_δ be as in Corollary 22.1.14. Assume that u_0 be an element of H such that $\|F(u_0) - f_\delta\| > C\delta^\zeta > \delta$. Assume in addition that u_0 satisfies either*

$$\|F(u_0) + a(0)(u_0 - \bar{u}) - f_\delta\| \leq p\|V_\delta(0)\|, \qquad 0 < p < 1 - \frac{q}{1 - 2q}, \quad (22.98)$$

or

$$\|F(u_0) + a(0)(u_0 - \bar{u}) - f_\delta\| \leq \theta\delta^\zeta, \qquad 0 < \theta < C. \qquad (22.99)$$

where $C > 0$ and $0 < \zeta \leq 1$ are constants from Corollary 22.1.15. Let t_δ be defined by (22.95). Then

$$\lim_{\delta \to 0} t_\delta = \infty. \qquad (22.100)$$

An example
Let

$$f(x) := \begin{cases} 1 - (1-x)^p & \text{if} \quad x \leq 1, \\ 1 & \text{if} \quad 1 \leq x \leq 2, \\ 1 + (x-2)^p & \text{if} \quad x \geq 2, \end{cases} \qquad p = const > \frac{1}{2}. \quad (22.101)$$

It follows from (22.101) that f is nondecreasing. From (22.101) it follows that $f'(x)$ is bounded for all $x \neq 1, 2$. Thus, $f(x)$ is Lipschitz continuous for all $x \neq 1, 2$. From (22.101) one gets $f'(1_+) = 0$. This and the fact that the function $g(x) = x^p, x \geq 0$, is Hölder continuous of order p imply that f is Hölder continuous of order p at $x = 1$. By similar arguments, one concludes that f is Hölder continuous of order p at $x = 2$. Therefore, f is monotone and Hölder continuous of order p on $\mathbb{R}$. It is clear that the equation $f(x) = 1$ has infinitely many solutions. Thus, the equation $f(x) = 1$ is ill-posed and $x^* = 1$ is the smallest solution to $f(x) = 1$. Consider the equation $f(x) = 1 + \delta$ where $\delta \neq 0$ and $|\delta|$ is small. Although equation $f(x) = 1 + \delta$ has a unique solution x_δ for all $\delta \neq 0$ and $|\delta| < 1$, the relation $\lim_{\delta \to 0} x_\delta = x^*$ does not hold. Indeed, it follows from (22.101) that $\lim_{\delta \to 0_+} x_\delta = 2 > 1 = x^*$.

Let us solve for x^* by the method given in this chapter.

Let u_δ solve the following Cauchy problem:

$$\dot{u}_\delta(t) = -[f(u_\delta(t)) + a(t)u_\delta(t) - 1 - \delta], \qquad t \geq 0, \quad u_\delta(0) = u_0, \quad (22.102)$$

where $a(t)$ and u_0 are suitably chosen. Then, by Theorem 22.1.9 and Theorem 22.1.12 one gets

$$\lim_{\delta \to 0} u_\delta(t_\delta) = x^*, \qquad (22.103)$$

if $a(t)$, u_0, and t_δ are chosen to satisfy conditions in these theorems.

22.2 AN ITERATIVE SCHEME FOR SOLVING EQUATIONS WITH σ-INVERSE MONOTONE OPERATORS

In this section we assume that F is a locally σ-inverse monotone operator in a Hilbert space H. An operator F is called locally σ-inverse monotone if for any $R > 0$ there exists a constant $\sigma_R > 0$ such that

$$\langle F(u) - F(v), u - v \rangle \geq \sigma_R \|F(u) - F(v)\|^2, \qquad \forall u, v \in B(0, R) \subset H. \quad (22.104)$$

Here, $\langle \cdot, \cdot \rangle$ denotes the inner product in H. If the constant σ_R in (22.104) is independent of R, then we call F a σ-inverse monotone operator.

A necessary condition for an operator F to be σ-inverse monotone is the following:

$$\|F(u) - F(v)\| \leq \sigma^{-1} \|u - v\|.$$

Indeed, inequality (22.104) and the Cauchy inequality imply the above estimate. If the σ-inverse monotone operator is a homeomorphism, then its inverse is strongly monotone:

$$\langle F^{-1}(u) - F^{-1}(v), u - v \rangle \geq \sigma \|u - v\|^2.$$

An example of σ-inverse operator is a linear self-adjoint compact nonnegative-definite operator A. Indeed, if $\lambda_1 \geq \lambda_2 \geq \ldots \geq 0$ are its eigenvalues and ϕ_j are the corresponding normalized eigenvectors, then

$$\langle Au - Av, u - v \rangle = \sum_j \lambda_j |\langle u - v, \phi_j \rangle|^2$$

$$\geq \sigma \sum_j \lambda_j^2 |\langle u - v, \phi_j \rangle|^2 = \sigma \|Au - Av\|^2,$$

where $\sigma = \lambda_1^{-1}$. An example of locally σ-inverse monotone operator is a nonlinear Fréchet differentiable monotone operator $F : H \rightarrow H$, provided that H is a complex Hilbert space and F' is locally bounded; that is, for any $R > 0$ there exists a constant $M(R)$ such that

$$\|F'(u)\| \leq M(R), \qquad \forall u \in B(0, R)$$

(see Lemma 22.2.8 in Section 22.2.1). In Lemma 22.2.8 we also prove that if H is a real Hilbert space, $F : H \rightarrow H$ is a Fréchet differentiable monotone operator, and F' is a self-adjoint locally bounded operator, then F is also a locally σ-inverse monotone operator. If (22.104) holds, then the operator $\sigma_R F$ satisfies (22.104) with $\sigma_R = 1$.

It is clear that if F is σ-inverse monotone, then it is monotone, that is,

$$\langle F(u) - F(v), u - v \rangle \geq 0, \qquad \forall u, v \in H. \tag{22.105}$$

If $F'(u)$ is not boundedly invertible, then solving equation (22.1) for u given noisy data f_δ is often (but not always) an ill-posed problem.

In [48], the following iterative scheme for solving equation (22.1) with monotone operators F was investigated:

$$u_{n+1} = u_n - \left(F'(u_n) + a_n I\right)^{-1}\left(F(u_n) + a_n u_n - f_\delta\right), \qquad u_0 = \tilde{u}_0. \tag{22.106}$$

The convergence of this method was justified with an a priori and an a posteriori choice of stopping rules (see [48]).

In this section we consider the following iterative for a stable solution to equation (22.1):

$$u_{n+1} = u_n - \gamma_n\left[F(u_n) + a_n u_n - f_\delta\right], \qquad u_0 = \tilde{u}_0, \tag{22.107}$$

where F is a locally σ-inverse monotone operator, $\gamma_n \in (0, 1)$, $n \geq 0$, and $\tilde{u}_0$ is a suitably chosen element in H which will be specified later.

The advantages of this iterative scheme compared with (22.106) are:

1. There is no inverse operator in the algorithm, which makes the algorithm (22.107) less expensive than (22.106).

2. One does not have to compute the Fréchet derivative of F.

3. The Fréchet differentiability of F is not required.

A more expensive algorithm (22.106) may converge faster than (22.107) in some cases.

In this section we investigate a stopping rule based on a *discrepancy principle* (DP) for the iterative scheme (22.107). Using the local σ-inverse monotonicity of F, we prove convergence of the method (22.107) to the minimal-norm solution to (22.1). The rate of decay of the regularizing sequence a_n in this section is also faster than the one in [50]. This saves the computer time and results in a faster convergence of our method. The main results of this section are Theorem 22.2.12, 22.2.14, and 22.2.16. In Theorem 22.2.12 a DP is formulated and the existence of a stopping time n_δ is proved. The convergence of the iterative scheme with the proposed DP to a solution to (22.1) is proved in Theorem (22.2.14). In Theorem (22.2.16), sufficient conditions for the convergence of the iterative scheme with the proposed DP to the minimal-norm solution to (22.1) is justified mathematically.

This section is based on paper [44].

22.2.1   Auxiliary results

Let us consider the following equation:

$$F(V_{\delta,n}) + a_n V_{\delta,n} - f_\delta = 0, \qquad a_n > 0, \qquad (22.108)$$

For simplicity let us denote $V_n := V_{\delta,n}$ when $\delta \neq 0$.

Remark 22.2.1 Let $V_{0,n} := V_{\delta,n}|_{\delta=0}$. Then $F(V_{0,n}) + a_n V_{0,n} - f = 0$. Note that

$$\|V_{\delta,n} - V_{0,n}\| \leq \frac{\delta}{a_n}. \qquad (22.109)$$

Indeed, from (19.3) one gets

$$F(V_{\delta,n}) - F(V_{0,n}) + a_n(V_{\delta,n} - V_{0,n}) = f_\delta - f.$$

Multiply this equality with $V_{\delta,n} - V_{0,n}$ and use (22.104) to get

$$\begin{aligned}
\delta\|V_{\delta,n} - V_{0,n}\| &\geq \langle f_\delta - f, V_{\delta,n} - V_{0,n} \rangle \\
&= \langle F(V_\delta, n) - F(V_{0,n}) + a_n(V_{\delta,n} - V_{0,n}), V_{\delta,n} - V_{0,n} \rangle \\
&\geq a_n\|V_{\delta,n} - V_{0,n}\|^2.
\end{aligned}$$

This implies (22.109). Similarly, from the equation

$$F(V_{0,n}) + a_n V_{0,n} - F(y) = 0,$$

one derives that

$$\|V_{0,n}\| \leq \|y\|. \qquad (22.110)$$

From (22.109) and (22.110), one gets the following estimate:

$$\|V_n\| \leq \|V_{0,n}\| + \frac{\delta}{a_n} \leq \|y\| + \frac{\delta}{a_n}. \qquad (22.111)$$

From equation (22.108) one gets

$$F(V_{n+1}) - F(V_n) = a_n V_n - a_{n+1} V_{n+1}.$$

This and the monotonicity of F imply

$$
\begin{aligned}
0 &\le \langle a_n V_n - a_{n+1} V_{n+1}, V_{n+1} - V_n \rangle \\
&= -a_n \|V_n - V_{n+1}\|^2 + (a_n - a_{n+1})\langle V_{n+1}, V_{n+1} - V_n \rangle \\
&\le -a_n \|V_n - V_{n+1}\|^2 + (a_n - a_{n+1})\|V_{n+1}\|\|V_{n+1} - V_n\|.
\end{aligned}
\tag{22.112}
$$

Thus, one gets

$$
\|V_n - V_{n+1}\| \le \frac{a_n - a_{n+1}}{a_n}\|V_{n+1}\|, \qquad \forall n \ge 0.
\tag{22.113}
$$

From Lemma 19.1.1 one gets the following result.

Lemma 22.2.2 *Assume* $\|F(0) - f_\delta\| > 0$. *Let* $0 < a_n \searrow 0$, F *be monotone, and*

$$
\ell_n := \|F(V_n) - f_\delta\|, \quad k_n := \|V_n\|, \qquad n = 0, 1, ...,
$$

where V_n *solves* (22.108). *Then* ℓ_n *is decreasing and* k_n *is increasing.*

Remark 22.2.3 From Lemma 19.1.2 and Lemma 22.2.2 one concludes that

$$
a_n \|V_n\| = \|F(V_n) - f_\delta\| \le \|F(0) - f_\delta\|, \qquad \forall n \ge 0.
\tag{22.114}
$$

Let $0 < a(t) \in C^1(\mathbb{R}_+)$ satisfy the following conditions:

$$
0 < a(t) \searrow 0, \qquad \nu(t) := \frac{|\dot{a}(t)|}{a^2(t)} \searrow 0, \qquad t \ge 0.
\tag{22.115}
$$

Let $0 < h = const$ and

$$
a_n := a(nh), \qquad n \ge 0.
$$

Remark 22.2.4 It follows from (22.115) that

$$
0 < \frac{1}{a_{n+1}} - \frac{1}{a_n} = -\int_{nh}^{(n+1)h} \frac{\dot{a}(s)}{a^2(s)}ds \le h\nu(nh) \le h\nu(0).
\tag{22.116}
$$

Inequalities (22.116) imply

$$
1 < \frac{a_n}{a_{n+1}} \le 1 + a_n h\nu(0).
\tag{22.117}
$$

From the relation $\lim_{n\to\infty} a_n = 0$ and (22.117) one gets

$$
\lim_{n\to\infty} \frac{a_n}{a_{n+1}} = 1.
\tag{22.118}
$$

From (22.115) and (22.116) one gets

$$
\lim_{n\to\infty} \frac{a_n - a_{n+1}}{a_n a_{n+1}} = 0.
\tag{22.119}
$$

Remark 22.2.5 Let $b \in (0,1)$, $c \geq 1$, $d > 0$ and

$$a(t) = \frac{d}{(c+t)^b}.$$

Then $a(t)$ satisfies (22.115).

Lemma 22.2.6 *Let $0 < h = const$ and $a(t)$ satisfy (22.115) and the following conditions*

$$a(0)h \leq 2, \qquad \nu(0) = \frac{|\dot{a}(0)|}{a^2(0)} \leq \frac{1}{10}. \tag{22.120}$$

Let $a_n := a(nh)$ and

$$\varphi_n := \sum_{i=1}^{n} \frac{a_i h}{2}, \qquad n \geq 1, \tag{22.121}$$

Then the following inequality holds:

$$e^{-\varphi_n} \sum_{i=0}^{n-1} e^{\varphi_{i+1}} (a_i - a_{i+1}) \|V_i\| \leq \frac{1}{2} a_n \|V_n\|, \qquad n \geq 1. \tag{22.122}$$

Proof: First, let us prove that

$$e^{\varphi_n} (a_{n-1} - a_n) \leq \frac{1}{2} (a_n e^{\varphi_n} - a_{n-1} e^{\varphi_{n-1}}), \qquad \forall n \geq 1. \tag{22.123}$$

Inequality (22.123) is equivalent to

$$\frac{3a_n}{a_{n-1}} \geq \frac{2e^{\varphi_n} + e^{\varphi_{n-1}}}{e^{\varphi_n}} = 2 + e^{-\frac{ha_n}{2}}, \qquad n \geq 1. \tag{22.124}$$

This inequality is equivalent to

$$\frac{a_{n-1} - a_n}{a_{n-1} a_n} \leq \frac{1 - e^{-\frac{ha_n}{2}}}{3a_n}, \qquad \forall n \geq 1. \tag{22.125}$$

Let us prove (22.125). From (22.115) and (22.120) one gets

$$\frac{a_{n-1} - a_n}{a_{n-1} a_n} = \int_{(n-1)h}^{nh} \frac{|\dot{a}(s)|}{a^2(s)} \, ds \leq h\nu((n-1)h) \leq h\nu(0) \leq \frac{h}{10}. \tag{22.126}$$

Note that the function $\tilde{f}(x) = \frac{1-e^{-x}}{x}$ is decreasing on $(0, \infty)$. Therefore, one gets

$$\frac{1 - e^{-\frac{ha_n}{2}}}{3a_n} = \frac{h\tilde{f}(\frac{ha_n}{2})}{6} \geq \frac{h\tilde{f}(\frac{ha_0}{2})}{6} \geq \frac{h\tilde{f}(1)}{6} \geq \frac{h\frac{6}{10}}{6} = \frac{h}{10}. \tag{22.127}$$

We have used the inequalities $a_n h \leq a_0 h \leq 2$, $\forall n \geq 1$, and $\tilde{f}(1) > \frac{6}{10}$ in (22.127). Inequality (22.125) follows from (22.126) and (22.127). Thus, (22.123) holds.

From inequality (22.123) one obtains

$$2\sum_{i=0}^{n-1} e^{\varphi_{i+1}}(a_i - a_{i+1}) \leq \sum_{i=0}^{n-1}(a_{i+1}e^{\varphi_{i+1}} - a_ie^{\varphi_i}) < e^{\varphi_n}a_n, \qquad n \geq 1. \quad (22.128)$$

Multiplying (22.128) by $\frac{1}{2}\|V_n\|e^{-\varphi_n}$ and recalling the fact that $\|V_i\|$ is increasing (see Lemma 22.2.2), one gets inequality (22.122). Lemma 22.2.6 is proved. ∎

Lemma 22.2.7 *Let R and σ_R be positive constants and F be an operator in a Hilbert space H satisfying the following inequality:*

$$\langle F(u) - F(v), u - v\rangle \geq \sigma_R\|F(u) - F(v)\|^2, \qquad \forall u, v \in B(0, R) \subset H. \quad (22.129)$$

Assume that

$$0 < \gamma \leq \frac{2}{\sigma_R^{-1} + 2a}, \qquad a = const \geq 0. \quad (22.130)$$

Then $\forall u, v \in B(0, R)$ the following inequality holds:

$$\mu(u, v) := \left\|u - v - \frac{\gamma}{1 - \gamma a}[F(u) - F(v)]\right\| \leq \|u - v\|, \quad (22.131)$$

Proof: Let us fix $R > 0$ and denote $\sigma := \sigma_R$ and $w := u - v$. From (22.129), one gets, $\forall u, v \in B(0, R)$, the following inequality:

$$\mu^2(u, v) = \|w\|^2 - \frac{2\gamma}{1 - \gamma a}\,\text{Re}\langle w, F(u) - F(v)\rangle + \frac{\gamma^2}{(1 - \gamma a)^2}\|F(u) - F(v)\|^2$$

$$\leq \|w\|^2 - \frac{2\gamma}{1 - \gamma a}\sigma\|F(u) - F(v)\|^2 + \frac{\gamma^2}{(1 - \gamma a)^2}\|F(u) - F(v)\|^2$$

$$= \|w\|^2 - \left(\frac{2\gamma\sigma}{1 - \gamma a} - \frac{\gamma^2}{(1 - \gamma a)^2}\right)\|F(u) - F(v)\|^2. \quad (22.132)$$

It follows from (22.130) that

$$\frac{2\gamma\sigma}{1 - \gamma a} - \frac{\gamma^2}{(1 - \gamma a)^2} = \frac{\gamma\sigma}{(1 - \gamma a)^2}[2(1 - \gamma a) - \sigma^{-1}\gamma]$$

$$= \frac{\gamma\sigma}{(1 - \gamma a)^2}(\sigma^{-1} + 2a)[\frac{2}{\sigma^{-1} + 2a} - \gamma] \geq 0. \quad (22.133)$$

Inequality (22.131) follows from inequalities (22.132) and (22.133). Lemma 22.2.7 is proved. ∎

Lemma 22.2.8 *Let $F : H \to H$ be a Fréchet differentiable monotone operator with locally bounded F', that is,*

$$\|F'(u)\| \leq M(R), \qquad \forall u \in B(u_0, R), \quad (22.134)$$

where H is a Hilbert space. Let one of the following assumptions hold:

1. H *is a real Hilbert space and F' is self-adjoint.*

2. H *is a complex Hilbert space.*

Then F is a locally σ-inverse monotone operator; that is, for all $R > 0$ there exists $\sigma_R > 0$ such that

$$\langle F(u) - F(v), u - v \rangle \geq \sigma_R \|F(u) - F(v)\|^2, \qquad \forall u, v \in B(0, R). \quad (22.135)$$

Moreover,

$$\sigma_R = \frac{1}{M(R)}, \qquad R > 0.$$

Proof: Fix $u, v \in B(0, R)$. One has

$$F(u) - F(v) = J(u - v), \qquad J := \int_0^1 F'(v + \xi(u - v)) \, d\xi. \quad (22.136)$$

By our assumption, J is a self-adjoint operator and

$$0 \leq J \leq M(R), \qquad M(R) := \sup_{w \in B(0, R)} \|F'(w)\|. \quad (22.137)$$

This and the self-adjointness of J imply

$$0 \leq J(I - \sigma_R J) = (I - \sigma_R J)J, \quad (22.138)$$

where I is the identity operator in H and σ_R is defined by (22.2.8). Thus,

$$
\begin{aligned}
\langle F(u) - F(v), u - v \rangle &= \langle J(u - v), u - v \rangle \\
&= \langle J(u - v), (I - \sigma_R J)(u - v) \rangle + \sigma_R \|J(u - v)\|^2 \\
&= \langle \big[(I - \sigma_R J)J\big](u - v), (u - v) \rangle + \sigma_R \|J(u - v)\|^2 \\
&\geq \sigma_R \|J(u - v)\|^2 = \sigma_R \|F(u) - F(v)\|^2. \quad (22.139)
\end{aligned}
$$

This implies (22.135). Lemma 22.2.8 is proved. ∎

Remark 22.2.9 It follows from the proof of Lemma 22.2.8 that if $F'(u)$ is self-adjoint and uniformly bounded, that is, the constant $M = M(R)$ in (22.134) is independent of R, then F is a σ-inverse monotone operator with $\sigma = \frac{1}{M}$.

Lemma 22.2.10 *Let $0 < h = const$, let $a(t)$ satisfy (22.115), let $a_n := a(nh)$, and let*

$$\phi_n = h \sum_{i=0}^n a_i, \qquad \phi(t) := \int_0^t a(s) \, ds. \quad (22.140)$$

Then $\forall n \geq 1$ the following inequalities hold:

$$e^{-\phi_n} \sum_{i=0}^{n-1} e^{\phi_{i+1}}(a_i - a_{i+1}) < e^{2a_0 h} e^{-\phi(nh)} \int_0^{nh} e^{\phi(s)} |\dot{a}(s)| \, ds, \qquad (22.141)$$

$$e^{-\phi_n} \sum_{i=0}^{n-1} e^{\phi_{i+1}} \frac{a_i - a_{i+1}}{a_i} < e^{2a_0 h} e^{-\phi(nh)} \int_0^{nh} e^{\phi(s)} \frac{|\dot{a}(s)|}{a(s)} \, ds. \qquad (22.142)$$

Proof: Let us prove (22.142). Since $a_n = a(nh)$ and $0 < a(t) \searrow 0$, one gets

$$\phi_n - \phi_{i+1} = \sum_{k=i+2}^{n} a_k h \geq \sum_{k=i+2}^{n} \int_{kh}^{(k+1)h} a(s) \, ds$$

$$= \int_{(i+2)h}^{(n+1)h} a(s) \, ds = \phi((n+1)h) - \phi((i+2)h)$$

$$> \phi(nh) - \phi((i+1)h) - a_0 h, \qquad 0 \leq i \leq n. \qquad (22.143)$$

Here, the inequalities $\phi((n+1)h) > \phi(nh)$ and $\phi(ih) + a_0 h > \phi((i+1)h)$ have been used. Inequality (22.143) and the inequalities

$$\phi((i+1)h) - \phi(s) = \int_s^{(i+1)h} a(s) \, ds \leq \int_s^{(i+1)h} a(0) \, ds \leq a(0)h,$$

for all $s \in [ih, (i+1)h]$, imply

$$-\phi_n + \phi_{i+1} \leq -\phi(nh) + \phi(s) + 2a_0 h, \qquad \forall s \in [ih, (i+1)h], \qquad (22.144)$$

where $0 \leq i \leq n - 1$.

Since $0 < a_n \searrow 0$ and $|\dot{a}(t)| = -a(t)$, one obtains

$$\frac{a_i - a_{i+1}}{a_i} = \int_{ih}^{(i+1)h} \frac{|\dot{a}(s)|}{a_i} \, ds \leq \int_{ih}^{(i+1)h} \frac{|\dot{a}(s)|}{a(s)} \, ds. \qquad (22.145)$$

It follows from (22.145) and (22.144) that

$$e^{-\phi_n} \sum_{i=0}^{n-1} e^{\phi_{i+1}} \frac{a_i - a_{i+1}}{a_i} < \sum_{i=0}^{n-1} \int_{ih}^{(i+1)h} e^{-\phi_n + \phi_{i+1}} \frac{|\dot{a}(s)|}{a(s)} \, ds$$

$$< e^{2a_0 h} e^{-\phi(nh)} \int_0^{nh} e^{\phi(s)} \frac{|\dot{a}(s)|}{a(s)} \, ds, \qquad (22.146)$$

for all $n \geq 1$. Thus, (22.142) is proved.

Let us prove inequality (22.141). From (22.144) one obtains

$$e^{-\phi_n} \sum_{i=0}^{n-1} e^{\phi_{i+1}}(a_i - a_{i+1}) = e^{-\phi_n} \sum_{i=0}^{n-1} \int_{ih}^{(i+1)h} e^{\phi_{i+1}} |\dot{a}(s)| \, ds$$

$$< e^{2a_0 h} e^{-\phi(nh)} \int_0^{nh} e^{\phi(s)} |\dot{a}(s)| \, ds,$$

for all $n \geq 1$. This implies (22.141).

Lemma 22.2.10 is proved.   ∎

Lemma 22.2.11 *Let $0 < h = const$, let $a(t)$ satisfy (22.115), let $a_n = a(nh)$ and let ϕ_n be as in (22.140). Then*

$$\lim_{n \to \infty} e^{\phi_{n-1}} a_n = \infty, \tag{22.147}$$

and

$$M := \lim_{n \to \infty} \frac{\sum_{i=0}^{n} e^{\phi_i} (a_i - a_{i+1})}{e^{\phi_n} a_{n+1}} = 0. \tag{22.148}$$

Proof: Let us first prove (22.147).

From (22.115) and (22.140), one gets

$$\lim_{n \to \infty} \phi_n = \lim_{n \to \infty} h \sum_{i=0}^{n} a_i \geq \int_0^{\infty} a(s)\, ds$$

$$\geq \int_0^{\infty} \frac{1}{\nu(0)} \frac{-\dot{a}(s)}{a(s)}\, ds = \frac{1}{\nu(0)} (\ln a(0) - \lim_{t \to \infty} \ln a(t)) = \infty. \tag{22.149}$$

We claim that if $n > 0$ is sufficiently large, then the following inequality holds:

$$\phi_{n-1} \geq \ln \frac{1}{a_n^2}. \tag{22.150}$$

Indeed, using a discrete analog of L'Hospital's rule, the relation $\ln(1 + x) = x + o(x)$, and (22.119), one gets

$$\lim_{n \to \infty} \frac{\phi_{n-1}}{\ln \frac{1}{a_n^2}} = \lim_{n \to \infty} \frac{\phi_n - \phi_{n-1}}{\ln \frac{1}{a_{n+1}^2} - \ln \frac{1}{a_n^2}} = \lim_{n \to \infty} \frac{a_n h}{2 \ln(1 + \frac{a_n - a_{n+1}}{a_{n+1}})}$$

$$= \lim_{n \to \infty} \frac{h}{2 \frac{a_n - a_{n+1}}{a_{n+1} a_n}} = \infty. \tag{22.151}$$

This implies that (22.150) holds for all $n \geq \tilde{N}$, provided that $\tilde{N} > 0$ is sufficiently large. It follows from inequality (22.150) that

$$\lim_{n \to \infty} a_n e^{\phi_{n-1}} \geq \lim_{n \to \infty} a_n e^{\ln \frac{1}{a_n^2}} = \lim_{n \to \infty} \frac{a_n}{a_n^2} = \infty. \tag{22.152}$$

Let us prove (22.148).

Since $a_n e^{\phi_{n-1}} \to \infty$ as $n \to \infty$, by (22.147), relation (22.148) holds if the numerator $\sum_{i=0}^{n} e^{\phi_i} (a_i - a_{i+1})$ in (22.148) is bounded. Otherwise, a discrete analog of L'Hospital's rule yields

$$M = \lim_{n \to \infty} \frac{e^{\phi_n} (a_n - a_{n+1})}{e^{\phi_n} a_{n+1} - e^{\phi_{n-1}} a_n} = \lim_{n \to \infty} \frac{a_n - a_{n+1}}{a_{n+1} - a_n e^{-h a_n}}$$

$$\leq \lim_{n \to \infty} \frac{1}{\frac{h a_n a_n}{(a_n - a_{n+1})} (1 - \frac{h a_n}{2}) - 1} = 0. \tag{22.153}$$

Here, we have used (22.118), (22.119), relation $\lim_{n\to\infty} a_n = 0$, and the following inequality:

$$e^{-ha_n} \le 1 - ha_n + \frac{(ha_n)^2}{2}, \qquad \forall n \ge 0.$$

Lemma 22.2.11 is proved. ∎

22.2.2 Main results

22.2.2.1 An iterative scheme Let $0 < a(t) \in C^1(\mathbb{R}_+)$ satisfy the following conditions (see also (22.115)):

$$0 < a(t) \searrow 0, \qquad \nu(t) := \frac{|\dot{a}(t)|}{a^2(t)} \searrow 0, \qquad t \ge 0. \qquad (22.154)$$

Let $0 < h = const \le 1$ and $a_n := a(nh), n \ge 0$. Consider the following iterative scheme:

$$u_{n+1} = u_n - \gamma_n[F(u_n) + a_n u_n - f_\delta], \qquad u_0 = \tilde{u}_0, \qquad (22.155)$$

where $\tilde{u}_0 \in H$ and

$$0 < h \le \gamma_n \le \frac{2}{\sigma_R^{-1} + 2a_n}, \qquad n \ge 0, \qquad (22.156)$$

where σ_R is the constant in (22.104) and $0 < R = const$.

Theorem 22.2.12 *Let $a(t)$ satisfy (22.154). Assume that $F : H \to H$ is a locally σ-inverse monotone operator. Assume that equation $F(u) = f$ has a solution, possibly nonunique. Let f_δ be such that $\|f_\delta - f\| \le \delta$ and let u_0 be an element of H satisfying the inequality:*

$$\|F(u_0) - f_\delta\| > C\delta^\zeta, \qquad (22.157)$$

where $C > 0$ and $\zeta \in (0,1]$ are constants satisfying $C\delta^\zeta > \delta$. Let $0 < R$ be sufficiently large and $0 < h$ and $0 < \gamma_n$ satisfy (22.156). Let u_n be defined by the iterative process (22.155). Then there exists a unique n_δ such that

$$\|F(u_{n_\delta}) - f_\delta\| \le C\delta^\zeta, \quad \|F(u_n) - f_\delta\| > C\delta^\zeta, \qquad 0 \le n < n_\delta, \qquad (22.158)$$

where C and ζ are constants from (22.157).
If

$$\lim_{\delta \to 0} n_\delta = \infty, \qquad (22.159)$$

and $\zeta \in (0,1)$, then

$$\lim_{\delta \to \infty} \|u_{n_\delta} - y\| = 0. \qquad (22.160)$$

Remark 22.2.13 In [50] the existence of n_δ was proved for the choice $a_n = d/(c+n)^b$, where $b \in (0, 1/2)$ and $d > 0$ is sufficiently large. However, it was not quantified in [50] how large d should be. In this section the existence of n_δ is proved for $a_n = d/(c+nh)^b$, for any $d > 0$, $c > 1$, $b \in (0,1)$, and $0 < h \leq \gamma_n$. This guarantees the existence of n_δ for small $a(0)$ or d. Moreover, our condition on b allows a_n to decay faster than the corresponding sequence a_n in [50] decays. Having smaller $a(0)$ and larger b reduces the cost of computations.

Inequality (22.157) is a very natural assumption. Indeed, if it does not hold and $\|u_0\|$ is not "large," then u_0 can be already considered an approximate solution to (22.1).

In general, if R in (22.156) is large, then the stepsize h in the iterative scheme (22.155) is small. Consequently, the computation time will be large since the rate of decay of $(a_n)_{n=1}^\infty$ is slow. However, if F is σ-inverse monotone (i.e., σ_R is independent of R), then it is easy to choose h and γ_n to satisfy (22.156).

Proof (Proof of Theorem 22.2.12):

Let us prove first that there exists $R > 0$ such that the sequence $(u_n)_{n=1}^{n_\delta}$ remains inside the ball $B(0, R)$.

We assume without loss of generality that $\delta \in (0, 1)$. It follows from (22.114) and the triangle inequality that

$$a_n \|V_n\| \leq \|F(0) - f_\delta\| \leq \|F(0) - f\| + \|f_\delta - f\| \leq \Gamma, \qquad \forall n \geq 0, \quad (22.161)$$

for all $\delta \in (0, 1)$, where

$$\Gamma := \|F(0) - f\| + 1.$$

From (22.161) one obtains

$$\|V_n\| \leq \frac{\Gamma}{a_n}, \qquad \forall n \geq 0. \quad (22.162)$$

Let

$$K = 1 + \sup_{t \geq 0} e^{a(0)} e^{-\phi(t)} \int_0^t e^{\phi(s)} \frac{|\dot{a}(s)|}{a(s)} \, ds, \quad (22.163)$$

where $\phi(t)$ is defined by (22.14). It follows from (22.24) that K is finite.

Let $V_\delta(t)$ solve the equation

$$F(V_\delta(t)) + a(t)V_\delta(t) - f_\delta = 0. \quad (22.164)$$

It follows from Lemma 22.2.2 that

$$\|F(V_\delta(t)) - f_\delta\| \leq \|F(V_n) - f_\delta\|, \qquad \forall t \geq nh. \quad (22.165)$$

Relation (22.24) and (21.7) imply that there exists $T > 0$ such that the following inequality holds $\forall t \geq T - 1$:

$$\|F(V_\delta(t)) - f_\delta\| + e^{a_0 - \phi(t)} h_0$$
$$+ e^{2a_0} e^{-\phi(t)} \int_0^t e^{\phi(s)} |\dot{a}(s)| \left(\frac{\Gamma K}{a(s)} + w_0 \right) ds < C\delta^\zeta, \quad (22.166)$$

where
$$h_0 = \|F(u_0) + a_0 u_0 - f_\delta\|, \qquad w_0 = \|u_0 - V_0\|. \qquad (22.167)$$

Let
$$R := \|V_\delta(T)\|K + w_0. \qquad (22.168)$$

Let
$$0 < h \le \min(1, \frac{2}{\sigma_R^{-1} + 2a(0)}).$$

Let N be the largest integer such that $Nh \le T$. Let us prove by induction that the sequence $(u_n)_{n=1}^N$ stays inside the ball $B(0, R)$. To prove this, it suffices to prove that

$$\|u_n - V_n\| \le w_0 + \|V_n\|(K-1), \qquad n = 0, 1, ..., N. \qquad (22.169)$$

Inequality (22.169) holds for $n = 0$, by (22.167). Assume that (22.169) holds for $0 \le n < N$. Let us prove that (22.169) also holds for $n + 1$.

It follows from equation (22.155) that

$$u_{n+1} - V_n = u_n - V_n - \gamma_n \left[F(u_n) + a_n u_n - F(V_n) - a_n V_n \right]$$
$$= (1 - \gamma_n a_n) \left[u_n - V_n - \frac{\gamma_n}{1 - \gamma_n a_n} \left(F(u_n) - F(V_n) \right) \right]. \qquad (22.170)$$

This and Lemma 22.2.7 imply

$$\|u_{n+1} - V_n\| \le \|u_n - V_n\|(1 - \gamma_n a_n). \qquad (22.171)$$

From (22.171), the triangle inequality, and (22.113), one gets

$$\|u_{n+1} - V_{n+1}\| \le \|u_n - V_n\|(1 - \gamma_n a_n) + \|V_{n+1} - V_n\|$$
$$\le \|u_n - V_n\| e^{-ha_n} + \frac{a_n - a_{n+1}}{a_n} \|V_{n+1}\|. \qquad (22.172)$$

Here we have used the inequality $1 - \gamma_n a_n \le e^{-ha_n}$, where $0 < h \le \gamma_n$ and $n \ge 0$. From (22.172) one gets by induction the following inequality:

$$\|u_{n+1} - V_{n+1}\| \le w_0 e^{-\phi_n} + e^{-\phi_n} \sum_{i=1}^{n} e^{\phi_i} \frac{a_i - a_{i+1}}{a_i} \|V_{i+1}\|, \qquad (22.173)$$

where
$$\phi_n = \sum_{i=0}^{n} ha_i, \qquad n \ge 0.$$

From (22.173), Lemma 22.2.2, Lemma 22.2.10, and (22.163), one obtains

$$\|u_{n+1} - V_{n+1}\| \le w_0 + \|V_{n+1}\| e^{-\phi_n} \sum_{i=1}^{n} e^{\phi_i} \frac{a_i - a_{i+1}}{a_i}$$
$$\le w_0 + \|V_{n+1}\| e^{a(0)} e^{-\phi((n+1)h)} \int_0^{(n+1)h} e^{\phi(s)} \frac{|\dot{a}(s)|}{a(s)} ds$$
$$\le w_0 + \|V_{n+1}\|(K-1), \qquad (22.174)$$

where $\phi(t)$ is defined by (22.14).

Hence, (22.169) holds for $n+1$. Thus, by induction, (22.169) holds for $0 \leq n \leq N$. Inequalities (22.169) and (22.168) and the inequality $\|V_n\| \leq \|V_\delta(T)\|$, $\forall n \leq N$ (see Lemma 22.2.2), imply that the sequence $(u_n)_{n=1}^N$ remains inside the ball $B(0, R)$

Let us prove the existence of n_δ.

Denote $g_n := g_{n,\delta} := F(u_n) + a_n u_n - f_\delta$. Equation (22.155) can be rewritten as

$$u_{n+1} - u_n = -\gamma_n g_n, \qquad n \geq 0. \tag{22.175}$$

This implies

$$
\begin{aligned}
g_{n+1} &= g_n + a_{n+1}(u_{n+1} - u_n) + F(u_{n+1}) - F(u_n) + (a_{n+1} - a_n)u_n \\
&= -\frac{u_{n+1} - u_n}{\gamma_n} + a_{n+1}(u_{n+1} - u_n) + F(u_{n+1}) - F(u_n) \\
&\quad + (a_{n+1} - a_n)u_n \\
&= -\frac{1 - \gamma_n a_{n+1}}{\gamma_n}\left[(u_{n+1} - u_n) - \frac{\gamma_n}{1 - \gamma_n a_{n+1}}(F(u_{n+1}) - F(u_n))\right] \\
&\quad + (a_{n+1} - a_n)u_n.
\end{aligned} \tag{22.176}
$$

Denote $h_n = \|g_n\|$. It follows from (22.176) that

$$
\begin{aligned}
h_{n+1} &\leq \frac{1 - \gamma_n a_{n+1}}{\gamma_n}\left\|(u_{n+1} - u_n) - \frac{\gamma_n}{1 - \gamma_n a_{n+1}}(F(u_{n+1}) - F(u_n))\right\| \\
&\quad + (a_n - a_{n+1})\|u_n\|.
\end{aligned} \tag{22.177}
$$

From Lemma 22.2.7 and (22.175) we get the following inequality:

$$
\left\|(u_{n+1} - u_n) - \frac{\gamma_n}{1 - \gamma_n a_{n+1}}(F(u_{n+1}) - F(u_n))\right\|
$$
$$
\leq \|u_{n+1} - u_n\| = \gamma_n h_n, \tag{22.178}
$$

for all $0 \leq n \leq N - 1$. From (22.178) and (22.177) one gets

$$h_{n+1} \leq (1 - \gamma_n a_{n+1})h_n + (a_n - a_{n+1})\|u_n\|. \tag{22.179}$$

Note that one has $1 - ha_{n+1} \leq e^{-a_{n+1}h}$, $\forall n \geq 0$. This, inequalities (22.179) and (22.169) imply

$$h_{n+1} \leq e^{-a_{n+1}h}h_n + (a_n - a_{n+1})(\|V_n\|K + w_0), \tag{22.180}$$

where $0 \leq n \leq N - 1$. From inequality (22.180) one gets by induction the following inequality:

$$h_n \leq h_0 e^{\phi_0 - \phi_n} + e^{-\phi_n}\sum_{i=0}^{n-1} e^{\phi_{i+1}}(a_i - a_{i+1})(\|V_i\|K + w_0), \tag{22.181}$$

where ϕ_n is defined by (22.140) and $1 \le n \le N$.

From the monotonicity of F one gets the following inequalities (see (20.135)):

$$a_n \|u_n - V_n\| \le h_n, \quad \|F(u_n) - F(V_n)\| \le h_n. \tag{22.182}$$

From (22.181) and (22.182), one gets, for $0 \le n \le N$, the following inequality:

$$\|F(u_n) - F(V_n)\| \le h_0 e^{\phi_0 - \phi_n}$$
$$+ e^{-\phi_n} \sum_{i=0}^{n-1} e^{\phi_{i+1}} (a_i - a_{i+1}) (\|V_i\| K + w_0). \tag{22.183}$$

This, the triangle inequality, and inequalities (22.162) imply

$$\|F(u_n) - f_\delta\| \le \|F(V_n) - f_\delta\| + h_0 e^{\phi_0 - \phi_n}$$
$$+ e^{-\phi_n} \sum_{i=0}^{n-1} e^{\phi_{i+1}} (a_i - a_{i+1}) \left(\frac{\Gamma K}{a_i} + w_0 \right), \tag{22.184}$$

where $0 \le n \le N$. From (22.14) and the fact that $a(t)$ is decreasing, one gets

$$\phi_n = \sum_{i=0}^{n} ha(ih) \ge \int_0^{(n+1)h} a(s)\,ds > \phi(nh), \quad n \ge 1. \tag{22.185}$$

Inequality (22.184) with $n = N$, Lemma 22.2.10, the inequality $T - 1 < Nh \le T$, by the definition of N, and (22.168) imply

$$\|F(u_N) - f_\delta\| \le \|F(V_\delta(Nh)) - f_\delta\| + h_0 e^{a_0 - \phi(Nh)}$$
$$+ e^{2a(0)} e^{-\phi(Nh)} \int_0^{Nh} e^{\phi(s)} |\dot{a}(s)| \left(\frac{\Gamma K}{a(s)} + w_0 \right) ds$$
$$< C\delta^\varsigma. \tag{22.186}$$

This implies the existence of n_δ.

The uniqueness of n_δ, satisfying (22.158), follows from its definition (22.158). *Let us prove* (22.160), *assuming that* (22.159) *holds.*

From (22.159), Lemma 22.2.11, and the fact that the sequence $(\|V_n\|)_{n=0}^{\infty}$ is increasing, one gets the following inequalities for sufficiently small $\delta > 0$:

$$h_0 e^{-\phi_{n_\delta - 2}} \le a_{n_\delta - 1} \|V_0\| < a_{n_\delta - 1} \|V_{n_\delta - 1}\| \tag{22.187}$$

and

$$e^{-\phi_{n_\delta - 2}} \sum_{i=0}^{n_\delta - 2} e^{\phi_i} (a_i - a_{i+1}) (\|V_i\| + K) \le a_{n_\delta - 1} \|V_{n_\delta - 1}\|. \tag{22.188}$$

From (22.158), (22.184) with $n = n_\delta - 1$, (22.187), and (22.188), one obtains

$$C\delta^\zeta < a_{n_\delta - 1} \|V_{n_\delta - 1}\| (1 + 1 + 1) \le 3\left(a_{n_\delta - 1}\|y\| + \delta\right), \qquad (22.189)$$

for all $0 < \delta$ sufficiently small. This and the relation $\lim_{\delta \to 0} \frac{\delta^\zeta}{\delta} = \infty$ for a fixed $\zeta \in (0,1)$ imply

$$\limsup_{\delta \to 0} \frac{\delta^\zeta}{a_{n_\delta}} \le \frac{3\|y\|}{C}. \qquad (22.190)$$

Inequalities (22.190), $\delta < C\delta^\zeta$, and (22.111) imply, for sufficiently small $\delta > 0$, the following inequality:

$$\|V_n\| \le \|y\| + \frac{\delta}{a_{n_\delta}} < \|y\| + \frac{C\delta^\zeta}{a_{n_\delta}} \le 4\|y\|, \qquad 0 \le n \le n_\delta. \qquad (22.191)$$

Using estimate (22.191), one obtains

$$\lim_{\delta \to 0} \frac{\sum_0^{n_\delta - 1} e^{\phi_i}(a_i - a_{i+1})\|V_i\|}{e^{\phi_{n_\delta - 1}} a_{n_\delta}} \le 4\|y\| \lim_{\delta \to 0} \frac{\sum_0^{n_\delta - 1} e^{\phi_i}(a_i - a_{i+1})}{e^{\phi_{n_\delta - 1}} a_{n_\delta}}. \qquad (22.192)$$

It follows from (22.148) and (22.192) that

$$\lim_{\delta \to 0} \frac{\sum_{i=0}^{n_\delta - 1} e^{\phi_i}(a_i - a_{i+1})\|V_i\|}{e^{\phi_{n_\delta - 1}} a_{n_\delta}} = 0. \qquad (22.193)$$

From (22.181) and (22.182), one gets

$$\|u_n - V_n\| \le \frac{e^{-\phi_{n-1}} h_0}{a_n} + \frac{e^{-\phi_{n-1}}}{a_n} \sum_{i=0}^{n-1} e^{-\phi_i}(a_i - a_{i+1})\left(\|V_i\| K + w_0\right). \qquad (22.194)$$

From (22.147), (22.159), (22.193), and (22.194), one obtains

$$\lim_{\delta \to 0} \|u_{n_\delta} - V_{n_\delta}\| = 0. \qquad (22.195)$$

It follows from (22.190) that

$$\lim_{\delta \to 0} \frac{\delta}{a_{n_\delta}} = 0. \qquad (22.196)$$

From the triangle inequality and inequality (22.109), one obtains

$$\|u_{n_\delta} - y\| \le \|u_{n_\delta} - V_{n_\delta}\| + \|V_{n_\delta} - V_{0,n_\delta}\| + \|V_{0,n_\delta} - y\|$$
$$\le \|u_{n_\delta} - V_{n_\delta}\| + \frac{\delta}{a_{n_\delta}} + \|V_{0,n_\delta} - y\|. \qquad (22.197)$$

From (22.195)–(22.197), (22.159), and Lemma 6.1.9, one obtains (22.160).

Theorem 22.2.12 is proved. ∎

Theorem 22.2.14 *Let F, f, f_δ, and u_δ be as in Theorem 22.2.12 and let y be the minimal-norm solution to the equation $F(u) = f$. Let $0 < (\delta_m)_{m=1}^\infty$ be a sequence such that $\delta_m \to 0$. If the sequence $\{n_{\delta_m}\}_{m=1}^\infty$ is bounded and $\{n_{\delta_{m_j}}\}_{j=1}^\infty$ is a convergent subsequence, then*

$$\lim_{j \to \infty} u_{n_{\delta_{m_j}}} = u^\star, \qquad (22.198)$$

where $u^\star$ is a solution to the equation $F(u) = f$.

Proof: Since the subsequence $n_{\delta_{m_j}}$ is a sequence of integers that is bounded and convergent all the members of this sequence for $j \geq j_0$, where j_0 is a sufficiently large integer, must be equal to a fixed integer, say, N. Therefore, the sequence $u_{n_{\delta_{m_j}}}$ converges to an element u_N.

If $n > 0$ is fixed, then u_n is a continuous function of f_δ. Denote

$$u^\star := u_N := \lim_{j \to \infty} u_{n_{\delta_{m_j}}}, \qquad (22.199)$$

where

$$\lim_{j \to \infty} n_{\delta_{m_j}} = N. \qquad (22.200)$$

From (22.199) and the continuity of F, one obtains

$$\|F(u^\star) - f\| = \lim_{j \to \infty} \|F(u_{n_{\delta_{m_j}}}) - f_{\delta_{m_j}}\| \leq \lim_{j \to \infty} C\delta_{m_j}^\zeta = 0. \qquad (22.201)$$

Thus, $u^\star$ is a solution to the equation $F(u) = f$, and (22.198) is proved.
Theorem 22.2.14 is proved. ∎

Let us assume in addition that $a(t)$ satisfies the following inequalities:

$$2 \geq a(0), \qquad \nu(0) = \frac{|\dot{a}(0)|}{a^2(0)} \leq \frac{1}{10}. \qquad (22.202)$$

Remark 22.2.15 Let $b \in (0, 1)$, $c \geq 5$, $d > 0$, and

$$a(t) = \frac{d}{(c + t)^b}, \qquad \frac{10b}{c^{1-b}} \leq d \leq 2c^b.$$

Then $a(t)$ satisfies (22.115) and (22.202).

We have the following result.

Theorem 22.2.16 *Let $a(t)$ satisfy (22.115) and (22.202). Let F, f, f_δ, n_δ, and u_δ be as in Theorem 22.2.12. Assume that u_0 satisfies either*

$$h_0 = \|F(u_0) + a_0 u_0 - f_\delta\| \leq \theta\delta^\zeta, \qquad 0 < \theta < C, \qquad (22.203)$$

or

$$\|F(u_0) + a_0 u_0 - f_\delta\| \leq \frac{1}{8} a_0 \|V_0\|. \tag{22.204}$$

Assume $\zeta \in (0,1)$. *Then*

$$\lim_{\delta \to 0} n_\delta = \infty. \tag{22.205}$$

Remark 22.2.17 The element u_0 satisfying (22.203) can be obtained easily by the following fixed point iterations:

$$v_{n+1} = v_n - \gamma(F(v_n) + a_0 v_n - f_\delta), \qquad n \geq 0, \tag{22.206}$$

where $v_0 \in B(0, R)$, $0 < R$ is sufficiently large, and γ is chosen so that

$$0 < \gamma < \frac{2}{\sigma_R^{-1} + 2a_0}. \tag{22.207}$$

Note that the operator $G(v) := v - \gamma(F(v) + a_0 v - f_\delta)$ is a contraction map as a consequence of estimate (22.131).

Inequality (22.204) is a sufficient condition for the following inequality to hold (see also (22.220) below):

$$e^{-\varphi_n} h_0 \leq \frac{1}{4} a_n \|V_n\|, \qquad t \geq 0. \tag{22.208}$$

By the arguments similar to the ones used in the proof of Lemma 22.2.11, one can prove that

$$\lim_{n \to \infty} e^{\varphi_n} a_n = \infty.$$

In the proof of Theorem 22.2.16, inequality (22.208) (or (22.220)) is used at $n = n_\delta$. The stopping time n_δ is often sufficiently large for the quantity $e^{\varphi_{n_\delta}} a_{n_\delta}$ to be large. In this case, inequality (22.220) with $n = n_\delta$ is satisfied for a wide range of u_0.

It is an *open problem* to choose ζ (see (22.157)) which is optimal in some sense. In practice it is *natural* to choose C and ζ so that $C\delta^\zeta$ is close to δ. Indeed, if v solves equation $F(u) = f$, then $\|F(v) - f_\delta\| = \|f - f_\delta\| \leq \delta$.

Proof: [Proof of Theorem 22.2.16]

Let us prove (22.205), assuming that (22.204) holds. When (22.203) holds, instead of (22.204), the proof follows similarly.

It follows from (22.179), the triangle inequality, and (22.182) that

$$h_{n+1} \leq (1 - ha_{n+1})h_n + (a_n - a_{n+1})\|u_n - V_n\| + (a_n - a_{n+1})\|V_n\|$$
$$\leq (1 - ha_{n+1})h_n + \frac{a_n - a_{n+1}}{a_n} h_n + (a_n - a_{n+1})\|V_n\|, \tag{22.209}$$

for $0 \leq n \leq n_\delta - 1$. From (22.116) and (22.202) one gets

$$1 - ha_{n+1} + \frac{a_n - a_{n+1}}{a_n} \leq 1 - \frac{ha_{n+1}}{2} \leq e^{-\frac{ha_{n+1}}{2}}. \tag{22.210}$$

From (22.209) and (22.210) one obtains

$$h_{n+1} \leq e^{-\frac{ha_{n+1}}{2}} h_n + (a_n - a_{n+1})\|V_n\|. \tag{22.211}$$

This implies

$$h_n \leq e^{-\varphi_n} h_0 + e^{-\varphi_n} \sum_{i=0}^{n-1} e^{-\varphi_i}(a_i - a_{i+1})\|V_i\|, \qquad 1 \leq n \leq n_\delta. \tag{22.212}$$

where

$$\varphi_n = \sum_{i=1}^{n} \frac{ha_n}{2}. \tag{22.213}$$

It follows from the triangle inequality, (22.108), (22.182), and (22.212) that

$$\|F(u_n) - f_\delta\| \geq \|F(V_n) - f_\delta\| - \|F(V_n) - F(u_n)\|$$
$$\geq a_n\|V_n\| - h_0 e^{-\varphi_n}$$
$$- e^{-\varphi_n} \sum_{i=0}^{n-1} e^{\varphi_{i+1}}(a_i - a_{i+1})\|V_i\|. \tag{22.214}$$

From Lemma 22.2.6 one obtains

$$\frac{1}{2}a_n\|V_n\| \geq e^{-\varphi_n} \sum_{i=0}^{n-1} e^{\varphi_{i+1}}(a_i - a_{i+1})\|V_i\|. \tag{22.215}$$

From the relation $h_n = \|g_n\| = \|F(u_n) + a_n u_n - f_\delta\|$ and (22.204) one gets

$$h_0 e^{-\varphi_n} \leq \frac{1}{8}a_0\|V_0\|e^{-\varphi_n}, \qquad n \geq 0. \tag{22.216}$$

It follows from (22.115) that

$$a_1 \leq a_{n+1}e^{\varphi_n}, \qquad \forall n \geq 0. \tag{22.217}$$

Indeed, inequality $a_1 \leq a_{n+1}e^{\varphi_n}$ is obviously true for $n = 0$, and $a_{n+1}e^{\varphi_n}$ is an increasing sequence because

$$a_{n+1}e^{\varphi_n} - a_n e^{\varphi_{n-1}} = e^{\varphi_{n-1}}\left(a_{n+1}e^{\frac{ha_n}{2}} - a_n\right)$$
$$\geq e^{\varphi_{n-1}}\left(a_{n+1} + \frac{ha_n}{2}a_{n+1} - a_n\right)$$
$$= e^{\varphi_{n-1}}a_n a_{n+1}\left(\frac{h}{2} - \frac{a_n - a_{n+1}}{a_n a_{n+1}}\right) \geq 0, \tag{22.218}$$

by (22.116) and (22.202). From (22.217), (22.202), and (22.117) one gets

$$e^{-\varphi_n}a_0 \leq a_{n+1}\frac{a_0}{a_1} < 2a_{n+1}, \qquad n \geq 0. \tag{22.219}$$

Inequalities (22.216) and (22.219) imply

$$e^{-\varphi_n} h_0 \le \frac{1}{4} a_{n+1} \|V_0\| \le \frac{1}{4} a_n \|V_n\|, \quad n \ge 0, \tag{22.220}$$

where we have used the inequality $\|V_{n'}\| \le \|V_n\|$ for $n' \le n$ (see Lemma 22.2.2). From (22.214), (22.215), and (22.220) one gets

$$\|F(u_{n_\delta}) - f_\delta\| \ge a_{n_\delta} \|V_{n_\delta}\| - \frac{1}{4} a_{n_\delta} \|V_{n_\delta}\| - \frac{1}{2} a_{n_\delta} \|V_{n_\delta}\| = \frac{1}{4} a_{n_\delta} \|V_{n_\delta}\|.$$

This and (22.158) imply

$$C\delta^\zeta \ge \|F(u_{n_\delta}) - f_\delta\| \ge \frac{1}{4} a_{n_\delta} \|V_{n_\delta}\|.$$

Thus,

$$\lim_{\delta \to 0} a_{n_\delta} \|V_{n_\delta}\| \le \lim_{\delta \to 0} 4C\delta^\zeta = 0. \tag{22.221}$$

From (22.109) and the triangle inequality we obtain

$$a_{n_\delta} \|V_{0,n_\delta}\| \le a_{n_\delta} \|V_{n_\delta}\| + a_{n_\delta} \|V_{n_\delta} - V_{0,n_\delta}\| \le a_{n_\delta} \|V_{n_\delta}\| + \delta. \tag{22.222}$$

This and (22.221) imply

$$\lim_{\delta \to 0} a_{n_\delta} \|V_{0,n_\delta}\| = 0. \tag{22.223}$$

Since $\|V_{0,n_\delta}\| \ge \|V_{0,0}\| > 0$, relation (22.223) implies $\lim_{\delta \to 0} a_{n_\delta} = 0$. Since $0 < a_n \searrow 0$, it follows that (22.205) holds.

Theorem 22.2.16 is proved. ∎

Instead of using iterative scheme (22.155), one may use the following iterative scheme:

$$u_{n+1} = u_n - \gamma_n [F(u_n) + a_n(u_n - \bar{u}) - f_\delta], \quad u_0 = \tilde{u}_0, \tag{22.224}$$

where $\bar{u}, \tilde{u}_0 \in H$. Denote $\tilde{F}(u) := F(u + \bar{u})$. If F is a locally σ-inverse monotone operator, then so is $\tilde{F}$. Using Theorem 22.2.12 with $F := \tilde{F}$, one gets the following corollary:

Corollary 22.2.18 *Let $a(t)$ satisfy (22.115) and (22.202). Let $0 < R = const$ be sufficiently large and let h and γ_n satisfy (22.156). Assume that $F : H \to H$ is a locally σ-inverse monotone operator, and u_0 is an element of H, satisfying inequality*

$$\|F(u_0) - f_\delta\| > C\delta^\zeta > \delta, \tag{22.225}$$

where $C > 0$ and $0 < \zeta \le 1$ are constants. Assume also that u_0 satisfy either

$$\|F(u_0) + a_0(u_0 - \bar{u}) - f_\delta\| \le \frac{1}{8} a_0 \|V_0\|,$$

or

$$\|F(u_0) + a_0(u_0 - \bar{u}) - f_\delta\| \le \theta\delta^\gamma, \qquad 0 < \theta = const < C.$$

Assume that equation $F(u) = f$ has a solution, possibly nonunique, and $z \in B(u_0, R)$ is the solution with minimal distance to $\bar{u}$. Let f_δ be such that $\|f_\delta - f\| \le \delta$. Let u_n be defined by (22.224). Then there exists a unique n_δ such that

$$\|F(u_{n_\delta}) - f_\delta\| \le C\delta^\zeta, \quad \|F(u_n) - f_\delta\| > C\delta^\zeta, \qquad 0 \le n < n_\delta, \qquad (22.226)$$

where C and ζ are constants from (22.157). If $\zeta \in (0,1)$ and n_δ satisfies (22.158), then

$$\lim_{\delta \to 0} \|u_{n_\delta} - z\| = 0. \qquad (22.227)$$

22.2.2.2 An algorithm for solving equations with σ-inverse operators   Let us formulate an algorithm for solving equations with σ-inverse operators.

Algorithm

1. Estimate the constant $\sigma = \sigma_R$ in (22.104).

2. Choose an $a(t)$ satisfying (22.115).

3. Choose $h = \frac{2}{\sigma^{-1}+2a(0)}$ and γ_n to satisfy conditions (22.156).

4. Find an initial approximation u_0 for y or simply set $u_0 = 0$.

5. Compute u_n by formula (22.119), use (22.158) to stop the iterations at n_δ, and use u_{n_δ} as an approximate solution to the equation $F(u) = f$.

Theorem 22.2.14 guarantees the convergence of u_{n_δ}, computed by the **Algorithm** above, to, at least, a solution to $F(u) = f$. If the equation $F(u) = f$ has a unique solution, then u_{n_δ} converges to this unique solution.

If one chooses $a(t)$ to satisfy (22.202) in addition, and u_0 to satisfy (22.203) or (22.204), then $n_\delta \to \infty$ as $\delta \to 0$, as proved in Theorem 22.2.16. Consequently, u_{n_δ} converges to the minimal-norm solution y as stated by Theorem 22.2.14.

Note that the element u_0 satisfying (22.203) can be found from iteration (22.206). Moreover, in practice n_δ, is often large when δ is sufficiently small. Thus, in practice one can also use $u_0 = 0$ as pointed out in Remark 22.2.17.

The **Algorithm** above can also be implemented for solving equations with locally σ-inverse operators. Since the constant σ_R depends on R, one should choose R sufficiently large so that the sequence $(u_n)_{n=1}^{n_\delta}$ remains inside the ball $B(0, R)$. However, if one chooses R too large, then h and γ_n satisfying (22.156) are small. Consequently, the computation cost will be large. Thus, R should be chosen not too small so that the sequence $(u_n)_{n=1}^{n_\delta}$ remains inside the ball $B(0, R)$, and not too large so that the computation cost is not large. The choice of R varies from problems to problems.

CHAPTER 23

DSM FOR SOLVING NONLINEAR OPERATOR EQUATIONS IN BANACH SPACES

Consider an operator equation

$$F(u) = f, \tag{23.1}$$

where F is an operator in a Banach space X with a Gateaux-differentiable norm.

A1. *Assume that F is continuously Fréchet differentiable, $F'(u) := A(u)$, and*

$$\|A(u) - A(v)\| \leq \omega(\|u - v\|), \quad \omega(r) = c_0 r^\kappa, \quad \kappa \in (0, 1], \tag{23.2}$$

where $c_0 > 0$ is a constant, and $\omega(r)$ is a continuous strictly growing function, $\omega(0) = 0$.

A2. *Assume that*

$$\|A_a^{-1}(u)\| \leq \frac{c_1}{|a|^b}; \ |a| > 0, \ A_a := A + aI, \ c_1 = const > 0, \ b > 0, \tag{23.3}$$

where a may be a complex number, $|a| > 0$, and there exists a smooth path L on the complex plane $\mathbb{C}$, such that for any $a \in L$, $|a| < \epsilon_0$, where $\epsilon_0 > 0$ is a

Dynamical Systems Method and Applications: Theoretical Developments and Numerical Examples, First Edition. A. G. Ramm and N. S. Hoang.
Copyright © 2012 John Wiley & Sons, Inc.

small fixed number independent of u, estimate (23.3) holds, and L joins the origin and some point a_0, $0 < |a_0| < \epsilon_0$.

Assumption (23.3) holds if there is a smooth path L on a complex a-plane, consisting of regular points of the operator $A(u)$, such that the norm of the resolvent $A_a^{-1}(u)$ grows, as $a \to 0$, not faster than a power $|a|^{-b}$. Thus, assumption (23.3) is a weak assumption. For example, assumption (23.3) is satisfied for the class of linear operators A, satisfying the spectral assumption, introduced in Chapter 8 in this book. This spectral assumption says that the set $\{a : |\arg a - \pi| \leq \phi_0, \ 0 < |a| < \epsilon_0\}$ consists of the regular points of the operator A. This assumption implies the estimate $\|A_a^{-1}\| \leq \frac{c_1}{a}$, $0 < a < \epsilon_0$, similar to estimate (23.3).

A3. *Assume additionally that the equation*

$$F(w_a) + aw_a - f = 0, \quad a \in L, \tag{23.4}$$

is uniquely solvable for any $f \in X$, and

$$\lim_{a \to 0, a \in L} \|w_a - y\| = 0, \quad F(y) = f. \tag{23.5}$$

We also assume that there exists a constant $c > 0$ such that

$$|\dot{a}(t)| \leq c|\dot{r}(t)|, \quad r(t) := |a(t)|. \tag{23.6}$$

In formula (23.15) inequality $|\dot{r}(t)| \leq |\dot{a}(t)|$ is established. Thus,

$$|\dot{r}(t)| \leq |\dot{a}(t)| \leq c|\dot{r}(t)|. \tag{23.7}$$

All the above assumptions are standing.

The result in this chapter has been published in [157]. We formulate the main result at the end of the chapter for convenience of the reader, because some additional assumptions, used in the proof of Theorem 23.2.1, are flexible and will arise naturally in the course of the proof. One of the goals in this chapter is to demonstrate the methodology for establishing the convergence results. In recent papers [162, 165] a new nonlinear differential inequality, generalizing inequality (23.10), is obtained.

These assumptions are satisfied, for example, if F is a monotone operator in a Hilbert space H and L is a segment $[0, \epsilon_0]$, in which case $c_1 = 1$ and $b = 1$. Sufficient conditions for (23.5) to hold are given in earlier chapters and in [157].

Every equation (23.1) with a linear, closed, densely defined in a Hilbert space H operator $F = A$ can be reduced to an equation with a monotone operator A^*A, where A^* is the adjoint to A. The operator $T := A^*A$ is selfadjoint and densely defined in H. If $f \in D(A^*)$, where $D(A^*)$ is the domain of A^*, then the equation $Au = f$ is equivalent to $Tu = A^*f$, provided that $Au = f$ has a solution, that is, $f \in R(A)$, where $R(A)$ is the range of A. Recall that $D(A^*)$ is dense in H if A is closed and densely defined in H. If

$f \in R(A)$ but $f \notin D(A^*)$, then equation $Tu = A^*f$ still makes sense and its normal solution y, that is, the solution with minimal norm, can be defined as

$$y = \lim_{a \to 0} T_a^{-1} A^* f. \qquad (23.8)$$

One proves that $Ay = f$, and $y \perp N(A)$, where $N(A)$ is the null-space of A. These results are proved in [152]–[158].

Our aim is to prove convergence of the DSM for solving equation (23.1):

$$\dot{u} = -A_{a(t)}^{-1}(u(t))[F(u(t)) + a(t)u(t) - f], \quad u(0) = u_0; \quad \dot{u} := \frac{du}{dt}, \qquad (23.9)$$

where $u_0 \in X$ is an initial element, $a(t) \in C^1[0, \infty)$, $a(t) \in L$. Our main result is formulated in Theorem 23.1.4, in Section 23.2.

The novel points in the current chapter include the larger class of the operator equations than earlier considered, along with the weakened assumptions on the smoothness of the nonlinear operator F. While in earlier chapters it was often assumed that $F''(u)$ is locally bounded, in the current chapter a weaker assumption (23.2) is used.

Our proof of Theorem 23.1.4 uses the following corollary which follows immediately from Theorem 5.6.1 when the function $\alpha(t, y)$ is of the form $\alpha(t)y^p$, where $p = const > 1$.

Corollary 23.0.19 *Assume that $g(t) \geq 0$ is continuously differentiable on any interval $[0, T)$, on which it is defined, and satisfies the following inequality:*

$$\dot{g}(t) \leq -\gamma(t)g(t) + \alpha(t)g^p(t) + \beta(t), \quad t \in [0, T), \qquad (23.10)$$

where $p > 1$ is a constant, $\alpha(t) > 0$, and $\gamma(t)$ and $\beta(t)$ are three continuous on $[0, \infty)$ functions. Suppose that there exists $\mu(t) > 0$, $\mu(t) \in C^1[0, \infty)$, such that

$$\alpha(t)\mu^{-p}(t) + \beta(t) \leq \mu^{-1}(t)[\gamma(t) - \dot{\mu}(t)\mu^{-1}(t)], \quad t \geq 0, \qquad (23.11)$$

and

$$\mu(0)g(0) < 1. \qquad (23.12)$$

Then $T = \infty$, that is, g exists on $[0, \infty)$, and

$$0 \leq g(t) < \mu^{-1}(t), \quad t \geq 0. \qquad (23.13)$$

In Section 23.2 a method is given for a proof of the following conclusions: *There exists a unique solution $u(t)$ to problem (23.9) for all $t \geq 0$, there exists $u(\infty) := \lim_{t \to \infty} u(t)$, and $F(u(\infty)) = f$, that is,*

$$\exists! u(t) \quad \forall t \geq 0; \ \exists u(\infty); \quad F(u(\infty)) = f. \qquad (23.14)$$

The assumptions on u_0 and $a(t)$ under which conclusions (23.14) hold for the solution to problem (23.9) are formulated in Theorem 23.1.4 in Section 23.2. Theorem 23.1.4 in Section 23.2 is our main result. Roughly speaking, this result says that conclusions (23.14) hold for the solution to problem (23.9), provided that $a(t)$ is suitably chosen.

Results in this chapter are taken from papers [58], [157], and [166].

23.1   PROOFS

Let $|a(t)| := r(t) > 0$. If $a(t) = a_1(t) + ia_2(t)$, where $a_1(t) = \operatorname{Re} a(t)$, $a_2(t) = \operatorname{Im} a(t)$, then

$$|\dot{r}(t)| \leq |\dot{a}(t)|. \qquad (23.15)$$

Indeed,

$$|\dot{r}(t)| = \frac{|a_1\dot{a}_1 + a_2\dot{a}_2|}{r(t)} \leq \frac{r(t)|\dot{a}(t)|}{r(t)}, \qquad (23.16)$$

and (23.16) implies (23.15).
   Let

$$g(t) := \|z(t)\|, \qquad z(t) := u(t) - w_a(t), \qquad (23.17)$$

where $u(t)$ solves (23.9) and $w_a(t)$ solves (23.4) with $a = a(t)$. By the assumption A3, $w_a(t)$ exists for every $t \geq 0$. The local existence of $u(t)$, the solution to (23.9), is the conclusion of Lemma 23.1.1. Let $\psi(t) \in C^1([0, \infty); X)$. In the following lemma a proof of local existence of the solution to problem (23.9) is given by a novel argument. The right-hand side of (23.9) is a nonlinear function of u, which does not satisfy the Lipschitz condition, which is the standard condition in the usual proofs of the local existence of the solution to an evolution problem. Our argument uses an abstract inverse function theorem. This argument is valid under the minimal assumption that $F'(u)$ depends continuously on u.

Lemma 23.1.1 *If assumption (23.3) holds and (23.4) is uniquely solvable for any $f \in X$, then the solution $u(t)$ to (23.9) exists locally.*

Proof: Differentiate equation (23.4) with $a = a(t)$ with respect to t. The result is

$$A_{a(t)}(w_a(t))\dot{w}_a(t) = -\dot{a}(t)w_a(t), \qquad (23.18)$$

or

$$\dot{w}_a(t) = -\dot{a}(t)A_{a(t)}^{-1}(w_a(t))w_a(t). \qquad (23.19)$$

Denote

$$\psi(t) := F(u(t)) + a(t)u(t) - f. \qquad (23.20)$$

For any $\psi \in H$, equation (23.20) is uniquely solvable for $u(t)$ by our assumption (23.4), which is used with $f + \psi(t)$ in place of f in (23.4). By the inverse function theorem, which holds due to our assumption (23.3), and by assumption (23.2), the solution $u(t)$ to (23.20) is continuously differentiable with respect to t if $\psi(t)$ is. One may solve (23.20) for u and write $u = G(\psi)$, where the map G is continuously Fréchet differentiable because F is.
   Differentiate (23.20) and get

$$\dot{\psi}(t) = A_{a(t)}(u(t))\dot{u}(t) + \dot{a}(t)u. \qquad (23.21)$$

If one wants the solution to (23.20) to be a solution to (23.9), then one has to require that

$$A_{a(t)}(u(t))\dot{u} = -\psi(t). \tag{23.22}$$

If (23.22) holds, then (23.21) can be written as

$$\dot{\psi}(t) = -\psi + \dot{a}(t)G(\psi), \quad G(\psi) := u(t), \tag{23.23}$$

where $G(\psi)$ is continuously Fréchet differentiable. Thus, equation (23.23) is equivalent to (23.9) for all $t \geq 0$ if

$$\psi(0) = F(u_0) + a(0)u_0 - f. \tag{23.24}$$

Indeed, if u solves (23.9), then ψ, defined in (23.20), solves the Cauchy problem (23.23)–(23.24). Conversely, if ψ solves (23.23)–(23.24), then $u(t)$, defined as the unique solution to (23.20), solves (23.9). Since the right-hand side of (23.23) is Fréchet differentiable, it satisfies a local Lipschitz condition. Thus, problem (23.23)–(23.24) is locally solvable. Therefore, problem (23.9) is locally solvable.
Lemma 23.1.1 is proved.   ∎

To prove that the solution $u(t)$ to (23.9) exists globally, it is sufficient to prove the following estimate:

$$\sup_{t \geq 0} \|u(t)\| < \infty. \tag{23.25}$$

Lemma 23.1.2 *Estimate* (23.25) *holds.*

Proof: Denote

$$z(t) := u(t) - w(t), \tag{23.26}$$

where $u(t)$ solves (23.9) and $w(t) = w_{a(t)}$ solves (23.4) with $a = a(t)$. When $t \to \infty$, the function $w(t)$ tends to the limit y by (23.5) and, therefore, is uniformly bounded. If one proves that

$$\lim_{t \to \infty} \|z(t)\| = 0, \tag{23.27}$$

then (23.25) follows from (23.27) and the boundedness of $w(t)$. Indeed,

$$\sup_{t \geq 0} \|u(t)\| \leq \sup_{t \geq 0} \|z(t)\| + \sup_{t \geq 0} \|w(t)\| < \infty. \tag{23.28}$$

                                                                           ∎

To prove (23.27), *we use Corollary 23.0.19.*
Rewrite (23.9) as

$$\dot{z} = -\dot{w} - A_{a(t)}^{-1}(u(t))[F(u(t)) - F(w(t)) + a(t)z(t)]. \tag{23.29}$$

Lemma 23.1.3 *If the norm* $\|w(t)\|$ *in* X *is differentiable, then*

$$\left| \frac{d\|w(t)\|}{dt} \right| \le \|\dot{w}(t)\|. \tag{23.30}$$

Proof: The triangle inequality implies

$$\frac{\|w(t+s)\| - \|w(t)\|}{s} \le \frac{\|w(t+s) - w(t)\|}{s}, \quad s > 0. \tag{23.31}$$

Passing to the limit $s \searrow 0$ and using the assumption concerning the differentiability of the norm in X, one gets $\frac{d\|w(t)\|}{dt} \le \|\dot{w}(t)\|$. Similarly, one gets $-\frac{d\|w(t)\|}{dt} \le \|\dot{w}(t)\|$. These two inequalities yield (23.30).
Lemma 23.1.3 is proved.   ∎

Various necessary and sufficient conditions for the Gateaux or Fréchet differentiability of the norm in Banach spaces are known in the literature (see, e.g., [22] and [29]), starting with Shmulian's paper of 1940 (see [178]).

Hilbert spaces, $L^p(D)$ and ℓ^p-spaces, $p \in (1, \infty)$, and Sobolev spaces $W^{\ell, p}(D)$, $p \in (1, \infty)$, $D \in \mathbb{R}^n$, is a bounded domain, have Fréchet differentiable norms. These spaces are uniformly convex and they have the following property: If $u_n \rightharpoonup u$ and $\|u_n\| \to \|u\|$ as $n \to \infty$, then $\lim_{n \to \infty} \|u_n - u\| = 0$.

From (23.19) and (23.7) one gets

$$\|\dot{w}\| \le c_1 |\dot{a}(t)| r^{-b}(t) \|w(t)\|, \quad r(t) = |a(t)|, \tag{23.32}$$

where $w(t) := w_a(t)$. Since we assume that $\lim_{t \to \infty} |a(t)| = 0$, one concludes that (23.5) and (23.32) imply the following inequality:

$$\|\dot{w}\| \le c_2 |\dot{a}(t)| r^{-b}(t), \quad c_2 = const > 0, \tag{23.33}$$

because (23.5) implies the following estimate:

$$c_1 \|w(t)\| \le c_2, \quad t \ge 0. \tag{23.34}$$

Inequalities (23.7) and (23.33) imply that

$$\|\dot{w}\| \le c_2 |\dot{r}(t)| r^{-b}(t), \quad t \ge 0. \tag{23.35}$$

Recall that $F'(u) := A(u)$ and note that

$$F(u) - F(w) = \int_0^1 F'(w + sz) \, dsz = A(u)z + \int_0^1 [A(w + sz) - A(u)] \, dsz. \tag{23.36}$$

Thus, one can write (23.29) as

$$\dot{z}(t) = -z(t) - \dot{w}(t) - A_{a(t)}^{-1}(u(t))\eta(t) := -z(t) + W, \tag{23.37}$$

$$\|\eta(t)\| = O(g^p(t)), \quad p = 1 + \kappa, \quad g(t) := \|z(t)\|, \tag{23.38}$$

where estimate (23.2) was used, and W is defined by the formula

$$W := -\dot{w}(t) - A_{a(t)}^{-1}(u(t))\eta(t). \tag{23.39}$$

Let

$$Z(t) := e^t z(t). \tag{23.40}$$

Then (23.37) yields

$$e^{-t}\dot{Z} = W. \tag{23.41}$$

Taking the norm of this equation yields

$$e^{-t}\|\dot{Z}\| = \|W\|. \tag{23.42}$$

One has

$$\|W\| \le c_2|\dot{r}(t)|r^{-b}(t) + c_3 r^{-b}(t)g^p(t), \qquad g(t) := \|z(t)\|, \quad p = 1+\kappa, \tag{23.43}$$

where $c_3 := c_0 c_1$, c_0 is the constant from (23.2) and c_1 is the constant from (23.3). Using estimate (23.30), one gets

$$\|\dot{Z}\| \ge \left|\frac{d\|Z(t)\|}{dt}\right| = \left|\frac{d(e^t g(t))}{dt}\right|. \tag{23.44}$$

Using formulas (23.41)–(23.44) one gets from (23.37) the following inequality:

$$\dot{g}(t) \le -g + c_2|\dot{r}(t)|r^{-b}(t) + c_3 r^{-b}(t)g^p, \qquad g(t) = \|z(t)\|, \quad p = 1+\kappa. \tag{23.45}$$

Inequality (23.45) is of the form (23.10) with

$$\gamma(t) = 1, \quad \alpha(t) = c_3 r^{-b}(t), \quad \beta(t) = c_2|\dot{r}(t)|r^{-b}(t). \tag{23.46}$$

Choose

$$\mu(t) = \lambda r^{-k}(t), \quad \lambda = const > 0, \quad k = const > 0. \tag{23.47}$$

Then

$$\dot{\mu}\mu^{-1} = -k\dot{r}r^{-1}. \tag{23.48}$$

Let us assume that

$$r(t) \searrow 0, \quad \dot{r} < 0, \quad |\dot{r}| \searrow 0. \tag{23.49}$$

Assumption (23.12) implies

$$g(0)\frac{\lambda}{r^k(0)} < 1, \tag{23.50}$$

and inequality (23.11) holds if

$$\frac{c_3 r^{-b}(t)r^{kp}}{\lambda^p} + c_2|\dot{r}(t)|r^{-b}(t) \le \frac{r^k(t)}{\lambda}\left(1 - k|\dot{r}(t)|r^{-1}(t)\right), \qquad t \ge 0. \tag{23.51}$$

Inequality (23.51) can be written as

$$\frac{c_3 r^{k(p-1)-b}(t)}{\lambda^{p-1}} + \frac{c_2\lambda|\dot{r}(t)|}{r^{k+b}(t)} + \frac{k|\dot{r}(t)|}{r(t)} \le 1, \qquad t \ge 0. \tag{23.52}$$

Let us choose k so that

$$k(p-1) - b = 1, \tag{23.53}$$

that is,

$$k = \frac{b+1}{p-1}. \tag{23.54}$$

Choose λ, for example, as follows:

$$\lambda := \frac{r^k(0)}{2g(0)}. \tag{23.55}$$

Then inequality (23.50) holds, and inequality (23.52) can be written as

$$c_3 \frac{r(t)[2g(0)]^{p-1}}{[r^k(0)]^{p-1}} + c_2\frac{r^k(0)}{2g(0)}\frac{|\dot{r}(t)|}{r^{k+b}(t)} + k\frac{|\dot{r}(t)|}{r(t)} \le 1, \qquad t \ge 0. \tag{23.56}$$

Note that (23.54) implies

$$k + b = kp - 1. \tag{23.57}$$

Choose $r(t)$ so that relations (23.49) hold and

$$k\frac{|\dot{r}(t)|}{r(t)} \le \frac{1}{2}, \qquad t \ge 0. \tag{23.58}$$

Since $r(0) \ge r(t)$ and (23.58) holds, then inequality (23.56) holds if

$$c_3 \frac{[2g(0)]^{p-1}}{r^b(0)} + c_2\frac{r^k(0)}{2g(0)}\frac{|\dot{r}(t)|}{r^{kp-1}} \le \frac{1}{2}, \qquad t \ge 0. \tag{23.59}$$

Denote

$$c_2\frac{r^k(0)}{2g(0)} = c_2\lambda := c_4. \tag{23.60}$$

Let

$$c_4\frac{|\dot{r}(t)|}{r^{kp-1}} = \frac{1}{4}, \qquad t \ge 0, \tag{23.61}$$

and $kp > 2$. Then equation (23.61) implies

$$r(t) = \left[1 + t\frac{4c_4}{kp-2}4c_4\right]^{-\frac{1}{kp-2}}. \tag{23.62}$$

This $r(t)$ satisfies conditions (23.49), and equation (23.61) implies

$$k\frac{|\dot{r}(t)|}{r(t)} = \frac{kr^{kp-2}(t)}{4c_4}, \qquad t \ge 0. \tag{23.63}$$

Recall that $r(t)$ decays monotonically. Therefore, inequality (23.58) holds if

$$\frac{kr^{kp-2}(0)}{4c_4} \le \frac{1}{2}.$$ (23.64)

Inequality (23.64) holds if

$$\frac{kg(0)}{c_2}r^{k(p-1)-2}(0) = \frac{kg(0)}{c_2}r^{b-1}(0) \le 1,$$ (23.65)

because (23.54) implies

$$k(p-1) - 2 = b - 1.$$ (23.66)

Condition (23.65) holds if $g(0)$ is sufficiently small or $r^{b-1}(0)$ is sufficiently large:

$$g(0) \le \frac{c_2}{k}r^{b-1}(0).$$ (23.67)

If $b > 1$, then condition (23.67) holds for any fixed $g(0)$ if $r(0)$ is sufficiently large. If $b = 1$, then (23.67) holds if $g(0) \le \frac{c_2}{k}$. If $b \in (0,1)$, then (23.67) holds if either $g(0)$ is sufficiently small or $r(0)$ is sufficiently small.

Consequently, if (23.62) and (23.67) hold, then (23.61) holds. Therefore, (23.59) holds if

$$c_3\frac{[2g(0)]^{p-1}}{r^b(0)} \le \frac{1}{4}.$$ (23.68)

It follows from (23.67) that (23.68) holds if

$$c_3 2^{p-1}\left(\frac{c_2}{k}\right)^{p-1}\frac{1}{r^{-1+p+2b-bp}(0)} \le \frac{1}{4}.$$ (23.69)

One has $p = 1 + \kappa$, and $\kappa \in (0,1]$. If $b > 0$ and $\kappa \in (0,1]$, then

$$-1 + p - pb + 2b = \kappa + (1 - \kappa)b > 0.$$ (23.70)

Thus, (23.69) always holds if $r(0)$ is sufficiently large, specifically, if

$$r(0) \ge [4c_3\left(2c_2k^{-1}\right)^{p-1}]^{\frac{1}{\kappa+(1-\kappa)b}}.$$ (23.71)

We have proved the following theorem.

Theorem 23.1.4 *Let the Assumptions A1, A2, and A3 hold. If $r(t) = |a(t)|$ is defined in (23.62), and inequalities (23.67) and (23.71) hold, then*

$$\|z(t)\| < r^k(t)\lambda^{-1}, \qquad \lim_{t\to\infty}\|z(t)\| = 0.$$ (23.72)

Thus, problem (23.9) has a unique global solution $u(t)$ and

$$\lim_{t\to\infty}\|u(t) - y\| = 0,$$ (23.73)

where $F(y) = f$.

23.2   THE CASE OF CONTINUOUS $F'(U)$

In this section the Hölder continuity of ω is replaced by a weaker assumption:
We only assume that $0 \leq \omega(r)$ is a strictly increasing continuous function and
$\omega(0) = 0$.

Assume that F is continuously Fréchet differentiable, $F'(u) := A(u)$, and
inequality (23.2) holds.

Assume

$$0 < \rho(t) \searrow 0, \qquad \int_0^\infty \rho(s)\,ds = \infty, \qquad \frac{|\dot{\rho}|}{\rho} \leq C < 1, \qquad \forall t \geq 0. \quad (23.74)$$

Theorem 23.2.1 *Assume that conditions* (23.3)–(23.5) *and* (23.74) *hold.*
Let

$$r(t) = \begin{cases} r_0 e^{-C_2 \int_0^t \rho(s)ds} & if & 0 < b \leq 1 \\ \left(r_0^{1-b} + (b-1)\int_0^t C_2\rho(s)ds\right)^{\frac{1}{1-b}} & if & b > 1 \end{cases}, \qquad t \geq 0,$$

$$(23.75)$$

*where C_2 and r_0 are positive constants. Assume that the function $\omega(r)$ satisfies
the following condition:*

$$\omega(\rho(t)) \leq C_1 r^b(t), \qquad t \geq 0, \qquad C_1 = const > 0. \quad (23.76)$$

Assume

$$c_2 C_2 + c_1 C_1 + C \leq 1, \qquad if \qquad b > 1, \quad (23.77)$$
$$c_2 C_2 r_0^{1-b} + c_1 C_1 + C \leq 1, \qquad if \qquad 0 < b \leq 1. \quad (23.78)$$

Let $u(t)$ solve (23.9) *with $a(t)$ chosen so that $r(t) = |a(t)|$. Let u_0 satisfy*

$$\|u_0 - w(0)\| < \rho(0). \quad (23.79)$$

Then

$$\lim_{t\to\infty} u(t) = y. \quad (23.80)$$

Proof: It follows from (23.75) that

$$\dot{r}(t) = \begin{cases} -C_2 r(t)\rho(t) & if & 0 < b \leq 1 \\ -C_2 r^b(t)\rho(t) & if & b > 1 \end{cases}, \qquad t \geq 0. \quad (23.81)$$

From equation (23.29) and (23.36) one gets

$$\dot{z} = \dot{w} - z + A_{a(t)}^{-1}(u(t)) \int_0^1 [A(tz + w) - A(u)]z\,dt. \quad (23.82)$$

Let $g(t) := \|z(t)\|$. From (23.2)–(23.3), (23.35), (23.82), and Lemma 23.1.3, one obtains

$$\dot{g} \leq -g(t) + \|\dot{w}\| + \frac{c_1\omega(g)}{r^b(t)}g(t) \leq -g(t) + \frac{c_2|\dot{r}(t)|}{r^b(t)} + \frac{c_1\omega(g)}{r^b(t)}g(t), \quad (23.83)$$

for $t \geq 0$. Let

$$\mu(t) := \frac{1}{\rho(t)}, \quad t \geq 0. \quad (23.84)$$

We claim that the following inequality holds:

$$\frac{1}{\mu(t)}\frac{c_1\omega(\frac{1}{\mu(t)})}{r^b(t)} + \frac{c_2|\dot{r}(t)|}{r^b(t)} \leq \frac{1}{\mu(t)}\left(1 - \frac{\dot{\mu}(t)}{\mu(t)}\right). \quad (23.85)$$

Let us prove this claim. From (23.81) and (23.84) one gets

$$\frac{|\dot{r}(t)|}{r^b(t)} = C_2 r^{1-b}(t)\rho(t) \leq \frac{C_2 r_0^{1-b}}{\mu(t)}, \quad 0 < b \leq 1, \quad (23.86)$$

$$\frac{|\dot{r}(t)|}{r^b(t)} = C_2\rho(t) = \frac{C_2}{\mu(t)}, \quad b > 1. \quad (23.87)$$

From (23.84) and (23.76) one obtains

$$\frac{\omega(\frac{1}{\mu(t)})}{r^b(t)} = \frac{\omega(\rho(t))}{r^b(t)} \leq \frac{C_1 r^b(t)}{r^b(t)} = C_1, \quad b > 0. \quad (23.88)$$

From (23.84) and (23.74) one gets

$$\frac{|\dot{\mu}(t)|}{\mu(t)} = \frac{|\dot{\rho}(t)|}{\rho(t)} \leq C. \quad (23.89)$$

It follows from (23.77), (23.78), and (23.87)–(23.89) that

$$\frac{1}{\mu}\frac{c_1\omega(\frac{1}{\mu})}{r^b(t)} + \frac{c_2|\dot{r}(t)|}{r^b(t)} \leq \frac{1}{\mu}(1-C) \leq \frac{1}{\mu}\left(1 - \frac{\dot{\mu}}{\mu}\right), \quad b > 0. \quad (23.90)$$

Thus, inequality (23.85) holds.

Note that inequality (23.83) is of the form (5.66). From Theorem 5.6.1, (23.79), (23.83), and (23.85), one obtains

$$\|u(t) - w(t)\| = g(t) < \frac{1}{\mu(t)} = \rho(t), \quad t \geq 0. \quad (23.91)$$

Since $\lim_{t\to\infty} \rho(t) = 0$, it follows from (23.91) that $\lim_{t\to\infty} \|u(t) - w(t)\| = 0$. This and the triangle inequality imply

$$\lim_{t\to\infty} \|u(t) - y\| \leq \lim_{t\to\infty} \|u(t) - w(t)\| + \lim_{t\to\infty} \|w(t) - y\| = 0. \quad (23.92)$$

Thus, Theorem 23.2.1 is proved.   ∎

Remark 23.2.2 In this remark we give an example which shows that the continuity modulus $\omega(r)$ may satisfy inequality (23.76) but does not satisfy (23.2).

From (23.75) it follows that $r(t)$ exists and is unique for all $t > 0$ and

$$\lim_{t \to \infty} r(t) = 0. \tag{23.93}$$

If $b = 1$, then (23.75) implies that $r(t) = r(0)e^{-C_2 \int_0^t \rho(s)ds}$. In this case, inequality (23.76) becomes

$$\omega(\rho(t)) \le r_0 e^{-C_2 \int_0^t \rho(s)ds}. \tag{23.94}$$

Let

$$\rho(t) = \frac{1}{(e + t)\ln(e + t)}, \qquad t \ge 0, \tag{23.95}$$

then (23.74) is satisfied. One has

$$e^{-C_2 \int_0^t \rho(s)ds} = e^{-C_2 \ln\ln(e+t)} = \frac{1}{\ln^{C_2}(e + t)}. \tag{23.96}$$

Thus, if

$$\omega\Big(\frac{1}{(e + t)\ln(e + t)}\Big) = \frac{C_3}{\ln^{C_2}(e + t)}, \qquad C_3 > 0, \qquad t \ge 0,$$

then inequality (23.94) is satisfied. One can see that for this $\omega(r)$ there does not exist $C_4 > 0$ and $\kappa > 0$ such that

$$\omega(r) \le C_4 r^\kappa, \qquad 0 \le r \ll 1.$$

NUMERICAL EXPERIMENTS

CHAPTER 24

SOLVING LINEAR OPERATOR EQUATIONS BY THE DSM

This chapter presents numerical solutions by various versions of the DSM to several linear problems of practical interest. These problems include numerical differentiation, Fredholm and Volterra integral equations of the first kind, and an image restoration problem. Some numerical experiments in this chapter are taken from papers [49] and [54].

24.1 NUMERICAL EXPERIMENTS WITH ILL-CONDITIONED LINEAR ALGEBRAIC SYSTEMS

24.1.1 Numerical experiments with Hilbert matrix

Consider a linear algebraic system

$$H_n u = f_\delta, \tag{24.1}$$

Dynamical Systems Method and Applications: Theoretical Developments and Numerical Examples, First Edition. A. G. Ramm and N. S. Hoang.
Copyright © 2012 John Wiley & Sons, Inc.

where

$$f_\delta = f + e, \quad f = H_n x, \quad H_n = \begin{bmatrix} 1 & \frac{1}{2} & \cdots & \frac{1}{n} \\ \frac{1}{2} & \frac{1}{3} & \cdots & \frac{1}{n+1} \\ \vdots & \vdots & \ddots & \vdots \\ \frac{1}{n} & \frac{1}{n+1} & \cdots & \frac{1}{2n-1} \end{bmatrix},$$

and $e \in \mathbb{R}^n$ is a random normally distributed vector such that $\|e\|_2 \le \delta_{rel}\|f\|_2$. The Hilbert matrix H_n has a very large condition number when n is large. If n is sufficiently large ($n > 8$), the system is severely ill-conditioned.

24.1.1.1 The condition numbers of Hilbert matrices   It is impossible to calculate the condition number of H_n by computing the ratio of the largest and the smallest eigenvalues of H_n because for large n the smallest eigenvalue of H_n is smaller than 10^{-16}. Note that singular values of H_n are its eigenvalues since H_n is self-adjoint and positive definite. Due to the limitation of machine precision, every value smaller than 10^{-16} is understood as 0. That is why if we use the function *cond* provided by MATLAB, the condition number of H_n for $n \ge 20$ is about $10^{16} \times \max |\lambda_i(H_n)|$. Since the largest eigenvalue of H_n grows very slowly, the condition numbers of H_n for $n \ge 20$ are all about 10^{20} according to the function *cond* of MATLAB, while, in fact, the condition number of H_{100} computed by the formula, given below, is about 10^{150} (see Table 24.1). In general, computing condition numbers of strongly ill-conditioned matrices is an open problem. The function *cond*, provided by MATLAB, is not always reliable for computing the condition number of ill-condition matrices. Fortunately, there is an analytic formula for the inverse of H_n. Indeed, one has (see [17]) $H_n^{-1} = (h_{ij})_{i,j=1}^n$, where

$$h_{ij} = (-1)^{i+j}(i + j - 1)\binom{n + i - 1}{n - j}\binom{n + j - 1}{n - i}\binom{i + j - 2}{i - 1}^2.$$

Thus, the condition number of the Hilbert matrix can be computed by the formula

$$cond(H_n) = \|H_n\| \|H_n^{-1}\|.$$

Here $cond(H_n)$ stands for the condition number of the Hilbert matrix H_n and $\|H_n\|$ and $\|H_n^{-1}\|$ are the largest eigenvalues of H_n and H_n^{-1}, respectively. Although MATLAB cannot compute values less than 10^{-16}, it can compute values up to 10^{200}. Therefore, it can compute $\|H_n^{-1}\|$ for n up to 100. In MATLAB, the matrices H_n and H_n^{-1} can be obtained by the syntax: $H_n =$ hilb(n) and $H_n^{-1} =$ invhilb(n), respectively.

The condition numbers of Hilbert matrices, computed by the above formula, are given in Table 24.1.

From Table 24.1 one can see that the computed condition numbers of the Hilbert matrix grow very fast as n grows.

Table 24.1   The condition number of Hilbert matrices

n	20	40	60	80	100
$cond(H_n)$	2.5×10^{28}	7.7×10^{58}	2.7×10^{89}	9.9×10^{119}	3.8×10^{150}

24.1.1.2 DSM compared to VR In this section, we test three methods: the DSM, the VR_i and the VR_n on linear algebraic systems with Hilbert matrices. These methods are developed and discussed in detail in [45]. The first linear system is obtained by taking H_{100} and $x = (x_1, ..., x_{100})^T$, where $x_i = \frac{i}{100}$. Numerical results for this system with $\delta_{rel} = 10^{-2}$ and $\delta_{rel} = 10^{-4}$ are shown in Figure 24.1.

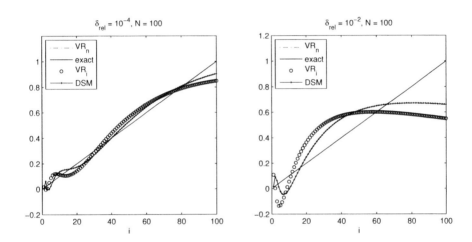

Figure 24.1   Plots of numerical solutions when $\delta_{rel} = 10^{-4}$ (left) and $\delta_{rel} = 10^{-2}$ (right).

Table 24.2 presents results with Hilbert matrices H_n for $n = 10, 20, ..., 100$, $\delta_{rel} = 0.01$, $x = (x_1, ..., x_n)^T$, $x_i = \sqrt{2 \frac{i-1}{n}} \pi$. Looking at this table it is clear that the results obtained by the DSM are slightly more accurate than those by the VR_n even in the cases when the VR_n requires much more work than the DSM. In this example, the DSM is better than the VR_n in both accuracy and time of computation.

24.2 NUMERICAL EXPERIMENTS WITH FREDHOLM INTEGRAL EQUATIONS OF THE FIRST KIND

Table 24.2   Numerical results for Hilbert matrix H_n for $\delta_{rel} = 0.01$, $n = 10, 20, ..., 100$

n	DSM N_{linsol}	DSM $\frac{\|u_\delta - y\|_2}{\|y\|_2}$	VR$_i$ N_{linsol}	VR$_i$ $\frac{\|u_\delta - y\|_2}{\|y\|_2}$	VR$_n$ N_{linsol}	VR$_n$ $\frac{\|u_\delta - y\|_2}{\|y\|_2}$
10	4	0.2368	1	0.3294	7	0.2534
20	5	0.1638	1	0.3194	7	0.1765
30	5	0.1694	1	0.3372	11	0.1699
40	5	0.1984	1	0.3398	8	0.2074
50	6	0.1566	1	0.3345	7	0.1865
60	5	0.1890	1	0.3425	8	0.1980
70	7	0.1449	1	0.3393	11	0.1450
80	7	0.1217	1	0.3480	8	0.1501
90	7	0.1259	1	0.3483	11	0.1355
100	6	0.1865	2	0.2856	9	0.1937

24.2.1   Numerical experiments for computing second derivative

Let us do some numerical experiments with linear algebraic systems arising in a numerical experiment of computing the second derivative of a noisy function.

The problem of finding second derivative of a twice differentiable function is reduced to a Fredholm integral equation of the first kind:

$$\int_0^1 K(s,t)u(t)\,dt = f(s), \qquad s \in [0,1], \tag{24.2}$$

where the kernel K is Green's function:

$$K(s,t) = \begin{cases} s(t-1), & \text{if } s < t \\ t(s-1), & \text{if } s \geq t \end{cases}, \qquad s,t \in [0,1].$$

Here $f(0) = 0$, $f(1) = 0$ and $f''(s) = u(s)$.

To solve equation (24.2) we use the Galerkin method. Let us look for for an approximation to $u(t)$ of the form

$$u_N(t) \approx \sum_{i=1}^N c_i \phi_i(t), \tag{24.3}$$

where

$$\phi_i(t) := \begin{cases} (\sqrt{h})^{-1} & \text{if } \frac{i-1}{N} \leq t \leq \frac{i}{N}, \\ 0 & \text{otherwise}, \end{cases} \qquad 1 \leq i \leq N. \tag{24.4}$$

The coefficients $x_N = (c_i)_{i=1}^N$ are found from the following linear algebraic system:

$$A_N x_N = b_N, \qquad A_N = (a_{ij})_{i,j=1}^N, \qquad b_N = (g_1, g_2, ..., g_N)^T, \tag{24.5}$$

where

$$a_{ij} = \int_0^1 \int_0^1 K(s,t)\phi_i(t)\phi_j(s)\,dt\,ds, \qquad g_i = \int_0^1 f(s)\phi_i(s)\,ds, \qquad 1 \le i \le N.$$

In numerical experiments, the following functions are used:

$$u_1(t) = t, \qquad f_1(s) = \frac{s^3 - s}{6}, \tag{24.6}$$

$$u_2(t) = \exp(t), \qquad f_2(s) = \exp(s) + (1-e)s - 1, \tag{24.7}$$

$$u_3(t) = \sin(\pi t), \qquad f_3(s) = \frac{1}{\pi^2}\sin(\pi t). \tag{24.8}$$

The functions (24.6) and (24.7) were also used in [26].

We use $N = 10, 20, ..., 100$ and $b_{N,\delta} = b_N + e_N$, where e_N is a random vector whose coordinates are independent and normally distributed, with mean 0 and variance 1, and scaled so that $\|e_N\| = \delta_{rel}\|b_N\|$. This linear algebraic system is mildly ill-posed: The condition number of A_{100} is 1.2158×10^4.

In this section we use the DSM version developed in Chapter 17. In all the experiments, we denote by DSMG the version of the DSM (17.1), by VR we denote the Variational Regularization, implemented using the discrepancy principle, by DSM we denote the method developed in [45], and by TSVD we denote the truncated SVD method.

Tables 24.3 and 24.4 present numerical results when $u(t) = u_1(t)$ and $u(t) = u_2(t)$, respectively. In these tables, N varies from 20 to 100. It can be seen from these tables that the DSM and the DSMG yield solutions with accuracy about the same as those of the VR. The DSM is the best method in these tables in terms of both accuracy and computational cost. The TSVD is the worst method in terms of accuracy in these experiments.

Table 24.3   Numerical results for computing second derivatives when $u(t) = u_1(t)$ and $\delta_{rel} = 0.01$

n	DSMG		DSM		VR		TSVD	
	n_{iter}	$\frac{\|u_n - y_n\|_2}{\|y_n\|_2}$	n_{iter}	$\frac{\|u_n - y_n\|_2}{\|y_n\|_2}$	n_{iter}	$\frac{\|u_n - y_n\|_2}{\|y_n\|_2}$	n_{trun}	$\frac{\|u_n - y_n\|_2}{\|y_n\|_2}$
20	6	0.3366	3	0.3130	5	0.3207	5	0.3603
30	6	0.3034	4	0.2822	6	0.2915	6	0.3347
40	6	0.3084	4	0.2995	6	0.3031	7	0.3107
50	8	0.2841	4	0.2639	7	0.2710	7	0.3111
60	9	0.2789	5	0.2687	7	0.2717	6	0.3312
70	6	0.2768	5	0.2720	7	0.2719	7	0.3056
80	8	0.3028	4	0.2958	7	0.2983	6	0.3317
90	5	0.2808	5	0.2762	7	0.2766	7	0.3063
100	9	0.2713	5	0.2664	7	0.2686	7	0.3067

Table 24.4   Numerical results for computing second derivatives when
$u(t) = u_2(t)$ and $\delta_{rel} = 0.01$

n	DSMG		DSM		VR		TSVD	
	n_{iter}	$\frac{\|u_n - y_n\|_2}{\|y_n\|_2}$	n_{linsol}	$\frac{\|u_n - y_n\|_2}{\|y_n\|_2}$	n_{linsol}	$\frac{\|u_n - y_n\|_2}{\|y_n\|_2}$	n_{trun}	$\frac{\|u_n - y_n\|_2}{\|y_n\|_2}$
20	7	0.2711	3	0.2710	4	0.2717	6	0.3046
30	6	0.3014	4	0.2948	4	0.2957	6	0.3104
40	7	0.2987	3	0.2736	6	0.2809	6	0.2997
50	8	0.2815	4	0.2766	7	0.2789	6	0.3000
60	8	0.2815	4	0.2684	7	0.2731	6	0.3009
70	8	0.2890	4	0.2801	6	0.2851	6	0.3055
80	8	0.2702	4	0.2662	7	0.2667	6	0.3009
90	8	0.2592	5	0.2417	7	0.2481	6	0.3019
100	7	0.2969	4	0.2833	6	0.2898	6	0.3011

Table 24.5 presents numerical results when N varies from 10 to 100, $u_3(t) = \sin(\pi t)$, and $t \in [0, 1]$. In this experiment the DSMG yields more accurate solutions than the DSM and the VR. The DSMG in this experiment takes more iterations than the DSM and the VR to get a solution.

Table 24.5   Numerical results for computing second derivatives with
$u_3(t) = \sin(\pi t)$ and $\delta_{rel} = 0.01$

N	DSM		DSM		VR	
	n_{iter}	$\frac{\|u_N - u\|_2}{\|u\|_2}$	n_{linsol}	$\frac{\|u_N - u\|_2}{\|u\|_2}$	n_{linsol}	$\frac{\|u_N - u\|_2}{\|u\|_2}$
20	9	0.0973	3	0.1130	6	0.1079
30	5	0.0831	4	0.1316	6	0.1160
40	7	0.0488	4	0.1150	6	0.1045
50	9	0.0614	4	0.1415	6	0.1063
60	6	0.0419	4	0.0919	6	0.0817
70	9	0.0513	4	0.0961	6	0.0842
80	6	0.0418	4	0.1225	6	0.0981
90	7	0.0287	4	0.0919	7	0.0840
100	7	0.0248	5	0.0778	7	0.0553

In Figure 24.2, the difference between the exact solution and solutions obtained by the DSMG, VR, and DSM are plotted. In these experiments, we used $N = 100$ and $u(t) = \sin(\pi t)$ with $\delta_{rel} = 0.05$ and $\delta_{rel} = 0.01$. Figure 24.2 shows that the results obtained by the VR and the DSM are close to each other. The results obtained by the DSMG are much better than those by the DSM and by the VR.

In this experiment the DSMG is implemented using the SVD of A obtained by the function *svd* in Matlab. As already mentioned, the SVD is a special case of the spectral decomposition (17.32). It is expensive to compute the SVD, in

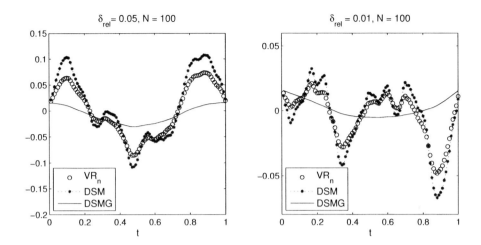

Figure 24.2   Plots of differences between the exact solution and solutions obtained by the DSMG, VR, and DSM.

general. However, there are practically important problems where the spectral decomposition (17.32) can be computed fast (see Section 24.3 below). These problems consist of deconvolution problems using the fast Fourier transform (FFT).

The conclusion from this experiment is: The DSMG may yield results with much better accuracy than the VR and DSM. Numerical experiments for various $u(t)$ show that the DSMG competes favorably with the VR and the DSM.

24.3 NUMERICAL EXPERIMENTS WITH AN IMAGE RESTORATION PROBLEM

In this section we present numerical experiments with the DSM version developed in Chapter 17. In all the experiments, by DSMG we denote the version of the DSM (17.1), by VR we denote the variational regularization, implemented using the discrepancy principle, and by DSM we denote the method developed in [45].

The image degradation process can be modeled by the following equation:

$$g_\delta = g + w, \quad g = h * f, \quad \|w\| \le \delta, \qquad (24.9)$$

where h represents a convolution function modelling the blurring that many imaging systems introduce. For example, camera defocus, motion blur, and imperfections of the lenses can be modeled by choosing a suitable h. The functions g_δ, f, and w are the observed image, the original signal, and the noise, respectively. The noise w can be due to the electronics used (thermal and shot noise), the recording medium (film grain), or the imaging process (photon noise).

In practice, g, h, and f in equation (24.9) are often given as functions of a discrete argument and equation (24.9) can be written in this case as

$$g_{\delta,i} = g_i + w_i = \sum_{j=-\infty}^{\infty} f_j h_{i-j} + w_i, \quad i \in \mathbb{Z}. \qquad (24.10)$$

Note that one (or both) signals f and h have compact support (finite length). Suppose that signal f is periodic with period N, that is, $f_{i+N} = f_i$, and $h_j = 0$ for $j < 0$ and $j \geq N$. Assume that f is represented by a sequence $f_0, ..., f_{N-1}$ and h is represented by $h_0, ..., h_{N-1}$. Then the convolution $h * f$ is periodic signal g with period N, and the elements of g are defined as

$$g_i = \sum_{j=0}^{N-1} h_j f_{(i-j) \bmod N}, \quad i = 0, 1, ..., N-1. \qquad (24.11)$$

Here $(i-j) \bmod N$ is $i-j$ modulo N. The discrete Fourier transform (DFT) of g is defined as the sequence

$$\mathcal{F}(g)_k := \hat{g}_k := \frac{1}{\sqrt{N}} \sum_{j=0}^{N-1} g_j e^{-i2\pi jk/N}, \quad k = 0, 1, ..., N-1.$$

Denote $\hat{g} = (\hat{g}_0, ..., \hat{g}_{N-1})^T$. Then equation (24.11) implies

$$\hat{g} = \hat{f}\hat{h}, \qquad \hat{f}\hat{g} := (\hat{f}_0\hat{h}_0, \hat{f}_1\hat{h}_1, ..., \hat{f}_{N-1}\hat{h}_{N-1})^T. \qquad (24.12)$$

Let $\text{Diag}(a)$ denote a diagonal matrix whose diagonal is $(a_0, ..., a_{N-1})$ and other entries are zeros. Then equation (24.12) can be rewritten as

$$g = \mathcal{F}^{-1}(S\mathcal{F}(f)), \qquad S := \text{Diag}(\hat{h}). \qquad (24.13)$$

Equation (24.10) is of the form $g = Af$ where A is of the form (17.32) with $U = \mathcal{F}^{-1}, V = \mathcal{F}$ and $S = \text{Diag}(\hat{h})$. Thus, one can use the DSMG method to solve equation (24.13) stably for f.

The image restoration test problem we use is taken from [85]. This problem has been widely used in the literature for testing image restoration algorithms. The original and blurred images have 256×256 pixels and are shown in Figure 24.3. These data have been widely used in the literature for testing image restoration algorithms.

Figure 24.4 plots the regularized images by the VR and the DSMG when $\delta_{rel} = 0.01$. Again, with an input value for δ_{rel}, the observed blurred noisy images are computed by

$$g_\delta = g + \delta_{rel} \frac{\|g\|}{\|err\|} err,$$

Original   Blurred noisy image

Figure 24.3   Original and blurred noisy images.

where err is a vector with random entries normally distributed with mean 0 and variance 1. In this experiment, it took 5 and 8 iterations for the DSMG and the VR, respectively, to yield numerical results. From Figure 24.4 one concludes that the DSMG is comparable to the VR in terms of accuracy. The time of computation in this experiment is about the same for the VR and DSMG.

VR   DSM

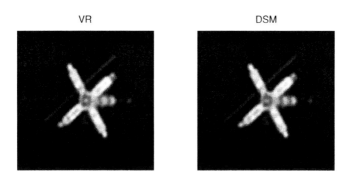

Figure 24.4   Regularized images when noise level is 1%.

Figure 24.5 plots the regularized images by the VR and the DSMG when $\delta_{rel} = 0.05$. It took 4 and 7 iterations for the DSMG and the VR, respectively, to yield numerical results. Figure 24.5 shows that the images obtained by the DSMG and the VR are about the same.

The conclusions from this experiment are: the DSMG yields results with the same accuracy as the VR, and requires less iterations than the VR. The restored images by the DSM are about the same as those by the VR.

VR   DSM

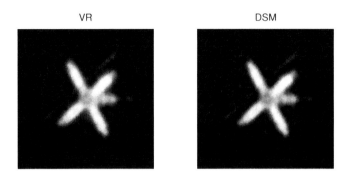

Figure 24.5 Regularized images when noise level is 5%.

Remark 24.3.1 Equation (24.9) can be reduced to equation (24.12) whenever one of the two functions f and h has compact support and the other is periodic.

24.4 NUMERICAL EXPERIMENTS WITH VOLTERRA INTEGRAL EQUATIONS OF THE FIRST KIND

24.4.1 Numerical experiments with an inverse problem for the heat equation

Consider the following heat equation (see [16]):

$$u_t = a^2 u_{xx}, \qquad 0 < x < \infty, \qquad 0 < t < \infty,$$
$$u(x,0) = 0, \qquad 0 \le x < \infty,$$
$$u(0,t) = f(t), \qquad 0 \le t < \infty, \tag{24.14}$$

where $a = const > 0$. The unique solution to (24.14) is

$$u(x,t) = \frac{x}{2a\sqrt{\pi}} \int_0^t \frac{f(\xi)}{(t-\xi)^{3/2}} \exp\left[\frac{-x^2}{4a^2(t-\xi)}\right] d\xi. \tag{24.15}$$

Let $g(t) := g_l(t) := u(l,t)$, $t \ge 0$, be the temperature at a fixed location $x = l > 0$.

The inverse problem is: *Find the temperature $f(t)$, $t \ge 0$, given the data $g(t)$, $t \ge 0$.*

The function $f(t)$ solves the following Volterra equation of the first kind:

$$(Kf)(t) = g(t), \qquad t \ge 0, \tag{24.16}$$

where

$$(Kf)(t) = \frac{l}{2a\sqrt{\pi}} \int_0^t \frac{f(\xi)}{(t-\xi)^{3/2}} \exp\left[\frac{-l^2}{4a^2(t-\xi)}\right] d\xi, \qquad 0 \le t < \infty. \tag{24.17}$$

In this section the DSM is applied for solving equation (24.16) numerically for $f(t)$, given noisy data $g_\delta(t)$.

We choose the following function as the exact solution:

$$f(t) = \begin{cases} 75t^2 & \text{if} \quad 0 \le t \le 0.1, \\ 0.75 + (20t - 2)(3 - 20t) & \text{if} \quad 0.1 \le t \le 0.15, \\ 0.75 \exp(-40(t - 1.5)) & \text{if} \quad 0.15 \le t \le 1. \end{cases} \qquad (24.18)$$

In numerical experiment we use $a = 1$ and $l = 1$. In this test the integral equation was discretized by the composite midpoint rule with n points and reduced to the following linear algebraic system:

$$A_n u_n = b_n. \qquad (24.19)$$

Here

$$u_n \approx \left(f\left(\frac{1}{2n}\right), f\left(\frac{3}{2n}\right), ..., f\left(\frac{2n-1}{2n}\right) \right)^T,$$

$$b_n = \left(g\left(\frac{1}{n}\right), g\left(\frac{2}{n}\right), ..., g\left(\frac{n}{n}\right) \right)^T,$$

and

$$A_n = \frac{1}{n} \begin{pmatrix} k\left(\frac{1}{2n}\right) & 0 & \cdots & 0 \\ k\left(\frac{3}{2n}\right) & k\left(\frac{1}{2n}\right) & \ddots & 0 \\ \vdots & \vdots & \ddots & \vdots \\ k\left(\frac{2n-1}{2n}\right) & k\left(\frac{2n-3}{2n}\right) & \cdots & k\left(\frac{1}{2n}\right) \end{pmatrix}, \quad k(t) = \frac{1}{2\sqrt{\pi}} \exp\left(-\frac{1}{4t}\right).$$

In our test, $n = 10, 20, ..., 100$ and $b_{n,\delta} = b_n + c e_n$, where e_n is a vector containing random entries, normally distributed with mean 0, variance 1, and $c = const > 0$ chosen so that

$$\|b_n - b_{n,\delta}\| = \delta_{rel} \|b_n\|. \qquad (24.20)$$

This linear system is ill-posed: The condition number of A_{100}, obtained by using the function *cond* provided in MATLAB, is 1.3717×10^{37}. This number shows that the corresponding linear algebraic system is severely ill-conditioned.

We apply iterative scheme (18.54) and the VR for solving linear algebraic system (24.19). We denote by DSM iterative scheme (18.54), by VR_i the Variational Regularization method (VR) with a as the regularization parameter, and by VR_n the VR in which Newton's method is used for finding the regularization parameter from a discrepancy principle. We compare these methods in terms of relative error and number of iterations, denoted by n_{iter}.

Table 24.6 shows that the results obtained by the DSM are comparable to those by the VR_n in terms of accuracy. The time of computation by the DSM

Table 24.6   Numerical results for the inverse heat equation with $\delta_{rel} = 0.05$, $n = 10i$, $i = \overline{1, 10}$

n	DSM		VR$_i$		VR$_n$	
	n_{iter}	$\frac{\|u_n - y_n\|_2}{\|y_n\|_2}$	n_{iter}	$\frac{\|u_n - y_n\|_2}{\|y_n\|_2}$	n_{iter}	$\frac{\|u_n - y_n\|_2}{\|y_n\|_2}$
10	3	0.1971	1	0.2627	5	0.2117
20	4	0.3359	1	0.4589	5	0.3551
30	4	0.3729	1	0.4969	5	0.3843
40	4	0.3856	1	0.5071	5	0.3864
50	5	0.3158	1	0.4789	6	0.3141
60	6	0.2892	1	0.4909	6	0.3060
70	7	0.2262	1	0.4792	8	0.2156
80	6	0.2623	1	0.4809	7	0.2600
90	5	0.2856	1	0.4816	7	0.2715
100	7	0.2358	1	0.4826	7	0.3405

is comparable to that by the VR$_n$. In some situations, the VR$_n$ and the DSM yield results with the same accuracy, although the VR$_n$ uses 3 more iterations than does the DSM. The conclusion from Table 24.6 is: The DSM competes favorably with the VR$_n$ in both accuracy and time of computation.

Figure 24.6 plots numerical solutions to the inverse heat equation for $\delta_{rel} = 0.05$ and $\delta_{rel} = 0.01$ when $n = 100$. From the figure one can see that the numerical solutions obtained by the DSM are about the same those by the VR$_n$. In these examples, the time of computation of the DSM is about the same as that of the VR$_n$.

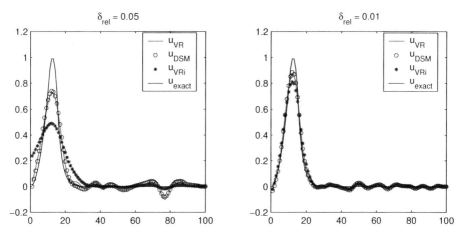

Figure 24.6   Plots of solutions obtained by DSM, VR for the inverse heat equation when $n = 100$, $\delta_{rel} = 0.05$ (left), and $\delta_{rel} = 0.01$ (right).

The conclusion is that the DSM competes favorably with the VR_n in this experiment.

24.5 NUMERICAL EXPERIMENTS WITH NUMERICAL DIFFERENTIATION

The problem of finding the derivative $u(t)$ of a function $f(t) \in C^1[0,1]$ can be reduced to the problem of solving the following Volterra equation:

$$(Au)(t) = \int_0^t u(\xi)\, d\xi = f(t) - f(0), \qquad t \in [0,1]. \qquad (24.21)$$

The operator A is called Volterra operator.

Let us discuss two approaches for finding u numerically from equation (24.21) when noisy data f_δ are available.

24.5.1 The first approach

Let us use the trapezoidal rule to discretize equation (24.21) and get the following algebraic system:

$$B_N u_N = g_N, \qquad (24.22)$$

where

$$
u_n = \begin{pmatrix} u_{n0} \\ u_{n1} \\ \vdots \\ u_{nn} \end{pmatrix} \approx \begin{pmatrix} u(0) \\ u(\frac{1}{n}) \\ \vdots \\ u(\frac{n}{n}) \end{pmatrix}, \qquad
B_n = \frac{1}{n} \begin{pmatrix} \frac{1}{2} & \frac{1}{2} & 0 & \cdots & 0 & 0 \\ \frac{1}{2} & 1 & \frac{1}{2} & \cdots & 0 & 0 \\ 0 & \frac{1}{2} & 1 & \cdots & 0 & 0 \\ \vdots & \vdots & \vdots & \ddots & & \vdots \\ 0 & 0 & 0 & \cdots & 1 & \frac{1}{2} \\ 0 & 0 & 0 & \cdots & \frac{1}{2} & \frac{1}{2} \end{pmatrix}, \qquad (24.23)
$$

and

$$
g_n = \begin{pmatrix} g_{n0} \\ g_{n1} \\ \vdots \\ g_{nn} \end{pmatrix}, \qquad
g_{nj} = \begin{cases} f(\frac{1}{n}) - f(0) & \text{if} & j = 0, \\ f(\frac{j+1}{n}) - f(\frac{j-1}{n}) & \text{if} & 1 \le j \le n-1, \\ f(1) - f(\frac{n-1}{n}) & \text{if} & j = n. \end{cases} \qquad (24.24)
$$

One can check that the matrix B_n is nonnegative definite.

We are going to find u_n numerically, assuming that g_n is not known but $g_{n,\delta}$ is known, $\|g_{n,\delta} - g_n\| \le \delta$. We use $g_{n,\delta} = g_n + ce_n$ where e_n is the noise function which will be chosen later. Here $c > 0$ is a constant chosen so that $\|g_{n,\delta} - g_n\| = \delta_{rel}\|g_n\|$, where $\delta_{rel} > 0$ is a relative noise level in our experiments.

Let us use the iterative scheme (22.155) in Section 22.2, that is,

$$u_{n+1} = u_n - h(Au_n + a_n u_n - f_\delta), \qquad n \ge 0, \qquad (24.25)$$

where

$$u_0 = 0, \qquad a_n = \frac{a_0}{1 + hn}. \tag{24.26}$$

We stop our iterations at $n = n_\delta$ using stopping rule (22.158), that is,

$$\|Au_{n_\delta} - f_\delta\| \le C\delta^\gamma, \qquad \|Au_n - f_\delta\| > C\delta^\gamma, \quad \forall n \le n_\delta. \tag{24.27}$$

We take $a_0 = 0.01$, $C = 1.1$, $h = 1$, and $\gamma = 0.99$.

In Figure 24.7 and 24.8 by *DSMS* we denote numerical results obtained by the method (24.25), by *VR* we denote the results obtained by the Variational Regularization method, and by *exact* we denote the exact solution.

Figure 24.7 plots the numerical results when the noisy function e_n is a vector with random entries normally distributed with mean zero and variance 1. Looking at Figure 24.7, one can see that the numerical results for the VR and the DSMS are about the same. In the implementation of the VR we computed

$$u = (A^*A + \alpha I)^{-1} A^* f_\delta, \qquad \alpha > 0, \tag{24.28}$$

and chose the best solution among several results for various $\alpha > 0$ of the form $n \times 10^{-i}$ where n and i are positive integers. When $\delta_{rel} = 0.03$ the DSMS used 90 iterations and 0.0015 seconds while the VR used 0.0006 seconds. When $\delta_{rel} = 0.01$ the DSMS used 121 iterations and 0.0021 seconds while the VR still used 0.0006 seconds. Thus, the computation time of the DSMS is about 4 times more than that of the VR. In practice, one may use Morozov's Discrepancy Principle to find the regularization parameter α. If the cost for finding the regularization parameter by the Morozov Discrepancy Principle is taken into account in these experiments, the VR takes more time than the DSMS.

Figure 24.8 plots the numerical results when the noise function e_n is defined by

$$e_n = \left(\tilde{f}(0), \tilde{f}\left(\frac{1}{n}\right), ..., \tilde{f}(1) \right)^T, \qquad \tilde{f}(x) = \sin(10\pi x). \tag{24.29}$$

From Figure 24.8 one can see that the DSMS yields numerical solutions which are of the same accuracy as those obtained by the VR. When $\delta_{rel} = 0.03$ the DSMS used 90 iterations and 0.0015 seconds. When $\delta_{rel} = 0.01$ the DSMS used 121 iterations and 0.0021 seconds. The VR in both cases used 0.006 seconds. Again, if the computational cost of finding regularization parameter α is taken into account, then it takes more computation time for the VR to get the solutions.

From the above results one can see that the DSMS works well in these experiments. The advantage of the DSMS is that one does not have to solve for the regularization parameter and the iterative scheme (24.25) is very easy to implement.

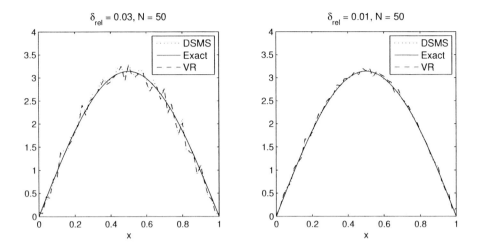

Figure 24.7 Plots of solutions of the DSMS and the VR when $N = 50$ and $\delta_{rel} = 0.05$ (left) and $\delta_{rel} = 0.01$ (right).

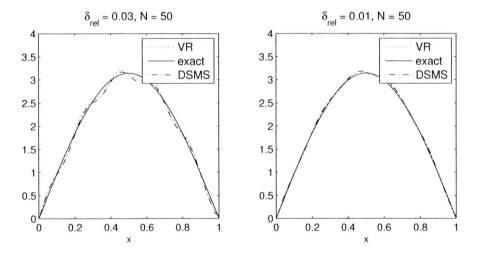

Figure 24.8 Plots of solutions of the DSMS and the VR when $N = 50$ and $\delta_{rel} = 0.05$ (left) and $\delta_{rel} = 0.005$ (right).

24.5.2 The second approach

Let us discretize (24.21) using trapezoidal rule, and get the following linear algebraic system

$$A_N u_N = f_N, \qquad (24.30)$$

where

$$
u_n = \begin{pmatrix} u_{n0} \\ u_{n1} \\ \vdots \\ u_{nn} \end{pmatrix} \approx \begin{pmatrix} u(0) \\ u(\frac{1}{n}) \\ \vdots \\ u(\frac{n}{n}) \end{pmatrix}, \qquad f_n = \begin{pmatrix} 0 \\ f(\frac{1}{n}) - f(0) \\ u(\frac{2}{n}) - f(0) \\ \vdots \\ u(\frac{n}{n}) - f(0). \end{pmatrix} \tag{24.31}
$$

and

$$
A_n = \frac{1}{n} \begin{pmatrix} 0 & 0 & 0 & \cdots & 0 \\ \frac{1}{2} & \frac{1}{2} & 0 & \cdots & 0 \\ \frac{1}{2} & 1 & \frac{1}{2} & \cdots & 0 \\ \vdots & \vdots & \vdots & \ddots & 0 \\ \frac{1}{2} & 1 & 1 & \cdots & \frac{1}{2} \end{pmatrix}. \tag{24.32}
$$

In this experiment, iterative scheme (16.75) is used, that is,

$$
u_{n+1} = (1 - h)u_n + h(A^*A + a_n I)^{-1} A^* f_\delta, \qquad n \geq 0, \tag{24.33}
$$

where

$$
u_0 = (A^*A + a_0 I)^{-1} A^* f_\delta. \tag{24.34}
$$

We used stopping rule (16.80), that is,

$$
\|Au_{n_\delta} - f_\delta\| \leq C\delta^\gamma, \qquad \|Au_n - f_\delta\| > C\delta^\gamma, \quad \forall n \leq n_\delta. \tag{24.35}
$$

Iterative scheme (24.33) is implemented with $a_n = a_0(1 - h)^n$, $h = 0.5$, $a_0 = 0.01$, $C = 1.1$, and $\gamma = 0.9$.

In Figures 24.9 and 24.10 we denote by DSMN the numerical results obtained by method (24.33).

Figure 24.9 plots the numerical results when the noise function e_n is a vector with random entries normally distributed with mean 0 and variance 1. The method used 8 and 11 iterations when $\delta_{rel} = 0.05$ and $\delta_{rel} = 0.01$, respectively.

Figure 24.10 plots the numerical results when the noise function e_n is defined in (24.29). In this experiment the method used 1 and 5 iterations when $\delta_{rel} = 0.05$ and $\delta_{rel} = 0.005$, respectively.

Note that the method DSMS in the previous section is not applicable for solving (24.30) directly since the matrix A_n is not nonnegative definite.

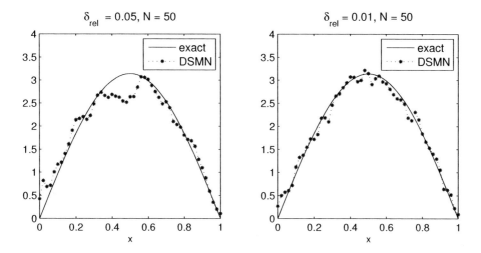

Figure 24.9   Plots of numerical solutions of the DSMN when $\delta_{rel} = 0.05$ and $\delta_{rel} = 0.01$.

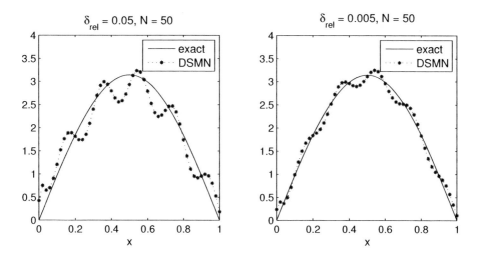

Figure 24.10   Plots of numerical solutions of the DSMN when $\delta_{rel} = 0.05$ (left) and $\delta_{rel} = 0.005$ (right).

CHAPTER 25

STABLE SOLUTIONS OF HAMMERSTEIN-TYPE INTEGRAL EQUATIONS

In this chapter we will carry out numerical experiments with iterative schemes developed in Chapter 20–22 for solving equations with monotone operators. Consider equation

$$F(u) = f \qquad (25.1)$$

where F is a monotone operator. We assume that (25.1) has a solution, possibly nonunique. Let y be the minimal norm solution to (25.1).

Numerical experiments in this chapter are taken from papers [42], [44], and [55].

25.1 DSM OF NEWTON TYPE

25.1.1 An experiment with an operator defined on $H = L^2[0, 1]$

Let us do a numerical experiment solving nonlinear equation (25.1) with

$$F(u) := B(u) + \arctan^3(u), \qquad (25.2)$$

Dynamical Systems Method and Applications: Theoretical Developments and Numerical Examples, First Edition. A. G. Ramm and N. S. Hoang.
Copyright © 2012 John Wiley & Sons, Inc.

where

$$B(u) := \int_0^1 e^{-|x-y|} u(y)\, dy.$$

Since the function $u \to \arctan^3 u$ is increasing on $\mathbb{R}$, one has

$$\langle \arctan^3(u) - \arctan^3(v), u - v \rangle \geq 0, \qquad \forall\, u, v \in H. \qquad (25.3)$$

Moreover,

$$e^{-|x|} = \frac{1}{\pi} \int_{-\infty}^{\infty} \frac{e^{i\lambda x}}{1 + \lambda^2}\, d\lambda. \qquad (25.4)$$

Therefore,

$$\langle B(u - v), u - v \rangle \geq 0, \qquad (25.5)$$

so

$$\langle F(u) - F(v), u - v \rangle \geq 0, \qquad \forall\, u, v \in H. \qquad (25.6)$$

Thus, F is a monotone operator. Note that

$$\langle \arctan^3(u) - \arctan^3(v), u - v \rangle = 0 \quad \text{iff} \quad u = v \quad a.e..$$

Therefore, the operator F, defined in (25.2), is injective and equation (25.1), with this F, has at most one solution.

The Fréchet derivative of F is

$$F'(u)w = \frac{3 \arctan^2(u)}{1 + u^2} w + \int_0^1 e^{-|x-y|} w(y)\, dy. \qquad (25.7)$$

If $u(x)$ vanishes on a set of positive Lebesgue measure, then $F'(u)$ is not boundedly invertible. If $u \in C[0, 1]$ vanishes even at one point x_0, then $F'(u)$ is not boundedly invertible in H.

In numerical implementation of the DSM, one often discretizes the Cauchy problem (20.129) and gets a system of ordinary differential equations (ODEs). Then, one can use numerical methods for solving ODEs to solve the system of ordinary differential equations obtained from discretization. There are many numerical methods for solving ODEs (see, e.g., [41]).

In practice one does not have to compute $u_\delta(t_\delta)$ exactly but can use an approximation to $u_\delta(t_\delta)$ as a stable solution to equation (25.1). To calculate such an approximation, one can use, for example, the iterative scheme

$$u_{n+1} = u_n - (F'(u_n) + a_n I)^{-1} (F(u_n) + a_n u_n - f_\delta),$$
$$u_0 = 0, \qquad (25.8)$$

and stop iterations at $n := n_\delta$ such that the following inequality holds:

$$\|F(u_{n_\delta}) - f_\delta\| < C\delta^\gamma, \quad \|F(u_n) - f_\delta\| \geq C\delta^\gamma, \quad n < n_\delta, \qquad (25.9)$$

where $C > 1$, $\gamma \in (0, 1)$.

The iterative scheme (25.8) can be considered a regularized Newton method. The existence of the stopping time n_δ is proved in [48, p. 733] and the choice $u_0 = 0$ is also justified in this paper. Iterative scheme (25.8) and stopping rule (25.9) are used in the numerical experiments. We proved in [48, p. 733] that u_{n_δ} converges to u^*, a solution of (25.1). Since F is injective as discussed above, we conclude that u_{n_δ} converges to the unique solution of equation (25.1) as δ tends to 0. The accuracy and stability are the key issues in solving the Cauchy problem. The iterative scheme (25.8) can be considered formally as the explicit Euler's method with the stepsize $h = 1$ (see, e.g., [41]). There might be other iterative schemes which are more efficient than scheme (25.8), but this scheme is simple and easy to implement.

Integrals of the form $\int_0^1 e^{-|x-y|} h(y) dy$ in (25.2) and (25.7) are computed by using the trapezoidal rule. The noisy function used in the test is

$$f_\delta(x) = f(x) + \kappa f_{noise}(x), \quad \kappa > 0. \tag{25.10}$$

The noise level δ and the relative noise level are defined by the formulas

$$\delta = \kappa \| f_{noise} \|, \quad \delta_{rel} := \frac{\delta}{\|f\|}. \tag{25.11}$$

In the test κ is computed in such a way that the relative noise level δ_{rel} equals some desired value, that is,

$$\kappa = \frac{\delta}{\| f_{noise} \|} = \frac{\delta_{rel} \|f\|}{\| f_{noise} \|}. \tag{25.12}$$

We have used the relative noise level as an input parameter in the test.

In all the figures the x-variable runs through the interval $[0, 1]$, and the graphs represent the numerical solutions $u_{DSM}(x)$ and the exact solution $u_{exact}(x)$.

In the test we took $h = 1$, $C = 1.01$, and $\gamma = 0.99$. The exact solution in the test is

$$u_e(x) = \begin{cases} 0 & \text{if } \frac{1}{3} \le x \le \frac{2}{3}, \\ 1 & \text{otherwise,} \end{cases} \tag{25.13}$$

here $x \in [0, 1]$, and the right-hand side is $f = F(u_e)$. As mentioned above, $F'(u)$ is not boundedly invertible in any neighborhood of u_e.

It is proved in [48] that one can take $a_n = \frac{d}{1+n}$ where d is sufficiently large. However, in practice, if we choose d too large, then the method will use too many iterations before reaching the stopping time n_δ in (25.9). This means that the computation time will be large in this case. Since

$$\| F(V_\delta) - f_\delta \| = a(t) \| V_\delta \|$$

and $\| V_\delta(t_\delta) - u_\delta(t_\delta) \| = O(a(t_\delta))$, we have

$$C\delta^\gamma = \| F(u_\delta(t_\delta)) - f_\delta \| \le a(t_\delta) \| V_\delta \| + O(a(t_\delta)),$$

and we choose

$$d = C_0 \delta^\gamma, \qquad C_0 > 0.$$

In the experiments our method works well with $C_0 \in [7, 10]$. In numerical experiments, we found out that the method diverged for smaller C_0. In the test we chose a_n by the formula $a_n := C_0 \frac{\delta^{0.99}}{n+1}$. The number of nodal points, used in computing integrals in (25.2) and (25.7), was $N = 100$. The accuracy of the solutions obtained in the tests with $N = 30$ and $N = 50$ was slightly less accurate than the one for $N = 100$.

Numerical results for various values of δ_{rel} are presented in Table 25.1. In this experiment, the noise function f_{noise} is a vector with random entries normally distributed, with mean value 0 and variance 1. Table 25.1 shows that the iterative scheme yields good numerical results.

Table 25.1   Results when $C_0 = 7$, $N = 100$, and $u = u_e$

δ_{rel}	0.02	0.01	0.005	0.003	0.001
Number of iterations	57	57	58	58	59
$\frac{\|u_{DSM} - u_{exact}\|}{\|u_{exact}\|}$	0.1437	0.1217	0.0829	0.0746	0.0544

Figure 25.1 presents the numerical results when $N = 100$ and $C_0 = 7$ with $\delta_{rel} = 0.01$ and $\delta_{rel} = 0.005$. The numbers of iterations for $\delta = 0.01$ and $\delta = 0.005$ were 57 and 58, respectively.

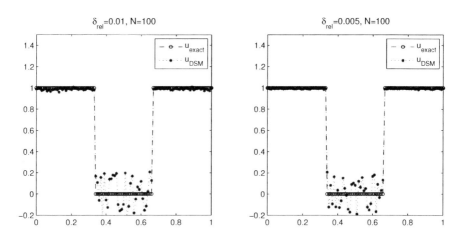

Figure 25.1   Plots of solutions obtained by the DSM when $N = 100$, $\delta_{rel} = 0.01$ (left), and $\delta_{rel} = 0.005$ (right).

Figure 25.2 presents the numerical results when $N = 100$ and $C_0 = 7$ with $\delta = 0.003$ and $\delta = 0.001$. In these cases, it took 58 and 59 iterations to get the numerical solutions for $\delta_{rel} = 0.003$ and $\delta_{rel} = 0.001$, respectively.

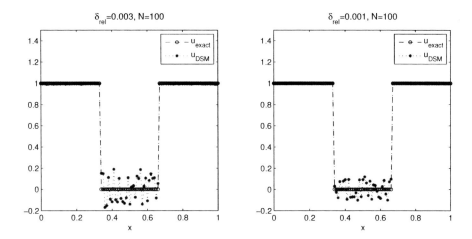

Figure 25.2   Plots of solutions obtained by the DSM when $N = 100$, $\delta_{rel} = 0.003$ (left), and $\delta_{rel} = 0.001$ (right).

We also carried out numerical experiments with $u(x) \equiv 1$, $x \in [0, 1]$, as the exact solution. Note that $F'(u)$ is boundedly invertible at this exact solution. However, in any arbitrarily small (in L^2 norm) neighborhood of this solution, there are infinitely many elements u at which $F'(u)$ is not boundedly invertible, because, as we have pointed out earlier, $F'(u)$ is not boundedly invertible if $u(x)$ is continuous and vanishes at some point $x \in [0, 1]$. In this case one cannot use the usual methods like Newton's method or the Newton–Kantorovich method. Numerical results for this experiment are presented in Table 25.2.

Table 25.2   Results when $C_0 = 4$, $N = 50$, and $u(x) \equiv 1$, $x \in [0, 1]$

δ_{rel}	0.05	0.03	0.02	0.01	0.003	0.001
Number of iterations	28	29	28	29	29	29
$\frac{\|u_{DSM} - u_{exact}\|}{\|u_{exact}\|}$	0.0770	0.0411	0.0314	0.0146	0.0046	0.0015

From Table 25.2 one concludes that the method works well in this experiment.

25.1.2   An experiment with an operator defined on a dense subset of $H = L^2[0,1]$

Our second numerical experiment with equation (25.1) deals with the operator F which is not defined on all of $H = L^2[0,1]$, but on a dense subset $D = C[0,1]$ of H:

$$F(u) := B(u) + u^3 := \int_0^1 e^{-|x-y|} u(y)\, dy + u^3. \qquad (25.14)$$

Therefore, the assumptions of Theorem 20.3.9 are not satisfied. Our goal is to show, by this numerical example, that numerically our method may work for an even wider class of problems than that covered by Theorem 20.3.9.

The operator B is compact in $H = L^2[0,1]$. The operator $u \longmapsto u^3$ is defined on a dense subset D of $L^2[0,1]$, for example, on $D := C[0,1]$. If $u, v \in D$, then

$$\langle u^3 - v^3, u - v \rangle = \int_0^1 (u^3 - v^3)(u - v)\, dx \geq 0. \qquad (25.15)$$

This and the inequality $\langle B(u - v), u - v \rangle \geq 0$, followed from equality (25.4), imply

$$\langle F(u) - F(v), u - v \rangle \geq 0, \qquad \forall u, v \in D.$$

Note that the equal sign of inequality (25.15) happens iff $u = v$ a.e. in Lebesgue measure. Thus, F is injective. Therefore, the element u_{n_δ} obtained from the iterative scheme (25.8) and the stopping rule (25.9) converges to the exact solution u_e as δ goes to 0.

Note that D does not contain subsets open in $H = L^2[0,1]$; that is, it does not contain interior points of H. This is a reflection of the fact that the operator $G(u) = u^3$ is unbounded on any open subset of H. For example, in any ball $\|u\| \leq C$, $C = const > 0$, where $\|u\| := \|u\|_{L^2[0,1]}$, there is an element u such that $\|u^3\| = \infty$. As such an element, one can take, for example, $u(x) = c_1 x^{-b}$, $\frac{1}{3} < b < \frac{1}{2}$. Here $c_1 > 0$ is a constant chosen so that $\|u\| \leq C$. The operator $u \longmapsto F(u) = G(u) + B(u)$ is maximal monotone on $D_F := \{u : u \in H, F(u) \in H\}$ (see [22, p. 102]), so that equation (25.1) with F defined by (25.14) is uniquely solvable for any $f_\delta \in H$.

The Fréchet derivative of F is

$$F'(u)w = 3u^2 w + \int_0^1 e^{-|x-y|} w(y)\, dy. \qquad (25.16)$$

If $u(x)$ vanishes on a set of positive Lebesgue measure, then $F'(u)$ is obviously not boundedly invertible. If $u \in C[0,1]$ vanishes even at one point x_0, then $F'(u)$ is not boundedly invertible in H.

We also use the iterative scheme (25.8) with the stopping rule (25.9).

We use the same exact solution u_e as in (25.13). The right-hand side f is computed by $f = F(u_e)$. Note that F' is not boundedly invertible in any neighborhood of u_e.

In experiments we found that our method works well with $C_0 \in [1, 4]$. Indeed, in the test we chose a_n by the formula $a_n := C_0 \frac{\delta^{0.9}}{n+6}$. The number of node points used in computing integrals in (25.2) and (25.7) was $N = 30$. In the test, the accuracy of the solutions obtained when $N = 30$, $N = 50$ were slightly less accurate than the one when $N = 100$.

Numerical results for various values of δ_{rel} are presented in Table 25.3. In this experiment, the noise function f_{noise} is a vector with random entries normally distributed of mean 0 and variance 1. Table 25.3 shows that the iterative scheme yields good numerical results.

Table 25.3   Results when $C_0 = 2$ and $N = 100$

δ_{rel}	0.02	0.01	0.005	0.003	0.001
Number of iterations	16	17	17	17	18
$\frac{\|u_{DSM} - u_{exact}\|}{\|u_{exact}\|}$	0.1387	0.1281	0.0966	0.0784	0.0626

Figure 25.3 presents the numerical results when $f_{noise}(x) = \sin(3\pi x)$ for $\delta_{rel} = 0.02$ and $\delta_{rel} = 0.01$. The number of iterations when $C_0 = 2$ for $\delta_{rel} = 0.02$ and $\delta_{rel} = 0.01$ were 16 and 17, respectively.

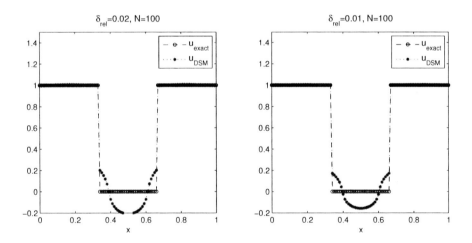

Figure 25.3   Plots of solutions obtained by the DSM with $f_{noise}(x) = \sin(3\pi x)$ when $N = 100$, $\delta_{rel} = 0.02$ (left), and $\delta_{rel} = 0.01$ (right).

Figure 25.4 presents the numerical results when $f_{noise}(x) = \sin(3\pi x)$ with $\delta_{rel} = 0.003$ and $\delta_{rel} = 0.001$. We also used $C_0 = 2$. In these cases, it took 17 and 18 iterations to give the numerical solutions for $\delta_{rel} = 0.003$ and $\delta_{rel} = 0.001$, respectively.

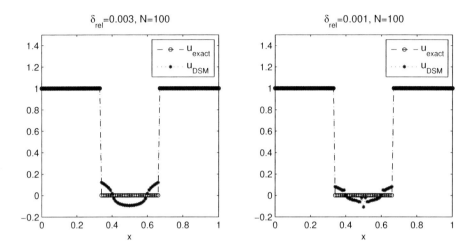

Figure 25.4   Plots of solutions obtained by the DSM with $f_{noise}(x) = \sin(3\pi x)$ when $N = 100$, $\delta_{rel} = 0.003$ (left), and $\delta_{rel} = 0.001$ (right).

We have included the results of the numerical experiments with $u(x) \equiv 1$, $x \in [0, 1]$, as the exact solution. The operator $F'(u)$ is boundedly invertible in $L^2([0, 1])$ at this exact solution. However, in any arbitrarily small L^2-neighborhood of this solution, there are infinitely many elements u at which $F'(u)$ is not boundedly invertible as was mentioned above. Therefore, even in this case one cannot use the usual methods such as Newton's method or the Newton–Kantorovich method. Numerical results for this experiment are presented in Table 25.4.

Table 25.4   Results when $C_0 = 1$, $N = 30$, and $u(x) = 1$, $x \in [0, 1]$

δ_{rel}	0.05	0.03	0.02	0.01	0.003	0.001
Number of iterations	7	8	8	9	10	10
$\frac{\|u_{DSM} - u_{exact}\|}{\|u_{exact}\|}$	0.0436	0.0245	0.0172	0.0092	0.0026	0.0009

From the numerical experiments we can conclude that the method works well in this experiment. Note that the function F used in this experiment is not defined on the whole space $H = L^2[0, 1]$ but defined on a dense subset $D = C[0, 1]$ of H.

25.2   DSM OF GRADIENT TYPE

We solve nonlinear equation (25.1) with F defined by (25.2). It follows from (25.6) that F is a monotone operator.

The Fréchet derivative of F is

$$F'(u)w = \frac{\arctan^2(u)}{1+u^2}w + \int_0^1 e^{-|x-y|}w(y)\,dy. \tag{25.17}$$

Thus, F' is self-adjoint. If $u(x)$ vanishes on a set of positive Lebesgue's measure, then $F'(u)$ is not boundedly invertible in H. If $u \in C[0,1]$ vanishes even at one point x_0, then $F'(u)$ is not boundedly invertible in H.

From (25.17) and (25.4) one can prove that

$$\|F'(u)\| \leq \sqrt{\frac{2}{\pi}} + \sup_{x \geq 0} \frac{\arctan^2 x}{1+x^2} < \sqrt{\frac{2}{\pi}} + \frac{1}{3}, \qquad \forall u \in H. \tag{25.18}$$

Moreover, one gets

$$\frac{2}{a_0^2 + \left(\sqrt{\frac{2}{\pi}} + \frac{1}{3} + a_0\right)^2} \geq 1 \quad \text{if} \quad a_0 \leq 0.5. \tag{25.19}$$

It follows from (21.83), (25.18), and (25.19) that we can use $h = 1$ if $a_0 \leq 0.5$. We use the iterative scheme

$$u_{n+1} = u_n - h(F'(u_n) + a_n I)^*(F(u_n) + a_n u_n - f_\delta), \tag{25.20}$$
$$u_0 = 0,$$

and stop iterations at $n := n_\delta$ such that the following inequality holds:

$$\|F(u_{n_\delta}) - f_\delta\| < C\delta^\zeta, \quad \|F(u_n) - f_\delta\| \geq C\delta^\zeta > \delta, \quad n < n_\delta, \tag{25.21}$$

where $\zeta \in (0,1)$.

Iterative scheme (25.20) can be considered a regularized Landweber iteration for equations with monotone operators. The existence of the stopping time n_δ is justified in Chapter 21. There might be other iterative schemes which are more efficient than scheme (25.20), but this scheme is simple and easy to implement.

Integrals of the form $\int_0^1 e^{-|x-y|}h(y)\,dy$ in (25.2) and (25.17) are computed by using the trapezoidal rule. The noisy function used in the test is generated as in (25.10)–(25.12).

In the test we took $h = 1$, $C = 1.01$, and $\zeta = 0.499$. The exact solution in the first and the second test is $u(x) = \sin(\pi x)$ and $u(x) \equiv 1$, respectively.

In the test we chose $a(t) = \frac{0.1}{(2+t)^{0.499}}$ and $h = 1$. Thus, $a_n = a(nh) = \frac{0.1}{(n+2)^{0.499}}$. This $a(t)$ satisfies condition (21.81) but does not satisfy condition (21.159). Note that condition (21.159) is one of sufficient conditions for relation (21.115) to hold. In our experiments we have found that relation (21.115) holds for our choice of $a(t)$ (see Table 25.5).

The number of nodal points, used in computing integrals in (25.2) and (25.17), was $N = 50$. The accuracy of the solutions obtained in the tests with $N = 70$ and $N = 80$ was about the same as for $N = 50$.

In all figures we denote by DSMG the numerical solutions obtained by iterative scheme (25.20).

Numerical results for various values of δ_{rel} are presented in Table 25.5. In this experiment, the noise function f_{noise} is a vector with random entries normally distributed, with mean value 0 and variance 1. Table 25.5 shows that the iterative scheme yields good numerical results.

Table 25.5   Results when $N = 50$

δ_{rel}	0.05	0.03	0.02	0.01	0.005	0.002
Number of iterations	8	15	31	126	519	3340
$\frac{\|u_{DSM}-u_{exact}\|}{\|u_{exact}\|}$	0.1187	0.0865	0.0533	0.0293	0.0184	0.0061

Figure 25.5 plots the numerical results when relative noise levels are $\delta_{rel} = 0.01$ and $\delta_{rel} = 0.001$. The exact solution in this experiment is $u(x) = \sin(\pi x)$, $x \in [0, 1]$.

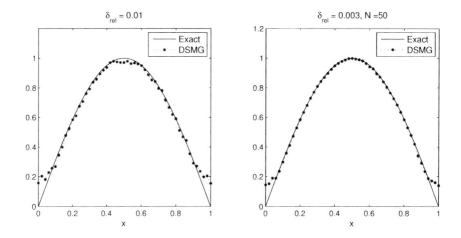

Figure 25.5   Plots of solutions obtained by the iterative scheme when $N = 50$, $\delta_{rel} = 0.01$ (left), and $\delta_{rel} = 0.003$ (right).

Figure 25.6 plots the numerical results when the noise levels are $\delta_{rel} = 0.01$ and $\delta_{rel} = 0.003$. The exact solution in this experiment is $u(x) = 1$, $x \in [0, 1]$.

25.3 DSM OF SIMPLE ITERATION TYPE

Let us do a numerical experiment solving nonlinear integral equation (25.1) with F defined by (25.2). It follows from (25.17) that F' is self-adjoint and

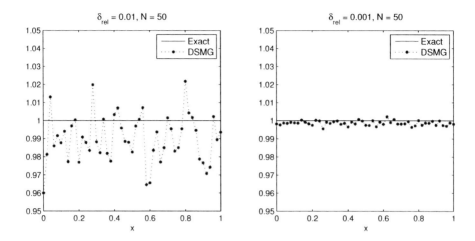

Figure 25.6 Plots of solutions obtained by the iterative scheme when $N = 50$, $\delta_{rel} = 0.01$ (left), and $\delta_{rel} = 0.001$ (right).

uniformly bounded. Thus, F is a σ-inverse operator. Inequality (25.18) and (22.2.8) imply that

$$\sigma_R^{-1} < 1 + \sqrt{\frac{2}{\pi}}, \qquad \forall R > 0.$$

Thus, if $a(0) < 1 - \sqrt{\frac{2}{\pi}}$, then (22.156) holds for $\gamma_n = h = 1$. Therefore, the existence of n_δ is guaranteed with $a_n = \frac{a(0)}{(5+n)^{0.99}}$ and $\gamma_n = 1$ by Theorem 22.2.12. It follows from (25.2) that equation $F(u) = f$ has not more than one solution for any $f \in H$. Thus if $(\delta_m)_{m=1}^\infty$ is a sequence decaying to 0 and $n_{\delta_{m_j}}$ is any convergent subsequence of n_{δ_m}, then one gets $u_{n_{\delta_{m_j}}} \to y$, the unique solution to $F(u) = f$, by Theorem 22.2.14.

If $u(x)$ vanishes on a set of positive Lebesgue's measure, then $F'(u)$ is not boundedly invertible. If $u \in C[0,1]$ vanishes even at one point x_0, then $F'(u)$ is not boundedly invertible in H.

Let us use the iterative process (22.155):

$$u_{n+1} = u_n - \gamma_n[F(u_n) + a_n u_n - f_\delta],$$
$$u_0 = 0. \tag{25.22}$$

We stop iterations at $n := n_\delta$ such that the following inequality holds:

$$\|F(u_{n_\delta}) - f_\delta\| < C\delta^\zeta, \quad \|F(u_n) - f_\delta\| \geq C\delta^\zeta, \quad n < n_\delta, \quad C > 1, \quad (25.23)$$

where $\zeta \in (0, 1)$.

Integrals of the form $\int_0^1 e^{-|x-y|}h(y)\,dy$ in (25.2) and (25.7) are computed by using the trapezoidal rule; the noisy function, used in the test, is generated by (25.10)–(25.12).

In all figures the x-variable runs through the interval $[0,1]$, and the graphs represent the numerical solutions $u_{DSM}(x)$ and the exact solution $u_{exact}(x)$.

As we have proved, the iterative scheme converges to the minimal-norm solution when $a_n = \frac{d}{(5+hn)^b}$, $b \in (0,1)$, $\frac{10b}{5^b} \le d \le 2 \times 5^{1-b}$, and γ_n are "sufficiently" small. The choice of γ_n depends on the problem one wants to solve because γ_n depends on σ_R which varies from problems to problems. Note that if one chooses γ_n to be too small, then one needs many iterations in order to reach the stopping time n_δ in (25.23). Consequently, the computation time will be large in this case. For σ-inverse problems where the constant $\sigma = \sigma_R$ can be estimated, it is not difficult to choose γ_n satisfying (22.156).

In the numerical experiments we found that our method works well with $a(0) \in [0.1, 1]$. In the test we chose a_n by the formula $a_n := \frac{a(0)}{(n+5)^\zeta}$ where $a(0) = 0.1$ and $\zeta = 0.99$. We carried out the experiments with $\gamma_n = h = const \in (0,1]$, and the method works well with this choice of γ_n. If one chooses $h > 0$ too small, then it takes more computer time for the method to converge. The number of node points, used in computing integrals (25.2) and (25.7), was $N = 100$. In all the experiments, the exact solution is chosen as follows:

$$u_{exact}(x) = \begin{cases} 0 & \text{if } x \in [0, 0.5) \\ 1 & \text{if } x \in (0.5, 1]. \end{cases}$$

As we have mentioned above, $F'(u)$ is not boundedly invertible in a neighborhood of u_{exact}. In particular, $F'(u_{exact})$ is not boundedly invertible. Thus, one cannot use classical methods such as Newton's method or Gauss–Newton method to solve for u_{exact}.

Numerical results for various values of δ_{rel} are presented in Table 25.6. From Table 25.6 one can see that the number of iterations n_δ tends to go to ∞ as δ goes to 0. Numerical experiments showed that $n_\delta \to \infty$ as $\delta \to 0$. Note that our choice of $a(t)$ in this experiment does not satisfy condition (22.2.15), which is a sufficient condition for having $n_\delta \to \infty$ as $\delta \to 0$. Table 25.6 shows that the iterative scheme yields good numerical results.

Table 25.6   Results when $a(0) = 0.1$ and $h = 1$

δ_{rel}	0.05	0.03	0.02	0.01	0.003	0.001
Number of iterations	5	6	8	13	39	104
$\frac{\|u_{DSM}-u_{exact}\|}{\|u_{exact}\|}$	0.166	0.111	0.108	0.076	0.065	0.045

Figure 25.7 plots the numerical results when relative noise levels are $\delta_{rel} = 0.01$ and $\delta_{rel} = 0.001$. The noise function in this example is a normally

distributed random vector of length N with mean 0 and variance 1. Here N is the number of nodal points used in discretizing the interval $[0, 1]$.

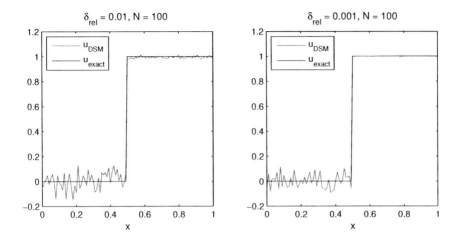

Figure 25.7   Plots of solutions obtained by the iterative scheme when $N = 100$, $\delta_{rel} = 0.01$ (left), and $\delta_{rel} = 0.001$ (right).

Figure 25.8 plots the numerical results when the noise levels are $\delta_{rel} = 0.01$ and $\delta_{rel} = 0.001$. In this experiment we choose the noise function by the formula $f_{noise}(x) = \sin(3\pi x), x \in [0, 1]$.

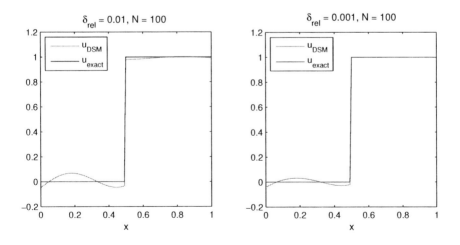

Figure 25.8   Plots of solutions obtained by the iterative scheme when $N = 100$, $\delta_{rel} = 0.01$ (left), and $\delta_{rel} = 0.001$ (right).

In computations the functions u, f and f_δ are vectors in $\mathbb{R}^N$ where N is the number of nodal points. The norm used in computations is the Euclidean length or L^2 norm of $\mathbb{R}^N$.

We have also carried out numerical experiments with $a_n = \frac{10}{(5+n)^{0.99}}$. For this choice of a_n the convergence of u_{n_δ} to the unique solution of the problem is guaranteed by Theorem 22.2.12–22.2.16. However, the numerical experiment showed that using this choice of a_n does not bring any improvement in accuracy while requiring more time for computation. Experiments also showed that for this problem it is better to use $a_n = \frac{a(0)}{(5+n)^{0.99}}$ with $a(0) \in [0.1, 1]$.

From the numerical results we conclude that the proposed stopping rule yields good results in this problem.

CHAPTER 26

INVERSION OF THE LAPLACE TRANSFORM FROM THE REAL AXIS USING AN ADAPTIVE ITERATIVE METHOD

In this chapter a new method for inverting the Laplace transform from the real axis is formulated. This method is based on a quadrature formula. We assume that the unknown function $f(t)$ is continuous with (known) compact support. An adaptive iterative method and an adaptive stopping rule, which yield the convergence of the approximate solution to $f(t)$, are proposed in this chapter. This chapter is based on paper [62].

26.1 INTRODUCTION

Consider the Laplace transform:

$$\mathcal{L}f(p) := \int_0^\infty e^{-pt} f(t)\, dt = F(p), \quad \operatorname{Re} p > 0, \tag{26.1}$$

where $\mathcal{L} : X_{0,b} \to L^2[0, \infty)$,

$$X_{0,b} := \{f \in L^2[0, \infty) \mid \operatorname{supp} f \subset [0, b)\}, \quad b > 0. \tag{26.2}$$

We assume in (26.2) that f has compact support. This is not a restriction practically. Indeed, if $\lim_{t \to \infty} f(t) = 0$, then $|f(t)| < \delta$ for $t > t_\delta$, where $\delta > 0$

Dynamical Systems Method and Applications: Theoretical Developments and Numerical Examples, First Edition. A. G. Ramm and N. S. Hoang.
Copyright © 2012 John Wiley & Sons, Inc.

is an arbitrary small number. Therefore, one may assume that supp $f \subset [0, t_\delta]$, and treat the values of f for $t > t_\delta$ as noise. One may also note that if $f \in L^1(0, \infty)$, then

$$F(p) := \int_0^\infty f(t)e^{-pt}dt = \int_0^b f(t)e^{-pt}dt + \int_b^\infty f(t)e^{-pt}dt := F_1(p) + F_2(p),$$

and $|F_2(p)| \le e^{-bp}\delta$, where $\int_b^\infty |f(t)|\,dt \le \delta$. Therefore, the contribution of the "tail" $f_b(t)$ of f,

$$f_b(t) := \left\{ \begin{array}{ll} 0, & t < b, \\ f(t), & t \ge b, \end{array} \right.$$

can be considered as noise if $b > 0$ is large and $\delta > 0$ is small. We assume in (26.2) that $f \in L^2[0, \infty)$. One may also assume that $f \in L^1[0, \infty)$, or that $|f(t)| \le c_1 e^{c_2 t}$, where c_1, c_2 are positive constants. If the last assumption holds, then one may define the function $g(t) := f(t)e^{-(c_2+1)t}$. Then $g(t) \in L^1[0, \infty)$, and its Laplace transform $G(p) = F(p + c_2 + 1)$ is known on the interval $[c_2 + 1, c_2 + 1 + b]$ of real axis if the Laplace transform $F(p)$ of $f(t)$ is known on the interval $[0, b]$. Therefore, our inversion methods are applicable to these more general classes of functions f as well.

The operator $\mathcal{L} : X_{0,b} \to L^2[0, \infty)$ is compact. Therefore, the inversion of the Laplace transform (26.1) is an ill-posed problem (see [83] and [137]). Since the problem is ill-posed, a regularization method is needed to obtain a stable inversion of the Laplace transform. There are many methods to solve equation (26.1) stably: variational regularization, quasisolutions, and iterative regularization (see, e.g, [60], [83], [137], and [151]). In this chapter we propose an adaptive iterative method based on the Dynamical Systems Method (DSM) developed in [137] and [151]. Some methods have been developed earlier for the inversion of the Laplace transform (see [18], [23], [28], and [63]). In many papers the data $F(p)$ are assumed exact and given on the complex axis. In [77] it is shown that the results of the inversion of the Laplace transform from the complex axis are more accurate than these of the inversion of the Laplace transform from the real axis. The reason is the ill-posedness of the Laplace transform inversion from the real axis. A survey regarding the methods of the Laplace transform inversion has been given in [23]. There are several types of the Laplace inversion method compared in [23]. The inversion formula for the Laplace transform is well known:

$$f(t) = \frac{1}{2\pi i} \int_{\sigma-i\infty}^{\sigma+i\infty} F(p)e^{pt}dp, \quad \sigma > 0, \tag{26.3}$$

is used in some of these methods, and then $f(t)$ is computed by some quadrature formulas, and many of these formulas can be found in [24] and [76]. Moreover, the ill-posedness of the Laplace transform inversion is not discussed in all the methods compared in [23]. The approximate $f(t)$, obtained by these methods when the data are noisy, may differ significantly from $f(t)$. There are some papers in which the inversion of the Laplace transform from the real axis was studied (see [2], [20], [27], [31], [77], [84], [108], [192], and [194]). In [2] and [108] a method based on the Mellin transform is developed. In this method the Mellin transform of the data $F(p)$ is calculated first and then inverted for $f(t)$. In [20] a Fourier series method for the inversion of Laplace transform from the real axis is developed. The drawback of this method comes from the ill-conditioning of the discretized problem. It is shown in [20] that if one uses some basis functions in $X_{0,b}$, the problem becomes extremely ill-conditioned if the number m of the basis functions exceeds 20. In [31] a reproducing kernel method is used for the inversion of the Laplace transform. In the numerical experiments in [31] the authors use double and multiple precision methods to obtain high accuracy inversion of the Laplace transform. The usage of the multiple precision increases the computation time significantly which is observed in [31], so this method may be not efficient in practice. A detailed description of the multiple precision technique can be found in [30] and [59]. Moreover, the Laplace transform inversion with perturbed data is not discussed in [31]. In [194] the authors develop an inversion formula, based on the eigenfunction expansion for the Laplace transform. The difficulties with this method are: (a) the inversion formula is not applicable when the data are noisy, (b) even for exact data the inversion formula is not suitable for numerical implementation.

The Laplace transform as an operator from C_{0k} into L^2, where $C_{0k} = \{f(t) \in C[0, +\infty) \mid \operatorname{supp} f \subset [0, k)\}$, $k = const > 0$, $L^2 := L^2[0, \infty)$, is considered in [27]. The finite difference method is used in [27] to discretize the problem, where the size of the linear algebraic system obtained by this method is fixed at each iteration, so the computation time increases if one uses large linear algebraic systems. The method of choosing the size of the linear algebraic system is not given in [27]. Moreover, the inversion of the Laplace transform when the data $F(p)$ is given only on a finite interval $[0, d]$, $d > 0$, is not discussed in [27].

The novel points in this chapter are:

(1) the representation of the approximation solution (26.73) of the function $f(t)$ which depends only on the kernel of the Laplace transform,

(2) the adaptive iterative scheme (26.76) and adaptive stopping rule (26.87), which generate the regularization parameter, the discrete data $F_\delta(p)$ and the number of terms in (26.73), needed for obtaining an approximation of the unknown function $f(t)$.

We study the inversion problem using the pair of spaces $(X_{0,b}, L^2[0, d])$, where $X_{0,b}$ is defined in (26.2), develop an inversion method, which can be easily implemented numerically, and demonstrate in the numerical experiments that our method yields the results comparable in accuracy with the results, presented in the literature—for example, with the double precision results given in paper [31].

The smoothness of the kernel allows one to use the compound Simpson's rule in approximating the Laplace transform. Our approach yields a representation (26.73) of the approximate inversion of the Laplace transform. The number of terms in approximation (26.73) and the regularization parameter are generated automatically by the proposed adaptive iterative method. Our iterative method is based on the iterative method proposed in [61]. The adaptive stopping rule we propose here is based on the discrepancy-type principle, established in [159]. This stopping rule yields convergence of the approximation (26.73) to $f(t)$ when the noise level $\delta \to 0$.

A detailed derivation of our inversion method is given in Section 26.2. In Section 26.3 some results of the numerical experiments are reported. These results demonstrate the efficiency and stability of the proposed method.

26.2   DESCRIPTION OF THE METHOD

Let $f \in X_{0,b}$. Then equation (26.1) can be written as

$$(\mathcal{L}f)(p) := \int_0^b e^{-pt} f(t)\, dt = F(p), \quad 0 \le p. \tag{26.4}$$

Let us assume that the data $F(p)$, the Laplace transform of f, are known only for $0 \le p \le d < \infty$. Consider the mapping $\mathcal{L}_m : L^2[0, b] \to \mathbb{R}^{m+1}$, where

$$(\mathcal{L}_m f)_i := \int_0^b e^{-p_i t} f(t) dt = F(p_i), \quad i = 0, 1, 2, \ldots, m, \tag{26.5}$$

$$p_i := ih, \quad i = 0, 1, 2, \ldots, m, \ h := \frac{d}{m}, \tag{26.6}$$

and m is an even number which will be chosen later. Then the unknown function $f(t)$ can be obtained from a finite-dimensional operator equation (26.5). Let

$$\langle u, v \rangle_{W^m} := \sum_{j=0}^m w_j^{(m)} u_j v_j \quad and \quad \|u\|_{W^m} := \langle u, u \rangle_{W^m} \tag{26.7}$$

be the inner product and norm in $\mathbb{R}^{m+1}$, respectively, where $w_j^{(m)}$ are the weights of the compound Simpson's rule (see [24, p. 58]), that is,

$$w_j^{(m)} := \begin{cases} h/3, & j = 0, m; \\ 4h/3, & j = 2l - 1, \ l = 1, 2, \ldots, m/2; \\ 2h/3, & j = 2l, \ l = 1, 2, \ldots, (m-2)/2, \end{cases} \quad h = \frac{d}{m}, \tag{26.8}$$

where m is an even number. Then

$$\langle \mathcal{L}_m g, v \rangle_{W^m} = \sum_{j=0}^{m} w_j^{(m)} \int_0^b e^{-p_j t} g(t)\, dt v_j$$

$$= \int_0^b g(t) \sum_{j=0}^{m} w_j^{(m)} e^{-p_j t} v_j\, dt = \langle g, \mathcal{L}_m^* v \rangle_{X_{0,b}}, \qquad (26.9)$$

where

$$\mathcal{L}_m^* v = \sum_{j=0}^{m} w_j^{(m)} e^{-p_j t} v_j, \quad v := \begin{pmatrix} v_0 \\ v_1 \\ \vdots \\ v_m \end{pmatrix} \in \mathbb{R}^{m+1}. \qquad (26.10)$$

and

$$\langle g, h \rangle_{X_{0,b}} := \int_0^b g(t) h(t)\, dt. \qquad (26.11)$$

It follows from (26.5) and (26.10) that

$$(\mathcal{L}_m^* \mathcal{L}_m g)(t) = \sum_{j=0}^{m} w_j^{(m)} e^{-p_j t} \int_0^b e^{-p_j z} g(z)\, dz := (T^{(m)} g)(t), \qquad (26.12)$$

and

$$\mathcal{L}_m \mathcal{L}_m^* v = \begin{pmatrix} \int_0^b e^{-p_0 t} \sum_{j=0}^{m} w_j^{(m)} e^{-p_j t} v_j dt \\ \int_0^b e^{-p_1 t} \sum_{j=0}^{m} w_j^{(m)} e^{-p_j t} v_j dt \\ \vdots \\ \int_0^b e^{-p_m t} \sum_{j=0}^{m} w_j^{(m)} e^{-p_j t} v_j dt \end{pmatrix} := Q^{(m)} v, \qquad (26.13)$$

where

$$(Q^{(m)})_{ij} := w_j^{(m)} \int_0^b e^{-(p_i + p_j)t}\, dt = w_j^{(m)} \frac{1 - e^{-b(p_i + p_j)}}{p_i + p_j}, \qquad (26.14)$$

and $i, j = 0, 1, 2, \ldots, m$.

Lemma 26.2.1 *Let $w_j^{(m)}$ be defined in (26.8). Then*

$$\sum_{j=0}^{m} w_j^{(m)} = d, \qquad (26.15)$$

for any even number m.

Proof: From definition (26.8) one gets

$$\sum_{j=0}^{m} w_j^{(m)} = w_0^{(m)} + w_m^{(m)} + \sum_{j=1}^{m/2} w_{2j-1}^{(m)} + \sum_{j=1}^{(m-2)/2} w_{2j}^{(m)}$$

$$= \frac{2h}{3} + \sum_{j=1}^{m/2} \frac{4h}{3} + \sum_{j=1}^{(m-2)/2} \frac{2h}{3}$$

$$= \frac{2h}{3} + \frac{2hm}{3} + \frac{h(m-2)}{3} = hm = \frac{d}{m} m = d. \tag{26.16}$$

Lemma 26.2.1 is proved.   ∎

Lemma 26.2.2 *The matrix $Q^{(m)}$, defined in (26.14), is positive semidefinite and self-adjoint in $\mathbb{R}^{m+1}$ with respect to the inner product (26.7).*

Proof: Let

$$(H_m)_{ij} := \int_0^b e^{-(p_i+p_j)t} dt = \frac{1 - e^{-b(p_i+p_j)}}{p_i + p_j}, \tag{26.17}$$

and

$$(D_m)_{ij} = \begin{cases} w_i^{(m)}, & i = j; \\ 0, & \text{otherwise,} \end{cases} \tag{26.18}$$

$w_j^{(m)}$ are defined in (26.8). Then $\langle D_m H_m D_m u, v \rangle_{\mathbb{R}^{m+1}} = \langle u, D_m H_m D_m v \rangle_{\mathbb{R}^{m+1}}$, where

$$\langle u, v \rangle_{\mathbb{R}^{m+1}} := \sum_{j=0}^{m} u_j v_j, \quad u, v \in \mathbb{R}^{m+1}. \tag{26.19}$$

We have

$$\langle Q^{(m)} u, v \rangle_{W^m} = \sum_{j=0}^{m} w_j^{(m)} (Q^{(m)} u)_j v_j = \sum_{j=0}^{m} (D_m H_m D_m u)_j v_j$$

$$= \langle D_m H_m D_m u, v \rangle_{\mathbb{R}^{m+1}} = \langle u, D_m H_m D_m v \rangle_{\mathbb{R}^{m+1}}$$

$$= \sum_{j=0}^{m} u_j (D_m H_m D_m v)_j = \sum_{j=0}^{m} u_j w_j^{(m)} (H_m D_m v)_j$$

$$= \langle u, Q^{(m)} v \rangle_{W^m}. \tag{26.20}$$

Thus, $Q^{(m)}$ is self-adjoint with respect to inner product (26.7). We have

$$(H_m)_{ij} = \int_0^b e^{-(p_i+p_j)t} dt = \int_0^b e^{-p_i t} e^{-p_j t} dt$$

$$= \langle \phi_i, \phi_j \rangle_{X_{0,b}}, \quad \phi_i(t) := e^{-p_i t}, \tag{26.21}$$

where $\langle \cdot, \cdot \rangle_{X_{0,b}}$ is defined in (26.11). This shows that H_m is a Gram matrix. Therefore,

$$\langle H_m u, u \rangle_{\mathbb{R}^{m+1}} \geq 0, \forall u \in \mathbb{R}^{m+1}. \tag{26.22}$$

This implies

$$\langle Q^{(m)} u, u \rangle_{W^m} = \langle Q^{(m)} u, D_m u \rangle_{\mathbb{R}^{m+1}} = \langle H_m D_m u, D_m u \rangle_{\mathbb{R}^{m+1}} \geq 0. \quad (26.23)$$

Thus, $Q^{(m)}$ is a positive semidefinite and self-adjoint matrix with respect to the inner product (26.7).   ∎

Lemma 26.2.3 *Let $T^{(m)}$ be defined in (26.12). Then $T^{(m)}$ is self-adjoint and positive semidefinite operator in $X_{0,b}$ with respect to inner product (26.11).*

Proof: From definition (26.12) and inner product (26.11) we get

$$\langle T^{(m)} g, h \rangle_{X_{0,b}} = \int_0^b \sum_{j=0}^m w_j^{(m)} e^{-p_j t} \int_0^b e^{-p_j z} g(z) \, dz h(t) \, dt$$

$$= \int_0^b g(z) \sum_{j=0}^m w_j^{(m)} e^{-p_j z} \int_0^b e^{-p_j t} h(t) \, dt dz$$

$$= \langle g, T^{(m)} h \rangle_{X_{0,b}}. \quad (26.24)$$

Thus, $T^{(m)}$ is a self-adjoint operator with respect to inner product (26.11). Let us prove that $T^{(m)}$ is positive semidefinite. Using (26.12), (26.8), (26.7), and (26.11), one gets

$$\langle T^{(m)} g, g \rangle_{X_{0,b}} = \int_0^b \sum_{j=0}^m w_j^{(m)} e^{-p_j t} \int_0^b e^{-p_j z} g(z) \, dz g(t) \, dt$$

$$= \sum_{j=0}^m w_j^{(m)} \int_0^b e^{-p_j z} g(z) \, dz \int_0^b e^{-p_j t} g(t) \, dt$$

$$= \sum_{j=0}^m w_j^{(m)} \left(\int_0^b e^{-p_j z} g(z) \, dz \right)^2 \geq 0. \quad (26.25)$$

Lemma 26.2.3 is proved.   ∎

From (26.10) we get $\text{Range}[\mathcal{L}_m^*] = span\{w_j^{(m)} k(p_j, \cdot, 0)\}_{j=0}^m$, where

$$k(p, t, z) := e^{-p(t+z)}. \quad (26.26)$$

Let us approximate the unknown $f(t)$ as follows:

$$f(t) \approx \sum_{j=0}^m c_j^{(m)} w_j^{(m)} e^{-p_j t} = T_{a,m}^{-1} \mathcal{L}_m^* F^{(m)} := f_m(t), \quad (26.27)$$

where p_j are defined in (26.6), $T_{a,m}$ is defined in (26.34), and $c_j^{(m)}$ are constants obtained by solving the linear algebraic system:

$$(aI + Q^{(m)}) c^{(m)} = F^{(m)}, \quad (26.28)$$

where $Q^{(m)}$ is defined in (26.13),

$$c^{(m)} := \begin{pmatrix} c_0^{(m)} \\ c_1^{(m)} \\ \vdots \\ c_m^{(m)} \end{pmatrix} \quad and \quad F^{(m)} := \begin{pmatrix} F(p_0) \\ F(p_1) \\ \vdots \\ F(p_m) \end{pmatrix}. \tag{26.29}$$

To prove the convergence of the approximate solution $f(t)$, we use the following estimates, which are proved in [151], so their proofs are omitted.

Lemma 26.2.4 *Let $T^{(m)}$ and $Q^{(m)}$ be defined in (26.12) and (26.13), respectively. Then, for $a > 0$, the following estimates hold:*

$$\|Q_{a,m}^{-1}\mathcal{L}_m\| \leq \frac{1}{2\sqrt{a}}, \tag{26.30}$$

$$a\|Q_{a,m}^{-1}\| \leq 1, \tag{26.31}$$

$$\|T_{a,m}^{-1}\| \leq \frac{1}{a}, \tag{26.32}$$

$$\|T_{a,m}^{-1}\mathcal{L}_m^*\| \leq \frac{1}{2\sqrt{a}}, \tag{26.33}$$

where

$$Q_{a,m} := Q^{(m)} + aI \quad T_{a,m} := T^{(m)} + aI, \tag{26.34}$$

I is the identity operator and $a = const > 0$.

Estimates (26.30) and (26.31) are used in proving inequality (26.92), while estimates (26.32) and (26.33) are used in the proof of Lemmas 26.2.9 and 26.2.10, respectively.

Let us formulate an iterative method for obtaining the approximation solution of $f(t)$ with the exact data $F(p)$. Consider the following iterative scheme:

$$u_n(t) = qu_{n-1}(t) + (1-q)T_{a_n}^{-1}\mathcal{L}^*F, \quad u_0(t) = 0, \tag{26.35}$$

where $\mathcal{L}^*$ is the adjoint of the operator $\mathcal{L}$, that is,

$$(\mathcal{L}^*g)(t) = \int_0^d e^{-pt}g(p)\,dp, \tag{26.36}$$

$$(Tf)(t) := (\mathcal{L}^*\mathcal{L}f)(t) = \int_0^b \int_0^d k(p,t,z)\,dpf(z)\,dz$$

$$= \int_0^b \frac{f(z)}{t+z}\left(1 - e^{-d(t+z)}\right)dz, \tag{26.37}$$

$k(p, t, z)$ is defined in (26.26),

$$T_a := aI + T, \quad a > 0, \tag{26.38}$$

$$a_n := q a_{n-1}, \quad a_0 > 0, \quad q \in (0, 1). \tag{26.39}$$

Lemma 26.2.5 *Let T_a be defined in (26.38), $\mathcal{L}f = F$, and $f \perp \mathcal{N}(\mathcal{L})$, where $\mathcal{N}(\mathcal{L})$ is the null space of $\mathcal{L}$. Then*

$$a\|T_a^{-1} f\| \to 0 \quad as \ a \to 0. \tag{26.40}$$

Proof: Since $f \perp \mathcal{N}(\mathcal{L})$, it follows from the spectral theorem that

$$\lim_{a \to 0} a^2 \|T_a^{-1} f\|^2 = \lim_{a \to 0} \int_0^\infty \frac{a^2}{(a+s)^2} d\langle E_s f, f \rangle = \|P_{\mathcal{N}(\mathcal{L})} f\|^2 = 0,$$

where E_s is the resolution of the identity corresponding to $\mathcal{L}^* \mathcal{L}$, and P is the orthogonal projector onto $\mathcal{N}(\mathcal{L})$.
Lemma 26.2.5 is proved. ∎

Theorem 26.2.6 *Let $\mathcal{L}f = F$, and u_n be defined in (26.35). Then*

$$\lim_{n \to \infty} \|f - u_n\| = 0. \tag{26.41}$$

Proof: By induction we get

$$u_n = \sum_{j=0}^{n-1} \omega_j^{(n)} T_{a_{j+1}}^{-1} \mathcal{L}^* F, \tag{26.42}$$

where T_a is defined in (26.38), and

$$\omega_j^{(n)} := q^{n-j-1} - q^{n-j}. \tag{26.43}$$

Using the identities

$$\mathcal{L}f = F, \tag{26.44}$$

$$T_a^{-1} \mathcal{L}^* \mathcal{L} = T_a^{-1}(T + aI - aI) = I - aT_a^{-1} \tag{26.45}$$

and

$$\sum_{j=0}^{n-1} \omega_j^{(n)} = 1 - q^n, \tag{26.46}$$

we get

$$f - u_n = f - \sum_{j=0}^{n-1} \omega_j^{(n)} f + \sum_{j=0}^{n-1} \omega_j^{(n)} a_{j+1} T_{a_{j+1}}^{-1} f$$

$$= q^n f + \sum_{j=0}^{n-1} \omega_j^{(n)} a_{j+1} T_{a_{j+1}}^{-1} f. \tag{26.47}$$

Therefore,

$$\|f - u_n\| \leq q^n \|f\| + \sum_{j=0}^{n-1} \omega_j^{(n)} a_{j+1} \|T_{a_{j+1}}^{-1} f\|. \tag{26.48}$$

To prove relation (26.41), the following lemma is needed:

Lemma 26.2.7 *Let $g(x)$ be a continuous function on $(0, \infty)$, let $c > 0$, and let $q \in (0, 1)$ be constants. If*

$$\lim_{x \to 0^+} g(x) = g(0) := g_0, \tag{26.49}$$

then

$$\lim_{n \to \infty} \sum_{j=0}^{n-1} \left(q^{n-j-1} - q^{n-j} \right) g(cq^{j+1}) = g_0. \tag{26.50}$$

Proof: Let

$$F_l(n) := \sum_{j=1}^{l-1} \omega_j^{(n)} g(cq^{j+1}), \tag{26.51}$$

where $\omega_j^{(n)}$ are defined in (26.43). Then

$$|F_{n+1}(n) - g_0| \leq |F_l(n)| + \left| \sum_{j=l}^{n} \omega_j^{(n)} g(cq^{j+1}) - g_0 \right|.$$

Take $\epsilon > 0$ arbitrarily small. For sufficiently large fixed $l(\epsilon)$ one can choose $n(\epsilon) > l(\epsilon)$, such that

$$|F_{l(\epsilon)}(n)| \leq \frac{\epsilon}{2}, \ \forall n > n(\epsilon),$$

because $\lim_{n \to \infty} q^n = 0$. Fix $l = l(\epsilon)$ such that $|g(cq^j) - g_0| \leq \frac{\epsilon}{2}$ for $j > l(\epsilon)$. This is possible because of (26.49). One has

$$|F_{l(\epsilon)}(n)| \leq \frac{\epsilon}{2}, \ n > n(\epsilon) > l(\epsilon)$$

and

$$\left| \sum_{j=l(\epsilon)}^{n} \omega_j^{(n)} g(cq^{j+1}) - g_0 \right| \leq \sum_{j=l(\epsilon)}^{n} \omega_j^{(n)} |g(cq^{j+1}) - g_0| + \left| \sum_{j=l(\epsilon)}^{n} \omega_j^{(n)} - 1 \right| |g_0|$$

$$\leq \frac{\epsilon}{2} \sum_{j=l(\epsilon)}^{n} \omega_j^{(n)} + q^{n-l(\epsilon)} |g_0|$$

$$\leq \frac{\epsilon}{2} + |g_0| q^{n-l(\epsilon)} \leq \epsilon,$$

if $n(\epsilon)$ is sufficiently large. Here we have used the relation

$$\sum_{j=l}^{n} w_j^{(n)} = 1 - q^{n-l}.$$

Since $\epsilon > 0$ is arbitrarily small, relation (26.50) follows.
Lemma 26.2.7 is proved. ∎

Lemma 26.2.5, together with Lemma 26.2.7 with $g(a) = a\|T_a^{-1}f\|$, yields

$$\lim_{n\to\infty} \sum_{j=0}^{n-1} w_j^{(n)} a_{j+1} \|T_{a_{j+1}}^{-1} f\| = 0. \tag{26.52}$$

This, together with estimate (26.48) and condition $q \in (0,1)$, yields relation (26.41).
Theorem 26.2.6 is proved. ∎

Lemma 26.2.8 *Let T and $T^{(m)}$ be defined in (26.37) and (26.12), respectively. Then*

$$\|T - T^{(m)}\| \le \frac{(2bd)^5}{540\sqrt{10}m^4}. \tag{26.53}$$

Proof: From definitions (26.37) and (26.12) we get

$$|(T - T^{(m)})f(t)| \le \int_0^b \left| \int_0^d k(p,t,z)\,dp - \sum_{j=0}^{m} w_j^{(m)} k(p_j,t,z) \right| |f(z)|\,dz$$

$$\le \int_0^b \left| \frac{d^5}{180m^4} \max_{p\in[0,d]} (t+z)^4 e^{-p(t+z)} \right| |f(z)|\,dz$$

$$= \int_0^b \frac{d^5}{180m^4} (t+z)^4 |f(z)|\,dz \le \frac{d^5}{180m^4} \left(\int_0^b (t+z)^8 dz \right)^{1/2} \|f\|_{X_{0,b}}$$

$$= \frac{d^5}{180m^4} \left[\frac{(t+b)^9 - t^9}{9} \right]^{1/2} \|f\|_{X_{0,b}}, \tag{26.54}$$

where the following upper bound for the error of the compound Simpson's rule was used (see [24, p. 58]): For $f \in C^{(4)}[x_0, x_{2l}]$, $x_0 < x_{2l}$,

$$\left| \int_{x_0}^{x_{2l}} f(x)\,dx - \frac{h}{3} \left[f_0 + 4\sum_{j=1}^{l} f_{2(j-1)} + 2\sum_{j=1}^{l-1} f_{2j} + f_{x_{2l}} \right] \right| \le R_l, \tag{26.55}$$

where

$$f_j := f(x_j), \quad x_j = x_0 + jh, \ j = 0,1,2,\ldots,2l, \ h = \frac{x_{2l} - x_0}{2l}, \tag{26.56}$$

and

$$R_l = \frac{(x_{2l} - x_0)^5}{180(2l)^4} |f^{(4)}(\xi)|, \quad x_0 < \xi < x_{2l}. \tag{26.57}$$

This implies

$$\|(T - T^{(m)})f\|_{X_{0,b}} \leq \frac{d^5}{540m^4} \left[\frac{(2b)^{10} - 2b^{10}}{10} \right]^{1/2} \|f\|_{X_{0,b}}$$

$$\leq \frac{(2bd)^5}{540\sqrt{10}m^4} \|f\|_{X_{0,b}}, \tag{26.58}$$

so estimate (26.53) is obtained.

Lemma 26.2.8 is proved. ∎

Lemma 26.2.9 *Let* $0 < a < a_0$,

$$m = \kappa \left(\frac{a_0}{a} \right)^{1/4}, \quad \kappa > 0. \tag{26.59}$$

Then

$$\|T - T^{(m)}\| \leq \frac{(2bd)^5}{540\sqrt{10}a_0\kappa^4} a, \tag{26.60}$$

where T *and* $T^{(m)}$ *are defined in* (26.37) *and* (26.12), *respectively.*

Proof: Inequality (26.60) follows from estimate (26.53) and formula (26.59). ∎

Lemma 26.2.9 leads to an adaptive iterative scheme:

$$u_{n,m_n}(t) = qu_{n-1,m_{n-1}} + (1 - q)T_{a_n,m_n}^{-1} \mathcal{L}_{m_n}^* F^{(m_n)}, \quad u_{0,m_0}(t) = 0, \tag{26.61}$$

where $q \in (0, 1)$, a_n are defined in (26.39), $T_{a,m}$ is defined in (26.34), $A_m\mathcal{L}$ is defined in (26.5), and

$$F^{(m)} := \begin{pmatrix} F(p_0) \\ F(p_1) \\ \dots \\ F(p_m) \end{pmatrix} \in \mathbb{R}^{m+1}, \tag{26.62}$$

p_j are defined in (26.6). In the iterative scheme (26.61) we have used the finite-dimensional-operator $T^{(m)}$ approximating the operator T. Convergence of the iterative scheme (26.61) to the solution f of the equation $\mathcal{L}f = F$ is established in the following lemma:

Lemma 26.2.10 *Let* $\mathcal{L}f = F$ *and* u_{n,m_n} *be defined in* (26.61). *If* m_n *are chosen by the rule*

$$m_n = \left[\left[\kappa \left(\frac{a_0}{a_n} \right)^{1/4} \right] \right], \quad a_n = qa_{n-1}, \ q \in (0, 1), \ \kappa, a_0 > 0, \tag{26.63}$$

where $[[x]]$ is the smallest even number not less than x, then

$$\lim_{n \to \infty} \|f - u_{n,m_n}\| = 0. \tag{26.64}$$

Proof: Consider the estimate

$$\|f - u_{n,m_n}\| \le \|f - u_n\| + \|u_n - u_{n,m_n}\| := I_1(n) + I_2(n), \tag{26.65}$$

where $I_1(n) := \|f - u_n\|$ and $I_2(n) := \|u_n - u_{n,m_n}\|$. By Theorem 26.2.6, we get $I_1(n) \to 0$ as $n \to \infty$. Let us prove that $\lim_{n\to\infty} I_2(n) = 0$. Let $U_n := u_n - u_{n,m_n}$. Then, from definitions (26.35) and (26.61), we get

$$U_n = qU_{n-1} + (1-q)\left(T_{a_n}^{-1}\mathcal{L}^*F - T_{a_n,m_n}^{-1}\mathcal{L}_{m_n}^*F^{(m_n)}\right), \quad U_0 = 0. \tag{26.66}$$

By induction we obtain

$$U_n = \sum_{j=0}^{n-1} \omega_j^{(n)}\left(T_{a_{j+1}}^{-1}\mathcal{L}^*F - T_{a_{j+1},m_{j+1}}^{-1}(\mathcal{L}_{m_{j+1}})^*F^{(m_{j+1})}\right), \tag{26.67}$$

where ω_j are defined in (26.43). Using the identities $\mathcal{L}f = F$, $\mathcal{L}_m f = F^{(m)}$,

$$T_a^{-1}T = T_a^{-1}(T + aI - aI) = I - aT_a^{-1}, \tag{26.68}$$
$$T_{a,m}^{-1}T^{(m)} = T_{a,m}^{-1}(T^{(m)} + aI - aI) = I - aT_{a,m}^{-1}, \tag{26.69}$$
$$T_{a,m}^{-1} - T_a^{-1} = T_{a,m}^{-1}(T - T^{(m)})T_a^{-1}, \tag{26.70}$$

one gets

$$U_n = \sum_{j=0}^{n-1} \omega_j^{(n)}a_{j+1}\left(T_{a_{j+1},m_{j+1}}^{-1} - T_{a_{j+1}}^{-1}\right)f$$

$$= \sum_{j=0}^{n-1} \omega_j^{(n)}a_{j+1}T_{a_{j+1},m_{j+1}}^{-1}\left(T - T^{(m_{j+1})}\right)T_{a_{j+1}}^{-1}f. \tag{26.71}$$

This, (26.63), estimate (26.32), and Lemma 26.2.8 yield

$$\|U_n\| \le \sum_{j=0}^{n-1} \omega_j^{(n)}a_{j+1}\|T_{a_{j+1},m_{j+1}}^{-1}\|\|T - T^{(m_{j+1})}\|\|T_{a_{j+1}}^{-1}f\|$$

$$\le \frac{(2bd)^5}{540\sqrt{10}a_0\kappa^4} \sum_{j=0}^{n-1} \omega_j^{(n)}a_{j+1}\|T_{a_{j+1}}^{-1}f\|. \tag{26.72}$$

Applying Lemma 26.2.5 and Lemma 26.2.7 with $g(a) = a\|T_a^{-1}f\|$, we obtain $\lim_{n\to\infty} \|U_n\| = 0$.
Lemma 26.2.10 is proved. ∎

26.2.1   Noisy data

When the data $F(p)$ are noisy, the approximate solution (26.27) is written as

$$f_m^\delta(t) = \sum_{j=0}^{m} w_j^{(m)} c_j^{(m,\delta)} e^{-p_j t} = T_{a,m}^{-1} \mathcal{L}_m^* F_\delta^{(m)}, \qquad (26.73)$$

where the coefficients $c_j^{(m,\delta)}$ are obtained by solving the following linear algebraic system:

$$Q_{a,m} c^{(m,\delta)} = F_\delta^{(m)}, \qquad (26.74)$$

$Q_{a,m}$ is defined in (26.34),

$$c^{(m,\delta)} := \begin{pmatrix} c_0^{(m,\delta)} \\ c_1^{(m,\delta)} \\ \dots \\ c_m^{(m,\delta)} \end{pmatrix}, \quad F_\delta^{(m)} := \begin{pmatrix} F_\delta(p_0) \\ F_\delta(p_1) \\ \dots \\ F_\delta(p_m) \end{pmatrix}, \qquad (26.75)$$

$w_j^{(m)}$ are defined in (26.8), and p_j are defined in (26.6).

To get the approximation solution of the function $f(t)$ with the noisy data $F_\delta(p)$, we consider the following iterative scheme:

$$u_{n,m_n}^\delta = q u_{n-1,m_{n-1}}^\delta + (1-q) T_{a_n,m_n}^{-1} \mathcal{L}_{m_n}^* F_\delta^{(m_n)}, \quad u_{0,m_0}^\delta = 0, \qquad (26.76)$$

where $T_{a,m}$ is defined in (26.34), a_n are defined in (26.39), $q \in (0,1)$, $F_\delta^{(m)}$ is defined in (26.75), and m_n are chosen by the rule (26.63). Let us assume that

$$F_\delta(p_j) = F(p_j) + \delta_j, \quad 0 < |\delta_j| \leq \delta, \quad j = 0, 1, 2, \dots, m, \qquad (26.77)$$

where δ_j are random quantities generated from some statistical distributions, e.g., the uniform distribution on the interval $[-\delta, \delta]$, and δ is the noise level of the data $F(p)$. It follows from assumption (26.77), definition (26.8), Lemma 26.2.1, and the inner product (26.7) that

$$\|F_\delta^{(m)} - F^{(m)}\|_{W^m}^2 = \sum_{j=0}^{m} w_j^{(m)} \delta_j^2 \leq \delta^2 \sum_{j=0}^{m} w_j^{(m)} = \delta^2 d. \qquad (26.78)$$

Lemma 26.2.11 *Let u_{n,m_n} and u_{n,m_n}^δ be defined in (26.61) and (26.76), respectively. Then*

$$\|u_{n,m_n} - u_{n,m_n}^\delta\| \leq \frac{\sqrt{d}\delta}{2\sqrt{a_n}}(1 - q^n), \quad q \in (0,1), \qquad (26.79)$$

where a_n are defined in (26.39).

Proof: Let $U_n^\delta := u_{n,m_n} - u_{n,m_n}^\delta$. Then, from definitions (26.61) and (26.76),

$$U_n^\delta = qU_{n-1}^\delta + (1-q)T_{a_n,m_n}^{-1}\mathcal{L}_{m_n}^*(F^{(m_n)} - F_\delta^{(m_n)}), \quad U_0^\delta = 0. \quad (26.80)$$

By induction we obtain

$$U_n^\delta = \sum_{j=0}^{n-1} \omega_j^{(n)} T_{a_{j+1},m_{j+1}}^{-1}(\mathcal{L}_{m_{j+1}})^*(F^{(m_{j+1})} - F_\delta^{(m_{j+1})}), \quad (26.81)$$

where $\omega_j^{(n)}$ are defined in (26.43). Using estimates (26.78) and inequality (26.33), one gets

$$\|U_n^\delta\| \le \sqrt{d}\sum_{j=0}^{n-1}\omega_j^{(n)}\frac{\delta}{2\sqrt{a_{j+1}}} \le \frac{\sqrt{d}\delta}{2\sqrt{a_n}}\sum_{j=0}^m \omega_j^{(n)} = \frac{\sqrt{d}\delta}{2\sqrt{a_n}}(1-q^n), \quad (26.82)$$

where ω_j are defined in (26.43).
Lemma 26.2.11 is proved. ∎

Theorem 26.2.12 *Suppose that conditions of Lemma 26.2.10 hold, and n_δ satisfies the following conditions:*

$$\lim_{\delta\to 0} n_\delta = \infty, \quad \lim_{\delta\to 0}\frac{\delta}{\sqrt{a_{n_\delta}}} = 0. \quad (26.83)$$

Then

$$\lim_{\delta\to 0}\|f - u_{n_\delta,m_{n_\delta}}^\delta\| = 0. \quad (26.84)$$

Proof: Consider the estimate

$$\|f - u_{n_\delta,m_{n_\delta}}^\delta\| \le \|f - u_{n_\delta,m_{n_\delta}}\| + \|u_{n_\delta,m_{n_\delta}} - u_{n_\delta,m_{n_\delta}}^\delta\|. \quad (26.85)$$

This, together with Lemma 26.2.11, yields

$$\|f - u_{n_\delta,m_{n_\delta}}^\delta\| \le \|f - u_{n_\delta,m_{n_\delta}}\| + \frac{\sqrt{d}\delta}{2\sqrt{a_{n_\delta}}}(1-q^n). \quad (26.86)$$

Applying relations (26.83) in estimate (26.86), one gets relation (26.84).
Theorem 26.2.12 is proved. ∎

In the following subsection we propose a stopping rule which implies relations (26.83).

26.2.2 Stopping rule

In this subsection a stopping rule which yields relations (26.83) in Theorem 26.2.12 is given. We propose the stopping rule

$$G_{n_\delta,m_{n_\delta}} \le C\delta^\varepsilon < G_{n,m_n}, \quad 1 \le n < n_\delta, \ C > \sqrt{d}, \ \varepsilon \in (0,1), \quad (26.87)$$

where

$$G_{n,m_n} = qG_{n-1,m_{n-1}} + (1-q)\|\mathcal{L}_{m_n} z^{(m_n,\delta)} - F_\delta^{(m_n)}\|_{W^{m_n}}, \quad G_{0,m_0} = 0, \quad (26.88)$$

$\|\cdot\|_{W^m}$ is defined in (26.7),

$$z^{(m,\delta)} := \sum_{j=0}^m c_j^{(m,\delta)} w_j^{(m)} e^{-p_j t}, \quad (26.89)$$

$w_j^{(m)}$ and p_j are defined in (26.8) and (26.6), respectively, and $c_j^{(m,\delta)}$ are obtained by solving linear algebraic system (26.74).

We observe that

$$\begin{aligned}
\mathcal{L}_{m_n} z^{(m_n,\delta)} - F_\delta^{(m_n)} &= Q^{(m_n)} c^{(m_n,\delta)} - F_\delta^{(m_n)} \\
&= Q^{(m_n)}(a_n I + Q^{(m_n)})^{-1} F_\delta^{(m_n)} - F_\delta^{(m_n)} \\
&= (Q^{(m_n)} + a_n I - a_n I)(a_n I + Q^{(m_n)})^{-1} F_\delta^{(m_n)} - F_\delta^{(m_n)} \\
&= -a_n(a_n I + Q^{(m_n)})^{-1} F_\delta^{(m_n)} = -a_n c^{(m_n,\delta)}.
\end{aligned}$$

$$(26.90)$$

Thus, the sequence (26.88) can be written in the following form:

$$G_{n,m_n} = qG_{n-1,m_{n-1}} + (1-q)a_n\|c^{(m_n,\delta)}\|_{W^{m_n}}, \quad G_{0,m_0} = 0, \quad (26.91)$$

where $\|\cdot\|_{W^m}$ is defined in (26.7), and $c^{(m,\delta)}$ solves the linear algebraic systems (26.74).

It follows from estimates (26.78), (26.30), and (26.31) that

$$\begin{aligned}
a_n\|c^{(m_n,\delta)}\|_{W^{m_n}} &= a_n\|(a_n I + Q^{(m_n)})^{-1} F_\delta^{(m_n)}\|_{W^{m_n}} \\
&\le a_n\|(a_n I + Q^{(m_n)})^{-1}(F_\delta^{(m_n)} - F^{(m_n)})\|_{W^{m_n}} \\
&\quad + a_n\|(a_n I + Q^{(m_n)})^{-1} F^{(m_n)}\|_{W^{m_n}} \\
&\le \|F_\delta^{(m_n)} - F^{(m_n)}\|_{W^{m_n}} \\
&\quad + a_n\|(a_n I + Q^{(m_n)})^{-1}\mathcal{L}_{m_n} f\|_{W^{m_n}} \\
&\le \delta\sqrt{d} + \sqrt{a_n}\|f\|_{X_{0,b}}.
\end{aligned}$$

$$(26.92)$$

This together with (26.91) yield

$$G_{n,m_n} \le qG_{n-1,m_{n-1}} + (1-q)(\delta\sqrt{d} + \sqrt{a_n}\|f\|_{X_{0,b}}), \quad (26.93)$$

or

$$G_{n,m_n} - \delta\sqrt{d} \le q(G_{n-1,m_{n-1}} - \delta\sqrt{d}) + (1-q)\sqrt{a_n}\|f\|_{X_{0,b}}. \quad (26.94)$$

Lemma 26.2.13 *The sequence* (26.91) *satisfies the following estimate:*

$$G_{n,m_n} - \delta\sqrt{d} \le \frac{(1-q)\sqrt{a_n}\|f\|_{X_{0,b}}}{1-\sqrt{q}}, \tag{26.95}$$

where a_n are defined in (26.39).

Proof: Define

$$\Psi_n := G_{n,m_n} - \delta\sqrt{d} \tag{26.96}$$

and

$$\psi_n := (1-q)\sqrt{a_n}\|f\|_{X_{0,b}}. \tag{26.97}$$

Then estimate (26.94) can be rewritten as

$$\Psi_n \le q\Psi_{n-1} + \sqrt{q}\psi_{n-1}, \tag{26.98}$$

where the relation $a_n = qa_{n-1}$ was used. Let us prove estimate (26.95) by induction. For $n = 0$ we get

$$\Psi_0 = -\delta\sqrt{d} \le \frac{(1-q)\sqrt{a_0}\|f\|_{X_{0,b}}}{1-\sqrt{q}}. \tag{26.99}$$

Suppose estimate (26.95) is true for $0 \le n \le k$. Then

$$\begin{aligned}
\Psi_{k+1} &\le q\Psi_k + \sqrt{q}\psi_k \le \frac{q}{1-\sqrt{q}}\psi_k + \sqrt{q}\psi_k \\
&= \frac{\sqrt{q}}{1-\sqrt{q}}\psi_k = \frac{\sqrt{q}}{1-\sqrt{q}}\frac{\psi_k}{\psi_{k+1}}\psi_{k+1} \\
&= \frac{\sqrt{q}}{1-\sqrt{q}}\frac{\sqrt{a_k}}{\sqrt{a_{k+1}}}\psi_{k+1} = \frac{1}{1-\sqrt{q}}\psi_{k+1},
\end{aligned} \tag{26.100}$$

where the relation $a_{k+1} = qa_k$ was used.
Lemma 26.2.13 is proved. ∎

Lemma 26.2.14 *Suppose*

$$G_{1,m_1} > \delta\sqrt{d}, \tag{26.101}$$

where G_{n,m_n} are defined in (26.91). *Then there exist a unique integer n_δ, satisfying the stopping rule* (26.87) *with $C > \sqrt{d}$.*

Proof: From Lemma 26.2.13 we get the estimate

$$G_{n,m_n} \le \delta\sqrt{d} + \frac{(1-q)\sqrt{a_n}\|f\|_{X_{0,b}}}{1-\sqrt{q}}, \tag{26.102}$$

where a_n are defined in (26.39). Therefore,

$$\limsup_{n\to\infty} G_{n,m_n} \le \delta\sqrt{d}, \tag{26.103}$$

where the relation $\lim_{n\to\infty} a_n = 0$ was used. This, together with condition (26.101), yields the existence of the integer n_δ. The uniqueness of the integer n_δ follows from its definition.

Lemma 26.2.14 is proved.   ∎

Lemma 26.2.15 *Suppose conditions of Lemma 26.2.14 hold and n_δ is chosen by the rule (26.87). Then*

$$\lim_{\delta\to 0} \frac{\delta}{\sqrt{a_{n_\delta}}} = 0. \qquad (26.104)$$

Proof: From the stopping rule (26.87) and estimate (26.102) we get

$$C\delta^\varepsilon \leq G_{n_\delta - 1, m_{n_\delta - 1}} \leq \delta\sqrt{d} + \frac{(1-q)\sqrt{a_{n_\delta - 1}}\|f\|_{X_{0,b}}}{1-\sqrt{q}}, \qquad (26.105)$$

where $C > \sqrt{d}$, $\varepsilon \in (0,1)$. This implies

$$\frac{\delta(C\delta^{\varepsilon - 1} - \sqrt{d})}{\sqrt{a_{n_\delta - 1}}} \leq \frac{(1-q)\|f\|_{X_{0,b}}}{1-\sqrt{q}}, \qquad (26.106)$$

so, for $\varepsilon \in (0,1)$, and $a_{n_\delta} = q a_{n_\delta - 1}$, one gets

$$\lim_{\delta\to 0} \frac{\delta}{\sqrt{a_{n_\delta}}} = \lim_{\delta\to 0} \frac{\delta}{\sqrt{q}\sqrt{a_{n_\delta - 1}}} \leq \lim_{\delta\to 0} \frac{(1-q)\delta^{1-\varepsilon}\|f\|_{X_{0,b}}}{(\sqrt{q} - q)(C - \delta^{1-\varepsilon}\sqrt{d})} = 0. \qquad (26.107)$$

Lemma 26.2.15 is proved.   ∎

Lemma 26.2.16 *Consider the stopping rule (26.87), where the parameters m_n are chosen by rule (26.63). If n_δ is chosen by the rule (26.87), then*

$$\lim_{\delta\to 0} n_\delta = \infty. \qquad (26.108)$$

Proof: From the stopping rule (26.87) with the sequence G_n defined in (26.91) one gets

$$qC\delta^\varepsilon + (1-q)a_{n_\delta}\|c^{(m_{n_\delta}, \delta)}\|_{W^{m_{n_\delta}}} \leq qG_{n_\delta - 1, m_{n_\delta - 1}}$$
$$+ (1-q)a_{n_\delta}\|c^{(m_{n_\delta}, \delta)}\|_{W^{m_{n_\delta}}} = G_{n_\delta, m_{n_\delta}} < C\delta^\varepsilon, \qquad (26.109)$$

where $c^{(m,\delta)}$ is obtained by solving linear algebraic system (26.74). This implies

$$0 < a_{n_\delta}\|c^{(m_{n_\delta}, \delta)}\|_{W^{m_{n_\delta}}} \leq C\delta^\varepsilon. \qquad (26.110)$$

Thus,

$$\lim_{\delta\to 0} a_{n_\delta}\|c^{(m_{n_\delta}, \delta)}\|_{W^{m_{n_\delta}}} = 0. \qquad (26.111)$$

If $F^{(m)} \neq 0$, then there exists a $\lambda_0^{(m)} > 0$ such that

$$E_{\lambda_0^{(m)}}^{(m)} F^{(m)} \neq 0, \quad \langle E_{\lambda_0}^{(m)} F^{(m)}, F^{(m)} \rangle_{W^m} := \xi^{(m)} > 0, \qquad (26.112)$$

where $E_s^{(m)}$ is the resolution of the identity corresponding to the operator $Q^{(m)} := \mathcal{L}_m \mathcal{L}_m^*$. Let

$$h_m(\delta, \alpha) := \alpha^2 \|Q_{m,\alpha}^{-1} F_\delta^{(m)}\|_{W^m}^2, \quad Q_{m,a} := aI + Q^{(m)}.$$

For a fixed number $a > 0$ we obtain

$$h_m(\delta, a) = a^2 \|Q_{m,a}^{-1} F_\delta^{(m)}\|_{W^m}^2$$

$$= \int_0^\infty \frac{a^2}{(a+s)^2} d\langle E_s^{(m)} F_\delta^{(m)}, F_\delta^{(m)}\rangle_{W^m}$$

$$\geq \int_0^{\lambda_0^{(m)}} \frac{a^2}{(a+s)^2} d\langle E_s^{(m)} F_\delta^{(m)}, F_\delta^{(m)}\rangle_{W^m}$$

$$\geq \frac{a^2}{(a+\lambda_0)^2} \int_0^{\lambda_0^{(m)}} d\langle E_s^{(m)} F_\delta^{(m)}, F_\delta^{(m)}\rangle_{W^m}$$

$$= \frac{a^2 \|E_{\lambda_0^{(m)}}^{(m)} F_\delta^{(m)}\|_{W^m}^2}{(a+\lambda_0^{(m)})^2}. \tag{26.113}$$

Since $E_{\lambda_0}^{(m)}$ is a continuous operator, and $\|F^{(m)} - F_\delta^{(m)}\|_{W^m} < \sqrt{d}\delta$, it follows from (26.112) that

$$\lim_{\delta \to 0} \langle E_{\lambda_0}^{(m)} F_\delta^{(m)}, F_\delta^{(m)}\rangle_{W^m} = \langle E_{\lambda_0}^{(m)} F^{(m)}, F^{(m)}\rangle_{W^m} > 0. \tag{26.114}$$

Therefore, for the fixed number $a > 0$ we get

$$h_m(\delta, a) \geq c_2 > 0 \tag{26.115}$$

for all sufficiently small $\delta > 0$, where c_2 is a constant which does not depend on δ. Suppose $\lim_{\delta \to 0} a_{n_\delta} \neq 0$. Then there exists a subsequence $\delta_j \to 0$ as $j \to \infty$, such that

$$a_{n_{\delta_j}} \geq c_1 > 0, \tag{26.116}$$

and

$$0 < m_{n_{\delta_j}} = \left[\left(\kappa(a_0/a_{n_{\delta_j}})^{1/4} \right) \right] \leq \left[\left(\kappa(a_0/c_1)^{1/4} \right) \right] := c_3 < \infty, \quad \kappa, a_0 > 0, \tag{26.117}$$

where the rule (26.63) was used to obtain the parameters $m_{n_{\delta_j}}$. This, (26.112), and (26.115) yield

$$\lim_{j \to \infty} h_{m_{n_{\delta_j}}}(\delta_j, a_{n_{\delta_j}}) \geq \lim_{j \to \infty} \frac{a_{n_{\delta_j}}^2 \|E_{\lambda_0^{(m_{n_{\delta_j}})}}^{(m_{n_{\delta_j}})} F_{\delta_j}^{(m_{n_{\delta_j}})}\|_{W^{m_{n_{\delta_j}}}}^2}{(a_{n_{\delta_j}} + \lambda_0^{(m_{n_{\delta_j}})})^2}$$

$$\geq \liminf_{j \to \infty} \frac{c_1^2 \|E_{\lambda_0^{(m_{n_{\delta_j}})}}^{(m_{n_{\delta_j}})} F^{(m_{n_{\delta_j}})}\|_{W^{m_{n_{\delta_j}}}}^2}{(c_1 + \lambda_0^{(m_{n_{\delta_j}})})^2} > 0. \tag{26.118}$$

This contradicts relation (26.111). Thus, $\lim_{\delta \to 0} a_{n_\delta} = \lim_{\delta \to 0} a_0 q^{n_\delta} = 0$, that is, $\lim_{\delta \to 0} n_\delta = \infty$.
Lemma 26.2.16 is proved.   ∎

It follows from Lemma 26.2.15 and Lemma 26.2.16 that the stopping rule (26.87) yields the relations (26.83). We have proved the following theorem:

Theorem 26.2.17 *Suppose all the assumptions of Theorem 26.2.12 hold, m_n are chosen by the rule (26.63), n_δ is chosen by the rule (26.87), and $G_{1,m_1} > C\delta$, where G_{n,m_n} are defined in (26.91), then*

$$\lim_{\delta \to 0} \| f - u^\delta_{n_\delta, m_{n_\delta}} \| = 0. \tag{26.119}$$

26.2.3   The algorithm

Let us formulate the algorithm for obtaining the approximate solution f^δ_m:

(1) The data $F_\delta(p)$ on the interval $[0, d]$, $d > 0$, the support of the function $f(t)$, and the noise level δ.

(2) Initialization: choose the parameters $\kappa > 0$, $a_0 > 0$, $q \in (0, 1)$, $\varepsilon \in (0, 1)$, $C > \sqrt{d}$, and set $u^\delta_{0,m_0} = 0$, $G_0 = 0$, $n = 1$.

(3) iterate, starting with $n = 1$, and stop when condition (26.126) (see below) holds,

   (a) $a_n = a_0 q^n$,

   (b) choose m_n by the rule (26.63),

   (c) construct the vector $F^{(m_n)}_\delta$:

$$(F^{(m_n)}_\delta)_l = F_\delta(p_l), \quad p_l = lh, \quad h = d/m_n, \quad l = \overline{0, m}. \tag{26.120}$$

   (d) Construct the matrices H_{m_n} and D_{m_n}:

$$(H_{m_n})_{ij} := \int_0^b e^{-(p_i + p_j)t} dt = \frac{1 - e^{-b(p_i + p_j)}}{p_i + p_j}, \quad i, j = \overline{1, m_n} \tag{26.121}$$

$$(D_{m_n})_{ij} = \begin{cases} w^{(m_n)}_i, & i = j; \\ 0, & \text{otherwise,} \end{cases} \tag{26.122}$$

   where $w^{(m)}_j$ are defined in (26.8).

   (e) Solve the following linear algebraic systems:

$$(a_n I + H_{m_n} D_{m_n}) c^{(m_n, \delta)} = F^{(m_n)}_\delta, \tag{26.123}$$

   where $(c^{(m_n, \delta)})_i = c^{(m_n, \delta)}_i$,

(f) Update the coefficient $c_j^{(m_n,\delta)}$ of the approximate solution $u_{n,m_n}^\delta(t)$ defined in (26.73) by the iterative formula:

$$u_{n,m_n}^\delta(t) = q u_{n-1,m_{n-1}}^\delta(t) + (1-q) \sum_{j=1}^{m_n} c^{(m_n,\delta)} w_j^{(m_n)} e^{-p_j t}, \quad (26.124)$$

where

$$u_{0,m_0}^\delta(t) = 0. \qquad (26.125)$$

Stop when for the first time the inequality

$$G_{n,m_n} = q G_{n-1,m_{n-1}} + a_n \| c^{(m_n,\delta)} \|_{W^{m_n}} \le C\delta^\varepsilon \qquad (26.126)$$

holds, and get the approximation $f^\delta(t) = u_{n_\delta,m_{n_\delta}}^\delta(t)$ of the function $f(t)$ by formula (26.124).

26.3   NUMERICAL EXPERIMENTS

26.3.1   The parameters κ, a_0, d

From definition (26.39) and the rule (26.63) we conclude that $m_n \to \infty$ as $a_n \to 0$. Therefore, one needs to control the value of the parameter m_n so that it will not grow too fast as a_n decreases. The role of the parameter κ in (26.63) is to control the value of the parameter m_n so that the value of the parameter m_n will not be too large. Since for sufficiently small noise level δ, namely $\delta \in (10^{-16}, 10^{-6}]$, the regularization parameter a_{n_δ}, obtained by the stopping rule (26.87), is at most $O(10^{-9})$, we suggest to choose κ in the interval $(0, 1]$. For the noise level $\delta \in (10^{-6}, 10^{-2}]$ one can choose $\kappa \in (1, 3]$. To reduce the number of iterations, we suggest to choose the geometric sequence $a_n = a_0 \delta^{\alpha n}$, where $a_0 \in [0.1, 0.2]$ and $\alpha \in [0.5, 0.9]$. One may assume without loss of generality that $b = 1$, because a scaling transformation reduces the integral over $(0, b)$ to the integral over $(0, 1)$. We have assumed that the data $F(p)$ are defined on the interval $J := [0, d]$. In the case of the interval $J = [d_1, d]$, $0 < d_1 < d$, the constant d in estimates (26.60), (26.78), (26.79), (26.82), (26.94), (26.95), and (26.102) are replaced with the constant $d - d_1$. If $b = 1$, that is, $f(t) = 0$ for $t > 1$, then one has to take d not too large. Indeed, if $f(t) = 0$ for $t > 1$, then an integration by parts yields $F(p) = [f(0) - e^{-p} f(1)]/p + O(1/p^2)$, $p \to \infty$. If the data are noisy and the noise level is δ, then the data becomes indistinguishable from noise for $p = O(1/\delta)$. Therefore it is useless to keep the data $F_\delta(p)$ for $d > O(1/\delta)$. In practice one may get a satisfactory accuracy of inversion by the method, proposed in this chapter, when one uses the data with $d \in [1, 20]$ when $\delta \le 10^{-2}$. In all the numerical examples we have used $d = 5$. Given the interval $[0, d]$, the proposed method generates automatically the discrete data $F_\delta(p_j)$, $j = 0, 1, 2, \ldots, m$, over the interval $[0, d]$ which are needed to get the approximation of the function $f(t)$.

26.3.2 Experiments

To test the proposed method we consider some examples proposed in [2], [18]–[20], [23], [28], [31], [77], [84], and [194]. To illustrate the numerical stability of the proposed method with respect to the noise, we use the noisy data $F_\delta(p)$ with various noise levels $\delta = 10^{-2}$, $\delta = 10^{-4}$ and $\delta = 10^{-6}$. The random quantities δ_j in (26.77) are obtained from the uniform probability density function over the interval $[-\delta, \delta]$. In Examples 1–12 we choose the value of the parameters as follows: $a_n = 0.1q^n$, $q = \delta^{1/2}$, and $d = 5$. The parameter $\kappa = 1$ is used for the noise levels $\delta = 10^{-2}$ and $\delta = 10^{-4}$. When $\delta = 10^{-6}$ we choose $\kappa = 0.3$ so that the value of the parameters m_n are not very large, namely $m_n \le 300$. Therefore, the computation time for solving linear algebraic system (26.123) can be reduced significantly. We assume that the support of the function $f(t)$ is in the interval $[0, b]$ with $b = 10$. In the stopping rule (26.87) the following parameters are used: $C = \sqrt{d} + 0.01$, $\varepsilon = 0.99$. In Example 13 the function $f(t) = e^{-t}$ is used to test the applicability of the proposed method to functions without compact support. The results are given in Table 26.13.

For a comparison with the exact solutions we use the mean absolute error:

$$MAE := \left[\frac{\sum_{j=1}^{100} (f(t_i) - f^\delta_{m_{n_\delta}}(t_i))^2}{100} \right]^{1/2}, \quad t_j = 0.01 + 0.1(j-1), \quad (26.127)$$

where $j = 1, \ldots, 100$, $f(t)$ is the exact solution and $f^\delta_{m_{n_\delta}}(t)$ is the approximate solution. The computation time (CPU time) for obtaining the approximation of $f(t)$, the number of iterations (Iter.), and the parameters m_{n_δ} and a_{n_δ} generated by the proposed method are given in each experiment (see Tables 26.1–26.12). All the calculations are done in double precision generated by MATLAB.

Example 1. (see [31])

$$f_1(t) = \begin{cases} 1, & 1/2 \le t \le 3/2, \\ 0, & otherwise, \end{cases} \qquad F_1(p) = \begin{cases} 1, & p = 0, \\ \dfrac{e^{-p/2} - e^{-3p/2}}{p}, & p > 0. \end{cases}$$

Table 26.1 Numerical results for example 1

δ	MAE	m_{n_δ}	Iter.	CPU time(second)	a_{n_δ}
10^{-2}	9.62×10^{-2}	30	3	3.13×10^{-2}	2×10^{-3}
10^{-4}	5.99×10^{-2}	32	4	6.25×10^{-2}	2×10^{-7}
10^{-6}	4.74×10^{-2}	54	5	3.28×10^{-1}	2×10^{-10}

The reconstruction of the exact solution for different values of the noise level δ is shown in Figure 26.1. When the noise level δ is 10^{-6}, our result

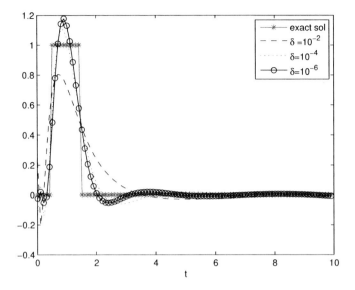

Figure 26.1   Plots of numerical results for example 1.

is comparable with the double precision results shown in [31]. The proposed method is stable with respect to the noise δ as shown in Table 26.1.

Example 2. (see [20] and [31])

$$
f_2(t) = \begin{cases} 1/2, & t = 1, \\ 1, & 1 < t < 10, \\ 0, & \text{elsewhere,} \end{cases} \qquad F_2(p) = \begin{cases} 9, & p = 0, \\ \dfrac{e^{-p} - e^{-10p}}{p}, & p > 0. \end{cases}
$$

Table 26.2   Numerical results for example 2

δ	MAE	m_{n_δ}	$Iter.$	CPU time (seconds)	a_{n_δ}
10^{-2}	1.09×10^{-1}	30	2	3.13×10^{-2}	2×10^{-3}
10^{-4}	8.47×10^{-2}	32	3	6.25×10^{-2}	2×10^{-6}
10^{-6}	7.410^{-2}	54	5	4.38×10^{-1}	2×10^{-12}

The reconstruction of the function $f_2(t)$ is plotted in Figure 26.2. In [31] a high-accuracy result is given by means of the multiple precision. But, as reported in [31], to get such high accuracy results, it takes 7 hours. From Table 26.2 and Figure 26.2 we can see that the proposed method yields stable solution with respect to the noise level δ. The reconstruction of the exact solution obtained by the proposed method is better than the reconstruction

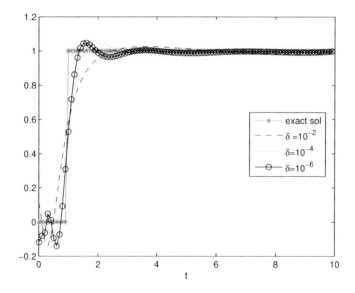

Figure 26.2   Plots of numerical results for example 2.

shown in [20]. The result is comparable with the double precision results given in [31]. For $\delta = 10^{-6}$ and $\kappa = 0.3$ the value of the parameter m_{n_δ} is bounded by the constant 54.

Example 3. (see [2], [20], [23], [84], and [194])

$$f_3(t) = \begin{cases} te^{-t}, & 0 \le t < 10, \\ 0, & \text{otherwise}, \end{cases} \qquad F_3(p) = \frac{1 - e^{-(p+1)10}}{(p+1)^2} - \frac{10e^{-(p+1)10}}{p+1}.$$

Table 26.3   Numerical results for example 3

δ	MAE	m_{n_δ}	$Iter.$	CPU time (seconds)	a_{n_δ}
10^{-2}	2.42×10^{-2}	30	2	3.13×10^{-2}	2×10^{-3}
10^{-4}	1.08×10^{-3}	30	3	3.13×10^{-2}	2×10^{-6}
10^{-6}	4.02×10^{-4}	30	4	4.69×10^{-2}	2×10^{-9}

We get an excellent agreement between the approximate solution and the exact solution when the noise level $\delta = 10^{-4}$ and 10^{-6} as shown in Figure 26.2. The results obtained by the proposed method are better than the results given in [20]. The mean absolute error MAE decreases as the noise level decreases, which shows the stability of the proposed method. Our results are more stable with respect to the noise δ than the results presented in [194]. The value of the

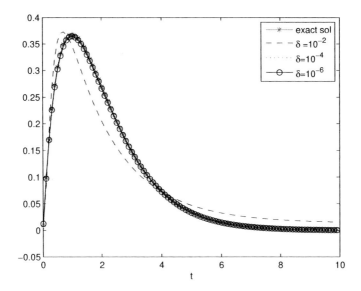

Figure 26.3   Plots of numerical results for example 3.

parameter m_{n_δ} is bounded by the constant 30 when the noise level $\delta = 10^{-6}$ and $\kappa = 0.3$.

Example 4. (see [20] and [31])

$$f_4(t) = \begin{cases} 1 - e^{-0.5t}, & 0 \leq t < 10, \\ 0, & \text{elsewhere.} \end{cases}$$

$$F_4(p) = \begin{cases} 8 + 2e^{-5}, & p = 0, \\ \dfrac{1-e^{-10p}}{p} - \dfrac{1-e^{-(p+1/2)10}}{p+0.5}, & p > 0. \end{cases}$$

As in our Example 3 when the noise $\delta = 10^{-4}$ and 10^{-6} are used, we get a satisfactory agreement between the approximate solution and the exact solution. Table 26.4 gives the results of the stability of the proposed method with respect to the noise level δ. Moreover, the reconstruction of the function $f_4(t)$ obtained by the proposed method is better than the reconstruction of $f_4(t)$ shown in [20] and is comparable with the double precision reconstruction obtained in [31].

In this example when $\delta = 10^{-6}$ and $\kappa = 0.3$ the value of the parameter m_{n_δ} is bounded by the constant 109 as shown in Table 26.4.

Example 5. (see [18], [20], and [28])

Table 26.4   Numerical results for example 4

δ	MAE	m_{n_δ}	$Iter.$	CPU time (seconds)	a_{n_δ}
10^{-2}	1.59×10^{-2}	30	2	3.13×10^{-2}	2×10^{-3}
10^{-4}	8.26×10^{-4}	30	3	9.400×10^{-2}	2×10^{-6}
10^{-6}	1.24×10^{-4}	30	4	1.250×10^{-1}	2×10^{-9}

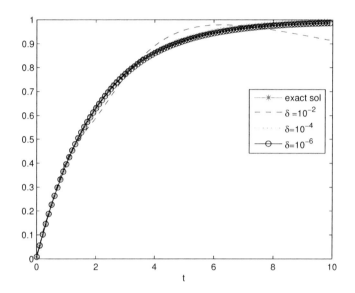

Figure 26.4   Plots of numerical results for example 4.

$$f_5(t) = 2/\sqrt{3}\,e^{-t/2}\sin(t\sqrt{3}/2)$$
$$F_5(p) = \frac{1 - \cos(10\sqrt{3}/2)e^{-10(p+0.5)}}{[(p+0.5)^2 + 3/4]} - \frac{2(p+0.5)e^{-10(p+0.5)}\sin(10\sqrt{3}/2)}{\sqrt{3}[(p+0.5)^2 + 3/4]}.$$

Table 26.5   Numerical results for example 5

δ	MAE	m_{n_δ}	$Iter.$	CPU time (seconds)	a_{n_δ}
10^{-2}	4.26×10^{-2}	30	3	6.300×10^{-2}	2×10^{-3}
10^{-4}	1.25×10^{-2}	30	3	9.38×10^{-2}	2×10^{-6}
10^{-6}	1.86×10^{-3}	54	4	3.13×10^{-2}	2×10^{-9}

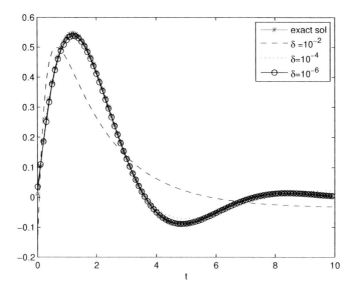

Figure 26.5   Plots of numerical results for example 5.

This is an example of the damped sine function. In [18] and [28] the knowledge of the exact data $F(p)$ in the complex plane is required to get the approximate solution. Here we only use the knowledge of the discrete perturbed data $F_\delta(p_j)$, $j = 0, 1, 2, \ldots, m$, and get a satisfactory result which is comparable with the results given in [18] and [28] when the noise level δ is 10^{-6}. The reconstruction of the exact solution $f_5(t)$ obtained by our method is better than this of the method given in [20]. Moreover, our method yields stable solution with respect to the noise level δ as shown in Figure 26.5 and Table 26.5 show. In this example when $\kappa = 0.3$ the value of the parameter m_{n_δ} is bounded by 54 for the noise level $\delta = 10^{-6}$ (see Table 26.5).

Example 6. (see [31])

$$f_6(t) = \begin{cases} t, & 0 \leq t < 1, \\ 3/2 - t/2, & 1 \leq t < 3, \\ 0, & \text{elsewhere.} \end{cases}$$

$$F_6(p) = \begin{cases} 3/2, & p = 0, \\ \frac{1 - e^{-p}(1+p)}{p^2} + \frac{e^{-3p} + e^{2p}(2p-1)}{2p^2}, & p > 0. \end{cases}$$

Example 26.6 represents a class of piecewise continuous functions. From Figure 26.6 the value of the exact solution at the points where the function is not differentiable can not be well approximated for the given levels of noise by the proposed method. When the noise level $\delta = 10^{-6}$, our result is comparable

Table 26.6   Numerical results for example 6

δ	MAE	m_{n_δ}	$Iter.$	CPU time (seconds)	a_{n_δ}
10^{-2}	4.19×10^{-2}	30	2	4.700×10^{-2}	2×10^{-3}
10^{-4}	1.64×10^{-2}	32	3	9.38×10^{-2}	2×10^{-6}
10^{-6}	1.22×10^{-2}	54	4	3.13×10^{-2}	2×10^{-9}

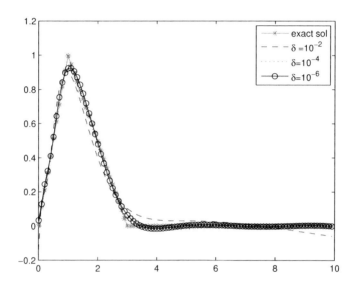

Figure 26.6   Plots of numerical results for example 6.

with the results given in [31]. Table 26.6 reports the stability of the proposed method with respect to the noise δ. It is shown in Table 26.6 that the value of the parameter m generated by the proposed adaptive stopping rule is bounded by the constant 54 for the noise level $\delta = 10^{-6}$ and $\kappa = 0.3$, which gives a relatively small computation time.

Example 7. (see [31])

$$f_7(t) = \begin{cases} -te^{-t} - e^{-t} + 1, & 0 \le t < 1, \\ 1 - 2e^{-1}, & 1 \le t < 10, \\ 0, & \text{elsewhere,} \end{cases}$$

$$F_7(p) = \begin{cases} 3/e - 1 + 9(1 - 2/e), & p = 0, \\ e^{-1-p}\frac{e^{1+p} - e(1+p)^2 + p(3+2p)}{p(p+1)^2} + (e-2)e^{-1-p-10p}\frac{e^{10p} - e^p}{p}, & p > 0. \end{cases}$$

Table 26.7   Numerical results for example 7

δ	MAE	m_{n_δ}	$Iter.$	CPU time (seconds)	a_{n_δ}
10^{-2}	1.52×10^{-2}	30	2	4.600×10^{-2}	2×10^{-3}
10^{-4}	2.60×10^{-3}	30	3	9.38×10^{-2}	2×10^{-6}
10^{-6}	2.02×10^{-3}	30	4	3.13×10^{-2}	2×10^{-9}

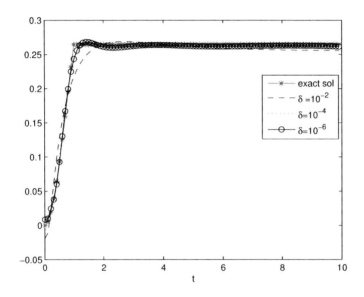

Figure 26.7   Plots of numerical results for example 7.

When the noise level δ is 10^{-4} and is 10^{-6}, we get numerical results which are comparable with the double precision results given in [31]. Figure 26.7 and Table 26.7 show the stability of the proposed method for decreasing δ.

Example 8. (see [19] and [20])

$$f_8(t) = \begin{cases} 4t^2 e^{-2t}, & 0 \le t < 10, \\ 0, & \text{elsewhere.} \end{cases}$$

$$F_8(p) = \frac{8 + 4e^{-10(2+p)}[-2 - 20(2+p) - 100(2-p)^2]}{(2+p)^3}.$$

The results of this example are similar to the results of Example 3. The exact solution can be well reconstructed by the approximate solution obtained by our method at the levels noise $\delta = 10^{-4}$ and $\delta = 10^{-6}$ (see Figure 26.8). Table 26.8 shows that the MAE decreases as the noise level decreases which shows the stability of the proposed method with respect to the noise. In all

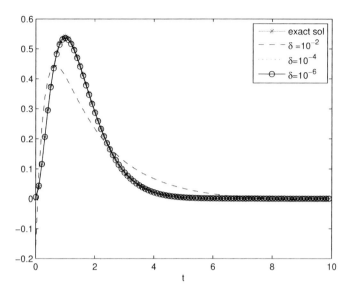

Figure 26.8 Plots of numerical results for example 8.

the levels of noise δ the computation time of the proposed method in obtaining the approximate solution are relatively small. We get better reconstruction results than the results shown in [20]. Our results are comparable with the results given in [19].

Table 26.8 Numerical results for example 8

δ	MAE	m_{n_δ}	$Iter.$	CPU time (seconds)	a_{n_δ}
10^{-2}	2.74×10^{-2}	30	2	1.100×10^{-2}	2×10^{-3}
10^{-4}	3.58×10^{-3}	30	3	3.13×10^{-2}	2×10^{-6}
10^{-6}	5.04×10^{-4}	30	4	4.69×10^{-2}	2×10^{-9}

Example 9. (see [84])

$$f_9(t) = \begin{cases} 5 - t, & 0 \leq t < 5, \\ 0, & \text{elsewhere}, \end{cases}$$

$$F_9(p) = \begin{cases} 25/2, & p = 0, \\ \frac{e^{-5p}+5p-1}{p^2}, & p > 0. \end{cases}$$

As in Example 6 the error of the approximate solution at the point where the function is not differentiable dominates the error of the approximation.

The reconstruction of the exact solution can be seen in Figure 26.9. The detailed results are presented in Table 26.9. When the double precision is used, we get comparable results with the results shown in [84].

Table 26.9 Numerical results for example 9

δ	MAE	m_{n_δ}	$Iter.$	CPU time (seconds)	a_{n_δ}
10^{-2}	2.07×10^{-1}	30	3	6.25×10^{-2}	2×10^{-6}
10^{-4}	7.14×10^{-2}	32	4	3.44×10^{-1}	2×10^{-9}
10^{-6}	2.56×10^{-2}	54	5	3.75×10^{-1}	2×10^{-12}

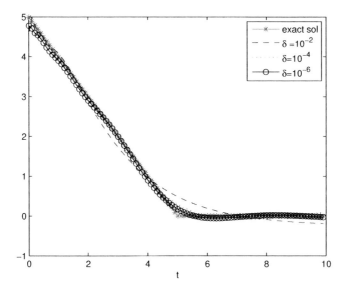

Figure 26.9 Plots of numerical results for example 9.

Example 10. (see [23])

$$f_{10}(t) = \begin{cases} t, & 0 \le t < 10, \\ 0, & \text{elsewhere,} \end{cases}$$

$$F_{10}(p) = \begin{cases} 50, & p = 0, \\ \frac{1-e^{-10p}}{p^2} - \frac{10e^{-10p}}{p}, & p > 0. \end{cases}$$

Table 26.10 shows the stability of the solution obtained by our method with respect to the noise level δ. We get an excellent agreement between the exact solution and the approximate solution for all the noise levels δ as shown in Figure 26.10.

Table 26.10   Numerical results for example 10

δ	MAE	m_{n_δ}	$Iter.$	CPU time (seconds)	a_{n_δ}
10^{-2}	2.09×10^{-1}	30	3	3.13×10^{-2}	2×10^{-6}
10^{-4}	1.35×10^{-2}	32	4	9.38×10^{-2}	2×10^{-9}
10^{-6}	3×10^{-3}	54	4	2.66×10^{-1}	2×10^{-9}

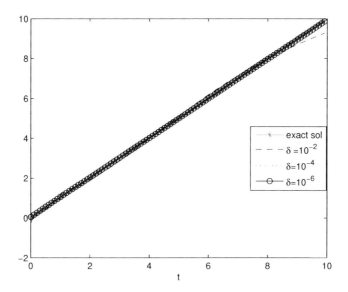

Figure 26.10   Plots of numerical results for example 10.

Example 11. (see [23] and [77])

$$f_{11}(t) = \begin{cases} \sin(t), & 0 \leq t < 10, \\ 0, & \text{elsewhere,} \end{cases}$$

$$F_{11}(p) = \frac{1 - e^{-10p}(p\sin(10) + \cos(10))}{1 + p^2}.$$

Here the function $f_{11}(t)$ represents the class of periodic functions. It is mentioned in [77] that oscillating function can be found with acceptable accuracy only for relatively small values of t. In this example the best approximation is obtained when the noise level δ is 10^{-6}, which is comparable with the results given in [23] and [77]. The reconstruction of the function $f_{11}(t)$ for various levels of the noise δ are given in Figure 26.11. The stability of the proposed method with respect to the noise δ is shown in Table 26.11. In this example the parameter m_{n_δ} is bounded by the constant 54 when the noise level δ is 10^{-6} and $\kappa = 0.3$.

Table 26.11   Numerical results for example 11

δ	MAE	m_{n_δ}	$Iter.$	CPU time (seconds)	a_{n_δ}
10^{-2}	2.47×10^{-1}	30	3	9.38×10^{-2}	2×10^{-6}
10^{-4}	4.910^{-2}	32	4	2.50×10^{-1}	2×10^{-9}
10^{-6}	2.46×10^{-2}	54	5	4.38×10^{-1}	2×10^{-12}

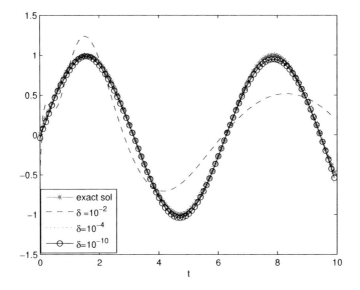

Figure 26.11   Plots of numerical results for example 11.

Example 12. (see [19] and [23])

$$f_{12}(t) = \begin{cases} t\cos(t), & 0 \le t < 10, \\ 0, & \text{elsewhere}, \end{cases}$$

$$F_{12}(p) = \frac{(p^2 - 1) - e^{-10p}(-1 + p^2 + 10p + 10p^3)\cos(10)}{(1 + p^2)^2}$$
$$+ \frac{e^{-10p}(2p + 10 + 10p^2)\sin(10)}{(1 + p^2)^2}.$$

Here we take an increasing function which oscillates as the variable t increases over the interval $[0, 10)$. A poor approximation is obtained when the

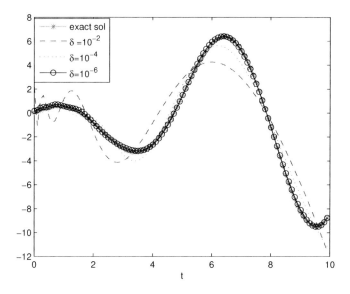

Figure 26.12 Plots of numerical results for example 12.

noise level δ is 10^{-2}. Figure 26.12 shows that the exact solution can be approximated very well when the noise level δ is 10^{-6}. The results of our method are comparable with these of the methods given in [19] and [23]. The stability of our method with respect to the noise level is shown in Table 26.12.

Table 26.12 Numerical results for example 12

δ	MAE	m_{n_δ}	$Iter.$	CPU time (seconds)	a_{n_δ}
10^{-2}	1.37×10^{0}	96	3	9.38×10^{-2}	2×10^{-6}
10^{-4}	5.98×10^{-1}	100	4	2.66×10^{-1}	2×10^{-9}
10^{-6}	2.24×10^{-1}	300	5	3.44×10^{-1}	2×10^{-12}

Example 13.

$$f_{13}(t) = e^{-t}, \quad F_{13}(p) = \frac{1}{1+p}.$$

Here the support of $f_{13}(t)$ is not compact. From the Laplace transform formula one gets

$$
\begin{aligned}
F_{13}(p) &= \int_0^\infty e^{-t} e^{-pt} dt = \int_0^b e^{-(1+p)t} dt + \int_b^\infty e^{-(1+p)t} dt \\
&= \int_0^b f_{13}(t) dt + \frac{e^{-(1+p)b}}{1+p} := I_1 + I_2,
\end{aligned}
$$

where $\delta(b) := e^{-b}$. Therefore, I_2 can be considered as noise of the data $F_{13}(p)$, that is,

$$
F_{13}^\delta(p) := F_{13}(p) + \delta(b), \tag{26.128}
$$

where $\delta(b) := e^{-b}$. In this example the following parameters are used: $d = 2$ and $\kappa = 10^{-1}$ for $\delta = e^{-5}$ and $\kappa = 10^{-5}$ for $\delta = 10^{-8}$, 10^{-20}, and 10^{-30}. Table 26.13 shows that the error decreases as the parameter b increases.

Table 26.13 Numerical results for example 13

b	MAE	m_δ	Iter	CPU time (seconds)
5	1.487×10^{-2}	2	4	3.125×10^{-2}
8	2.183×10^{-4}	2	4	3.125×10^{-2}
20	4.517×10^{-9}	2	4	3.125×10^{-2}
30	1.205×10^{-13}	2	4	3.125×10^{-2}

26.4 CONCLUSION

We have tested the proposed algorithm on many examples considered in the literature. Using the rule (26.63) and the stopping rule (26.87), the number of terms in representation (26.73), the discrete data $F_\delta(p_j)$, $j = 0, 1, 2, \ldots, m$, and regularization parameter a_{n_δ}, which are used in computing the approximation $f_m^\delta(t)$ (see (26.73)) of the unknown function $f(t)$, are obtained automatically. Our numerical experiments show that the computation time (CPU time) for approximating the function $f(t)$ is small, namely CPU time ≤ 1 seconds, and the proposed iterative scheme and the proposed adaptive stopping rule yield stable solution with respect to the noise level δ. The proposed method also works for f without compact support as shown in Example 13. Moreover, in the proposed method we only use a simple representation (26.73) which is based on the kernel of the Laplace transform integral, so it can be easily implemented numerically.

APPENDIX A

AUXILIARY RESULTS FROM ANALYSIS

A.1 CONTRACTION MAPPING PRINCIPLE

Let F be a mapping in a Banach space X. Assume that there is a closed set $D \subset X$ such that

$$F(D) \subset D, \quad \|F(u) - F(v)\| \leq q\|u - v\|, \quad u, v \in D, \quad q \in (0, 1). \quad (A.1)$$

Theorem A.1.1 *If (A.1) holds, then equation*

$$u = F(u) \quad (A.2)$$

has a unique solution in D. This solution can be obtained by the iterative method

$$u_{n+1} = F(u_n), \quad u_0 \in D, \quad u = \lim_{n \to \infty} u_n, \quad (A.3)$$

where $u_0 \in D$ is arbitrary, and

$$\|u_n - u\| \leq \frac{q^n}{1 - q}\|u_1 - u_0\|. \quad (A.4)$$

Dynamical Systems Method and Applications: Theoretical Developments and Numerical Examples, First Edition. A. G. Ramm and N. S. Hoang.
Copyright © 2012 John Wiley & Sons, Inc.

Proof: One has

$$\|u_{n+1}-u_n\| = \|F(u_n)-F(u_{n-1})\| \le q\|u_n-u_{n-1}\| \le \cdots \le q^n\|u_1-u_0\|. \quad (A.5)$$

Thus

$$\|u_n - u_p\| \le \sum_{j=p+1}^{n} \|u_j - u_{j-1}\|$$

$$\le \sum_{j=p+1}^{n} q^{j-1}\|u_1 - u_0\| \le \frac{q^p}{1-q}\|u_1 - u_0\|. \quad (A.6)$$

By the Cauchy test there exists the limit

$$\lim_{n\to\infty} u_n = u, \quad (A.7)$$

and

$$\|u - u_n\| \le \frac{q^n}{1-q}\|u_1 - u_0\|.$$

Passing to the limit in (A.3) yields (A.2). Uniqueness of the solution to (A.2) in D is immediate: If u and v solve (A.2), then

$$\|u - v\| = \|F(u) - F(v)\| \le q\|u - v\|. \quad (A.8)$$

If $0 < q < 1$, then (A.8) implies $\|u - v\| = 0$, so $u = v$.
    Theorem A.1.1 is proved.   ∎

Remark A.1.2 If

$$\|F^m(u) - F^m(v)\| \le q\|u - v\|, \quad u,v \in D, \quad q \in (0,1), \quad (A.9)$$

where $m > 1$ is an integer, and $F(D) \subset D$, then equation (A.2) has a unique solution in D.
    Indeed, the equation

$$u = F^m(u) \quad (A.10)$$

has a unique solution in D by Theorem A.1.1 due to assumption (A.9). Therefore $F(u) = F^{m+1}(u) = F^m(F(u))$. Since the solution to (A.10) is unique, it follows that $u = F(u)$, so u solves equation (A.2). The solution to (A.2) is unique if assumption (A.9) holds. Indeed, if $v = F(v)$, then $v = F^m(v)$, and therefore $v = u$, because the solution to (A.10) is unique. One can also prove that (A.3) holds.

Remark A.1.3 The contraction assumption (A.1) cannot be replaced by a weaker one:

$$\|F(u) - F(v)\| < \|u - v\|. \quad (A.11)$$

For example, if $X = \mathbb{R}^1$, $D = X$ and $F(u) = \frac{\pi}{2} + u - \arctan u$, then $u = F(u)$ implies $\frac{\pi}{2} = \arctan u$, and there is no real number u such that $\arctan u = \frac{\pi}{2}$. On the other hand,

$$|F(u) - F(v)| = |u - v - (\arctan u - \arctan v)| < |u - v|.$$

Remark A.1.4 If F maps D into a precompact subset of D and (A.11) holds, then equation (A.2) has a solution in D and this solution is unique.

Uniqueness of the solution is obvious: If v and u solve (A.2), then by (A.11) we have

$$\|u - v\| = \|F(u) - F(v)\| < \|u - v\|. \tag{A.12}$$

Thus, $u = v$.

Let us prove the existence of the solution to (A.2). Consider F on $F(D)$. The set $F(D)$ is precompact by the assumption. It is closed because F is continuous and D is closed. Thus the set $F(D)$ is is compact.

The continuous function $\|u - F(u)\|$ on the compact set $F(D)$ attains its minimum at some point y, that is,

$$\|y - F(y)\| \leq \|u - F(u)\|, \quad \forall u \in F(D). \tag{A.13}$$

If $\|y - F(y)\| > 0$, then

$$\|F(y) - F^2(y)\| < \|y - F(y)\|,$$

which is a contradiction to (A.13). Therefore $y = F(y)$, and equation (A.2) has a solution.

Assume that F depends continuously on a parameter $z \in Z$, where Z is a Banach space, that is,

$$\lim_{\|z - \xi\| \to 0} \|F(u, z) - F(u, \xi)\| = 0, \quad \forall u \in D \subset X. \tag{A.14}$$

Consider the equation

$$u = F(u, z). \tag{A.15}$$

Theorem A.1.5 *Assume that for every*

$$z \in B(z_0, \epsilon) := \{z : z \in Z, \|z - z_0\| \leq \epsilon\}, \qquad \epsilon = const > 0,$$

one has

$$\|F(u, z) - F(v, z)\| \leq q\|u - v\|, \tag{A.16}$$

where $q \in (0, 1)$ does not depend on $z \in \overline{B(z_0, \epsilon)}$,

$$\lim_{\|z - z_0\| \to 0} \|F(u, z) - F(u, z_0)\| = 0, \quad \forall u \in D, \tag{A.17}$$

and $F(\cdot, z)$ *maps a closed set $D \subset X$ into itself for every $z \in B(z_0, \epsilon)$. Then equation (A.15) has a unique solution $u(z)$, which is continuous as $z \to z_0$:*

$$\lim_{\|z-z_0\| \to 0} \|u(z) - u(z_0)\| = 0. \tag{A.18}$$

If $F(\cdot, z) \in C^m(B(z_0, \epsilon))$, then $u \in C^m(B(z_0, \epsilon))$, $m \geq 1$.

Proof: For each $z \in B(z_0, \epsilon)$, equation (A.15) has a unique solution $u(z)$. Moreover,

$$\begin{aligned}
\|u(z) - u(z_0)\| &= \|F(u(z), z) - F(u(z_0), z_0)\| \\
&\leq \|F(u(z), z) - F(u(z_0), z)\| \\
&\quad + \|F(u(z_0), z) - F(u(z_0), z_0)\| \\
&\leq q\|u(z) - u(z_0)\| + \|F(u(z_0), z) - F(u(z_0), z_0)\|.
\end{aligned}$$

Therefore

$$\|u(z) - u(z_0)\| \leq \frac{1}{1-q}\|F(u(z_0), z) - F(u(z_0), z_0)\|. \tag{A.19}$$

By assumption (A.17) one gets from (A.19) the desired conclusion (A.18). If $m = 1$, then differentiate (A.15) with respect to z and get the equation for $u'_z := u'$:

$$u' = F_u(u, z)u' + F_z(u, z).$$

From (A.16) it follows that

$$\|F'_u(u, z)\| \leq q < 1,$$

so that the above equation has a unique solution and this implies u is in $C^1(B(z_0, \epsilon))$. Similarly, one treats the case $m > 1$.

Theorem A.1.5 is proved. ∎

Remark A.1.6 Under the assumption of Theorem A.1.1, the sequence u_n, defined in (A.3), is a minimizing sequence for the functional $g(u) := \|u - F(u)\|$.

Remark A.1.7 Suppose that there is a norm $\|\cdot\|$ on X equivalent to the original norm, that is,

$$c_1\|u\|_1 \leq \|u\| \leq c_2\|u\|_1, \quad \forall u \in X. \tag{A.20}$$

It may happen that the map F is not a contraction on D with respect to the original norm, but is a contraction with respect to an equivalent norm. A standard example deals with the equation

$$u(t) = \int_0^t f(s, u(s))\, ds := F(u), \tag{A.21}$$

where $u(t)$ is a continuous vector-function with values in $\mathbb{R}^n$. Assume that

$$|f(t, u) - f(t, v)| \leq k|u - v|, \tag{A.22}$$

where

$$|u| := \left(\sum_{j=1}^{n} |u_j|^2 \right)^{\frac{1}{2}}$$

is the length of the vector u, and $k > 0$ is a constant independent of u, v and t. Let X be the space $C(0, T)$ of continuous vector-functions with the norm

$$\|u\| = \max_{0 \leq t \leq T} |u(t)|. \tag{A.23}$$

Define an equivalent norm:

$$\|u\|_1 = \max_{0 \leq t \leq T} \{e^{-\gamma t}|u(t)|\}, \quad \gamma = const > 0. \tag{A.24}$$

We have

$$
\begin{aligned}
\|F(u) - F(v)\| &\leq \max_{0 \leq t \leq T} \left\{ e^{-\gamma t} k \int_0^t |u - v|\, ds \right\} \\
&\leq \max_{0 \leq t \leq T} \left\{ e^{-\gamma t} k \int_0^t e^{\gamma s}\, ds \right\} \|u - v\|_1 \\
&\leq k \frac{1 - e^{-\gamma T}}{\gamma} \|u - v\|_1.
\end{aligned} \tag{A.25}
$$

One can always choose $\gamma > 0$ such that

$$q := k \frac{1 - e^{-\gamma T}}{\gamma} < 1 \tag{A.26}$$

no matter how large the fixed k and T are. It is easy to see that in the original norm (corresponding to $\gamma = 0$) the map F is not necessarily a contraction if k and T are sufficiently large.

A.2   EXISTENCE AND UNIQUENESS OF THE LOCAL SOLUTION TO THE CAUCHY PROBLEM

Let

$$\dot{u} = F(t, u), \quad u(0) = u_0, \tag{A.27}$$

where $F : X \rightarrow X$ is a map in a Banach space. Assume that

$$\|F(t, u) - F(t, v)\| \leq M_1(R)\|u - v\|, \quad u, v \in B(u_0, R), \tag{A.28}$$

where

$$B(u_0, R) = \{u : \|u - u_0\| \leq R, \, u \in X\},$$

and

$$\|F(t, u)\| \leq M_0(R), \qquad \forall u, v \in B(u_0, R). \tag{A.29}$$

For any fixed $u \in B(u_0, R)$ the element $F(t, u)$ is assumed continuous with respect to $t \in [0, T]$. Consider the equation

$$u(t) = u_0 + \int_0^t F(s, u(s)) \, ds := G(u) \tag{A.30}$$

in the space $C([0, T]); X)$ of continuous functions with values in X and the norm

$$|u| = \max_{0 \leq t \leq T} \|u(t)\|. \tag{A.31}$$

Let us check that the map G is a contraction on the set

$$D = \{u(t) : u \in B(u_0, R), \, t \in [0, T]\},$$

provided that T is sufficiently small, and $GD \subset D$.

Using (A.29), we get

$$|u(t) - u_0| \leq M_0(R)T \leq R \qquad \text{if} \qquad T \leq \frac{R}{M_0(R)}, \tag{A.32}$$

so

$$GD \subset D \qquad \text{if} \quad T \leq \frac{R}{M_0(R)}. \tag{A.33}$$

Furthermore, using (A.28), we get

$$|G(u) - G(v)| \leq M_1(R)T|u - v| := q|u - v|, \tag{A.34}$$

where

$$q = M_1(R)T < 1 \qquad \text{if} \qquad T \leq \frac{1}{M_1(R)}. \tag{A.35}$$

From Theorem A.1.1 we obtain the following result.

Theorem A.2.1 *Assume (A.28), (A.29), (A.32), and (A.35). Then problem (A.27) has a unique solution $u = u(t, u_0) \in B(u_0, R)$ defined on $[0, T]$, where*

$$T = \min\left(\frac{1}{M_1(R)}, \frac{R}{M_0(R)}\right). \tag{A.36}$$

Remark A.2.2 By Theorem A.1.5 the solution $u(t, u_0)$ depends continuously on the initial approximation u_0 in the following sense.

If

$$\|u_0 - v_0\| \leq \epsilon, \quad T = \min\left(\frac{1}{M_1(R + \epsilon)}, \frac{R}{M_0(R + \epsilon)}\right),$$

then
$$\lim_{\|v_0 - u_0\| \to 0} |u(t, u_0) - u(t, v_0)| = 0.$$

Theorem A.2.1 allows one to introduce the notion of maximal interval of the existence of the solution. Namely, if the solution $u(t)$ exists on the interval $[0, T)$ and does not exist on the interval $[0, \tau]$ for any $\tau > T$, then we say that $[0, T]$ is the maximal interval of the existence of the solution to (A.27). We have defined the maximal interval of the form $[0, T)$ of the existence of the solution. Usually the definition does not fix the lower point of the maximal interval of the existence of the solution, which is zero in our case (see, e.g., [40] for the standard definition of the maximal interval of the existence of the solution).

Under the assumptions of Theorem A.2.1, we can prove that if $[0, T]$ is the maximal interval of the existence of the solution to (A.27), then

$$\lim_{t \to T^-} \|u(t)\| = \infty, \tag{A.37}$$

where $t \to T^-$ denotes convergence from the left.

Indeed, assuming that (A.37) fails, we have

$$\sup_{0 \le t \le T} \|u(t)\| \le c < \infty. \tag{A.38}$$

Thus

$$\|u(t_n)\| \le c, \qquad t_n \to T, \quad t_n < T. \tag{A.39}$$

Consider the problem

$$u(t) = u(t_n) + \int_{t_n}^{t} F(s, u(s)) \, ds, \tag{A.40}$$

and let

$$D = \{u(t) : u(t) \in B(u(t_n), R), \qquad t \in [t_n, t_n + \tau]\}.$$

We choose t_n such that

$$t_n > T - \tau \tag{A.41}$$

and

$$\tau < \min \left(\frac{1}{M_1(R)}, \frac{1}{M_0(R)} \right), \tag{A.42}$$

where

$$\|u(t) - u(t_n)\| \le R, \qquad t_n \le t \le t_n + \tau. \tag{A.43}$$

If the constant c, independent of n, is given in (A.39), then we can find R and $\tau = \tau(R) > 0$, independent of n, so that the inequalities (A.41) and (A.43) hold. Therefore the unique solution $u(t)$ to problem (A.27) is defined on the interval $[0, T_1]$, where $T_1 = t_n + \tau > T$. This contradicts the assumption that

$T < \infty$ and $[0, T]$ is the maximal interval of the existence of the solution $u(t)$. We have proved the following result.

Theorem A.2.3 *Assume that conditions (A.28) and (A.29) hold with some $R > 0$ for any u_0 such that $\|u_0\| \leq c$, where $c > 0$ is an arbitrary constant and $R = R(c)$. Then either the maximal interval $[0, T]$ of the existence of the solution to (A.27) is finite, and then (A.37) holds, or it is infinite, and then (A.38) holds for any $T < \infty$ with a constant $c = c(T) > 0$.*

Remark A.2.4 Theorems A.2.1 and A.2.3 imply that a unique local solution to problem (A.27) is a global one provided that a uniform with respect to time a priori estimate

$$\sup_t \|u(t)\| \leq c < \infty \tag{A.44}$$

is established for the solution to problem (A.27) and $F(t, u)$ satisfies the assumptions of Theorem A.2.3.

This argument was used often in Chapters 3, 6–11, and 13.

Remark A.2.5 We give two standard examples of the finite maximal interval of the existence of the solution to problem (A.27).

Example A.1
Let

$$\dot{u} = 1 + u^2, \quad u(0) = 1. \tag{A.45}$$

Then

$$u = \tan\left(\frac{\pi}{4} + t\right). \tag{A.46}$$

Thus, $u(t)$ does not exist on any interval of length greater than π, and

$$\lim_{t \to \frac{\pi}{4}} u(t) = \infty. \tag{A.47}$$

Example A.2
Let

$$u_t - \Delta u = u^2, \quad t \geq 0, \quad x \in D \subset R^n, \tag{A.48}$$

$$u_N|_S = 0, \quad t \geq 0, \tag{A.49}$$

$$u|_{t=0} = u_0(x). \tag{A.50}$$

Here, D is a bounded domain with a sufficient smooth boundary S, N is the exterior normal to S, and

$$u_0(x) \geq 0, \quad \int_D u_0(x)\, dx > 0. \tag{A.51}$$

Let us verify that the local solution to problem (A.48)–(A.51) has to blow up in a finite time; that is, (A.37) holds for some $T < \infty$. Indeed, integrate (A.48) over D, and use formula

$$\int_D \Delta u\, dx = \int_S u_N\, ds = 0,$$

and get

$$\frac{\partial}{\partial t} \int_D u \, dx = \int_D u^2 \, dx. \tag{A.52}$$

Moreover, assuming that u is real-valued, we get

$$g(t) := \int_D u \, dx \le \left(\int_D u^2 dx \right)^{\frac{1}{2}} |D|^{\frac{1}{2}}, \quad |D| := \text{meas } D, \tag{A.53}$$

where meas D is the volume of the domain D.

Thus,

$$\frac{dg}{dt} \ge cg^2, \quad c := \frac{1}{|D|}, \quad g(0) = \int_D u_0 \, dx > 0. \tag{A.54}$$

Integrate (A.54) and get

$$g(t) \ge \frac{1}{\frac{1}{g(0)} - ct}. \tag{A.55}$$

Therefore

$$\lim_{t \to T^-} g(t) = +\infty, \quad T := \frac{1}{cg(0)}. \tag{A.56}$$

If (A.56) holds, then

$$\lim_{t \to T} \|u\|_{L^2(D)} = \infty, \tag{A.57}$$

because of the estimate (A.53).

A.3 DERIVATIVES OF NONLINEAR MAPPINGS

Let $F : X \to Y$ be a mapping from a Banach space X into a Banach space Y.

Definition A.3.1 *If there exists a bounded linear map $A = A(u)$, such that*

$$F(u + h) = F(u) + A(u)h + o(\|h\|), \quad \|h\| \to 0, \tag{A.58}$$

then the map F is called F-differentiable (Fréchet differentiable) at the point u and one writes $A(u) := F'(u)$. If $A(u)$ depends continuously on u in the sense

$$\lim_{\|v-u\| \to 0} \|A(u) - A(v)\| = 0, \quad u, v \in D, \tag{A.59}$$

then we write $F \in C^1(D)$. Here D is an open set in X, and $\|A(u)\|$ is the norm of the linear operator from X to Y.

If $F'(u)$ exists, then it is unique.

Definition A.3.2 *The map $F : X \to Y$ is called G-differentiable (Gâteaux differentiable) at a point u if*

$$F(u + th) = F(u) + tA(u)h + o(t), \quad t \to 0, \tag{A.60}$$

where t is a number, $h \in X$ is any arbitrary element, and $A(u)$ is a bounded linear operator $A(u) : X \rightarrow Y$.

The term $o(t)$ in (A.60) depends on h. Clearly, if F is F-differentiable, then F is G-differentiable.

If F is F-differentiable at a point u, then F is continuous at this point. One has

$$F(u + h) - F(u) = \int_0^1 \frac{d}{dt} F(u + th) \, dt = \int_0^1 F'(u + th)h \, dt.$$

Thus

$$\|F(u + h) - F(u)\| \leq \sup_{v \in B(u,R)} \|F'(v)\| \|h\|, \quad \|h\| \leq R,$$

where $B(u, R) : \{u : \|u - v\| \leq R\}$. The first derivative $F'(u)$ is a map $X \rightarrow L(X, Y)$, where $L(X, Y)$ is the space of bounded linear operators from X into Y. If this map is F-differentiable, then its derivative is called the second derivative of F, $F''(u)$. The map $F''(u)$ maps X into a set $L(X, L(X, Y))$. This set can be identified with the set of $L(X, X; Y)$ of bilinear mappings. An m-linear mapping $G : X_1 \times X_2 \times \cdots \times X_m \rightarrow Y$ is a mapping which is a bounded linear mapping with respect to each of the variables x_j, assuming that all the other variables are fixed, thus

$$\|G(x_1, x_2, \cdots, x_m)\| \leq c \prod_{j=1}^m \|x_j\|.$$

One can define higher-order derivatives inductively. Then $F^{(n)}(u)$ is the first derivative of $F^{(n-1)}(u)$. If $F \in C^{n+1}(D)$, that is, F is $n + 1$ times F-differentiable in an open set $D \subset X$, then an analog of Taylor's formula holds:

$$F(u + h) = \sum_{j=0}^n \frac{F^{(j)}(u)h^j}{j!} + R_n, \tag{A.61}$$

where

$$R_n = \int_0^1 \frac{(1 - s)^n}{n!} F^{(n+1)}(u + sh)h^{n+1} ds. \tag{A.62}$$

Here

$$F^{(n)}(u)h^n = \frac{\partial^n}{\partial t_1 \cdots \partial t_n} F\left(u + \sum_{j=1}^n t_j h_j\right) \Bigg|_{\substack{t_1 = \cdots = t_n = 0 \\ h_1 = \cdots = h_n = h}} \cdot \tag{A.63}$$

Example A.3

Uryson operators are defined by the formula

$$F(u) = \int_D K(x, y, u(y)) \, dy, \quad D \subset \mathbb{R}^n. \tag{A.64}$$

They are considered in $X = C(D)$ or $X = L^p(D)$, where $C(D)$ is the space of continuous in a bounded domain D the function with sup-norm and $L^p(D)$, $p \geq 1$, are the Lebesgue spaces. If the function $K(x, y, u(y))$ is continuous in the region $D \times D \times B_R$, where $B = \{u : |u| < R\}$, and $K_u = \frac{\partial K}{\partial u}$ is continuous in $D \times D \times B_R$, then the operator (A.64) is F-differentiable in $C(D)$ and in $L^p(D)$ at any point u such that $\|u\| \leq R$. One has

$$F'(u)h = \int_D K_u(x, y, u(y))h(y)\, dy. \tag{A.65}$$

If one wishes that the operator F, defined in (A.64), be defined on all of the space $X = C(D)$, then the function $K(x, y, u)$ has to be defined in $D \times D \times \mathbb{R}$. If $X = L^p(D)$, then it is necessary to restrict the growth of K with respect to variable u in order that the domain of definition of F in X be nontrivial.

For example, for F, defined in (A.64), to act from $L^{p_1}(D)$ int $L^{p_2}(D)$, it is sufficient that

$$|K(x, y, u)| \leq c_1 + c_2 |u|^{\frac{p_1}{p_2}},$$

where c_1 and c_2 are positive constants. For F, defined in (A.64) to act in $L^p(D)$, it is sufficient that

$$|K(x, y, u)| \leq a(x, y)(c_1 + c_2 |u|^\gamma),$$

$$\int_D \int_D |a(x, y)|^b dx dy < \infty, \quad \gamma \leq b - 1,$$

$$\frac{\gamma b}{b - 1} \leq p \leq b,$$

where c_1, c_2, γ, and b are positive constants. This is verified by using Hölder inequality.

A particular case of the Uryson operator (A.64) is the Hammerstein operator

$$F(u) = \int_D K(x, y) f(y, u(y))\, dy, \tag{A.66}$$

where the function $f(y, u)$ is continuous with respect to u and integrable with respect to y. The operator

$$(Nu)(y) := f(y, u(y))$$

is called Nemytskij operator. It acts from $L^{p_1}(D)$ into $L^{p_2}(D)$ if

$$|f(y, u(y))| \leq a(y) + b|u(y)|^{\frac{p_1}{p_2}}, \quad a \in L^{p_2}(D). \tag{A.67}$$

The operator

$$Ku := \int_D K(x, y) u(y)\, dy \tag{A.68}$$

acts from $L^p(D)$ into $L^q(D)$, $p, q \geq 1$, if

$$\int_D \int_D |K(x,y)| \, dx dy < \infty, \quad r' = \frac{r}{r-1}, \quad r := \min(p, q'). \tag{A.69}$$

Indeed

$$|Ku| \leq \left(\int_D |K(x,y)|^{r'} \right)^{\frac{1}{q'}} \left(\int_D |u(y)|^q \right)^{\frac{1}{q'}}, \tag{A.70}$$

and $r \leq q'$ implies $r' \geq q$. Note that since $|D| := \text{meas } D < \infty$ and $r \leq p$, one has

$$\|u\|_{L^{r'}(D)} \leq \|u\|_{L^p(D)}. \tag{A.71}$$

Furthermore,

$$\left(\int_D |Ku|^q dx \right)^{\frac{1}{q}} \leq \left[\int_D dx \left(\int_D |K(x,y)|^{r'} dy \right)^{\frac{q}{r'}} \right]^{\frac{1}{q}} \|u\|_{L^p(D)}$$

$$\leq |D|^{\frac{1}{q}\frac{r'}{q}} \|u\|_{L^p(D)} \left(\int_D \int_D |K(x,y)|^{r'} dy dx \right)^{\frac{1}{r'}}. \tag{A.72}$$

From (A.70)–(A.72) the desired conclusion follows and

$$\|K\|_{L^p(D) \to L^q(D)} \leq \|K(x,y)\|_{L^{r'}(D \times D)} |D|^{\frac{1}{q}\left(\frac{r'}{q}\right)'}. \tag{A.73}$$

A.4 IMPLICIT FUNCTION THEOREM

Assume that X, Y, and Z are Banach spaces $U \subset X$ and $V \subset Y$ are open sets, $u_0 \in U$, $v_0 \in V$, $F : X \times Y \to Z$ is a $C^1(U \times V)$ map, and $F(u_0, v_0) = 0$.

Theorem A.4.1 *If $[F'_v(u_0, v_0)]^{-1} := L$ is a bounded linear operator, then there exists a unique map $v = f(u)$, such that $F(u, f(u)) = 0$, $u \in U_1 \subset U$, $v_0 = f(u_0)$, and f is continuous in U_1. If $F \in C^m(U \times V)$, then $f \in C^m(U_1)$.*

Proof: Without loss of generality assume that $u_0 = 0$, $v_0 = 0$. Consider the equation

$$v = v - LF(u,v) := T(v,u). \tag{A.74}$$

Let us check that the operator T maps a ball $B_\epsilon := \{v : \|v\| \leq \epsilon\}$ into itself and is a contraction on B_R for any $u \in U_1$. If this is checked, then Theorem A.1.1 implies the existence and uniqueness of the solution to $v = f(u)$ to the equation (A.74), and Theorem A.1.5 implies that $f \in C^m(U_1)$ if $F \in C^m(U \times V)$.

We have

$$\|v - LF(u,v)\| = \|v - LF_u(0,0)u - v - LR\|$$
$$\leq \|L\| \|F_u(0,0)\| \|u\| + \|L\| o(\|v\| + \|u\|), \tag{A.75}$$

where $R = R(u, v)$ is the remainder in the formula

$$F(u, v) = F_u(0, 0)u + F_v(0, 0)v + R, \quad R = o(\|u\| + \|v\|), \quad F_v(0, 0) = L^{-1}. \tag{A.76}$$

If

$$\|u\| \leq \delta, \quad \|L\|\|F_u(0, 0)\|\delta \leq \frac{\epsilon}{2}, \tag{A.77}$$

and

$$o(\|v\| + \|u\|) \leq \frac{\epsilon}{2}, \tag{A.78}$$

then

$$T(v, u)B_\epsilon \subset B_\epsilon, \quad \forall u \in \{u : \|u\| \leq \delta\} := \tilde{B}_\delta \subset U. \tag{A.79}$$

Let $v, w \in B_\epsilon$ and $u \in \tilde{B}_\delta$. Then

$$T_v'(v, u) = I - LF_v'(u, v), \tag{A.80}$$

and, by the continuity of $F_v'(u, v)$,

$$\|F_v'(u, v) - F_v(0, 0)\| \leq \alpha(\delta, \epsilon), \tag{A.81}$$

where

$$\lim_{\delta, \epsilon \to 0} \alpha(\delta, \epsilon) = 0. \tag{A.82}$$

Since $LF_v'(0, 0) = I$, formulas (A.80)–(A.82) imply

$$\|T_v'(v, u)\| \leq \|L\|\|F_v'(u, v) - F_v'(0, 0)\| \leq q < 1, \tag{A.83}$$

provided that

$$0 \leq \epsilon \leq \epsilon_0, \quad 0 \leq \delta \leq \delta_0, \tag{A.84}$$

where ϵ_0 and δ_0 are sufficiently small. The number $q < 1$ in (A.83) depends on ϵ_0 and δ_0 but not on $u \in \tilde{B}_{\epsilon_0}$ or $v \in \tilde{B}_{\delta_0}$.

Thus, Theorems A.1.1 and A.1.5 imply the conclusions of Theorem A.4.1. ∎

A particular case of Theorem A.4.1 is the equation for the inverse function

$$v = f(u), \quad v_0 = f(u_0). \tag{A.85}$$

Theorem A.4.2 *Assume that $f : Y \to X$,*

$$L := [f'(v_0)]^{-1}, \tag{A.86}$$

is a bounded linear operator, $u_0 = f(v_0)$, and $f \in C^1(V)$, $v_0 \in V$, where V is an open set in Y. Then there exists and is unique the inverse function $u = f^{-1}(v)$, such that $u = f^{-1}(f(u))$, $u \in U$, $u_0 \in U$, where $U \subset X$ is a neighborhood of u_0, and $f^{-1}(v)$ is C^m if $f \in C^m$.

Proof: Let $F(u, v) = f(v) - u$, then

$$F_v(u_0, v_0) = f'(v_0), \tag{A.87}$$

the operator $f'(v_0)$ has a bounded inverse operator (A.86), and Theorem A.4.2 follows from Theorem A.4.1. ∎

A.5 AN EXISTENCE THEOREM

Consider the Cauchy problem:

$$\dot{u} = F(u,t), \quad u(0) = u_0. \tag{A.88}$$

Let

$$J = [0,a] \subset \mathbb{R}, \quad \Lambda = [\alpha, \beta] \subset \mathbb{R}_+,$$

where X_λ is a scale of Banach spaces, satisfying the following assumptions:

$$X_\lambda \subset X_\mu \quad \text{if} \quad \mu < \lambda, \quad \lambda, \mu \in \Lambda, \tag{A.89}$$

$$\|u\|_\mu \le \|u\|_\lambda \quad \text{if} \quad \mu < \lambda, \quad \lambda, \mu \in \Lambda. \tag{A.90}$$

Theorem A.5.1 *Assume that*

$$F : J \times X_\lambda \to X_\mu \quad \text{is continuous if} \quad \mu < \lambda, \quad \lambda, \mu \in \Lambda, \tag{A.91}$$

$$F(t,0) \in X_\beta, \tag{A.92}$$

$$\|F(t,u) - F(t,v)\|_\mu \le \frac{M}{\lambda - \mu} \|u - v\|_\lambda, \quad \mu < \lambda, \quad t \in J \quad \lambda, \mu \in \Lambda. \tag{A.93}$$

Then problem (A.88) with $u_0 \in X_\beta$ has a unique solution $u(t) \in X_\lambda$ for every $\lambda \in (\alpha, \beta)$, and this solution is defined on the interval $t \in [0, \delta(\beta - \lambda))$, where $\delta := \min\left(a, \frac{1}{Me}\right)$.

Proof: Let

$$u_{n+1}(t) = u_0 + \int_0^t F(s, u_n(s)) \, ds. \tag{A.94}$$

Assumptions (A.89)–(A.93) imply that $u_n(t) \in X_\lambda$ for every $\lambda \in [\alpha, \beta]$, and the map $u_n : J \to X_\lambda$ is continuous.

Denote $J(t) = [0,t]$ and

$$m(t) := \|u_0\|_\beta + \frac{\beta - \alpha}{M} \max_{s \in J(t)} \|F(s,0)\|_\beta. \tag{A.95}$$

Let us check that

$$\|u_{n+1}(t) - u_n(t)\|_\lambda \le m(t) \left(\frac{tMe}{\beta - \lambda}\right)^{n+1}, \quad t \in J. \tag{A.96}$$

We have

$$\|u_1(t) - u_0(t)\|_\lambda \le t \left[\frac{M}{\beta - \lambda} \|u_0\|_\beta + F(s,0)\|_\beta\right]$$

$$\le \frac{Mt}{\beta - \lambda} m(t). \tag{A.97}$$

If (A.96) holds for $n = j - 1$, then it holds for $n = j$. Indeed, with $\gamma < \beta - \lambda$ we have

$$
\begin{aligned}
\|u_{j+1}(t) - u_j(t)\|_\lambda &\leq \int_0^t \|F(s, u_j(s)) - F(s, u_{j-1}(s))\|_\lambda\, ds \\
&\leq \frac{M}{\gamma} \int_0^t \|u_j(s) - u_{j-1}(s)\|_{\lambda+\gamma}\, ds \\
&\leq \frac{M}{\gamma} \int_0^t m(s) \left(\frac{sMe}{\beta - \lambda - \gamma}\right)^j ds.
\end{aligned}
\tag{A.98}
$$

Choose

$$
\gamma = \frac{\beta - \lambda}{j + 1}.
\tag{A.99}
$$

Then

$$
\begin{aligned}
\|u_{j+1}(t) - u_j(t)\|_\lambda &\leq \frac{M(j+1)}{\beta - \lambda} m(t) \frac{(Me)^j t^{j+1}}{(\beta - \lambda)^j \left(1 - \frac{1}{j+1}\right)^j (j+1)} \\
&= m(t) \left(\frac{tMe}{\beta - \lambda}\right)^{j+1} \frac{1}{e\left(1 - \frac{1}{j+1}\right)^j} \\
&\leq m(t) \left(\frac{tMe}{\beta - \lambda}\right)^{j+1},
\end{aligned}
\tag{A.100}
$$

because

$$
\left(1 + \frac{1}{j}\right)^j \leq e.
\tag{A.101}
$$

Thus (A.96) is proved.

If $t \in [0, \delta(\beta - \lambda))$, where $\delta = \min\left(a, \frac{1}{Me}\right)$, then

$$
\frac{tMe}{\beta - \lambda} := q < 1.
\tag{A.102}
$$

Therefore (A.96) implies

$$
\lim_{n \to \infty} u_n(t) = u(t),
\tag{A.103}
$$

where convergence is in X_λ uniform in $t \in [0, \delta(\beta - \lambda))$, and $u(t)$ solves the equation

$$
u(t) = u_0 + \int_0^t F(s, u(s))\, ds
\tag{A.104}
$$

and problem (A.88).

If $v(t)$ is another solution to (A.88) in X_λ, then $w := u(t) - v(t)$ solves the problem:

$$
\dot{w} = F(t, u) - F(t, v), \quad w(0) = 0.
\tag{A.105}
$$

Let

$$u_{n+1}(t) = u_0 + \int_0^t F(s, u_n(s))\, ds,$$

$$v_{n+1}(t) = v_0 + \int_0^t F(s, v_n(s))\, ds,$$

$$w_n(t) = u_n(t) - v_n(t).$$

Then, as in (A.100), we have

$$\|w_n(t)\|_\lambda \leq m(t)\left(\frac{tMe}{\beta - \lambda}\right),$$

and if (A.102) holds, then

$$\lim_{n \to \infty} \|w_n(t)\|_\lambda = 0. \tag{A.106}$$

Thus $w(t) = 0$ for $t \in [0, t_0]$, where $t_0 < \delta(\beta - \lambda)$.

Theorem A.5.1 is proved. ∎

A.6 CONTINUITY OF SOLUTIONS TO OPERATOR EQUATIONS WITH RESPECT TO A PARAMETER

Let X and Y be Banach spaces, let $k \in \Delta \subset \mathbb{C}$ be a parameter, let Δ be an open bounded set on a complex plane $\mathbb{C}$, let $A(k) : X \to Y$ be a map, possibly nonlinear, and let $f := f(k) \in Y$ be a function.

Consider an equation

$$A(k)u(k) = f(k). \tag{A.107}$$

We are interested in conditions, sufficient for the continuity of $u(k)$ with respect to $k \in \Delta$. The novel points in our presentation include necessary and sufficient conditions for the continuity of the solution to equation (A.107) with linear operator $A(k)$ and sufficient conditions for its continuity when the operator $A(k)$ is nonlinear.

Consider separately the cases when $A(k)$ is a linear map and when $A(k)$ is a nonlinear map.

Assumptions 1. $A(k) : X \to Y$ is a linear bounded operator, and

(a) equation (A.107) is uniquely solvable for any

$$k \in \Delta_0 := \{k : |k - k_0| \leq r\}, \quad k_0 \in \Delta, \quad \Delta_0 \subset \Delta,$$

(b) $f(k)$ is continuous with respect to $k \in \Delta_0$, $\sup_{k \in \Delta_0} \|f(k)\| \leq c_0$,

(c) $\lim_{h \to 0} \sup_{\substack{k \in \Delta_0 \\ v \in M}} \|[A(k + h) - A(k)]v\| = 0$, where $M \subset X$ is an arbitrary bounded set,

(d) $\sup_{\substack{k\in\Delta_0 \\ f\in N}} \|A^{-1}(k)f\| \leq c_1$, where $N \subset Y$ is an arbitrary bounded set and c_1 may depend on N.

Theorem A.6.1 *If Assumptions 1 hold, then*

$$\lim_{h\to 0} \|u(k+h) - u(k)\| = 0. \tag{A.108}$$

Proof: One has

$$
\begin{aligned}
u(k+h) - u(k) &= A^{-1}(k+h)f(k+h) - A^{-1}(k)f(k) \\
&= A^{-1}(k+h)f(k+h) - A^{-1}(k)f(k+h) \\
&\quad + A^{-1}(k)f(k+h) - A^{-1}(k)f(k). \tag{A.109}
\end{aligned}
$$

Furthermore, we have

$$\left\|A^{-1}(k)\big[f(k+h) - f(k)\big]\right\| \leq c_1\|f(k+h) - f(k)\| \to 0 \text{ as } h \to 0. \tag{A.110}$$

Using the relation

$$A^{-1}(k+h) - A^{-1}(k) = -A^{-1}(k+h)\big[A(h+k) - A(k)\big]A^{-1}(k)$$

and estimating the norms of inverse operators, we obtain

$$\left\|A^{-1}(k+h) - A^{-1}(k)\right\| \leq c_1^2\|A(k+h) - A(k)\| \to 0 \text{ as } h \to 0. \tag{A.111}$$

From (A.109)–(A.111) and Assumptions 1 the conclusion of Theorem A.6.1 follows. ∎

Remark A.6.2 Assumptions 1 are not only sufficient for the continuity of the solution to (A.107), but also necessary if one requires the continuity of $u(k)$ uniform with respect to f running through arbitrary bounded sets. Indeed, the necessity of the assumption (a) is clear; that of the assumption (b) follows from the case $A(k) = I$, where I is the identity operator; that of the assumption (c) follows from the case $A(k) = I$, $A(k+h) = 2I$, $\forall h \neq 0$, $f(k) = g \neq 0$, $\forall k \in \Delta_0$. Indeed, in this case assumption (c) fails and one has $u(k) = g$, $u(k+h) = \frac{g}{2}$, so $\|u(k+h) - u(k)\| = \frac{\|g\|}{2}$ does not tend to zero as $h \to 0$.

To prove the necessity of the assumption (d), suppose that

$$\sup_{k\in\Delta_0} \|A^{-1}(k)\| = \infty.$$

Then, by the Banach–Steinhaus theorem, there is an element f such that $\sup_{k\in\Delta_0} \|A^{-1}(k)f\| = \infty$, so that

$$\lim_{j\to\infty} \|A^{-1}(k_j)f\| = \infty, \qquad k_j \to k \in \Delta_0.$$

Then

$$\|u_j\| := \|u(k_j)\| = \|A^{-1}(k_j)f\| \to \infty,$$

so u_j does not converge to $u := u(k) = A^{-1}(k)f$, although $k_j \to k$.

Assumptions 2. $A(k) : X \to Y$ is a nonlinear map, and (a), (b), (c), and (d) of Assumptions 1 hold, and the following assumption holds:

(e) $A^{-1}(k)$ *is a homeomorphism of X onto Y for each $k \in \Delta_0$.*

Remark A.6.3 *Assumption (e) is included in (d) in the case of a linear operator $A(k)$ because if $\|A(k)\| \leq c_2$ and $\|A^{-1}(k)\| \leq c_1$, then $A(k)$, $k \in \Delta_0$, is an isomorphism of X onto Y.*

Theorem A.6.4 *If Assumptions 2 hold, then (A.108) holds.*

Let us make the following Assumption A_d:
Assumptions A_d. Assumptions 2 hold and

(f) $\dot{f}(k) := \frac{df(k)}{dk}$ is continuous in Δ_0,

(g) $\dot{A}(u, k) := \frac{\partial A(u,k)}{\partial k}$ is continuous with respect to (wrt) k in Δ_0 and wrt $u \in X$,

(j) $\sup_{k \in \Delta_0} \|[A'(u, k)]^{-1}\| \leq c_3$, where $A'(u, k)$ is the Fréchet derivative of $A(u, k)$ and $[A'(u, k)]^{-1}$ is continuous with respect to u and k. $\dot{f}(k) := \frac{df(k)}{dk}$ is continuous in Δ_0.

Remark A.6.5 *If Assumption A_d holds, then*

$$\lim_{h \to 0} \|\dot{u}(k + h) - \dot{u}(k)\| = 0. \tag{A.112}$$

Remark A.6.6 *If Assumptions 1 hold except one: $A(k)$ is not necessarily a bounded linear operator, $A(k)$ may be unbounded, closed, densely defined operator-function, then the conclusion of Theorem A.6.4 still holds and its proof is the same. For example, let*

$$A(k) = L + B(k),$$

where $B(k)$ is a bounded linear operator continuous with respect to $k \in \Delta_0$, and L is a closed, linear, densely defined operator from $D(L) \subset X$ into Y. Then

$$\|A(k + h) - A(k)\| = \|B(k + h) - B(k)\| \to 0, \quad as \quad h \to 0,$$

although $A(k)$ and $A(k + h)$ are unbounded.

Proofs of Theorem A.6.4 and of Remark A.6.5 are given below.

Proof of Theorem A.6.4. One has

$$A(k + h)u(k + h) - A(k)u(k) = f(k + h) - f(k) = o(1) \quad as \quad h \to 0.$$

Thus

$$A(k)u(k+h) - A(k)u(k) = o(1) - [A(k+h)u(k+h) - A(k)u(k+h)].$$

Since

$$\sup_{\{u(k+h): \|u(k+h)\| \leq c\}} \|A(k+h)u(k+h) - A(k)u(k+h)\| \underset{h \to 0}{\to} 0,$$

one gets

$$A(k)u(k+h) \to A(k)u(k) \text{ as } h \to 0. \tag{A.113}$$

By Assumptions 2, item (e), the operator $A(k)$ is a homeomorphism. Thus (A.113) implies (A.108).

Theorem A.6.4 is proved. ∎

Proof of Remark A.6.5. First, assume that $A(k)$ is linear. Then

$$\frac{d}{dk}A^{-1}(k) = -A^{-1}(k)\dot{A}(k)A^{-1}(k), \quad \dot{A} := \frac{dA}{dk}. \tag{A.114}$$

Indeed, differentiate the identity $A^{-1}(k)A(k) = I$ and get

$$\frac{dA^{-1}(k)}{dk}A(k) + A^{-1}(k)\dot{A}(k) = 0.$$

This implies (A.114).

This argument proves also the existence of the deriviative $\frac{dA^{-1}(k)}{dk}$. Formula $u(k) = A^{-1}(k)f(k)$ and the continuity of $\dot{f}$ and of $\frac{dA^{-1}(k)}{dk}$ yield the existence and continuity of $\dot{u}(k)$. Remark A.6.5 is proved for linear operators $A(k)$.

Assume now that $A(k)$ is nonlinear, $A(k)u := A(k, u)$. Then one can differentiate (A.107) with respect to k and get

$$\dot{A}(k, u) + A'(k, u)\dot{u} = \dot{f}, \tag{A.115}$$

where A' is the Fréchet derivative of $A(k, u)$ with respect to u. Formally one assumes that $\dot{u}$ exists, when one writes (A.115), but in fact (A.115) proves the existence of $\dot{u}$, because $\dot{f}$ and $\dot{A}(k, u) := \frac{\partial A(k, u)}{\partial k}$ exist by Assumption A_d and $[A'(k, u)]^{-1}$ exists and is an isomorphism by Assumption A_d, item (j). Thus, (A.115) implies

$$\dot{u} = [A'(k, u)]^{-1}\dot{f} - [A'(k, u)]^{-1}\dot{A}(k, u). \tag{A.116}$$

Formula (A.116) and Assumption A_d imply (A.112).

Remark A.6.5 is proved. ∎

Consider some application of the above results to Fredholm equations depending on a parameter.

Let

$$Au := u - \int_D b(x, y, k) u(y)\, dy := [I - B(k)]u = f(k), \qquad (A.117)$$

where $D \subset R^n$ is a bounded domain, $b(x, y, k)$ is a function on $D \times D \times \Delta_0$, $\Delta_0 := \{|k - k_0| < r\}$, $k_0 > 0$, and $r > 0$ is a sufficiently small number. Assume that $A(k_0)$ is an isomorphism of $H := L^2(D)$ onto H, for example, $\int_D \int_D |b(x, y, k_0)|^2 dx dy < \infty$ and $\mathcal{N}(I - B(k_0)) = \{0\}$, where $\mathcal{N}(A)$ is the null-space of A. Then, $A(k_0)$ is an isomprohism of H onto H by the Fredholm alternative, and Assumptions 1 hold if $f(k)$ is continuous with respect to $k \in \Delta_0$ and

$$\lim_{h \to 0} \int_D \int_D |b(x, y, k + h) - b(x, y, k)|^2 dx\, dy = 0, \qquad k \in \Delta_0. \qquad (A.118)$$

Condition (A.118) implies that if $A(k_0)$ is an isomorphism of H onto H, then so is $A(k)$ for all $k \in \Delta_0$ if $|k - k_0|$ is sufficiently small.

Remark A.6.5 applies to (A.117) if $\dot{f}$ is continuous with respect to $k \in \Delta_0$, and $\dot{b} := \frac{\partial b}{\partial k}$ is continuous with respect to $k \in \Delta_0$ as an element of $L^2(D \times D)$. Indeed, under these assumptions one has

$$\dot{u} = [I - B(k)]^{-1} (\dot{f} - \dot{B}(k)u),$$

and the right-hand side of this formula is continuous in Δ_0.

A.7   MONOTONE OPERATORS IN BANACH SPACES

Suppose that $F : X \to X^*$ is a map from a real Banach space X into its adjoint.

Definition A.7.1   • X is *uniformly convex* if $\|u\| = \|v\| = 1$ and $\|u - v\| \geq \epsilon$ imply $\|\frac{u+v}{2}\| \leq 1 - \delta$, $\forall \epsilon (0, 2]$ and $\delta = \delta(\epsilon) > 0$.

• X is *locally uniformly convex* if $\|u\| = \|v\| = 1$ and $\|u - v\| \geq \epsilon$ imply $\|\frac{u+v}{2}\| \leq 1 - \delta$, $\forall \epsilon (0, 2]$ and $\delta = \delta(u, \epsilon) > 0$.

• X is *strictly convex* if $\|u\| = \|v\| = 1$ and $u \neq v$ implies

$$\|su + (1 - s)v\| < 1, \quad \forall s \in (0, 1).$$

Theorem A.7.1 *(a) If X is uniformly convex, then X is reflexive and locally uniformly convex.*

(b) If X is locally uniformly convex, then X is strictly convex and

$$\{u_n \rightharpoonup u, \|u_n\| \to \|u\|\} \quad implies \quad u_n \to u.$$

(c) X is strictly convex if and only if $\|u + v\| = \|u\| + \|v\|$ implies $v = 0$ or $u = \lambda v$, $\lambda = const \geq 0$.

(d) X is strictly convex if and only if every $u \in X$, $u \neq 0$, considered as an element of X^{**}, attains its norm at exactly one element $u^* \in X^*$, $\|u\| = \sup_{\|u^*\|=1} |u^*(u)|$.

(e) X is reflexive if and only if every $x^* \in X^*$ attains it norm on an element u, $\|u\| = 1$.

(f) X^* is uniformly convex if and only if the norm $\|\cdot\|_X$ is uniformly differentiable on the set $\partial B_1 := \{u : \|u\| = 1, u \in X\}$.

(g) $\|\cdot\|_X$ is Gâteaux-differentiable on $X\backslash\{0\}$ if and only if X^* is strictly convex.

Proofs of Theorem A.7.1 and of many other results in this section are omitted. One can find them in [25].

Definition A.7.2 *The map*

$$j : X \to 2^{X^*}, \quad ju = \{u^* \in X^* : \|u^*\| = \|u\|, u^*(u) = \|u\|^2\} \tag{A.119}$$

is called the duality map.

Theorem A.7.2 *(a) The set $\{ju\}$ is convex, $j(\lambda u) = \lambda j(u)$, $\forall \lambda \in \mathbb{R}$.*

(b) The set $\{ju\}$ consists of one element if and only if X^* is strictly convex; $j = I$, the identity operator, if $X = H$, the Hilbert space.

Semi-inner products in a Banach space are defined by the formulas

$$\|v\| \lim_{t \to 0, t > 0} t^{-1}(\|v + tu\| - \|v\|) := (u, v)_+. \tag{A.120}$$

$$\|v\| \lim_{t \to 0, t > 0} t^{-1}(\|v\| - \|v - tu\|) := (u, v)_-. \tag{A.121}$$

The existence of the limit in (A.120) follows from the properties of the norm:

$$\|v + su\| - \|v\| = \|\frac{s}{t}v + \frac{s}{t}tu + \left(1 - \frac{s}{t}\right)v\| - \|v\|$$
$$\leq \frac{s}{t}(\|v + tu\| - \|v\|), \quad 0 < s < t. \tag{A.122}$$

Thus the function

$$\frac{\|v + su\| - \|v\|}{s} \tag{A.123}$$

is nondecreasing. This function is bounded from below as $s \to 0$, $s > 0$:

$$\frac{\|v + su\| - \|v\|}{s} \geq -\|u\|. \tag{A.124}$$

This inequality follows from the triangle inequality:

$$\|v + su\| \geq |v\| - s\|u| \geq \|v\| - s\|u\|. \tag{A.125}$$

Therefore the limits in (A.120) and (A.121) exist.

Theorem A.7.3 *One has*

$$(u, v)_+ = (u, v)_- \tag{A.126}$$

if and only if X^* is strictly convex.

If $u(t) \in C^1(a, b; X)$, then $\|u(t)\| := g(t)$ satisfies the equations

$$g(t)D_+g = (\dot{u}(t), u(t))_+, \quad g(t)D_-g = (\dot{u}(t), u(t))_-, \tag{A.127}$$

where

$$D_+g = \varlimsup_{h \to 0,\, h>0} \frac{g(t+h) - g(t)}{h}; \quad D_-g = \varlimsup_{h \to 0,\, h>0} \frac{g(t) - g(t-h)}{h}. \tag{A.128}$$

Definition A.7.3 *An operator $F : D \subset X \to X$ is called accretive if*

$$\langle F(u) - F(v), u - v \rangle_+ \geq 0, \quad \forall u, v \in D, \tag{A.129}$$

strictly accretive if

$$\langle F(u) - F(v), u - v \rangle_+ > 0, \quad \forall u, v \in D, \quad u \neq v, \tag{A.130}$$

strongly accretive if

$$\langle F(u) - F(v), u - v \rangle_+ \geq c\|u - v\|^2, \quad \forall u, v \in D, \tag{A.131}$$

maximal accretive (or m-accretive) if

$$\langle F(u) - f, u - v \rangle_+ \geq 0, \quad \forall u \in D \tag{A.132}$$

implies $v \in D$ and $F(v) = f$, and hyperaccretive if (A.129) holds and

$$R(F + \lambda I) = X, \tag{A.133}$$

for some $\lambda = const > 0$.

Theorem A.7.4 *Assume that $F : D \to X$ is hyperaccretive. Then $R_\lambda := (I + \lambda F)^{-1}$, $\lambda > 0$, is nonexpansive, $\lim_{\lambda \to 0} R_\lambda(u) = u$, $\|R_\lambda(u)\| \leq \|F(u)\|$. If X and X^* are uniformly convex, then $\lim_{\lambda \to 0} \|FR_\lambda(u) - F(u)\| = 0$ and the problem*

$$\dot{u} = -F(u), \quad u(0) = u_0, \quad u_0 \in D, \tag{A.134}$$

has a unique solution on $[0, \infty)$. The solution $u(t)$ is continuous and weakly differentiable, and $\|\dot{u}(t)\|$ is decreasing.

Consider the operator

$$U(t)u_0 = u(t; u_0), \tag{A.135}$$

where $u(t; u_0)$ solves (A.134). We have

$$U(0) = I, \quad \lim_{t \to 0,\, t>0} \|U(t)u_0 - u_0\| = 0, \tag{A.136}$$

$$U(t + s) = U(t)U(s). \tag{A.137}$$

The family $\{U(t)\}$ forms a semigroup. In [82] one finds a presentation of the nonlinear semigroup theory, and in [89] the linear semigroup theory is presented.

A.8 EXISTENCE OF SOLUTIONS TO OPERATOR EQUATIONS

In this section we mention briefly the methods for proving the existence results for solutions to operator equations. The reader will find much more in [14], [32], [69], [73], [74], [75], and [196].

For linear equations

$$u = Au + f \qquad (A.138)$$

in a Banach space X, assuming that A is a linear operator, one has the existence of a unique solution for any f, provided that $\|A\| < 1$. This follows from the contraction mapping principle. The solution can be calculated by the iterative method

$$u_{n+1} = Au_n + f, \quad u_0 = u_0, \qquad (A.139)$$

where $u_0 \in X$ is arbitrary. The method converges at the rate of geometrical series with the denominator $q = \|A\| < 1$.

If A is compact, then the Fredholm–Riesz theory applies, (see, e.g., [195]). This theory yields the existence of a unique solution to (A.138) for any $f \in X$, provided that $\mathcal{N}(I-A) = \{0\}$, where $\mathcal{N}(I-A)$ is the null-space of the operator $I - A$. If $\mathcal{N}(I - A) \neq \{0\}$, then equation (A.138) is solvable if and only if $< u^*, f > \; > 0$ for all $u^* \in \mathcal{N}(I-A^*)$, where A^* is the adjoint to A operator. This is the well-known Fredholm alternative.

If A is nonlinear, then the existence of a solution to the equation

$$u = A(u) \qquad (A.140)$$

is often proved by applying topological methods, such as degree theory. Typical examples are Schauder's principle, Rothe's theorem, and Leray–Schauder's principle.

Theorem A.8.1 *(Schauder) If a continuous in X operator A maps a convex closed set D into its compact subset, then equation (A.140) has a solution in D.*

Theorem A.8.2 *(Rothe) If A is compact on $\bar{D}$, where $D \subset X$ is a bounded convex domain, and $A(\partial D) \subset \bar{D}$, where ∂D is the boundary of D, then equation (A.139) has a solution in $\bar{D}$.*

Let $A(\cdot, \lambda)$ be a parametric family of compact operators, $0 \leq \lambda \leq 1$, $A(u, 1) = A(u)$, and equation $u = A(u, 0)$ has a solution,

$$\|A(u, 0)\| \leq b, \quad \forall u \in \{u : \|u\| = b\}. \qquad (A.141)$$

Theorem A.8.3 *(Leray–Schauder). Under the above assumptions, equation (A.140) has a solution, provided that*

$$\|u(\lambda)\| \leq a < b, \qquad \forall \lambda \in [0, 1], \qquad (A.142)$$

where $\{u(\lambda)\}$ *is the set of all solutions to the equation*

$$u = A(u, \lambda), \quad \lambda \in [0, 1]. \tag{A.143}$$

Theorem A.8.3 is often used in the following form:
If all the solutions $u(\lambda)$ *to the equation*

$$u = \lambda A(u), \quad \lambda \in [0, 1] \tag{A.144}$$

satisfy the a priori estimate

$$\|u(\lambda)\| \le a, \tag{A.145}$$

then equation (A.140) has a solution in the ball $\{u : \|u\| \le a\}$.

Definition A.8.1 *A map* A *in a Banach space* X *is called a generalized contraction mapping if*

$$\|A(u) - A(v)\| \le q(a, b)\|u - v\|, \quad 0 < a \le \|u - v\| \le b, \tag{A.146}$$

where $q(a, b) < 1$.

For example, A is a generalized contraction mapping if

$$\|A(u) - A(v)\| \le \|u - v\| - g(\|u - v\|), \tag{A.147}$$

where $g(s) > 0$ if $s > 0$, $g(0) = 0$, and g is a continuous function on $[0, \infty)$.

Theorem A.8.4 *(Krasnoselsky) If* A *is a generalized contraction mapping on a closed set* D *and* $AD \subset D$, *then equation (A.140) has a solution in* D.

Proof: The proof follows [75]. Consider the sequence

$$u_{n+1} = A(u_n), \quad u_0 \in D. \tag{A.148}$$

We have

$$\begin{aligned} \|u_{n+2} - u_{n+1}\| &= \|A(u_{n+1} - A(u_n)\| \\ &\le q(a, b)\|u_{n+1} - u_n\| < \|u_{n+1} - u_n\|. \end{aligned} \tag{A.149}$$

Therefore there exists

$$\lim_{n \to \infty} \|u_{n+1} - u_n\| = s < \infty. \tag{A.150}$$

We claim that $s = 0$. Indeed, if $s > 0$, then

$$s \le \|u_{n+1} - u_n\| \le s + \epsilon, \quad \forall n > n(\epsilon), \tag{A.151}$$

and

$$s \le \|u_{n+m} - u_{n+m-1}\| \le q^m(s, s + \epsilon)(s + \epsilon), \quad \forall n > n(\epsilon). \tag{A.152}$$

If m is sufficiently large, then $q^m(s, s+\epsilon)$ is as small as one wishes, so (A.152) yields a contradiction which proves that $s = 0$.

Let us prove that

$$AB(u_p, \epsilon) \subset B(u_p, \epsilon), \tag{A.153}$$

where $\epsilon > 0$ is an arbitrary small fixed number, and u_p is chosen so that

$$d_p := \|u_{p+1} - u_p\| \leq \frac{\epsilon}{2}\left[1 - q\left(\frac{\epsilon}{2}, \epsilon\right)\right]. \tag{A.154}$$

Such a number p exists because

$$\lim_{n\to\infty} d_n := \lim_{n\to\infty} \|u_{n+1} - u_n\| = 0, \tag{A.155}$$

as we have proved.

If (A.153) is verified, then the sequence $\{u_n\}$ is a Cauchy sequence, so it has a limit

$$\lim_{n\to\infty} u_n = u. \tag{A.156}$$

The limit u solves equation (A.140) as one can see by passing to the limit $n \to \infty$ in equation (A.148). Uniqueness of the solution (A.140) follows from (A.146). Thus, the proof is completed if (A.153) is verified.

Let us verify (A.153) . Let $\|u - u_p\| \leq \frac{\epsilon}{2}$. Then

$$\|A(u) - u_p\| \leq \|A(u) - A(u_p)\| + \|A(u_p) - u_p\|$$
$$\leq \|u - u_p\| + d_p \leq \frac{\epsilon}{2} + \frac{\epsilon}{2} = \epsilon. \tag{A.157}$$

Let $\frac{\epsilon}{2} < \|u - u_p\| \leq \epsilon$. Then, using (A.154), one gets

$$\|A(u) - u_p\| \leq \|u - u_p\| + d_p \leq q\left(\frac{\epsilon}{2}, \epsilon\right)\epsilon + d_p$$
$$\leq q\left(\frac{\epsilon}{2}, \epsilon\right)\epsilon + \frac{\epsilon}{2} - \frac{\epsilon}{2}q\left(\frac{\epsilon}{2}, \epsilon\right) \leq \epsilon. \tag{A.158}$$

Thus, (A.153) is verified, and Theorem A.8.4 is proved. ∎

Theorem A.8.4 generalizes Theorem A.1.1, because if condition (A.1) holds, then condition (A.146) holds.

Theorem A.8.5 (*Krasnoselsky*). *Let $A : D \to D$, where D is a convex bounded, closed subset of a Banach space X. Assume that $A = B + T$, where B is a generalized contraction map and T is compact. Then equation (A.140) has a solution $u \in D$.*

Definition A.8.2 *A map $A : X \to X$ is called nonexpansive if*

$$\|A(u) - A(v)\| \leq \|u - v\|. \tag{A.159}$$

Let H denote a Hilbert space.

Theorem A.8.6 *If $A : D \to D$ is nonexpansive and $D \subset H$ is a bounded, convex, and closed set, then equation (A.140) has a solution $u \in D$.*

Definition A.8.3 *A closed convex set $K \subset X$ is called a cone if $u \in K$ and $u \neq 0$ imply that $\lambda u \in K$, $\forall \lambda \geq 0$ and $\lambda u \notin K$, $\forall \lambda \in \mathbb{R}$.*

We write $u \geq v$ if $u - v \in K$. Elements of K are called positive elements. A cone is called solid if there is an element $u \in K$ such that $B(u, r) \subset K$ for some $r > 0$, where $B(u, r) := \{v : \|u - v\| \leq r\}$. A cone is called reproducing if every element $w \in X$ can be represented as $w = u - v$, where $u, v \in K$. A cone is called normal iff $0 \leq u \leq v$ implies $\|u\| \leq N\|v\|$, where the constant N does not depend on u and v.

Definition A.8.4 *An operator $A : X \to X$ in a Banach space X with a cone K is called K-monotone if $u \leq v$ implies $A(u) \leq A(v)$.*

If A is K-monotone and $u \leq v$, $A(u) \geq u$, $A(v) \leq v$, then $u \leq w \leq v$ implies $u \leq A(w) \leq v$.

A cone is called strongly minihedral if any bounded set $U = \{u\} \in K$ has a supremum—that is, an element which is the minimal element in the set $W = \{w\} \in K$ of the elements such that $u \leq w$, $\forall u \in U$.

Let $[u_1, u_2] := \{u : u_1 \leq u \leq u_2, \ u \in K\}$.

Theorem A.8.7 *If K is strongly minihedral, and $A[u_1, u_2] \subset [u_1, u_2]$. Then equation (A.140) has a solution $u \in [u_1, u_2]$.*

Proof: The set W of elements $w \in [_1, u_2]$ such that $A(w) \geq w$ is non-void: It contains u_1. Also, $AW \subset W$. Let $s = \sup W$. Then $w \in W$ implies $w \leq A(w) \leq A(s)$, so $s \leq A(s)$ and $s \in W$. Therefore $A(s) \in W$. Consequently $A(s) \leq s$. Thus, $A(s) = s$.

Theorem A.8.7 is proved.   ∎

Example A.4

Consider the problem

$$\dot{u} = f(t, u), \quad u(0) = u_0 \tag{A.160}$$

in a Banach space X. Assume that

$$f(t, u) = g(t, u) + h(t, u), \tag{A.161}$$

where

$$\|g(t, u_1) - g(t, u_2)\| \leq L\|u_1 - u_2\|, \quad L = const > 0, \tag{A.162}$$

and $h(t, \cdot)$ is a compact operator. Problem (A.160) can be written as

$$u(t) = u_0 + \int_0^t f(s, u(s)) \, ds = Bu + Tu, \tag{A.163}$$

where

$$Bu := u_0 + \int_0^t g(s, u(s))\, ds,$$

$$Tu := \int_0^t h(s, u(s))\, ds. \tag{A.164}$$

Consider equation (A.163) in the space $C([0, \delta], X)$ of continuous on $[0, \delta]$ functions $u(t)$ with values in X. Then the operator B is a contraction mapping in this space if $\delta > 0$ is sufficiently small, and T is compact in this space. By Theorem A.8.5, problem (A.163) has a solution in $C([0, \delta], X)$ for sufficiently small $\delta > 0$.

A.9 COMPACTNESS OF EMBEDDINGS

The basic result of this section is:

Theorem A.9.1 *Let $X_1 \subset X_2 \subset X_3$ be Banach spaces,*

$$\|u\|_1 \geq \|u\|_2 \geq \|u\|_3,$$

that is, the norms are comparable, and if $\|u_n\|_3 \to 0$ as $n \to \infty$ and u_n is fundamental in X_2, then $\|u_n\|_2 \to 0$ (i.e., the norms in X_2 and X_3 are compatible). Under the above assumptions the embedding operator $i : X_1 \to X_2$ is compact if and only if the following two conditions are valid:

(a) The embedding operator $j : X_1 \to X_3$ is compact,
and
(b) The following inequality holds:

$$\|u\|_2 \leq s\|u\|_1 + c(s)\|u\|_3, \qquad \forall u \in X_1, \quad \forall s \in (0,1),$$

where $c(s) > 0$ is a constant.

This result is an improvement of the author's old result [92]. We follow [154]. We construct a counterexample to a theorem in [13, p. 35], where the validity of the inequality (b) in Theorem A.9.1 is claimed without the assumption of the compatibility of the norms of X_2 and X_3; see Remark A.9.2 at the end of this section.

Proof of Theorem A.9.1.
1. *The sufficiency of conditions (a) and (b) for compactness of $i : X_1 \to X_2$.*

Assume that (a) and (b) hold, and let us prove the compactness of i. Let $S = \{u : u \in X_1, \|u\|_1 = 1\}$ be the unit sphere in X_1. Using assumption (a), select a sequence u_n which converges in X_3. We claim that this sequence converges also in X_2. Indeed, since $\|u_n\|_1 = 1$, one uses assumption (b) to get

$$\|u_n - u_m\|_2 \leq s\|u_n - u_m\|_1 + c(s)\|u_n - u_m\|_3 \leq 2s + c(s)\|u_n - u_m\|_3.$$

Let $\eta > 0$ be an arbitrary small given number. Choose $s > 0$ such that

$$2s < \frac{1}{2}\eta,$$

and for a fixed s choose n and m so large that

$$c(s)\|u_n - u_m\|_3 < \frac{1}{2}\eta.$$

This is possible because the sequence u_n converges in X_3. Consequently,

$$\|u_n - u_m\|_2 \leq \eta$$

if n and m are sufficiently large. This means that the sequence u_n converges in X_2. Thus, the embedding $i : X_1 \to X_2$ is compact. In the above argument, that is, in the proof of the sufficiency, the compatibility of the norms was not used.

2. *The necessity of the compactness of $i : X_1 \to X_2$ for conditions (a) and (b) to hold.*

Assume now that i is compact. Let us prove that conditions (a) and (b) hold. In the proof of the necessity of these conditions the assumption about the compatibility of the norms of X_2 and X_3 is used essentially. Without this assumption, one cannot prove that conditions (a) and (b) hold. This is demonstrated in Remark A.9.2 after the end of the proof of the Theorem.

If i is compact, then assumption (a) holds because $\|u\|_2 \geq \|u\|_3$. Suppose that assumption (b) fails. Then there is a sequence u_n and a number $s_0 > 0$ such that $\|u_n\|_1 = 1$ and

$$\|u_n\|_2 \geq s_0 + n\|u_n\|_3. \qquad (A.165)$$

If the embedding operator i is compact and $\|u_n\|_1 = 1$, then one may assume that the sequence u_n converges in X_2. Its limit cannot be equal to zero, because, by (A.165),

$$\|u_n\|_2 \geq s_0 > 0.$$

The sequence u_n converges in X_3 because of the inequality

$$\|u_n - u_m\|_2 \geq \|u_n - u_m\|_3$$

and because the sequence u_n converges in X_2.

Its limit in X_3 is not zero, because the norms in X_3 and in X_2 are compatible.

Thus,

$$\lim_{n \to \infty} \|u_n\|_3 > 0.$$

Consequently, inequality (A.165) implies

$$\|u_n\|_3 = O(\frac{1}{n}) \to 0 \quad \text{as} \quad n \to \infty,$$

while

$$\lim_{n\to\infty} \|u_n\|_3 > 0.$$

This is a contradiction, which proves that (b) holds.

Theorem A.9.1 is proved.   ∎

Remark A.9.2 In [13], p. 35, the following claim is stated:
Claim. Let $X_1 \subset X_2 \subset X_3$ be three Banach spaces. Suppose the embedding $X_1 \to X_2$ is compact. Then given any $\epsilon > 0$, there is a $K(\epsilon) > 0$, such that

$$\|u\|_2 \le \epsilon\|u\|_1 + K(\epsilon)\|u\|_3$$

for all $u \in X_1$.

In this claim there is no explicit assumption about the compatibility of the norms of X_2 and X_3. Let us discuss the necessity to have this assumption.

For example, let $L^2(0,1)$ be the usual Lebesgue space of square integrable functions, $X_3 = L^2(0,1)$, and X_2 be a Banach space of $L^2(0,1)$ functions with a finite value at a fixed point $y \in [0,1]$ and with the norm

$$\|u\|_2 := \|u\|_{L^2(0,1)} + |u(y)| = \|u\|_3 + |u(y)|.$$

The space X_2 is complete because X_3 is complete and the one-dimensional space, consisting of numbers $u(y)$ with the usual norm $|u(y)|$, is complete. A function $u_0(x) = 0$ for $x \neq 0$ and $u_0(y) = 1$ has the properties

$$\|u_0\|_3 = 0, \quad \|u_0\|_2 = 1.$$

One has $X_2 \subset X_3$. The norms in X_2 and X_3 are *comparable*, that is, $\|u\|_3 \le \|u\|_2$. However, these norms are *not compatible*: There is a convergent to zero sequence $\lim_{n\to\infty} u_n = 0$ in X_3 such that it does not converge to zero in X_2, for example, $\lim_{n\to\infty}\|u_n\|_2 = 1$ in X_2. For instance, one may take $u_n(x) = u_0(x)$ for all $n = 1, 2, \ldots$, and an arbitrary fixed $y \in [0,1]$. Then $\|u_n\|_2 = 1$ and $\|u_n\|_3 = 0$, $\lim_{n\to\infty}\|u_n\|_2 = 1$, and $\lim_{n\to\infty}\|u_n\|_3 = 0$. The sequence u_n converges to zero in X_3 and to a nonzero element u_0 in X_2. In this case, inequality (A.165) holds for any fixed $s_0 \in (0,1)$ and any n, but the contradiction, which was used in the proof of the necessity in Theorem A.9.1, cannot be obtained because $\|u_n\|_3 = 0$ for all n.

Let us construct a counterexample which shows that the Claim, mentioned above, is not correct. Fix a $y \in [0,1]$. Choose the one-dimensional space of functions $\{u : u = \lambda u_0(x)\}$ as X_1, where $\lambda = const$ and $u_0(x)$ was defined above, and define the norm in X_1 by the formula $\|u\|_1 = |\lambda|$. Let $X_3 = L^2(0,1)$. The space X_1 is a one-dimensional Banach space. Therefore, bounded sets in X_1 are precompact. Note that $|\lambda| = \|\lambda u_0\|_1 = \|\lambda u_0\|_2 \ge \|\lambda u_0\|_3 = 0$ because $\|u_0\|_3 = 0$. Here the Banach space X_2 is defined as above with the norm $\|u\|_2 := \|u\|_{L^2(0,1)} + |u(y)|$, and the equalities $\|u_0\|_2 = 1$ and $\|u_0\|_3 = 0$ are used.

Consequently,

$$X_1 \subset X_2 \subset X_3, \quad \|u\|_1 \geq \|u\|_2 \geq \|u\|_3,$$

and the embedding $i : X_1 \to X_2$ is compact because bounded sets in finite-dimensional spaces are precompact and X_1 is a one-dimensional space. Thus, all the assumptions of the Claim are satisfied. However, the inequality of the Claim

$$\|u\|_2 \leq \epsilon \|u\|_1 + K(\epsilon)\|u\|_3, \quad \forall u \in X_1$$

does not hold for any fixed $\epsilon \in (0, 1)$. In our counterexample we have

$$u = \lambda u_0, \quad \|u_0\|_3 = 0,$$

and the above inequality takes the form

$$|\lambda| \leq \epsilon |\lambda|.$$

Clearly, this inequality does not hold for a fixed $\epsilon \in (0, 1)$ unless $\lambda = 0$.

APPENDIX B

BIBLIOGRAPHICAL NOTES

The contents of this book is based predominantly on the authors' papers cited in the bibliography. The most important of the preceding papers was the paper by M. Gavurin [34], who deals with a continuous analog of Newton's method. There is a large literature on solving ill-posed problems (see, e.g., [10], [64], [83], [137], [186], [191], and [193], to mention a few). The DSM as a tool for solving operator equations, especially ill-posed, linear and nonlinear, is developed in this book systematically.

The examples of inverse and ill-posed problems, mentioned in Section 2.1, are partly taken from [114] and [137]. Some of these examples are discussed in more detail in many books and papers. We mention the books [9, 81] and the papers [96, 97, 99, 100] on antenna synthesis. Variational regularization (Section 2.2) is discussed in the books [10, 64, 83, 137, 186, 191, 193], to mention a few. Our presentation contains several new points. We deal with unbounded, densely defined, closed linear operators and define the operator $(A^*A + aI)^{-1}A^*$ for $a = \text{const} > 0$ on the whole Hilbert space in the case when the domain $D(A^*)$ is dense in H, but is not the whole space. This allows us to extend to the case of unbounded operators the usual theory of variational regularization without requiring compactness properties from the stabilizing

Dynamical Systems Method and Applications: Theoretical Developments and Numerical Examples, First Edition. A. G. Ramm and N. S. Hoang.
Copyright © 2012 John Wiley & Sons, Inc.

functional [151, 152, 155]. We give a new discrepancy principle which does not require us to solve theoretically exactly the usual discrepancy principle equation, but rather to find an approximate minimizer of the functional which is used in the standard theory of variational regularization (Theorem 2.2.6, see also [137] and [141]). We formulate a new notion of regularizer [128].

Section 2.3, Quasisolutions, contains some well-known material (see, e.g., [64]).

Section 2.4, Iterative regularization, contains a proof of the following general result: Every solvable linear equation in a Hilbert space is solvable by a convergent iterative process (Theorem 2.4.1). There are many papers and books on iterative methods (see, e.g., [10], [191], and [193]).

Section 2.5, Quasi-inversion, presents very briefly the idea of the quasi-inversion method for solving ill-posed problems. A simple example of the application of this method is given in Section 2.5. More material on the quasi-inversion methods is found in [78].

In Section 2.6 the idea of the dynamical systems method (DSM) is discussed.

In Section 2.7 variational regularization for nonlinear operator equations is discussed (cf. [126] and [141]). Nonlinear ill-posed problems are discussed in [186].

Section 3.1 follows [133]. The results presented in this section generalize some results from [3]. Example 3.1 was not published earlier. Sections 3.2–3.7 contain applications of the results obtained in Section 3.1. Various versions of the DSM are constructed in these Sections for continuous analogs of classical methods, including Newton's method, modified Newton's method, Gauss–Newton's method, gradient method, simple iteration method, and a minimization method.

The results of Chapter 4 are based on papers [122], [153], and [155]. Section 4.4 uses some ideas from [104].

In Section 4.5 a new approach is given to stable calculation of values of unbounded operators. This problem has been earlier treated by the variational regularization method in [64] and [83].

In Chapter 5 some auxiliary inequalities are presented. Theorems 5.1.1 and 5.2.1 are used in the following chapters. The first version of Theorem 5.1.1 appeared first in [4], and its refinement was given in [3], [130], and [137].

An erroneous version of Theorem 5.3.1 appeared in [6]. A counterexample to the claim in [6] was given in [137], where a corrected version of the theorem has been proved and the basic idea of the proof from[6] was used (see also [4]). Theorem 5.3.2 appeared in [197]. Our proof is shorter. The results in Section 5.4 can be found, for example, in [184].

The results of Section 6.1, Chapter 6, are known, but our presentation is self-contained. The main results of Chapter 6 are given in Sections 6.2 and 6.3. Theorem 6.2.1 is taken from [130]. Its earlier versions appeared in [4] and [3]; see also [137].

The results of Section 6.3 are taken from [137].

Sections 7.1 and 7.2 of Chapter 7 are based on the results from [137] and [156], and Section 7.3, Theorem 7.3.1, is taken from [156] and [136].

Chapter 8 is based on papers [142] and [149].

Chapter 9 is based on papers [145] and [147], see also [137] and [142].

Chapter 10 is based on the results from [137] in the case of well-posed problems and from [125] in the case of ill-posed problems.

Chapter 11 is based on [148].

Chapter 12 is based on [131] and [137].

Chapter 13 is based on [132], [135], and [146].

There is a long history of the results preceding Theorem 13.2.2 which gives a sufficient condition for a local homeomorphism to be a global one. It starts with the Hadamard's paper of 1906 [38], and its further developments are described in [87]. Our approach is purely analytical. There are also approaches based on algebraic topology [180].

Sections 14.2 of Chapter 14 is based on [137]. Originally these results appeared in [3]. Our presentation is slightly different. Section 14.3 is based on [136] and Section 14.4 is based on [156].

Section 15.1 is based on [98], where for the first time the idea of using the stepsize as a regularizing parameter was proposed and implemented in a solution of the stable numerical differentiation of noisy data. This idea has been used in many papers of various authors and in many applications (see also [101], [104], [105], and [106]; Appendix 4 in [107]; [121]; Section 7.3 in [143]; [123], and [127].

Section 15.2 is based on [138] and [150]. It follows closely [150].

Section 15.3 follows closely [102].

Section 15.4 was not published earlier in this form, but it is based on the known ideas. Section 15.5 is based on [123] and [1].

Section 15.6 follows closely [113]. Part I of the book follows closely the earlier book of the first author [151]. The authors thank Elsevier for the permission to use the material from this book.

Chapter 19 is based on [51].

Chapter 20 is based on [58].

Chapter 21 is based on [42].

Chapter 22 is based on [43] and [44].

Chapter 23 is based on [157].

Chapter 17 is based on [54].

Chapter 18 is based on [49].

Section 25.1 is based on [55].

Section 25.2 is based on [42].

Section 25.3 is based on [44].

Section 24.1 is based on [45].

Section 24.2 is based on [54].

Section 24.3 is based on [54].

Section 24.4 is based on [49].

Chapter 26 is based on [62].

Appendix A contains auxiliary material. The material from Sections A.1–A.5 is well known and can be found, for example, in [25].

The material from Sections A.7 and A.8 can be found in [25], [73], [74], and in other references, cited in these sections.

Section A.6 is based on [134].

Section A.9 is based on [92] and follows closely [154].

REFERENCES

1. S. Ahn, A. G. Ramm, and U J. Choi, A scheme for a stable numerical differentiation, *J. Comp. Appl. Math.*, 186, N2 (2006), 325-334.

2. R. G. Airapetyan and A. G. Ramm, Numerical inversion of the Laplace transform from the real axis, *J. Math. Anal. Appl.*, 248 (2000), 572-587.

3. R. Airapetyan, A. G. Ramm, Dynamical systems and discrete methods for solving nonlinear ill-posed problems, in *Applied Mathematical Reviews*, vol. 1, G. Anastassiou, ed., World Scientific Publishers, Singapore, 2000, pp. 491-536.

4. R. Airapetyan, A. Smirnova, A. G. Ramm, Continuous methods for solving nonlinear ill-posed problems, in *Operator theory and applications*, American Mathematical Society, Fields Institute Communications, Providence, RI, 2000, pp. 111-138.

5. N. Akhiezer, *Lectures on approximation theory*, Nauka, Moscow, 1965.

6. Y. Alber, A new approach to the investigation of evolution differential equations in Banach spaces, *Nonlin. Anal.: Theory Methods Appl.*, 23, N9 (1994), 1115-1134.

7. S. Alinhac, and P. Gerard, *Pseudo-differential Operators and the Nash-Moser Theorem*, American Mathematical Society, Providence, RI, (2007).

8. V. Ambarzumian, *Ueber eine Frage der Eigenwerttheorie*, Zeitschrift für Physik, 53 (1929), 690-695.

9. M. Andrijchuk, N. Voitovich, D. Savenko, V. Tkachuk, *Synthesis of antenna*, Naukova Dumka, Kiev, 1993.

10. A. Bakushinsky, A. Goncharsky, *Iterative methods for solving ill-posed problems*, Nauka, Moscow, 1989 (Russian).

11. E. F. Beckenbach and R Bellman, *Inequalities*, Springer-Verlag, Berlin, 1961.

12. J. Bergh and J. Löfstrom, *Interpolation spaces*, Springer, New York, 1976.

13. M. Berger, *Nonlinearity and functional analysis*, Academic Press, New York, 1977.

14. F. Browder, *Nonlinear operators and nonlinear equations of evolution in Banach spaces*, American Mathematical Society, Providence, RI, 1976.

15. P. S. Bullen, *A dictionary of inequalities*, Longman, Essex, 1998.

16. A. S. Carasso, Determining surface temperatures from interior observations, *SIAM J. Appl. Math.*, 42 (1982), 558-574.

17. M. D. Choi, *Tricks or treats with the Hilbert matrix*, Am. Math. Monthly, 90 (1983), pp. 301-312.

18. K. S. Crump, Numerical inversion of Laplace transforms using a Fourier series approximation, *J. Assoc. Comput. Mach.*, 23, N1 (1976), 89-96.

19. S. Cuomo, L. D'Amore, A. Murli, and M. Rizzardi, Computation of the inverse Laplace transform based on a collocation method which uses only real values, *J. Comput. Appl. Math.*, 198 (2007), 98-115.

20. L. D'Amore and A. Murli, Reguarization of a Fourier series method for the Laplace transform inversion with real data, *Inverse Problems*, 18, (2002), 1185-1205.

21. Yu. L. Daleckii and M. G. Krein, *Stability of solutions of differential equations in Banach spaces*, American Mathematical Society, Providence, RI, 1974.

22. I. Daubechies, *Ten lectures on wavelets*, SIAM, Philadelphia, 1992.

23. B. Davies and B. Martin, Numerical inversion of the Laplace transform: A survey and comparison of methods, *J. Comp. Phys.*, 22 (1979), 1-32.

24. P. Davis and P. Rabinowitz, *Methods of numerical integration*, Academic Press, New York, 1989.

25. K. Deimling, *Nonlinear functional analysis*, Springer, New York, 1985.

26. L. M. Delves and J. L. Mohamed, *Computational Methods for Integral Equations*, Cambridge University Press, New York, 1985.

27. C. W. Dong, A regularization method for the numerical inversion of the Laplace transform, *SIAM J. Numer. Anal.*, 30, N3 (1993), 759-773.

28. H. Dubner and J. Abate, Numerical inversion of Laplace transforms by relating them to the finite Fourier cosine transform, *J. Assoc. Comput. Mach.*, 15, N1 (1968), 115-123.

29. N. Dunford and J. Schwartz, *Linear operators, Interscience*, New York, 1958.

30. H. Fujiwara, exflib, a multiple precision arithmetic software, *http://www-an.acs.i.kyoto-u.ac.jp/fujiwara/exflib*.

31. H. Fujiwara, T. Matsura, S. Saitoh, and Y. Sawano, Numerical real inversion of the Laplace transform by using a high-accurate numerical method *(private communication)*.

32. H. Gajewski, K. Gröger, and K. Zacharias, *Nichtlineare operatorgleichungen und operator differentialgleichungen*, Akad Verlag, Berlin, 1974.

33. A. Galperin and Z. Waksman, Ulm's method under regular smoothness, *Num. Funct. Anal. Optim.*, 19 (1998), 285-307.

34. M. Gavurin, Nonlinear functional equations and continuous analysis of iterative methods, *Izvestiya Vusov, Math.*, 5 (1958), 18-31 (Russian).

35. I. Glazman, *Direct methods of qualitative spectral analysis of differential operators*, Moscow, Nauka, 1963.

36. V. Gol'dshtein and A. G. Ramm, Embedding operators for rough domains, *Math. Ineq. Appl.*, 4, N1 (2001), 127-141.

37. V. Gol'dshtein and A. G. Ramm, Embedding operators and boundary-value problems for rough domains, *Intern. J. Appl. Math. Mech.*, 1 (2005), 51-72.

38. J. Hadamard, Sur les transformations ponctuelles, *Bull. Soc. Math. France*, 34 (1906), 71-84.

39. G. Hall and J. Watt, ed., *Modern numerical methods for ordinary differential equations*, Clarendon Press, Oxford, 1976.

40. P. Hartman, *Ordinary differential equations*, Wiley, New York, 1964.

41. Hairer, E., and Nørsett, S. P., and Wanner, G., *Solving ordinary differential equation I, nonstiff problems*, Springer, Berlin, 1987.

42. N. S. Hoang, Dynamical Systems Method of gradient type for solving nonlinear equations with monotone operators, *BIT Numer. Math.*, 50, N4 (2010), 751-780.

43. N. S. Hoang, Dynamical System Method for solving nonlinear equations with locally Hölder continuous monotone operators, *Int. J. Comput. Sci. Math.*, 3, N1/2 (2010), 5675.

44. N. S. Hoang, An iterative scheme for solving equations with locally σ-inverse monotone operators, *submitted for publication*.

45. N. S. Hoang and A. G. Ramm, Solving ill-conditioned linear algebraic systems by the dynamical systems method (DSM), *Inverse Problems Sci. Eng.*, 16, N5 (2008), 617-630.

46. N. S. Hoang and A. G. Ramm, Dynamical Systems Gradient Method for solving nonlinear operator equations with monotone operators, *Acta Appl. Math.*, 106 (2009), 473-499.

47. N. S. Hoang and A. G. Ramm, A nonlinear inequality, *J. Math. Ineq.*, 2, N4 (2008), 459-464.

48. N. S. Hoang and A. G. Ramm, An iterative scheme for solving nonlinear equations with monotone operators, *BIT Numer. Math.*, 48, N4 (2008), 725-741.

49. N. S. Hoang and A. G. Ramm, Dynamical systems method for solving linear finite-rank operator equations, *Ann. Polon. Math.*, 95, N1 (2009), 77–93.

50. N. S. Hoang and A. G. Ramm, A new version of the Dynamical Systems Method (DSM) for solving nonlinear equations with monotone operators, *Diff. Eqns Appl.*, 1, N1 (2009), 1–25.

51. N. S. Hoang and A. G. Ramm, A discrepancy principle for equations with monotone continuous operators, *Nonlinear Anal.: Theory Methods Appl.*, 70 (2009), 4307-4315.

52. N. S. Hoang and A. G. Ramm, A nonlinear inequality and applications, *Nonlinear Anal.: Theory Methods Appl.*, 71 (2009), 2744-2752.

53. N. S. Hoang and A. G. Ramm, Existence of solution to an evolution equation and a justification of the DSM for equations with monotone operators, *Comm. Math. Sci.*, 7, N4 (2009), 1073-1079.

54. N. S. Hoang and A. G. Ramm, Dynamical systems gradient method for solving ill-conditioned linear algebraic systems, *Acta Appl. Math.*, 111, N2 (2010), 189-204.

55. N. S. Hoang and A. G. Ramm, Dynamical systems method for solving nonlinear equations with monotone operators, *Math. Comput.*, 79, 269 (2010), 239-258.

56. N. S. Hoang and A. G. Ramm, The Dynamical Systems Method for solving nonlinear equations with monotone operators, *Asian Eur. Math. J.*, 3, N1 (2010), 57-105.

57. N. S. Hoang and A. G. Ramm, Dynamical Systems Method (DSM) for solving equations with monotone operators without smoothness assumptions on $F'(u)$, *J. Math. Anal. Appl.*, 367, N2 (2010), 508-515.

58. N. S. Hoang and A. G. Ramm, DSM of Newton-type for solving operator equations $F(u) = f$ with minimal smoothness assumptions on F, *Int. J. Comput. Sci. Math.* (IJCSM), 3, N1/2 (2010), 3-55.

59. K.M. Howell, Multiple precision arithmetic techniques, *Comput. J.*, 9, 4 (1967), 383-387.

60. S. Indratno and A. G. Ramm, Dynamical Systems Method for solving ill-conditioned linear algebraic systems, *Int. J. Comput. Sci. Math.* (IJCSM), 2, N4 (2009), 308-333.

61. S. Indratno and A. G. Ramm, An iterative method for solving Fredholm integral equations of the first kind, *Int. J. Comput. Sci. Math.* (IJCSM), 2, N4 (2009), 354-379.

62. S. Indratno and A. G. Ramm, Inversion of the Laplace transform from the real axis using an adaptive iterative method, *Int. J. Math. Math. Sci.* (IJMMS), 2009, Article 898195, 38 pages.

63. P. Iseger, Numerical transform inversion using Gaussian quadrature, *Prob. Eng. Inform. Sci.*, 20 (2006), 1–44.

64. V. Ivanov, V. Tanana, and V. Vasin, *Theory of ill-posed problems and its applications*, VSP, Utrecht, 2002.

65. J. Jerome, *Approximation of nonlinear evolution systems*, Academic Press, New York, 1983.

66. B. Kaltenbacher, A. Neubauer, A. G. Ramm, Convergence rates of the continuous regularized Gauss–Newton method, *J. Inv. Ill-Posed Probl.*, 10, N3 (2002), 261-280.

67. B. Kaltenbacher, A. Neubauer and O. Scherzer, *Iterative regularization methods for nonlinear ill-posed problems*, Walter de Gruyter, Berlin, 2008.

68. E. Kamke, *Differential Gleichungen. Lösungsmethoden und Lösungen*, Akademie Verlag., Leipzig, 1959.

69. L. Kantorovich and G. Akilov, *Functional analysis*, Pergamon Press, New York, 1982.

70. T. Kato, *Perturbation theory for linear operators*, Springer Verlag, New York, 1984.

71. A. I. Katsevich and A. G. Ramm, Multidimensional algorithm for finding discontinuities of functions from noisy data. *Math. Comp. Modelling*, 18, N1 (1993), 89-108.

72. A. I. Katsevich and A. G. Ramm, Nonparametric estimation of the singularities of a signal from noisy measurements, *Proc. AMS*, 120, N8 (1994), 1121-1134.

73. M. Krasnosel'skii, *Positive solutions of operator equations*, Groningen, Noordhoff, 1964.

74. M. Krasnosel'skii and P. Zabreiko, *Geometrical methods of nonlinear analysis*, Springer, New York, 1984.

75. M. Krasnosel'skii *et al.*, *Approximate solutions of operator equations*, Groningen, Noordhoff, 1972.

76. V. I. Krylov and N. Skoblya, *Reference book on numerical inversion of the Laplace transform*, Nauka i technika, Minsk, 1968 (in Russian).

77. V. V. Kryzhniy, Numerical inversion of the Lapace transform: Analysis via regularized analytic continuation, *Inverse Problem*, 22 (2006), 579-597.

78. J. Lattes and J. Lions, *Méthode de quasi-réversibilité et applications*, Dunod, Paris, 1967.

79. J. Lions, *Quelques méthodes de résolution des problemes aux limites nonlineatres*, Dunod, Paris, 1969.

80. O. Liskovetz, Regularization of equations with a closed linear operator, *Diff. equations*, 7 (1970), 972-976 (in Russian).

81. B. Minkovich, V. Yakovlev, *Theory of antenna synthesis*, Soviet Radio, Moscow, 1969.

82. I. Miyadera, *Nonlinear semigroups*, American Mathematical Society, Providence, RI, 1977.

83. V. Morozov, *Methods of solving incorrectly posed problems*, Springer Verlag, New York, 1984.

84. A. Murli, S. Cuomo, L. D'Amore, and A. Galleti, Numerical regularization of a real inversion formula based on the Laplace transform's eigenfunction expansion of the inverse function, *Inverse Problems*, 23 (2007), 713-731.

85. J. G. Nagy and K. M. Palmer, Steepest descent, CG, and iterative regularization of ill-posed problems, *BIT Numer. Math.*, 43 (2003), 1003-1017.

86. M. Naimark, *Linear differential operators*, Ungar, New York, 1969.

87. J. Ortega, W. Rheinboldt, *Iterative solution of nonlinear equations in several variables*, SIAM, Philadelphia, 2000.

88. D. Pascali and S. Sburlan, *Nonlinear Mappings of Monotone Type*, Noordhoff, Leyden, 1978.

89. A. Pazy, *Semigroups of linear operators and applications to partial differential equations*, Springer, New York, 1983.

90. G. Polya, G. Szego, *Problems and theorems in analysis*, Springer Verlag, New York, 1983.

91. M. Powell, *Approximation theory and methods*, Cambridge University Press, Cambridge, 1981.

92. A. G. Ramm, A necessary and sufficient condition for compactness of embedding. *Vestnik Lenigr. Univ. (Vestnik)*, N1 (1963), 150-151.

93. A. G. Ramm, *Theory and applications of some new classes of integral equations*, Springer-Verlag, New York, 1980.

94. A. G. Ramm, Iterative solution of linear equations with unbounded operators, *J. Math. Anal. Appl.*, 330, N2 (2007), 1338-1346

95. A. G. Ramm, On unbounded operators and applications, *Appl. Math. Lett.*, 21 (2008), 377-382.

96. A. G. Ramm, Antenna synthesis with the prescribed pattern, 22nd *science session dedicated the day of radio*, Moscow, 1966, Section of antennas, pp. 9–13.

97. A. G. Ramm, Optimal solution of the antenna synthesis problem, *Doklady Acad. Sci.* USSR, 180 (1968), 1071-1074.

98. A. G. Ramm, On numerical differentiation, *Math. Izvest. Vuzov*, 11 (1968), 131-135.

99. A. G. Ramm, Nonlinear antenna synthesis problems, *Doklady Acad. Sci.* USSR, 186 (1969), 1277-1280.

100. A. G. Ramm, Optimal solution of the linear antenna synthesis problem, *Radiofisika*, 12 (1969), 1842-1848.

101. A. G. Ramm, Simplified optimal differentiators, *Radiotech. Electron.* 17 (1972), 1325-1328.

102. A. G. Ramm, On simultaneous approximation of a function and its derivative by interpolation polynomials, *Bull. Lond. Math. Soc.*, 9 (1977), 283-288.

103. A. G. Ramm, Stationary regimes in passive nonlinear networks, in *Nonlinear Electromagnetics*, P. L. E. Uslenghi, ed., Academic Press, New York, 1980, pp. 263-302.

104. A. G. Ramm, Stable solutions of some ill-posed problems, *Math. Meth. Appl. Sci.*, 3 (1981), 336-363.

105. A. G. Ramm, Estimates of the derivatives of random functions. *J. Math. Anal. Appl.*, 102 (1984), 244-250.

106. A. G. Ramm, T. Miller, Estimates of the derivatives of random functions II, *J. Math. Anal. Appl.* 110 (1985), 429-435.

107. A. G. Ramm, *Scattering by obstacles*, D. Reidel, Dordrecht, 1986, 1-442.

108. A. G. Ramm, Inversion of the Laplace transform, *Inverse Problems*, 2 (1986), 55-59.

109. A. G. Ramm, Characterization of the scattering data in multidimensional inverse scattering problem, in *Inverse Problems: An Interdisciplinary Study*, P. Sabatier, ed., Academic Press, New York, 1987, 153-167.

110. A. G. Ramm, Necessary and sufficient conditions for a function to be the scattering amplitude corresponding to a reflecting obstacle, *Inverse Problems*, 3 (1987), L53-L57.

111. A. G. Ramm, Completeness of the products of solutions to PDE and uniqueness theorems of inverse scattering, *Inverse Problems*, 3 (1987), L77-L82.

112. A. G. Ramm, Recovery of the potential from fixed energy scattering data. *Inverse Problems*, 4 (1988), 877-886; 5 (1989), 255.

113. A. G. Ramm and A. van der Sluis, Calculating singular integrals as an ill-posed problem, *Numer. Math.*, 57 (1990), 139-145.

114. A. G. Ramm, *Multidimensional inverse scattering problems*, Longman/Wiley, New York, 1992.

115. A. G. Ramm and A. Zaslavsky, Reconstructing singularities of a function given its Radon transform, *Math. Comput. Modelling*, 18, N1 (1993), 109-138.

116. A. G. Ramm, Optimal local tomography formulas, *Pan Am. Math. J.*, 4, N4 (1994), 125-127.

117. A. G. Ramm, Finding discontinuities from tomographic data, *Proc. Am. Math. Soc.*, 123, N8 (1995), 2499-2505.

118. A. G. Ramm and A. I. Katsevich, *The Radon Transform and Local Tomography*, CRC Press, Boca Raton, FL, 1996.

119. A. G. Ramm and A. B. Smirnova, A numerical method for solving nonlinear ill-posed problems, *Numer. Funct. Anal. Optim.*, 20, N3 (1999), 317-332.

120. A. G. Ramm, A numerical method for some nonlinear problems, *Math. Models Meth. Appl. Sci.*, 9, N2 (1999), 325-335.

121. A. G. Ramm, Inequalities for the derivatives, *Math. Ineq. Appl.*, 3, N1 (2000), 129-132.

122. A. G. Ramm, Linear ill-posed problems and dynamical systems, *J. Math. Anal. Appl.*, 258, N1 (2001), 448-456.

123. A. G. Ramm and A. B. Smirnova, On stable numerical differentiation, *Math. Comput.*, 70 (2001), 1131-1153.

124. A. G. Ramm, Stability of solutions to inverse scattering problems with fixed-energy data, *Milan J. Math.*, 70 (2002), 97-161.

125. A. G. Ramm and A. B. Smirnova, Continuous regularized Gauss-Newton-type algorithm for nonlinear ill-posed equations with simultaneous updates of inverse derivative, *Int. J. Pure Appl. Math.*, 2, N1 (2002), 23-34.

126. A. G. Ramm, Regularization of ill-posed problems with unbounded operators, *J. Math. Anal. Appl.*, 271 (2002), 547-550.

127. A. G. Ramm and A. Smirnova, Stable numerical differentiation: When is it possible? *J. Korean SIAM*, 7, N1 (2003), 47-61.

128. A. G. Ramm, On a new notion of regularizer, *J. Phys A*, 36 (2003), 2191-2195.

129. A. G. Ramm, On the discrepancy principle, *Nonlinear Funct. Anal. Appl.*, 8, N2 (2003), 307-312.

130. A. G. Ramm, Global convergence for ill-posed equations with monotone operators: The dynamical systems method, *J. Phys A*, 36 (2003), L249-L254.

131. A. G. Ramm, Dynamical systems method for solving nonlinear operator equations, *Int. J. Appl. Math. Sci.*, 1, N1 (2004), 97-110.

132. A. G. Ramm, Dynamical systems method for solving operator equations, *Commun. Nonlinear Sci. Numer. Simulat.*, 9, N2 (2004), 383-402.

133. A. G. Ramm, Inequalities for solutions to some nonlinear equations, *Nonlinear Funct. Anal. Appl.*, 9, N2 (2004), 233-243.

134. A. G. Ramm, Continuity of solutions to operator equations with respect to a parameter, *Int. J. Pure Appl. Math. Sci.*, 1, N1 (2004), 1-5.

135. A. G. Ramm, Dynamical systems method and surjectivity of nonlinear maps, *Commun. Nonlinear Sci. Numer. Simulat.*, 10, N8 (2005), 931-934.

136. A. G. Ramm, DSM for ill-posed equations with monotone operators, *Commun. Nonlinear Sci. Numer. Simulat.*, 10, N8 (2005), 935-940.

137. A. G. Ramm, *Inverse problems*, Springer, New York, 2005.

138. A. G. Ramm, Inequalities for the derivatives and stable differentiation of piecewise-smooth discontinuous functions, *Math. Ineq. Appl.*, 8, N1 (2005), 169-172.

139. A. G. Ramm, Discrepancy principle for the dynamical systems method, *Commun. Nonlinear Sci. Numer. Simulat.*, 10, N1 (2005), 95-101.

140. A. G. Ramm, *Wave scattering by small bodies of arbitrary shapes*, World Scientific Publishers, Singapore, 2005.

141. A. G. Ramm, A new discrepancy principle, *J. Math. Anal. Appl.*, 310 (2005), 342-345.

142. A. G. Ramm, Dynamical systems method (DSM) and nonlinear problems, in *Spectral Theory and Nonlinear Analysis*, J. Lopez-Gomez, ed., World Scientific Publishers, Singapore, 2005, pp. 201-228.

143. A. G. Ramm, *Random fields estimation*, World Scientific Publishers, Singapore, 2005.

144. A. G. Ramm, Uniqueness of the solution to inverse obstacle scattering problem, *Phys. Lett. A*, 347, N4-6 (2005), 157-159.

145. A. G. Ramm, Dynamical systems method for nonlinear equations in Banach spaces, *Commun. Nonlinear Sci. Numer. Simulat.*, 11, N3 (2006), 306-310.

146. A. G. Ramm, Dynamical systems method and a homeomorphism theorem, *Am. Math. Monthly*, 113, N10 (2006), 928-933.

147. A. G. Ramm, A nonlinear singular perturbation problem, *Asymptotic Anal.*, 47, N1-2 (2006), 49-53.

148. A. G. Ramm, Dynamical systems method (DSM) for unbounded operators, *Proc. Am. Math. Soc.*, 134, N4 (2006), 1059-1063.

149. A. G. Ramm, Existence of a solution to a nonlinear equation, *J. Math. Anal. Appl.*, 316 (2006), 764-767.

150. A. G. Ramm, Finding discontinuities of piecewise-smooth functions, *J. Inequalities Pure Appl. Math. (JIPAM)*, 7, N2 (2006), Article 55, 1-7.

151. A. G. Ramm, *Dynamical systems method for solving operator equations*, Elsevier, Amsterdam, 2007.

152. A. G. Ramm, Ill-posed problems with unbounded operators, *J. Math. Anal. Appl.*, 325 (2007), 490-495.

153. A. G. Ramm, Iterative solution of linear equations with unbounded operators, *J. Math. Anal. Appl*, 330, N2 (2007), 1338-1346.

154. A. G. Ramm, Compactness of embeddings, *Nonlinear Funct. Anal. Appl.*, 11, N4 (2006), 655-658.

155. A. G. Ramm, Dynamical systems method for solving linear ill-posed problems, *Ann. Polon. Math.*, 95, N3 (2009), 253-272.

156. A. G. Ramm, Dynamical systems method (DSM) for general nonlinear equations, *Nonlinear Anal.: Theory, Methods Appl.*, 69, N7 (2008), 1934-1940.

157. A. G. Ramm, How large is the class of operator equations solvable by a DSM Newton-type method? *Appl. Math. Lett*, 24, N6 (2011), 860-865.

158. A. G. Ramm, On unbounded operators and applications, *Appl. Math. Lett.*, 21 (2008), 377-382.

159. A. G. Ramm, Discrepancy principle for DSM2, *Commun. Nonlin. Sci. Numer. Simulat.*, 13 (2008), 1256-1263.

160. A. G. Ramm, Dynamical systems method for solving linear ill-posed problems, *Ann. Polon. Math.*, 95, N3 (2009), 253-272.

161. A. G. Ramm, A DSM proof of surjectivity of monotone nonlinear mappings, *Ann. Polon. Math.*, 95, N2 (2009), 135-139.

162. A. G. Ramm, Asymptotic stability of solutions to abstract differential equations, *J. Abstract Diff. Equations and Appl. (JADEA)*, 1, N1 (2010), 27-34.

163. A. G. Ramm, Implicit Function Theorem via the DSM, *Nonlinear Anal.: Theory, Methods Appl.*, 72, N3-4 (2010), 1916-1921.

164. A. G. Ramm, On a new notion of the solution to an ill-posed problem, *J. Comp. Appl. Math.*, 234 (2010), 3326-3331.

165. A. G. Ramm, A nonlinear inequality and evolution problems, *J. Ineq. Special Funct., (JIASF)*, 1, N1 (2010), 1-9.

166. A. G. Ramm, Dynamical Systems Method (DSM) for solving non-linear operator equations in Banach spaces *(submitted for publication)*.

167. A. G. Ramm, Distribution of particles which produces a "smart" material, *J. Stat. Phys.*, 127, N5 (2007), 915-934.

168. A. G. Ramm, Inverse scattering problem with data at fixed energy and fixed incident direction, *Nonlinear Anal.: Theory, Methods Appl.*, 69, N4 (2008), 1478-1484.

169. A. G. Ramm, Fixed-energy inverse scattering, *Nonlinear Anal.: Theory, Methods Appl.*, 69, N3 (2008), 971-978.

170. A. G. Ramm, Inverse scattering with non-overdetermined data, *Phys. Lett. A*, 373 (2009), 2988-2991.

171. A. G. Ramm, Uniqueness theorem for inverse scattering problem with non-overdetermined data, *J. Phys. A, Fast Track Commun.*, 43, (2010), 112001.

172. A. G. Ramm, Uniqueness of the solution to inverse scattering problem with backscattering data, *Eurasian Math. J. (EMJ)*, 1, N3 (2010), 97-111.

173. A. G. Ramm, Justification of the Dynamical Systems Method (DSM) for global homeomorphisms, *Eurasian Math. J. (EMJ)*, 1, N4 (2010), 116-123.

174. A.G. Ramm, On the DSM Newton-type method, *J. Appl. Math. Comput.*, *doi:10.1007/s12190-011-0494-z.* (to appear)

175. A.G. Ramm, On the DSM version of Newton's method, *Eurasian Math. J. (EMJ).* (to appear)

176. E. Rothe, *Introduction to various aspects of degree theory in Banach spaces*, American Mathematical Society, Providence, RI, 1986.

177. J. Schröder, *Operator inequalities*, Academic Press, New York, 1980.

178. R. Showalter, *Monotone operators in Banach space and nonlinear partial differential equations*, American Mathematical Society, Providence, RI, 1997.

179. I. V. Skrypnik, *Methods for analysis of nonlinear elliptic boundary value problems*, American Mathematical Society, Providence, RI, 1994.

180. E. Spanier, *Algebraic topology*, McGraw-Hill, New York, 1966.

181. J. E. Slotine and W. Li, *Applied nonlinear control*, Prentice Hall, Englewood Cliffs, 1991.

182. J. Szarski, *Differential inequalities*, PWN, Warszawa, 1967.

183. U. Tautenhahn, On the method of Lavrentiev regularization for nonlinear ill-posed problems, *Inverse Problems*, 18 (2002), 191-207.

184. R. Temam, *Infinite-dimensional dynamical systems in physics and mechanics*, Springer, New York, 1997.

185. V. Tikhomirov, *Some questions in approximation theory*, Nauka, Moscow, 1976.

186. A. Tikhonov, A. Leonov, and A. Yagola, *Nonlinear ill-posed problems*, Chapman and Hall, London, 1998.

187. A. Timan, *Theory of approximation of functions of a real variable*, Dover, Mineola, New York, 1993.

188. A. Turetsky, *Interpolation theory in problems*, High School, Minck, 1968. (Russian)

189. S. Ulm, On iterative methods with successive approximation of the inverse operator, *Izv. Acad. Nauk Eston SSR*, 16 (1967), 403-411.

190. M. Vainberg, *Variational method and method of monotone operators*, Wiley, New York, 1973.

191. G. Vainikko, A. Veretennikov, *Iterative procedures in ill-posed problems*, Nauka, Moscow, 1986. (Russian)

192. J. Varah, Pitfalls in the numerical solution of linear ill posed problems, *SIAM J. Stat. Comput.*, 4 (1983), 164-76 .

193. V. Vasin, A. Ageev, *Ill-posed problems with a priori information*, Nauka, Ekaterinburg, 1993. (Russian)

194. J. G. Whirter and E. R. Pike, Laplace transform and other similar Fredholm integral equations of the first kind, *J. Phys. A: Math. Gen.*, 11 (1978), 1729-1745.

195. K. Yosida, *Functional analysis*, Springer, New York, 1980.

196. E. Zeidler, *Nonlinear functional analysis, I–V*, Springer, New York, 1985.

197. S. Zheng, *Nonlinear evolution equations*, Chapman and Hall, Boca Raton, FL, 1994.

INDEX

Dynamical Systems Method and Applications: Theoretical Developments and
Numerical Examples, First Edition. A. G. Ramm and N. S. Hoang.
Copyright © 2012 John Wiley & Sons, Inc.